NANOTECHNOLOGY IN DIAGNOSIS, TREATMENT AND PROPHYLAXIS OF INFECTIOUS DISEASES

NANOTECHNOLOGY IN DIAGNOSIS, TREATMENT AND PROPHYLAXIS OF INFECTIOUS DISEASES

Edited by

MAHENDRA RAI
Biotechnology Department, SGB Amravati University, Amravati, Maharashtra, India

KATERYNA KON
Department of Microbiology, Virology and Immunology, Kharkiv National Medical University, Kharkiv, Ukraine

ELSEVIER

AMSTERDAM • BOSTON • HEIDELBERG • LONDON
NEWYORK • OXFORD • PARIS • SAN DIEGO
SAN FRANCISCO • SINGAPORE • SYDNEY • TOKYO
Academic Press is an imprint of Elsevier

Academic Press is an imprint of Elsevier
32 Jamestown Road, London NW1 7BY, UK
525 B Street, Suite 1800, San Diego, CA 92101-4495, USA
225 Wyman Street, Waltham, MA 02451, USA
The Boulevard, Langford Lane, Kidlington, Oxford OX5 1GB, UK

ISBN: 978-0-12-801317-5

British Library Cataloguing-in-Publication Data
A catalogue record for this book is available from the British Library.

Library of Congress Cataloging-in-Publication Data
A catalog record for this book is available from the Library of Congress.

For Information on all Academic Press publications
visit our website at http://store.elsevier.com/

Typeset by MPS Limited, Chennai, India
www.adi-mps.com

Contents

List of Contributors

Ravikumar Aalinkeel Department of Medicine, Division of Allergy, Immunology, and Rheumatology, State University of New York at Buffalo, Clinical Translational Research Center, Buffalo, NY, USA

Vojtech Adam Central European Institute of Technology, Brno University of Technology, Technicka, Brno, Czech Republic, European Union; Department of Chemistry and Biochemistry, Faculty of Agronomy, Mendel University in Brno, Zemedelska, Brno, Czech Republic, European Union

Vipul Bansal Ian Potter NanoBioSensing Facility and NanoBiotechnology Research Laboratory (NBRL), School of Applied Sciences, RMIT University, Melbourne, Australia

Sunita Bansod Nanobiotechnology Laboratory, Department of Biotechnology, Sant Gadge Baba Amravati University, Amravati, Maharashtra, India

Pedro V. Baptista CIGMH, Departamento de Ciências da Vida, Faculdade de Ciências e Tecnologia, Universidade NOVA de Lisboa, Faculdade de Ciências e Tecnologia Caparica, Portugal

Debora Barros Barbosa Department of Dental Materials and Prosthodontics, Araçatuba Dental School, Univ Estadual Paulista (UNESP), São Paulo, Brazil

Juan Bueno Bioprospecting Development and Consulting, Bogotá, Colombia

Dagmar Chudobova Department of Chemistry and Biochemistry, Faculty of Agronomy, Mendel University in Brno, Zemedelska, Brno, Czech Republic, European Union

Kristyna Cihalova Department of Chemistry and Biochemistry, Faculty of Agronomy, Mendel University in Brno, Zemedelska, Brno, Czech Republic, European Union

Nicola Cioffi Department of Chemistry, University of Bari "Aldo Moro," Bari, Italy

Lili Dai Department of Medicine, Division of Allergy, Immunology, and Rheumatology, State University of New York at Buffalo, Clinical Translational Research Center, Buffalo, NY, USA

Hemant Kumar Daima Department of Biotechnology, Siddaganga Institute of Technology, Tumkur, Karnataka, India

Emerson Rodrigues de Camargo Department of Chemistry, Federal University of São Carlos (UFSCar), São Paulo, Brazil

Luiz Fernando Gorup Department of Chemistry, Federal University of São Carlos (UFSCar), São Paulo, Brazil

Anna M. Grudniak Department of Bacterial Genetics, Institute of Microbiology, Faculty of Biology, University of Warsaw, Warsaw, Poland

Alexandru Mihai Grumezescu AMG Transcend, Bucharest, Romania; Department of Science and Engineering of Oxide Materials and Nanomaterials, Faculty of Applied Chemistry and Materials Science, University Politehnica of Bucharest, Bucharest, Romania

Michael R Hamblin Department of Dermatology, Harvard Medical School, Boston, MA, USA; Wellman Center for Photomedicine, Massachusetts General Hospital, Boston, MA, USA; Harvard-MIT Division of Health Sciences and Technology, Cambridge, MA, USA

Mariana Henriques CEB—Center of Biological Engineering, LIBRO—Laboratório de Investigação em Biofilmes, Rosário Oliveira, University of Minho, Braga, Portugal

Alina Maria Holban AMG Transcend, Bucharest, Romania; Microbiology Immunology Department, Faculty of Biology, University of Bucharest, Bucharest, Romania; Department of Science and Engineering of Oxide Materials and Nanomaterials, Faculty of Applied Chemistry and Materials Science, University Politehnica of Bucharest, Bucharest, Romania

Nabil A. Ibrahim Textile Research Division, National Research Centre, Giza, Egypt

Avinash P. Ingle Nanobiotechnology Laboratory, Department of Biotechnology, Sant Gadge Baba Amravati University, Amravati, Maharashtra, India

Florin Iordache Institute of Cellular Biology and Pathology of Romanian Academy, "Nicolae Simionescu," Department of Fetal and Adult Stem Cell Therapy, Bucharest, Romania

Zeyd Issa University of Exeter Medical School, Exeter, Devon, UK; Wellman Center for Photomedicine, Massachusetts General Hospital, Boston, MA, USA

Konrad Kamiński Department of Bacterial Genetics, Institute of Microbiology, Faculty of Biology, University of Warsaw, Warsaw, Poland

Rene Kizek Central European Institute of Technology, Brno University of Technology, Technicka, Brno, Czech Republic, European Union; Department of Chemistry and Biochemistry, Faculty of Agronomy, Mendel University in Brno, Zemedelska, Brno, Czech Republic, European Union

Kateryna Kon Department of Microbiology, Virology and Immunology, Kharkiv National Medical University, Kharkiv, Ukraine

Pavel Kopel Central European Institute of Technology, Brno University of Technology, Technicka, Brno, Czech Republic, European Union; Department of Chemistry and Biochemistry, Faculty of Agronomy, Mendel University in Brno, Zemedelska, Brno, Czech Republic, European Union

Aydan Ayse Kose Department of Plastic and Reconstructive Surgery, Eskisehir Osmangazi University, Eskisehir, Turkey

Nusret Kose Department of Orthopedics and Traumatology, Eskisehir Osmangazi University, Eskisehir, Turkey

Dong Gun Lee School of Life Sciences and Biotechnology, College of Natural Sciences, Kyungpook National University, Daegu, Republic of Korea

Won Young Lee School of Life Sciences and Biotechnology, College of Natural Sciences, Kyungpook National University, Daegu, Republic of Korea

Supriya D. Mahajan Department of Medicine, Division of Allergy, Immunology, and Rheumatology, State University of New York at Buffalo, Clinical Translational Research Center, Buffalo, NY, USA

Manoj J. Mammen Department of Medicine, Division of Allergy, Immunology, and Rheumatology, State University of New York at Buffalo, Clinical Translational Research Center, Buffalo, NY, USA

Katarzyna Markowska Department of Bacterial Genetics, Institute of Microbiology, Faculty of Biology, University of Warsaw, Warsaw, Poland

Lukas Melichar Department of Chemistry and Biochemistry, Faculty of Agronomy, Mendel University in Brno, Zemedelska, Brno, Czech Republic, European Union

Douglas Roberto Monteiro Department of Pediatric Dentistry and Public Health, Araçatuba Dental School, Univ Estadual Paulista (UNESP), São Paulo, Brazil

Maria Jose Morilla Programa de Nanomedicinas, Departamento de Ciencia y Tecnología, Universidad Nacional de Quilmes, Buenos Aires, Argentina

Bindukumar B. Nair Department of Medicine, Division of Allergy, Immunology, and Rheumatology, State University of New York at Buffalo, Clinical Translational Research Center, Buffalo, NY, USA

Melyssa Negri Faculdade INGÁ, Maringá, Paraná, Brazil

Maria Angela Nitti Department of Physics "M. Merlin", University of Bari "Aldo Moro," Bari, Italy

Noha Nafee Department of Pharmaceutics and Biopharmacy, Philipps University, Marburg, Germany; Department of Pharmaceutics, Faculty of Pharmacy, Alexandria University, Alexandria, Egypt

Federica Paladini Department of Engineering for Innovation, University of Salento, Lecce, Italy

Pedro Pedrosa CIGMH, Departamento de Ciências da Vida, Faculdade de Ciências e Tecnologia, Universidade NOVA de Lisboa, Faculdade de Ciências e Tecnologia Caparica, Portugal

Rosaria Anna Picca Department of Chemistry, University of Bari "Aldo Moro," Bari, Italy

Mauro Pollini Department of Engineering for Innovation, University of Salento, Lecce, Italy

Paras N. Prasad Institute for Laser, Photonics and Biophotonics, State University of New York at Buffalo, Buffalo, NY, USA

Mahendra Rai Nanobiotechnology Laboratory, Department of Biotechnology, Sant Gadge Baba Amravati University, Amravati, Maharashtra, India

Jessica L. Reynolds Department of Medicine, Division of Allergy, Immunology, and Rheumatology, State University of New York at Buffalo, Clinical Translational Research Center, Buffalo, NY, USA

Eder Lilia Romero Programa de Nanomedicinas, Departamento de Ciencia y Tecnología, Universidad Nacional de Quilmes, Buenos Aires, Argentina

Branislav Ruttkay-Nedecky Central European Institute of Technology, Brno University of Technology, Technicka, Brno, Czech Republic, European Union; Department of Chemistry and Biochemistry, Faculty of Agronomy, Mendel University in Brno, Zemedelska, Brno, Czech Republic, European Union

Alessandro Sannino Department of Engineering for Innovation, University of Salento, Lecce, Italy

Stanley A. Schwartz Department of Medicine, Division of Allergy, Immunology, and Rheumatology, State University of New York at Buffalo, Clinical Translational Research Center, Buffalo, NY, USA

Ranjita Shegokar Freie Universität Berlin, Institute of Pharmacy Department of Pharmaceutics, Biopharmaceutics & NutriCosmetics, Kelchstraße, Berlin, Germany

Sónia Silva CEB—Center of Biological Engineering, LIBRO—Laboratório de Investigação em Biofilmes, Rosário Oliveira, University of Minho, Braga, Portugal

Maria Chiara Sportelli Department of Chemistry, University of Bari "Aldo Moro," Bari, Italy

Marketa Vaculovicova Central European Institute of Technology, Brno University of Technology, Technicka, Brno, Czech Republic, European Union; Department of Chemistry and Biochemistry, Faculty of Agronomy, Mendel University in Brno, Zemedelska, Brno, Czech Republic, European Union

Antonio Valentini Department of Physics "M. Merlin", University of Bari "Aldo Moro," Bari, Italy

Marco Valentini Department of Physics "M. Merlin", University of Bari "Aldo Moro," Bari, Italy

Krystyna I. Wolska Department of Bacterial Genetics, Institute of Microbiology, Faculty of Biology, University of Warsaw, Warsaw, Poland

Preface

Resistance to antimicrobial agents has been reaching high levels among all types of microorganisms. Bacteria constantly demonstrate growing rates of resistance to classical and newly introduced antibiotics, fungi increase rates of resistance to antimycotics, viruses increase rates of resistance to antiviral agents, and even insect vectors carrying microorganisms have been acquiring the ability to develop resistance to the most common insecticidal agents. Because of this, the efforts of scientists all over the world are being directed to the search for new and effective methods to cope with drug resistance. One promising approach is the application of nanotechnology in the battle against microorganisms.

Nanotechnology is being applied not only to the treatment of infectious diseases but also to diagnostics of infections and to prophylaxis by reducing the number of insecticidal vectors spreading microorganisms. This book discusses the potential of nanotechnology for fighting all common types of infective agents (bacteria, viruses, fungi, protozoa) and their vectors (ticks, mosquitoes, flies, etc.), as well as recent advances in diagnostics of infectious diseases and nanotechnology techniques.

Potential readers include researchers in applied microbiology, biotechnology, pharmacology, nanotechnology, and infection control, students of medical and biological faculties, and clinicians dealing with infectious diseases.

The editors thank Elizabeth Gibson, Editorial Project Manager, Academic Press/Elsevier S&T Books, Waltham, MA, USA, for her constant help and valuable suggestions, and the contributors for devoting their time to this book.

CHAPTER

1

Gold and Silver Nanoparticles for Diagnostics of Infection

Pedro Pedrosa and Pedro V. Baptista

CIGMH, Departamento de Ciências da Vida, Faculdade de Ciências e Tecnologia, Universidade NOVA de Lisboa, Faculdade de Ciências e Tecnologia Caparica, Portugal

1.1 NANOTECHNOLOGY AND INFECTION

Infectious diseases are responsible for approximately 16.2% of total deaths per year, with children being the most affected group and accounting for a disproportionate number of deaths (WHO, 2014). For each agent and disease, great effort has been directed toward the identification of associated molecular characteristics and other biomarkers, such as toxins, antigens, or nucleic acids. Pathogens that can cause disease have a complex and broad range; some of them have prolonged asymptomatic incubation times, and diagnosis of infection is a challenging task (Kaittanis et al., 2010). Molecular medicine has improved our knowledge of disease mechanisms and has brought clinicians a set of diagnostic tools that may help reduce the clinical uncertainty. Among these tools are nanosystems, which bring innovative molecular recognition tools and point-of-care (POC) diagnostic solutions. These new technologies can be integrated into a multitude of platforms allowing for enhancement of current techniques (McGlennen, 2001).

Nanotechnology has received international attention in the past decade, with more than 3 million patents registered with the term "nano" (Alharbi and Al-Sheikh, 2014). Nanotechnology involves the understanding, design, and fabrication of materials at the atomic and molecular scale, where applications to medicine are commonly related to diagnostics and treatment. Because of their size (1–100 nm), nanomaterials have the advantage of interacting in a one-on-one basis with biomolecules, potentiating accuracy and effectiveness (Cabral and Baptista, 2013). Nanotechnology has brought forth a plethora of new materials suitable for application in biosensing that greatly boost current methodologies for clinic diagnostics, including gene expression profiling, biomarker quantification, and imaging. Such strategies have focused on the development of nanoscale devices and platforms that can be used for single molecule characterization (nucleic acids, peptides, proteins, antibodies, small molecules)

Nanotechnology in Diagnosis, Treatment and Prophylaxis of Infectious Diseases.
DOI: http://dx.doi.org/10.1016/B978-0-12-801317-5.00001-3

at an increased rate when compared with conventional systems. Interestingly, the majority of proposed platforms greatly focus on demonstrating increased sensitivity by lowering the limits of detection of standard laboratory strategies such as real-time polymerase chain reaction (RT-PCR). Most platforms for protein detection have been designed to include antigens and antibodies for molecular recognition (e.g., sandwich immunoassay) that can be coupled to distinct detection strategies (reporter molecules and tags). Systems for nucleic acid sensing, whether for screening of nucleotide sequences or single base mismatch discrimination (e.g., mutations and single nucleotide polymorphisms [SNPs]), are usually based on differential hybridization stringency because of mismatch that causes conformational stress to the DNA duplex that enables detection.

Molecular diagnostics requires highly paralleled and miniaturized assays that make use of the immense available information on pathogen biomarkers. Molecular diagnostics of pathogen infection may be broadly divided in terms of the target for detection: the pathogen itself (considering the whole or biomolecular fragments such as DNA, RNA, and peptides) or the host response to the presence of the pathogen (this includes the detection of antibodies and other immune response molecules and cell stress molecules).

Over the past decades, noble metal NPs, because of their optical and physicochemical properties, have been used in proof-of-concept demonstrations of a wide range of biosensing tools for selective and specific identification of DNA/RNA sequences associated with infection and pathogens. These tools have been used as either substitutes or coupled to fluorescence/chemiluminescence signal transduction. Nanoparticles (NPs) are an optimal tool for tagging biomolecular probes because of the ease of synthesis and functionalization with DNA/RNA molecules, proteins, and other biomolecules. Currently used bioassays for detection of known biomarkers or nucleotide sequences have progressively been integrated into NP-based systems, increasing sensitivity and lowering costs. Because of their nanosize scale, they present a high ratio of surface area to volume, with the capability of interacting in the same scale as target biological molecules. Diagnostics strategies using NPs have been reported for the detection of nucleic acids, proteins, pH variations, or small analytes via colorimetric, fluorescence, mass spectrometry, electrochemical, and scattering approaches. Nanodiagnostic systems have the ability to perform molecular tests quicker with more sensitivity, and with increased flexibility at reduced costs (Azzazy et al., 2006). Most of the reported systems are still in the preclinical stage, with few commercially available products being transposed to the clinic. In fact, the majority of these new nanoplatforms still need further evaluation and validation with clinical samples before they can be fully translated into the clinics.

In this chapter, we focus on the methods used for diagnosing infectious diseases that take advantage of noble metal NPs, detailing their use in biomolecular recognition and their most promising approaches, and discussing their advantages and disadvantages.

1.2 GOLD AND SILVER NPs FOR MOLECULAR DIAGNOSTICS

Biomolecular detection systems based on noble metal NPs, particularly gold and silver, have been widely used because of their unique optical and physicochemical properties (Goluch et al., 2006). They may be synthesized as single inorganic compounds, alloys, or core shells, all of which have different sizes and shapes (i.e., spheres, rods, prisms, and stars) and convey different properties that can be used for tagging biomolecules and promoting biorecognition (for a thorough review of the synthesis and properties of gold and silver

NPs, see Sau et al., 2010; Zhang et al., 2012 and references therein). The nanosized particles present a high ratio of area to volume that favors functionalization with (bio)molecules.

Of particular relevance for pathogen identification and molecular characterization, gold NPs (AuNPs) and silver NPs (AgNPs) present extraordinary optical properties, such as their bright intense colors in solution because of their localized surface plasmon resonance (LSPR) (Doria et al., 2012). This phenomenon is explained by the interaction of light with the electrons at the surface of the metal. The electric field intensity and the scattering and absorption cross-sections are all strongly enhanced at the LSPR frequency, which lies in the visible region of the electromagnetic spectrum. Because of this surface plasmon enhancement, optical cross-sections are at least five orders of magnitude larger than those of common dyes used to tag biomolecules (Dreaden et al., 2012). The LSPR band varies with the size, shape, and environment of NPs. AuNPs and AgNPs also present other interesting properties that are exploited in molecular detection strategies, such as the capability to quench or enhance fluorescence of a fluorophore in the vicinity in a distance-dependent manner. Fluorescence modulation depends on NP size, coating, wavelengths of the incident light, emitted light, and intrinsic quantum yield of the fluorophore (Kang et al., 2011).

Nanosurfaces are also optimal for the enhancement of Raman spectrum, with increments of 10^5-times to 10^6-times the surface enhancement Raman spectroscopy (SERS) (Le Ru et al., 2007; Doria et al., 2012). This allows the use of organic dyes with a defined Raman spectrum to provide specific fingerprints that are strongly enhanced by the NPs, permitting the construction of multi-labeled detection systems. Further applications of AuNPs and AgNPs include conductivity of these metals, which take advantage of the electrical signal enhancement provided by the NPs (Jain et al., 2006).

Several protocols have been described for the synthesis of AuNPs and AgNPs using chemical, physical, photochemical, and biological routes. Each method possesses its advantages and disadvantages with varying costs, scalability, particle sizes, and size distribution (see Tran et al., 2013; Zhao et al., 2013, and references therein). The reaction principle is basically the reduction of an Ag or Au salt inducing the nucleation of neutral charged NPs; afterward, a capping agent "blocks" NP growth, stabilizing the NPs as specific shapes and sizes. The capping agent used during synthesis can leave functional groups available for subsequent functionalization (e.g., carboxylic groups).

Functionalization of AuNPs and AgNPs with biomolecules is generally through thiol–metal quasi-covalent bonds, taking advantage of the strong affinity between the metal and sulfur that either exist in the biomolecule (e.g., proteins) or can be easily added (e.g., oligonucleotides) (Zhao et al., 2013). For functionalization with proteins, sugars, and other polymers, direct amine–metal bonding or carboxyl groups of the coating agent is also used. This simple procedure renders AuNPs (and silver to a lesser extent) suitable for simple and easy biofunctionalization (Tran et al., 2013).

1.2.1 Biomarkers

A biomarker is referred to a measurable characteristic used as an indicator of a biological state or condition. Infection biomarkers are generally molecules of the infecting agent and can be nucleic acids, proteins, cell wall, or metabolites, or a host response to infection (e.g., antibodies) that can also be used as biomarkers (Kaittanis et al., 2010). Microbial pathogens are usually identified by performing classic techniques, such as microscopy, culture on selective or differential media, or biochemical or serological tests, or performing molecular biology techniques. Classical techniques rely on time-consuming and laborious practices, whereas molecular biological techniques tend to be expensive and unsuitable for field application (Patrick Boisseau and Loubaton, 2011). Also, biomarker detection in physiological fluid samples is difficult because most specimens (e.g., blood) contain components that interfere with the biorecognition and/or with the sensor itself. For example, lateral flow cytometry and other colorimetric systems face limitations imposed by the characteristics of the sample, because opaque or viscous samples can interfere with the readout. This situation is usually overcome by a preliminary sample purification step (Wei et al., 2010). Although molecular diagnostic technologies are commonly associated with disease diagnostics, it is important to clarify that they only detect the presence of the molecular "fingerprints" of the pathogen. Hence, these assays do not provide any direct information about the presence or metabolic state of the microorganism in the analyzed samples.

Systems using noble metal NP detection may be based on several techniques, namely, colorimetric, fluorescence, electrochemical, SERS, lateral flow, and others, with each showing strengths and weaknesses. For example, colorimetric systems are an asset for fast and easy screening, SERS and fluorescence are advantageous for multiplex assays, electrochemical systems have high sensitivity, and lateral flow assays (LFAs) are portable.

1.3 NANODIAGNOSTICS FOR NUCLEIC ACIDS

With whole genome sequencing of pathogens, it is now possible to identify specific DNA fingerprints for individual organisms and strains, whose sequence variations can be associated with particular phenotypes (Boyce et al., 2004; Wei et al., 2010). Single nucleobase variations have been extensively mapped and associated with phenotypic characteristics, such as antibiotic resistance and virulence (Kaittanis et al., 2010; Veigas et al., 2012a). In clinical samples, DNA/RNA target sequences are usually present at low copy numbers, making it difficult to directly detect nucleic acids without a previous step of amplification. This DNA amplification step, requiring specialized equipment and approximately 2–3 h, is typically performed via the PCR or loop-mediated isothermal amplification (LAMP) (Notomi, 2000), whereas for RNA the choice relies on quantitative RT-PCR (qRT-PCR) (Freeman et al., 1999) and nucleic acid sequence–based amplification (NASBA) (Compton, 1991).

1.3.1 Homogeneous Colorimetric Assays

UV–Vis spectral behavior of noble metal NPs strongly depends on size, shape, and interparticle distance. Typically, the LSPR peak of AuNPs may be strongly red-shifted when the distance between NPs decreases, that is, NPs aggregate in solution. This is the basis of a plethora of colorimetric detection schemes. Aggregation of NPs may be induced by hybridization of DNA cross-linkers, protein scaffolding, and/or increasing the ionic strength (for details refer to Baptista et al., 2011).

1.3.1.1 Unmodified NPs

In 2004, Li and Rothberg (2004) described the different interaction between ssDNA and dsDNA at the surface of AuNPs. ssDNA sequences nonspecifically adsorb to the surface of the NPs, increasing stability, which is in contrast to dsDNA sequences that form rigid non-adsorbent structures. This allowed the creation of a detection system based on the aggregation of unmodified AuNPs, where an ssDNA complementary to the target is added to the AuNP solution, increasing its stability due to surface adsorption. After target and salt addition, a DNA duplex is formed, unprotecting the NPs that immediately aggregate. With some variations, this method was used for the detection of hepatitis B virus (HBV) (Liu et al., 2011), human immunodeficiency virus type 1 (HIV-1) (Xie et al., 2011), *Bacillus anthracis* (Deng et al., 2013), and members of the *Mycobacterium tuberculosis* (Hussain et al., 2013).

Later, Fu et al. (2013) described a disruptive approach using unmodified AuNPs coupled with rolling circle amplification and thiol-modified primers. In case of amplification, NPs bind to the thiol primers and do not aggregate on salt addition. After isothermal amplification, it was possible to detect with a limit of 0.1 fM of DNA (Fu et al., 2013).

1.3.1.2 Cross-Linking

In 1996, Mirkin et al. described the specific DNA target recognition using oligonucleotide functionalized AuNPs as gold nanoprobes in a cross-linking approach, where the target DNA acts as a linker between two different Au nanoprobes. Sequence complementarity between the target and the two probes forces NP aggregation and consequent visual change from red to blue (Mirkin et al., 1996).

Based on this concept, Storhoff et al. (2004a) developed the first method for pathogen detection with a spot-and-read platform for detection of *mecA* in methicillin-resistant *Staphylococcus aureus* directly from genomic DNA samples with a limit of detection (LOD) of 66 ng/μL (Storhoff et al., 2004a). Subsequently, several cross-linking protocols were described for the detection of pathogens where the main difference between approaches, besides the target sequence and organism, is the need for prior nucleic acid amplification (i.e., PCR, LAMP, and NASBA). Some protocols claim a LOD that permits direct nucleic acid detection. These include the detection of HIV-1 (He et al., 2008), *Cryptosporidium parvum* (Javier et al., 2009), *M. tuberculosis* complex strains (Soo et al., 2009; Gill et al., 2009), and *Salmonella* sp. (Kalidasan et al., 2013). Mancuso et al. (2013) described a cross-linking multiplex assay for the detection of Kaposi's sarcoma-associated herpes virus (KSHV) and *Bartonella*. This assay used both Au and Ag nanoprobes, each detecting a specific target. The different spectral profiles of gold and silver allow discrimination between silver or gold NPs aggregation in solution (Mancuso et al., 2013).

1.3.1.3 Noncross-Linking

Another colorimetric detection strategy for nucleic acids based on AuNPs was described in 2006 by Baptista et al., who reported the direct detection of genomic DNA of *M. tuberculosis* from clinical samples (Baptista et al., 2006). The method was named noncross-linking because it only needs one Au nanoprobe for molecular detection, which constitutes one advantage. Aggregation of Au nanoprobes is induced by an increase in the ionic strength of the media (i.e., salt addition), and colorimetric discrimination is possible because of the differential behavior of Au nanoprobes in the presence (red due to stabilization of Au-nanoprobe system) or absence (blue due to extensive aggregation) of a specific complementary target (Figure 1.1A). Additional developments were introduced to the protocol for single base substitutions (mutations) associated with resistance to antibiotics (Veigas et al., 2010, 2013; Pedrosa et al., 2014). Moreover, Liandris et al. (2009) described application of this approach for the detection of unamplified genomic DNA of *Mycobacterium* spp. Recently, the noncross-linking approach was also integrated into a POC platform and a mobile phone for image analysis, simplifying the detection process and reducing dependency of equipment (Veigas et al., 2012b).

Other protocols have introduced slight variations toward identification of different pathogens. Bakthavathsalam et al. (2012) reported on the visual direct detection of *Escherichia coli* genomic DNA using HCl to promote aggregation, whereas Chen, S.H. et al. (2009) used salt and cooling of the reaction to promote aggregation of

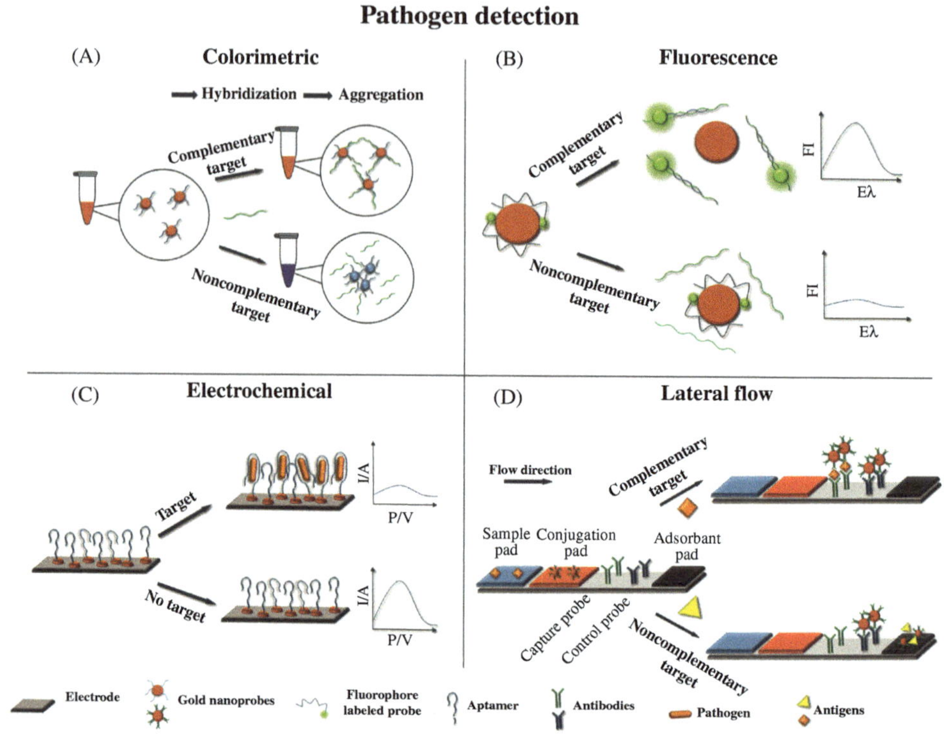

FIGURE 1.1 **Schematics of pathogen nanodetection.** (A) DNA colorimetric detection based on noncross-linking method. (B) DNA fluorescence detection based on NSET. Fluorophore-labeled oligonucleotide adsorbs to the surface of AuNPs, leading to fluorescence quenching. In the presence of the complementary target, dsDNA does not adsorb to the NPs surface, lengthening fluorophore–NP distance and increasing fluorescence signal. (C) Aptamer detection. Aptamer and target binding increases resistance at the surface of the electrode decreasing current signal. (D) LFA. Sample flows through the strip and the specific antigen forms a sandwich between probes, immobilizing the AuNPs.

Au nanoprobes for the detection of human papillomavirus (HPV) type 16 and type 18.

1.3.2 Heterogeneous Detection

1.3.2.1 *Microarrays*

Despite the higher extinction coefficient of silver NPs that confers higher spectral signal, the majority of colorimetric systems rely on AuNPs because they are more stable in solution and have higher functionalization efficiencies (Love et al., 2005). In some cases AuNPs are used for molecular detection and biofunctionalization; later, a silver staining is performed, creating core shell structures that share the best properties of both metals (Agasti et al., 2010).

Storhoff et al. (2004b) described a silver staining protocol in a microarray sandwich approach, where specific nucleotides are fixed on a platform and then the target and probe hybridization are promoted. After a washing step, AuNPs are immobilized to the corresponding spots of each target and silver staining is performed. By analyzing the scattering, it was possible to specifically detect *S. aureus mecA* gene (LOD = 100 fM). Based on this method, Zhao et al. (2010) also suggested the use a microarray approach for the detection of influenza A virus. They were able to detect and discriminate H5N1 from H1N1 and H3N2 strains with an LOD of 100 fM after PCR. Li, X.Z. et al. (2013) described the same system for detection of HPV integrated with an optical detection system and software for automated data analysis.

In terms of application in the field, microarray-based approaches present the advantage of high throughput, allowing screening for a high number of samples for a few pathogens or a few samples for a wide range of pathogens. However, to do so, microarray technologies require specialized equipment operated by dedicated technicians who must be capable of handling high-density data and complex analysis.

1.3.2.2 Lateral Flow Assays

The LFA provides a disposable platform for POC. Most of these devices rely on sandwich approaches, where a capture probe captures the analyte of interest and a second probe reports the event that occurs on a suitable matrix (e.g., nitrocellulose, paper, and plastic) (Posthuma-Trumpie et al., 2009; Anfossi et al., 2013a). In terms of performance, LFAs are a simple operation: a liquid sample is placed in a sample pad that, because of the capillarity forces, migrates to a conjugation pad and binds to the probe; a second probe is immobilized and captures the target and the first probe like a sandwich. Commonly, the second probe is immobilized on the platform using impregnated streptavidin-coated microspheres or anti-streptavidin antibodies and/or biotin-modified oligonucleotides forming streptavidin–biotin bonds (Hu et al., 2014). An adsorbent pad is always placed on the extremity of the platform to guarantee vector migration of the sample and a control line is used to discard recognition/hybridization failure. Although fast and easy to perform and requiring minimal technical experience, the limitation of LFAs is sensitivity, which in the case of nucleic acids commonly requires previous amplification (Posthuma-Trumpie et al., 2009). Other systems rely on postdetection signal enhancement (e.g., silver enhancement) that allows improvement of LOD (~5 ng of DNA) (Chua et al., 2011). Many authors described the use of this technology to detect DNA/RNA related to food-borne pathogens, for example, *Vibrio cholerae* (Chua et al., 2011; Ang et al., 2012), *Salmonella* sp. (Liu et al., 2013), and *E. coli* (Pohlmann et al., 2014).

Rastogi et al. (2012) reported the use of locked nucleic acids (LNA) as probes for improved detection of hemolysin-A gene of *E. coli* O157:H7 DNA (LOD ~ 0.4 nM). Hu et al. (2010) used a cross-linking method based on the formation of AuNP aggregates that detect single targets, enhancing the visual signal per target (LOD ~0.1 nM RNA of HIV-1 virus).

1.3.3 Electrochemical Assays

Electrochemical sensors offer simplicity in operation and sample manipulation, provide high sensitivity and specificity, and have the potential to be easily transposed from laboratorial settings to POC devices (Luo and Davis, 2013). Electrochemical detection systems transduce biomolecular recognition into detectable changes of electrochemical properties, such as redox kinetics or electrical impedance or amperage (Doria et al., 2012). A typical electrochemical DNA sensor is composed by electrode, capture probe, reporter probe, and the target (e.g., DNA). Capture probes are usually immobilized onto the electrode surface (e.g., gold, carbon electrodes) and recognize the target DNA; however, they can also be immobilized on other nanomaterials (e.g., gold or magnetic NPs) (Thiruppathiraja et al., 2011; Low et al., 2013). The reporter probe generates an electrochemical signal in response to a reaction. Both probes have complementarity to the target and hybridize in a sandwich method (Wang, Y. et al., 2008). Because of recent advances in nanofabrication and microfluidics, miniaturization of sensors has allowed for sample and reagent volume reduction, improving sensitivity and efficiency. In some sensors, the capture and reporter probes are combined, like the system described by Li et al. (2012). In this configuration, a capture probe in the shape of a hairpin DNA immobilized on a glassy carbon electrode via the 3′ end and reporter probe (Au nanoprobe functionalized at the 5′ end) are used to detect the presence of RNA from hepatitis C virus (HCV). The change in the conformational structure of the hairpin approaches the AuNP and the electrode, leading to a change in current/potential.

One of the advantages of using AuNPs in electrochemical sensors was the possibility of functionalizing enzymes to the surface via simple protocols, and the nanoconjugates can be used to recognize specific sequences or catalyze reactions originating an electrochemical signal (Wang, Y. et al., 2008; Luo and Davis, 2013). Thiruppathiraja et al. (2011) used dual functionalized NPs for the detection of the genomic DNA of several species of *Mycobacterium*. A capture oligonucleotide probe is immobilized on the surface of the electrode, and the reporter Au nanoprobe is bifunctionalized with oligonucleotides and an alkaline phosphatase. The target forms a sandwich with the capture and reporter probes, and paranitrophenol (p-NP) is added as substrate. In the presence of the target, the enzyme hydrolyzes the substrate to paranitrophenyl phosphate (p-NPP), which is detected via differential pulse voltammetry measurements (LOD of circa 1.25 ng/mL of genomic DNA). Taking advantage of magnetic NPs for target separation, electrochemical detection of *V. cholerae* and *Salmonella enteritidis* was also shown to be possible (Vetrone et al., 2012; Low et al., 2013). After magnetic separation, hydrochloric acid promotes gold dissolution on a surface plasmon-coupled emission (SPCE) plate that measures $Au3^{+}$ ions. The authors reported a detection limit of 5 and 100 ng/mL, respectively.

Recently, Russell et al. (2014) reported a method for detection of *E. coli* genomic DNA with a detection limit of only 10 ng of DNA. With a capture oligo probe immobilized on the surface of a gold electrode, a target DNA triggers a rolling circle amplification—with the help of a padlock probe—that produces a long, repetitive ssDNA fragment. Au nanoprobes specifically hybridize along the DNA fragment, forming a bridge to a second electrode. After silver or gold enhancement, a nanowire is generated that connects both electrodes and drastically reduces current resistance.

1.3.4 Fluorescence Assays

The phenomenon called nanosurface energy transfer (NSET) is commonly associated with methods coupling fluorescence to metal nanosurfaces (Swierczewska et al., 2011). Similar to more common Förster resonance energy transfer (FRET), NSET reflects the modulation of fluorescence at the vicinity of metal surfaces (ruled by the LSPR of the NPs) (Griffin et al., 2009). This interaction is distance-dependent, occurring in a range from 2 to 30 nm, twice as that of FRET (Yun et al., 2005). Nucleic acid detection systems based on NSET are typically composed by a fluorophore restrained to the proximity of an NP that highly quenches its fluorescence; the presence of the target induces a conformation modification that increases the distance between fluorophore and NP, thus decreasing the quenching intensity and fluorescence signal can be detected (Swierczewska et al., 2011) (see also gold nanobeacons, Rosa et al., 2012; Conde et al., 2013).

The use of the difference in adsorption to AuNPs experienced by ssDNA versus dsDNA has been previously described for colorimetric DNA detection (see Section 1.3.1.1). NPs are capable of modulating fluorescence emitted from a fluorophore located in close vicinity to the surface of the NPs. This property has been used for the development of simple sensing protocols where detection of the target is based on differential adsorption of an oligonucleotide labeled with a fluorophore onto the surface of the NP (Darbha et al., 2008a). When adsorbed to the surface, fluorescence is quenched. In presence of a complementary target, hybridization (i.e., dsDNA) causes de-adsorption from the surface of the AuNPs and the fluorophore moves away, reducing quenching efficiency (see Figure 1.1B) (Darbha et al., 2008a). The authors used this approach for the detection of *Campylobacter*, *C. perfringens*, and *S. aureus* DNA. Using spectrally different dyes, they were able to specifically detect the presence of each target individually with discrimination of single base mismatches and an LOD of 600 fM. Ganbolda et al. (2012) used a similar method to detect single base mutations in influenza A H1N1 DNA. They also showed that discrimination of single base mutations can be attained if the protocol is performed under near-nonaggregative conditions. Griffin et al. (2009) studied the variations between quenching efficiency and NP size for the detection of HCV RNA. According to the authors, the quenching efficiency of fluorescence increases by three orders of magnitude with NP size. Therefore, NPs with a diameter of 110 nm allow for LOD of 300 fM of RNA.

The use of fluorescence for detection shows a crucial advantage over colorimetric assays by allowing conjugation of several chromophores for several pathogens in multiplex assays, each one having specific spectral signatures. Also, several studies report a direct correlation between the signal intensity and target concentration, which allows the quantification of the infection load (Darbha, et al. 2008a; Griffin et al., 2009; Zhang et al., 2009). A major disadvantage is the need for specific detection equipment and costly dye molecules.

1.3.5 Raman and SERS

It has been shown that the junctions between noble metal NP aggregates are hotspots of SERS because of the overlap of LSPR fields (Park et al., 2008). Cao et al. (2002) described a multiplex approach using five different probes labeled with five different Raman dyes for the detection of five viral agents: hepatitis A (HVA), hepatitis B (HVB), HIV, Ebola, and variola virus, together with bacterium *B. anthracis*. Based on a sandwich assay, the target hybridizes to an immobilized probe and to a Raman-labeled Au nanoprobe, which, after silver deposition for signal enhancement, allows sensitive detection of the SERS signal. To increase signal intensity, Hu et al. (2010) added a second Au nanoprobe that cross-links to the first, forming an NP network that confines dyes and NPs into a small space. Using HIV-1 as the target, they reported an LOD of 0.1 aM.

Kang et al. (2010) used Au nanowires and Raman-labeled Au nanoprobes for detection of *Enterococcus faecium*, *S. aureus*, *Stenotrophomonas maltophilia*, and *Vibrio vulnificus*. Presence of DNA target confines the Raman dye between the AuNP and Au nanowire, thus enhancing the SERS signal. This approach describes a multiplex system with a detection limit of 10 pM. Other authors have used NP aggregation to increase signal in several ways, either by addition of a salt (Ganbold et al., 2012; Papadopoulou and Bell, 2012) or by magnet NPs (Zhang et al., 2011) for the detection of influenza virus A H1N1 DNA (Ganbold et al., 2012), *E. coli* (Papadopoulou and Bell, 2012), and West Nile virus (Zhang et al., 2011).

Another method using the SERS-based approach for detection of infection conditions was described by Wang et al. (2013). Using AgNPs functionalized with a dye-labeled oligonucleotide, the system works like a beacon. The presence of the target opens the beacon and leads to a decrease in SERS signal. The target is a radical *S*-adenosyl methionine domain-containing two (RSAD2) RNA whose product is involved in antiviral defense. RSAD2 has been recognized as one of the most highly induced genes on interferon stimulation or on infection with various viruses, including human cytomegalovirus, influenza virus, HCV, dengue virus, alpha viruses, and retroviruses such HIV. Using this system it was possible to distinguish individuals with symptomatic acute respiratory infections from uninfected individuals with more than 95% accuracy (Wang et al., 2013).

The main advantage of SERS-based assays is clearly the sensitivity and multiplexing capability, whereas the need for specialized equipment and complexity of data analysis are major drawbacks.

1.3.6 Other

Other noticeable approaches using noble metal NP detection have been reported. These are based on coupling the biobarcode assay with mass spectrometry detection (Yang et al., 2010), piezoelectric sensors (Chen et al., 2008; Wang, L. et al., 2008; Hao et al., 2011), hyper-Rayleigh scattering (Darbha et al., 2008b), and acoustic detection (Uludağ et al., 2010).

Since Nam et al. (2002) first described the use of the biobarcode assay for detection of nucleic acids, several protocols followed for a plethora of targets. DNA-functionalized magnetic NPs are used for target separation, and then bifunctionalized AuNPs hybridize to the target. These AuNPs are functionalized with both a specific sequence that recognizes the desired target and a signature sequence in a proportion of 1:100. After separation using the magnetic NPs, thiol–Au bonds are cleaved, releasing the signature sequences and resulting in signal amplification. The method reported by Yang et al. (2010) uses of MALDI TOF mass spectrometry for detection of different signature sequences. Because each oligonucleotide has its characteristic profile, they were able to detect DNA from HIV, HBV, HCV, and *Treponema pallidum* in a multiplex assay with LOD of 0.5 aM.

1.4 APTAMERS AND ANTIBODIES

Ag and AuNP-mediated identification of pathogens can be achieved via detection of specific proteins or carbohydrates using aptamers and/or antibodies (Kaittanis et al., 2010). Methods for detection of these moieties include lateral flow stripes, fluorescent and electrochemical assays, and enzyme-linked immunosorbent assays (ELISAs). Aptamers are small oligonucleotides or peptides capable of molecular detection of specific targets via electrostatic and conformation recognition/binding (Zhou et al., 2014). Aptamers are usually selected via repeated rounds *in vitro* and can be easily functionalized on the surface of noble metal NPs (Zhang et al., 2013). Antibodies are large glycoproteins produced by the immune system to identify and neutralize nonself-moieties, such as bacteria and viruses (Tauran et al., 2013). Both aptamers and antibodies have been described for the detection of specific proteins that are biomarkers of specific pathogens. However, they can also be used for detection of whole organisms. Generally, in these cases, antibodies or aptamers recognize the carbohydrates or proteins from the cellular wall. Toxins may also be detected through antibodies or aptamers (Tauran et al., 2013; Zhang et al., 2013). For instance, detection of Ochratoxin A, produced by *Aspergillus ochraceus*, *Aspergillus carbonarius*, and *Penicillium verrucosum*, is one of the most abundant food-contaminating toxins and has been widely used for the proof-of-concept demonstration of AuNP-based detection. The bottlenecks in the development of these assays are the lack of proper antibodies, the elevated costs, and need for specific storage conditions. Aptamers present cost, manufacture, manipulation, and storage advantages over antibodies (Song et al., 2012). Aptamers are mostly reported in electrochemical assays, whereas antibodies are mostly described in LFAs.

1.4.1 Colorimetric

Colorimetric systems based on noble metal NPs for detection of proteins are similar to those used for nucleic acids detection. Based on the previously described approach of unmodified NPs, some authors described the use of aptamers for detection of Ochratoxin A (Yang et al., 2011) and *Plasmodium falciparum* lactate dehydrogenase (Jeon et al., 2013; Cheung et al., 2013). In these cases, binding of the target to the aptamer results in the deprotection AuNP from aggregation. A cross-linking approach has also been described for detection of *Salmonella* spp. in milk samples (Wang et al., 2010). The method is similar to the previously described cross-linking method, where the presence of a target promotes NP aggregation. In this case, the AuNPs are functionalized with antibodies instead of oligonucleotides.

Another interesting approach uses NP growth after selective separation of pathogens bonded to the immunoprobes by a filter membrane (Li et al., 2009; Sung et al., 2013). Li et al. then proposed centrifugation for separation of bonded and nonbonded immunoprobes for detection of *Giardia lamblia* cysts, whereas Sung et al. (2013) used magnetic NPs for detection of *S. aureus*. Cho and Irudayaraj (2013) described another strategy for signal enhancement for detection of *E. coli* O157:H7 and *Salmonella typhimurium* (Cho and Irudayaraj, 2013). Three immunoprobes were used: a primary goat antibacterium antibody, which varies according to the detection target; a secondary rabbit antigoat antibody; and a tertiary goat antirabbit antibody. Because the secondary and tertiary antibodies can detect each other, presence of a target cross-links AuNP and allows colorimetric detection.

ELISA tests use enzymes that catalyze reactions responsible for a colorimetric change or via a fluorescence signal. In the presence of the target, the enzymes are immobilized in an immuno sandwich assay on top of the target and probe. Plasmonic ELISA has also been reported to benefit from the intense colors of noble metal NPs. Rica and Stevens (2012) used the conventional ELISA method, substituting the dye with AuNPs. In the presence of HIV-1 capsid antigen p24, a catalase-mediated reaction promotes NPs aggregation, which permits an ultrasensitive detection of HIV-infected patients with the naked eye and of viral loads undetectable by a gold standard nucleic acid-based test.

1.4.2 Electrochemical

Most electrochemical detection sensors described for the detection of pathogens membranes, proteins, or toxins using noble metal NPs are based on aptamers. Generally, they follow the same protocol: an aptamer is immobilized on the surface of an electrode and if the target binds to the aptamer, then an alteration in the electrochemical signal is produced (Wang, Y. et al., 2008; Wei et al., 2010; Luo and Davis, 2013). The electrodes used are generally gold or screen-printed carbon with AuNPs on the surface. Both these structures allow for easy functionalization with aptamers via simple thiol chemistry; then, while a quasi-reversible redox reaction is occurring on the surface of the electrode, the binding of the target to the aptamer raises electrochemical resistance due to the steric hindrance of the target (see Figure 1.1C). This principle was described for the detection of several targets, such as viable *S. typhimurium* (Labib et al., 2012), vaccinia virus (Labib et al., 2013), norovirus (Giamberardino et al., 2013), *Salmonella* (Ma et al., 2014), and Ochratoxin A (Evtugyn et al., 2013).

A second strategy for electrochemical detection using aptamers and AuNPs was described for the detection of Ochratoxin A (Kuang et al., 2010; Yang et al., 2014) and toxin A of

Clostridium difficile (Luo et al., 2014). A gold electrode functionalized with a capture oligonucleotide probe complementary to the aptamer is used in a sandwich-like system using Au nanoprobes complementary to different regions of the aptamer. Proximity of the AuNPs to the electrode gives an electrical signal. Target-binding to the aptamer leads to release of both probes, leading to an increase of distance between Au nanoprobes and electrode with a consequent decrease in electric signal.

Wu et al. (2013) used a similar approach for detection of *Brucella melitensis* but instead of aptamers, antibodies were used where presence of the pathogen originates an increase in electrical resistance.

1.4.3 Lateral Flow Assays

The majority of gold and silver immunoprobe assays are based on lateral flow chromatography and the working principle is the same as that for nucleic acid detection, but with capture by an immobilized antibody and detection by gold immunoprobe in a sandwich approach (see Figure 1.1D). *V. cholerae* O1 serotype Ogawa (Chen et al., 2014) and *Paragonimiasis skrjabini* (Wang et al., 2014) are examples of pathogens detected by LFA aided by NPs. However, for increased sensitivity, gold enhancement (Li, J. et al., 2013), silver enhancement (Anfossi et al., 2013b), and the use of gold-functionalized secondary antibodies (Laderman et al., 2008) are some of the strategies used. Li et al. (2011) further reported an interesting approach for the detection of several pathogens in a multiplex assay using fork, star, and cross-like shaped stripes; they described the possibility of detecting several pathogens in a single sample.

Mdluli et al. (2014) described a lateral flow-based method for the detection *M. tuberculosis* 38 kDa monoclonal antibody using an AuNP conjugate functionalized with protein A as the reporter and an antigen as the immobilizer. This system allowed detection of only 5 ng/mL of the target in a sandwich-like method. Because protein A can bind to several antibodies, authors also described the use of dual pathway in the immune chromatographic test strips for reduction of cross-reactivity with other antibodies in serum. It is possible to assess host response to infection with this method.

1.5 iPCR AND OTHER METHODS

Gold spheres, rods, and silver NPs have been reported for pathogen immunodetection. Based on fluorescence (Guirgis et al., 2012; Tan et al., 2012; Chen et al., 2014a), piezoelectric (Shen et al., 2011), SERS (Wang et al., 2011; Temur et al., 2012), and Rayleigh scattering (Singh et al., 2009) assays, these systems have taken advantage of antibodies or aptamers for the detection of several pathogens.

Sano et al. (1992) introduced immune PCR (iPCR) in 1992 as a highly sensitive technique that combines both antibodies and PCR for protein detection. Conjugated with specific DNA sequences, antibodies are used to target the antigen of interest followed by PCR amplification or LAMP (Pourhassan-Moghaddam et al., 2013). Amplification of the DNA probe bound to the antibodies allows for quantitative determination of antigen. iPCR presents itself as an extremely sensitive method for protein detection, deferring from other Au and AgNP methods by not using optical or electrochemical properties; methods using iPCR are sensitive because Au and AgNP are excellent carriers of biomolecules (Syed and Bokhari, 2011). Conjugation of antibodies and DNA with AuNPs is easier when compared with direct conjugation of antibodies with DNA. Furthermore, by using several DNA molecules per AuNP, signal amplification and sensitivity increment are obtained (Chen, L. et al., 2009). This has been used for the detection of Hantaan virus nucleocapsid protein. In this assay, an ELISA plate coated with polyclonal antibodies and AuNPs bearing a signature DNA and monoclonal antibody against the target were used. The signature sequence was released by heating and was amplified using real-time PCR. Later, Perez et al. used magnetic NPs instead of ELISA plates for the separation of respiratory syncytial virus (Perez et al., 2011).

1.6 CONCLUSION

We have provided an overview of existing strategies relying on the use of Au and AgNPs for the detection of biomarkers of infection. Rather than providing an exhaustive list of applications, we aimed to introduce the variety and scope of combinations and modalities that can be easily coupled with current molecular diagnostics technologies to facilitate integration into laboratory and clinical settings.

Noble metal NPs along with their unique electrical, optical, and chemical properties allow faster, sensitive, and cheaper diagnostic assays that can assist medical procedures. Besides sensitivity and speed, nanotechnology has also been focused on affordable and reproducible systems that can be used in remote areas. In fact, there has been increasing interest in and application of traditional molecular diagnostic techniques in clinical settings all over the world, mainly for detection of pathogens. Recent advances in nanotechnology-based detection systems applied to pathogen infection biomarkers have increased over the past decade and have now matured to the point where they can be translated to the clinical setting. These systems have relied strongly on the use of noble metal NPs (mainly gold and silver) because of the inherent simplicity of use associated with the flexibility of functionalization and interaction with biomolecules of interest. Also, detection strategies based on AuNPs and AgNPs provide detection capabilities comparable with that of standard techniques but at a fraction of cost and time without the need for cumbersome sample preparation or equipment. It is expected that, because of these characteristics, NP-based approaches will be incrementally applied to pathogen detection with particular emphasis on POC and remote locations.

One should recognize that despite the tremendous investment in these technologies, translation to the clinics has not yet been achieved. This is clearly observed from the number of different strategies that have been proposed and are found in the literature that lack validation in clinical samples. The next step is focusing on the some of the strategies that exist in the laboratory.

Acknowledgments

The authors thank Fundação para Ciência e Tecnologia, MEC for financial support through CIGMH (Pest-OE/SAU/UI0009/2011-14) and PTDC/BBB-NAN/1812/2012.

References

Agasti, S.S., Rana, S., Park, M.H., Kim, C.K., You, C.C., Rotello, V.M., 2010. Nanoparticles for detection and diagnosis. Adv. Drug Deliv. Rev. 62, 316–328.

Alharbi, K.K., Al-Sheikh, Y.A., 2014. Role and implications of nanodiagnostics in the changing trends of clinical diagnosis. Saudi J. Biol. Sci. 21, 109–117.

Anfossi, L., Baggiani, C., Giovannoli, C., D'arco, G., Giraudi, G., 2013a. Lateral-flow immunoassays for mycotoxins and phycotoxins: a review. Anal. Bioanal. Chem. 405, 467–480.

Anfossi, L., Di Nardo, F., Giovannoli, C., Passini, C., Baggiani, C., 2013b. Increased sensitivity of lateral flow immunoassay for Ochratoxin A through silver enhancement. Anal. Bioanal. Chem. 405, 9859–9867.

Ang, G.Y., Yu, C.Y., Yean, C.Y., 2012. Ambient temperature detection of PCR amplicons with a novel sequence-specific nucleic acid lateral flow biosensor. Biosens. Bioelectron. 38, 151–156.

Azzazy, H.M., Mansour, M.M., Kazmierczak, S.C., 2006. Nanodiagnostics: a new frontier for clinical laboratory medicine. Clin. Chem. 52, 1238–1246.

Bakthavathsalam, P., Rajendran, V.K., Mohammed, J.A., 2012. A direct detection of *Escherichia coli* genomic DNA using gold nanoprobes. J. Nanobiotechnol. 10, 8.

Baptista, P.V., Koziol-Montewka, M., Paluch-Oles, J., Doria, G., Franco, R., 2006. Gold-nanoparticle-probe–based assay for rapid and direct detection of *Mycobacterium tuberculosis* DNA in clinical samples. Clin. Chem. 52, 1433–1434.

Baptista, P.V., Doria, G., Quaresma, P., Cavadas, M., Neves, C.S., Gomes, I., et al., 2011. Nanoparticles in molecular diagnostics. Prog. Mol. Biol. Transl. Sci. 104, 427–488.

Boisseau, P., Loubaton, B., 2011. Nanomedicine, nanotechnology in medicine. C.R. Phys. 12, 620–636.

Boyce, J.D., Cullen, P.A., Adler, B., 2004. Genomic-scale analysis of bacterial gene and protein expression in the host. Emerg. Infect. Dis. 10, 1357–1362.

Cabral, R.M., Baptista, P.V., 2013. The chemistry and biology of gold nanoparticle-mediated photothermal therapy: promises and challenges. Nano. LIFE. 3 (3), 1330001.

Cao, Y.C., Jin, R., Mirkin, C.A., 2002. Nanoparticles with Raman spectroscopic fingerprints for DNA and RNA detection. Science 297, 1536–1540.

Chen, J., Zhang, X., Cai, S., Wu, D., Chen, M., Wang, S., et al., 2014a. A fluorescent aptasensor based on DNA-scaffolded silver-nanocluster for Ochratoxin A detection. Biosens. Bioelectron. 57, 226–231.

Chen, L., Wei, H., Guo, Y., Cui, Z., Zhang, Z., Zhang, X.-E., 2009. Gold nanoparticle enhanced immuno-PCR for ultrasensitive detection of Hantaan virus nucleocapsid protein. J. Immunol. Methods 346, 64–70.

Chen, S.H., Wu, V.C., Chuang, Y.C., Lin, C.S., 2008. Using oligonucleotide-functionalized Au nanoparticles to rapidly detect foodborne pathogens on a piezoelectric biosensor. J. Microbiol. Methods 73, 7–17.

Chen, S.H., Lin, K.I., Tang, C.Y., Peng, S.L., Chuang, Y.C., Lin, Y.R., et al., 2009. Optical detection of human papillomavirus type 16 and type 18 by sequence sandwich hybridization with oligonucleotide-functionalized Au nanoparticles. IEEE Trans. Nanobiosci. 8, 120–131.

Chen, W., Zhang, J., Lu, G., Yuan, Z., Wu, Q., Li, J., et al., 2014. Development of an immunochromatographic lateral flow device for rapid diagnosis of *Vibrio cholerae* O1 serotype Ogawa. Clin. Biochem. 47, 448–454.

Cheung, Y.-W., Kwok, J., Law, A.W.L., Watt, R.M., Kotaka, M., Tanner, J.A., 2013. Structural basis for discriminatory recognition of *Plasmodium* lactate dehydrogenase by a DNA aptamer. Proc. Natl. Acad. Sci. U.S.A. 110, 15967–15972.

Cho, I.-H., Irudayaraj, J., 2013. *In-situ* immuno-gold nanoparticle network ELISA biosensors for pathogen detection. Int. J. Food Microbiol. 164, 70–75.

Chua, A., Yean, C.Y., Ravichandran, M., Lim, B., Lalitha, P., 2011. A rapid DNA biosensor for the molecular diagnosis of infectious disease. Biosens. Bioelectron. 26, 3825–3831.

Compton, J., 1991. Nucleic acid sequence-based amplification. Nature 350, 91–92.

Conde, J., Rosa, J., De La Fuente, J.M., Baptista, P.V., 2013. Gold-nanobeacons for simultaneous gene specific silencing and intracellular tracking of the silencing events. Biomaterials 34, 2516–2523.

Darbha, G.K., Lee, E., Anderson, Y.R., Fowler, P., Mitchell, K., Ray, P.C., 2008a. Miniaturized sensor for microbial pathogens DNA and chemical toxins. IEEE Sensors J. 8, 693–700.

Darbha, G.K., Rai, U.S., Singh, A.K., Ray, P.C., 2008b. Gold-nanorod-based sensing of sequence specific HIV-1 virus DNA by using hyper-Rayleigh scattering spectroscopy. Chemistry 14, 3896–3903.

Deng, H., Zhang, X., Kumar, A., Zou, G., Zhang, X., Liang, X. J., 2013. Long genomic DNA amplicons adsorption onto unmodified gold nanoparticles for colorimetric detection of *Bacillus anthracis*. Chem. Commun. (Camb). 49, 51–53.

Doria, G., Conde, J., Veigas, B., Giestas, L., Almeida, C, Assunção, M., et al., 2012. Noble metal nanoparticles for biosensing applications. Sensors (Basel). 12, 1657–1687.

Dreaden, E.C., Alkilany, A.M., Huang, X., Murphy, C.J., El-Sayed, M.A., 2012. The golden age: gold nanoparticles for biomedicine. Chem. Soc. Rev. 41, 2740–2779.

Evtugyn, G., Porfireva, A., Stepanova, V., Kutyreva, M., Gataulina, A., Ulakhovich, N., et al., 2013. Impedimetric aptasensor for Ochratoxin a determination based on au nanoparticles stabilized with hyper-branched polymer. Sensors (Basel). 13, 16129–16145.

Freeman, W.M., Walker, S.J., Vrana, K.E., 1999. Quantitative RT-PCR: pitfalls and potential. Biotechniques 26, 112–122, 124–5.

Fu, Z., Zhou, X., Xing, D., 2013. Sensitive colorimetric detection of *Listeria monocytogenes* based on isothermal gene amplification and unmodified gold nanoparticles. Methods 64, 260–266.

Ganbold, E.O., Kang, T., Lee, K., Lee, S.Y., Joo, S.W., 2012. Aggregation effects of gold nanoparticles for single-base mismatch detection in influenza A (H1N1) DNA sequences using fluorescence and Raman measurements. Colloids Surf B Biointerfaces 93, 148–153.

Giamberardino, A., Labib, M., Hassan, E.M., Tetro, J.A., Springthorpe, S., Sattar, S.A., et al., 2013. Ultrasensitive norovirus detection using DNA aptasensor technology. PLoS One 8, e79087.

Gill, P., Ghalami, M., Ghaemi, A., Mosavari, N., Abdul-Tehrani, H., Sadeghizadeh, M., 2009. Nanodiagnostic method for colorimetric detection of *Mycobacterium tuberculosis* 16S rRNA. NanoBiotechnol. 4, 28–35.

Goluch, E.D., Nam, J.M., Georganopoulou, D.G., Chiesl, T. N., Shaikh, K.A., Ryu, K.S., et al., 2006. A bio-barcode assay for on-chip attomolar-sensitivity protein detection. Lab Chip 6, 1293–1299.

Griffin, J., Singh, A.K., Senapati, D., Rhodes, P., Mitchell, K., Robinson, B., et al., 2009. Size- and distance-dependent nanoparticle surface-energy transfer (NSET) method for selective sensing of HCV RNA. Chemistry 15, 342–351.

Guirgis, B.S., Sa e Cunha, C., Gomes, I., Cavadas, M., Silva, I., Doria, G., et al., 2012. Gold nanoparticle-based fluorescence immunoassay for malaria antigen detection. Anal. Bioanal. Chem. 402, 1019–1027.

Hao, R.Z., Song, H.B., Zuo, G.M., Yang, R.F., Wei, H.P., Wang, D.B., et al., 2011. DNA probe functionalized QCM biosensor based on gold nanoparticle amplification for *Bacillus anthracis* detection. Biosens. Bioelectron. 26, 3398–3404.

He, W., Huang, C.Z., Li, Y.F., Xie, J.P., Yang, R.G., Zhou, P. F., et al., 2008. One-step label-free optical genosensing system for sequence-specific DNA related to the human immunodeficiency virus based on the measurements of light scattering signals of gold nanorods. Anal. Chem. 80, 8424–8430.

Hu, J., Zheng, P.C., Jiang, J.H., Shen, G.L., Yu, R.Q., Liu, G.K., 2010. Sub-attomolar HIV-1 DNA detection using surface-enhanced Raman spectroscopy Analyst 135, 1084–1089.

Hu, J., Wang, S., Wang, L., Li, F., Pingguan-Murphy, B., Lu, T.J., et al., 2014. Advances in paper-based point-of-care diagnostics. Biosens. Bioelectron. 54, 585–597.

Hussain, M.M., Samir, T.M., Azzazy, H.M., 2013. Unmodified gold nanoparticles for direct and rapid detection of *Mycobacterium tuberculosis* complex. Clin. Biochem. 46, 633–637.

Jain, P.K., Lee, K.S., El-Sayed, I.H., El-Sayed, M.A., 2006. Calculated absorption and scattering properties of gold nanoparticles of different size, shape, and composition: applications in biological imaging and biomedicine. J. Phys. Chem. B. 110, 7238–7248.

Javier, D.J., Castellanos-Gonzalez, A., Weigum, S.E., White Jr., A.C., Richards-Kortum, R., 2009. Oligonucleotide-gold nanoparticle networks for detection of *Cryptosporidium parvum* heat shock protein 70 mRNA. J. Clin. Microbiol. 47, 4060–4066.

Jeon, W., Lee, S., Manjunatha, D.H., Ban, C., 2013. A colorimetric aptasensor for the diagnosis of malaria based on cationic polymers and gold nanoparticles. Anal. Biochem. 439, 11–16.

Kaittanis, C., Santra, S., Perez, J.M., 2010. Emerging nanotechnology-based strategies for the identification of microbial pathogenesis. Adv. Drug Deliv. Rev. 62, 408–423.

Kalidasan, K., Neo, J.L., Uttamchandani, M., 2013. Direct visual detection of *Salmonella* genomic DNA using gold nanoparticles. Mol. Biosyst. 9, 618–621.

Kang, K.A., Wang, J., Jasinski, J.B., Achilefu, S., 2011. Fluorescence manipulation by gold nanoparticles: from complete quenching to extensive enhancement. J. Nanobiotechnol. 9, 16.

Kang, T., Yoo, S.M., Yoon, I., Lee, S.Y., Kim, B., 2010. Patterned multiplex pathogen DNA detection by Au particle-on-wire SERS sensor. Nano. Lett. 10, 1189–1193.

Kuang, H., Chen, W., Xu, D., Xu, L., Zhu, Y., Liu, L., et al., 2010. Fabricated aptamer-based electrochemical "signal-off" sensor of Ochratoxin A. Biosens. Bioelectron. 26, 710–716.

Labib, M., Zamay, A.S., Kolovskaya, O.S., Reshetneva, I.T., Zamay, G.S., Kibbee, R.J., et al., 2012. Aptamer-based viability impedimetric sensor for bacteria. Anal. Chem. 84, 8966–8969.

Labib, M., Zamay, A.S., Berezovski, M.V., 2013. Multifunctional electrochemical aptasensor for aptamer clones screening, virus quantitation in blood and viability assessment. Analyst 138, 1865–1875.

Laderman, E.I., Whitworth, E., Dumaual, E., Jones, M., Hudak, A., Hogrefe, W., et al., 2008. Rapid, sensitive, and specific lateral-flow immunochromatographic point-of-care device for detection of herpes simplex virus type 2-specific immunoglobulin g antibodies in serum and whole blood. Clin. Vaccine Immunol. 15, 159–163.

Le Ru, E.C., Blackie, E., Meyer, M., Etchegoin, P.G., 2007. Surface enhanced Raman scattering enhancement factors: a comprehensive study. J. Phys. Chem. C 111, 13794–13803.

Li, C.-Z., Vandenberg, K., Prabhulkar, S., Zhu, X., Schneper, L., Methee, K., et al., 2011. Paper based point-of-care testing disc for multiplex whole cell bacteria analysis. Biosens. Bioelectron. 26, 4342–4348.

Li, H., Rothberg, L., 2004. Colorimetric detection of DNA sequences based on electrostatic interactions with unmodified gold nanoparticles. Proc. Natl. Acad. Sci. U.S.A. 101, 14036–14039.

Li, J., Zou, M., Chen, Y., Xue, Q., Zhang, F., Li, B., et al., 2013. Gold immunochromatographic strips for enhanced detection of Avian influenza and Newcastle disease viruses. Anal. Chim. Acta. 782, 54–58.

Li, W., Wu, P., Zhang, H., Cai, C., 2012. Catalytic signal amplification of gold nanoparticles combining with conformation-switched hairpin DNA probe for hepatitis C virus quantification. Chem. Commun. (Camb). 48, 7877–7879.

Li, X.X., Cao, C., Han, S.J., Sim, S.J., 2009. Detection of pathogen based on the catalytic growth of gold nanocrystals. Water Res. 43, 1425–1431.

Li, X.Z., Kim, S., Cho, W., Lee, S.Y., 2013. Optical detection of nanoparticle-enhanced human papillomavirus genotyping microarrays. Biomed. Opt. Express 4, 187–192.

Liandris, E., Gazouli, M., Andreadou, M., Čomor, M., Abazovic, N., Sechi, L.A., et al., 2009. Direct detection of unamplified DNA from pathogenic mycobacteria using DNA-derivatized gold nanoparticles. J. Microbiol. Methods 78, 260–264.

Liu, C.C., Yeung, C.Y., Chen, P.H., Yeh, M.K., Hou, S.Y., 2013. Salmonella detection using 16S ribosomal DNA/RNA probe-gold nanoparticles and lateral flow immunoassay. Food Chem. 141, 2526–2532.

Liu, M., Yuan, M., Lou, X., Mao, H., Zheng, D., Zou, R., et al., 2011. Label-free optical detection of single-base mismatches by the combination of nuclease and gold nanoparticles. Biosens. Bioelectron. 26, 4294–4300.

Love, J.C., Estroff, L.A., Kriebel, J.K., Nuzzo, R.G., Whitesides, G.M., 2005. Self-assembled monolayers of thiolates on metals as a form of nanotechnology. Chem. Rev. 105, 1103–1169.

Low, K.F., Karimah, A., Yean, C.Y., 2013. A thermostabilized magnetogenosensing assay for DNA sequence-specific detection and quantification of *Vibrio cholerae*. Biosens. Bioelectron. 47, 38–44.

Luo, P., Liu, Y., Xia, Y., Xu, H., Xie, G., 2014. Aptamer biosensor for sensitive detection of toxin A of *Clostridium difficile* using gold nanoparticles synthesized by *Bacillus stearothermophilus*. Biosens. Bioelectron. 54, 217–221.

Luo, X., Davis, J.J., 2013. Electrical biosensors and the label free detection of protein disease biomarkers. Chem. Soc. Rev. 42, 5944–5962.

Ma, X., Jiang, Y., Jia, F., Yu, Y., Chen, J., Wang, Z., 2014. An aptamer-based electrochemical biosensor for the detection of *Salmonella*. J. Microbiol. Methods 98, 94–98.

Mancuso, M., Jiang, L., Cesarman, E., Erickson, D., 2013. Multiplexed colorimetric detection of Kaposi's sarcoma associated herpesvirus and Bartonella DNA using gold and silver nanoparticles. Nanoscale 5, 1678–1686.

Mcglennen, R.C., 2001. Miniaturization technologies for molecular diagnostics. Clin. Chem. 47, 393–402.

Mdluli, P., Tetyana, P., Sosibo, N., Van Der Walt, H., Mlambo, M., Skepu, A., et al., 2014. Gold nanoparticle based Tuberculosis immunochromatographic assay: the quantitative ESE Quanti analysis of the intensity of test and control lines. Biosens. Bioelectron. 54, 1–6.

Mirkin, C.A., Letsinger, R.L., Mucic, R.C., Storhoff, J.J., 1996. A DNA-based method for rationally assembling nanoparticles into macroscopic materials. Nature 382, 607–609.

Nam, J.M., Park, S.J., Mirkin, C.A., 2002. Bio-barcodes based on oligonucleotide-modified nanoparticles. J. Am. Chem. Soc. 124, 3820–3821.

Notomi, T., 2000. Loop-mediated isothermal amplification of DNA. Nucleic Acids Res. 28, 63e-63.

Papadopoulou, E., Bell, S.E., 2012. Label-free detection of nanomolar unmodified single- and double-stranded DNA by using surface-enhanced Raman spectroscopy on Ag and Au colloids. Chemistry 18, 5394–5400.

Park, W.-H., Ahn, S.-H., Kim, Z.H., 2008. Surface-enhanced Raman scattering from a single nanoparticle–plane junction. ChemPhysChem. 9, 2491–2494.

Pedrosa, P., Veigas, B., Machado, D., Couto, I., Viveiros, M., Baptista, P.V., 2014. Gold nanoprobes for multi loci assessment of multi-drug resistant tuberculosis. Tuberculosis 94, 332–337.

Perez, J.W., Vargis, E.A., Russ, P.K., Haselton, F.R., Wright, D.W., 2011. Detection of respiratory syncytial virus using nanoparticle amplified immuno-polymerase chain reaction. Anal. Biochem. 410, 141–148.

Pohlmann, C., Dieser, I., Sprinzl, M., 2014. A lateral flow assay for identification of *Escherichia coli* by ribosomal RNA hybridisation. Analyst 139, 1063–1071.

Posthuma-Trumpie, G., Korf, J., Amerongen, A., 2009. Lateral flow (immuno)assay: its strengths, weaknesses, opportunities and threats. A literature survey. Anal. Bioanal. Chem. 393, 569–582.

Pourhassan-Moghaddam, M., Rahmati-Yamchi, M., Akbarzadeh, A., Daraee, H., Nejati-Koshki, K., Hanifehpour, Y., et al., 2013. Protein detection through different platforms of immuno-loop-mediated isothermal amplification. Nanoscale Res. Lett. 8, 485.

Rastogi, S.K., Gibson, C.M., Branen, J.R., Aston, D.E., Branen, A.L., Hrdlicka, P.J., 2012. DNA detection on lateral flow test strips: enhanced signal sensitivity using LNA-conjugated gold nanoparticles. Chem. Commun. (Camb). 48, 7714–7716.

Rica, R., Stevens, M.M., 2012. Plasmonic ELISA for the ultrasensitive detection of disease biomarkers with the naked eye. Nat. Nanotechnol. 7, 821–824.

Rosa, J., Conde, J., De La Fuente, J.M., Lima, J.C., Baptista, P.V., 2012. Gold-nanobeacons for real-time monitoring of RNA synthesis. Biosens. Bioelectron. 36, 161–167.

Russell, C., Welch, K., Jarvius, J., Cai, Y., Brucas, R., Nikolajeff, F., et al., 2014. Gold nanowire based electrical DNA detection using rolling circle amplification. ACS Nano. 8, 1147–1153.

Sano, T., Smith, C.L., Cantor, C.R., 1992. Immuno-PCR: very sensitive antigen detection by means of specific antibody-DNA conjugates. Science 258, 120–122.

Sau, T.K., Rogach, A.L., Jackel, F., Klar, T.A., Feldmann, J., 2010. Properties and applications of colloidal nonspherical noble metal nanoparticles. Adv. Mater. 22, 1805–1825.

Shen, Z.-Q., Wang, J.-F., Qiu, Z.-G., Jin, M., Wang, X.-W., Chen, Z.-L., et al., 2011. QCM immunosensor detection of *Escherichia coli* O157:H7 based on beacon immunomagnetic nanoparticles and catalytic growth of colloidal gold. Biosens. Bioelectron. 26, 3376–3381.

Singh, A.K., Senapati, D., Wang, S., Griffin, J., Neely, A., Candice, P., et al., 2009. Gold nanorod based selective identification of *Escherichia coli* bacteria using two-photon Rayleigh scattering spectroscopy. ACS Nano. 3, 1906–1912.

Song, K.-M., Lee, S., Ban, C., 2012. Aptamers and their biological applications. Sensors (Basel). 12, 612–631.

Soo, P.C., Horng, Y.T., Chang, K.C., Wang, J.Y., Hsueh, P. R., Chuang, C.Y., et al., 2009. A simple gold nanoparticle probes assay for identification of *Mycobacterium tuberculosis* and *Mycobacterium tuberculosis* complex from clinical specimens. Mol. Cell. Probes 23, 240–246.

Storhoff, J.J., Lucas, A.D., Garimella, V., Bao, Y.P., Muller, U.R., 2004a. Homogeneous detection of unamplified genomic DNA sequences based on colorimetric scatter of gold nanoparticle probes. Nat. Biotechnol. 22, 883–887.

Storhoff, J.J., Marla, S.S., Bao, P., Hagenow, S., Mehta, H., Lucas, A., et al., 2004b. Gold nanoparticle-based detection of genomic DNA targets on microarrays using a novel optical detection system. Biosens. Bioelectron. 19, 875–883.

Sung, Y.J., Suk, H.-J., Sung, H.Y., Li, T., Poo, H., Kim, M.-G., 2013. Novel antibody/gold nanoparticle/magnetic nanoparticle nanocomposites for immunomagnetic separation and rapid colorimetric detection of *Staphylococcus aureus* in milk. Biosens. Bioelectron. 43, 432–439.

Swierczewska, M., Lee, S., Chen, X., 2011. The design and application of fluorophore-gold nanoparticle activatable probes. Phys. Chem. Chem. Phys. 13, 9929–9941.

Syed, M.A., Bokhari, S.H.A., 2011. Gold nanoparticle based microbial detection and identification. J. Biomed. Nanotechnol. 7, 229–237.

Tan, L., Neoh, K.G., Kang, E.-T., Choe, W.-S., Su, X., 2012. Affinity analysis of DNA aptamer–peptide interactions using gold nanoparticles. Anal. Biochem. 421, 725–731.

Tauran, Y., Brioude, A., Coleman, A.W., Rhimi, M., Kim, B., 2013. Molecular recognition by gold, silver and copper nanoparticles. World J. Biol. Chem. 4, 35–63.

Temur, E., Zengin, A., Boyacı, İ.H., Dudak, F.C., Torul, H., Tamer, U., 2012. Attomole sensitivity of staphylococcal enterotoxin B detection using an aptamer-modified surface-enhanced Raman scattering probe. Anal. Chem. 84, 10600–10606.

Thiruppathiraja, C., Kamatchiammal, S., Adaikkappan, P., Santhosh, D.J., Alagar, M., 2011. Specific detection of *Mycobacterium* sp. Genomic DNA using dual labeled gold nanoparticle based electrochemical biosensor. Anal. Biochem. 417, 73–79.

Tran, Q.H., Nguyen, V.Q., Le, A.-T., 2013. Silver nanoparticles: synthesis, properties, toxicology, applications and perspectives. Adv. Natl. Sci. Nanosci. Nanotech. 4, 033001.

Uludağ, Y., Hammond, R., Cooper, M.A., 2010. A signal amplification assay for HSV type 1 viral DNA detection using nanoparticles and direct acoustic profiling. J. Nanobiotechnol. 8, 3.

Veigas, B., Machado, D., Perdigão, J., Portugal, I., Couto, I., Viveiros, M., et al., 2010. Au-nanoprobes for detection of SNPs associated with antibiotic resistance in *Mycobacterium tuberculosis*. Nanotechnology 21, 415101.

Veigas, B., Doria, G., Baptista, P.V., 2012a. In: Cardona, P. (Ed.), Nanodiagnostics for Tuberculosis. InTech—Open Access Publishers, Croatia, pp. 257–276 (Chapter 12).

Veigas, B., Jacob, J.M., Costa, M.N., Santos, D.S., Viveiros, M., Inácio, J., et al., 2012b. Gold on paper-paper platform for Au-nanoprobe TB detection. Lab Chip 12, 4802–4808.

Veigas, B., Pedrosa, P., Couto, I., Viveiros, M., Baptista, P., 2013. Isothermal DNA amplification coupled to Au-nanoprobes for detection of mutations associated to Rifampicin resistance in *Mycobacterium tuberculosis*. J. Nanobiotechnol. 11, 38.

Vetrone, S.A., Huarng, M.C., Alocilja, E.C., 2012. Detection of non-PCR amplified *S. enteritidis* genomic DNA from food matrices using a gold-nanoparticle DNA biosensor: a proof-of-concept study. Sensors (Basel) 12, 10487–10499.

Wang, H.-N., Fales, A.M., Zaas, A.K., Woods, C.W., Burke, T., Ginsburg, G.S., et al., 2013. Surface-enhanced Raman scattering molecular sentinel nanoprobes for viral infection diagnostics. Anal. Chim. Acta 786, 153–158.

Wang, L., Wei, Q., Wu, C., Hu, Z., Ji, J., Wang, P., 2008. The *Escherichia coli* O157:H7 DNA detection on a gold nanoparticle-enhanced piezoelectric biosensor. Chin. Sci. Bull. 53, 1175–1184.

Wang, S., Singh, A.K., Senapati, D., Neely, A., Yu, H., Ray, P.C., 2010. Rapid colorimetric identification and targeted photothermal lysis of *Salmonella* bacteria by using bioconjugated oval-shaped gold nanoparticles. Chemistry 16, 5600–5606.

Wang, Y., Xu, H., Zhang, J., Li, G., 2008. Electrochemical sensors for clinic analysis. Sensors (Basel) 8, 2043–2081.

Wang, Y., Ravindranath, S., Irudayaraj, J., 2011. Separation and detection of multiple pathogens in a food matrix by magnetic SERS nanoprobes. Anal. Bioanal. Chem. 399, 1271–1278.

Wang, Y., Wang, L., Zhang, J., Wang, G., Chen, W., Chen, L., et al., 2014. Preparation of colloidal gold immunochromatographic strip for detection of *Paragonimiasis skrjabini*. PLoS One 9, e92034.

Wei, F., Lillehoj, P.B., Ho, C.M., 2010. DNA diagnostics: nanotechnology-enhanced electrochemical detection of nucleic acids. Pediatr. Res. 67, 458–468.

WHO, 2014. The top 10 causes of death. Fact sheet N°310. Available from: <http://www.who.int/mediacentre/factsheets/fs310/en>.

Wu, H., Zuo, Y., Cui, C., Yang, W., Ma, H., Wang, X., 2013. Rapid quantitative detection of *Brucella melitensis* by a label-free impedance immunosensor based on a gold nanoparticle-modified screen-printed carbon electrode. Sensors (Basel) 13, 8551–8563.

Xie, X., Xu, W., Li, T., Liu, X., 2011. Colorimetric detection of HIV-1 ribonuclease H activity by gold nanoparticles. Small 7, 1393–1396.

Yang, B., Gu, K., Sun, X., Huang, H., Ding, Y., Wang, F., et al., 2010. Simultaneous detection of attomolar pathogen DNAs by Bio-MassCode mass spectrometry. Chem. Commun. (Camb). 46, 8288–8290.

Yang, C., Wang, Y., Marty, J.-L., Yang, X., 2011. Aptamer-based colorimetric biosensing of Ochratoxin A using unmodified gold nanoparticles indicator. Biosens. Bioelectron. 26, 2724–2727.

Yang, X., Qian, J., Jiang, L., Yan, Y., Wang, K., Liu, Q., et al., 2014. Ultrasensitive electrochemical aptasensor for Ochratoxin A based on two-level cascaded signal amplification strategy. Bioelectrochemistry 96, 7–13.

Yun, C.S., Javier, A., Jennings, T., Fisher, M., Hira, S., Peterson, S., et al., 2005. Nanometal surface energy transfer in optical rulers, breaking the FRET barrier. J. Am. Chem. Soc. 127, 3115–3119.

Zhang, D., Carr, D.J., Alocilja, E.C., 2009. Fluorescent bio-barcode DNA assay for the detection of *Salmonella enterica* serovar Enteritidis. Biosens. Bioelectron. 24, 1377–1381.

Zhang, H., Harpster, M.H., Park, H.J., Johnson, P.A., Wilson, W.C., 2011. Surface-enhanced Raman scattering detection of DNA derived from the West Nile virus genome using magnetic capture of Raman-active gold nanoparticles. Anal. Chem. 83, 254–260.

Zhang, J., Liu, B., Liu, H., Zhang, X., Tan, W., 2013. Aptamer-conjugated gold nanoparticles for bioanalysis. Nanomedicine (Lond). 8, 983–993.

Zhang, Y., Huang, R., Zhu, X., Wang, L., Wu, C., 2012. Synthesis, properties, and optical applications of noble metal nanoparticle-biomolecule conjugates. Chin. Sci. Bull. 57, 238–246.

Zhao, J., Tang, S., Storhoff, J., Marla, S., Bao, Y.P., Wang, X., et al., 2010. Multiplexed, rapid detection of H5N1 using a PCR-free nanoparticle-based genomic microarray assay. BMC Biotechnol. 10, 74.

Zhao, P., Li, N., Astruc, D., 2013. State of the art in gold nanoparticle synthesis. Coord. Chem. Rev. 257, 638–665.

Zhou, W., Jimmy Huang, P.-J., Ding, J., Liu, J., 2014. Aptamer-based biosensors for biomedical diagnostics. Analyst 139, 2627–2640.

CHAPTER

2

Antimicrobial Models in Nanotechnology: From the Selection to Application in the Control and Treatment of Infectious Diseases

Juan Bueno

Bioprospecting Development and Consulting, Bogotá, Colombia

2.1 INTRODUCTION

The emergence of antibiotic resistance produces approximately 2,049,442 diseases and causes 23,000 deaths in the United States every year, as well as 25,000 deaths in Europe annually (Cars et al., 2011; Hecker et al., 2014; Fauci and Marston, 2014). Also, it has been estimated that 70% of nosocomial infections present resistance to one of the antimicrobials used for treatment (Mishra et al., 2012). Currently, a large number of microorganisms associated with human diseases have become multidrug-resistant (MDR), including *Mycobacterium tuberculosis, Acinetobacter baumannii, Burkholderia cepacia, Campylobacter jejuni, Citrobacter freundii, Clostridium difficile, Enterobacter* spp., *Enterococcus faecium, Enterococcus faecalis, Escherichia coli, Haemophilus influenzae, Klebsiella pneumoniae, Proteus mirabilis, Pseudomonas aeruginosa, Salmonella* spp., *Serratia* spp., *Staphylococcus aureus, Staphylococcus epidermidis, Stenotrophomonas maltophilia*, and *Streptococcus pneumoniae*. All these microorganisms are grouped under the term "superbugs," which possess various mutations that cause an advanced level of antimicrobial resistance and, in some cases, increased virulence and transmissibility (Davies and Davies, 2010).

By definition, an antimicrobial agent is referred to as a compound that has the ability to kill or inhibit the growth of microorganisms, in most cases by blocking the metabolism and altering the synthesis of useful substances for cellular activities. Thus, it is necessary to increase the intracellular concentration of the same for more microbicidal activity and further reduction of tissue toxicity, which limits their use (Talbot et al., 2006). To tackle antimicrobial resistance and decrease the toxicity of antibiotics, there has

Nanotechnology in Diagnosis, Treatment and Prophylaxis of Infectious Diseases.
DOI: http://dx.doi.org/10.1016/B978-0-12-801317-5.00002-5

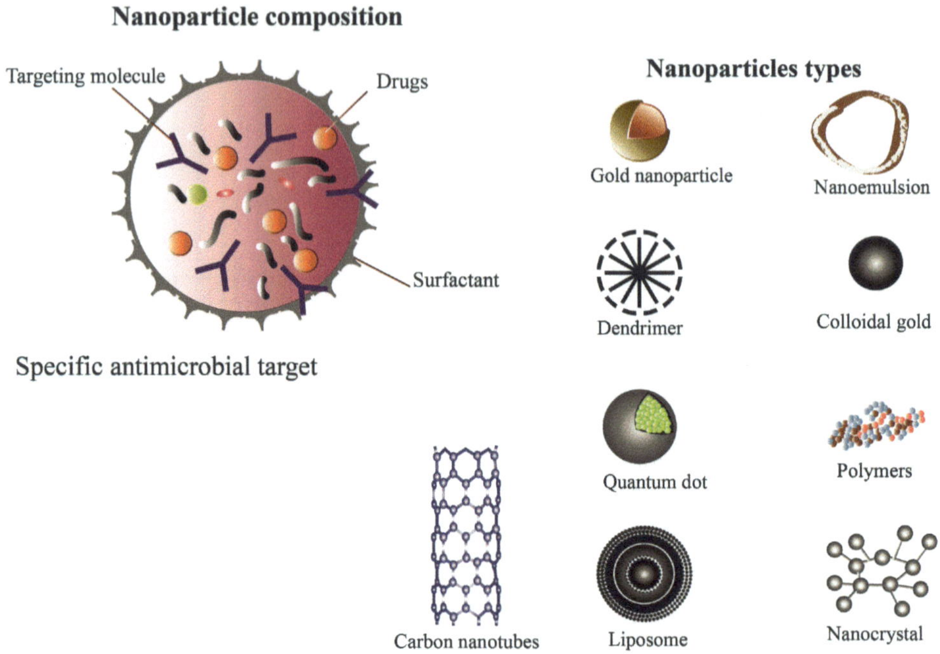

FIGURE 2.1 NMs with antimicrobial activity.

been an increasing interest in the discovery of new drugs and the development of new formulations contemplating the use of nanotechnology (Allahverdiyev et al., 2011). Nanotechnology is defined as the technology that uses nanomaterials (NMs) (particles, fibers, tubes, grains, etc.) in their design and applications (Figure 2.1); these have the main characteristics of being 1–100 nm in size and having novel chemical and physical properties such as high reactivity and ability to selectively destroy the cell membranes of pathogenic bacteria (Table 2.1) (Veerapandian and Yun, 2011). Because of their smaller size, they have a large surface area and can modify many properties such as conductivity, magnetism, as well as mechanical and catalytic properties of materials (Wijnhoven et al., 2009; Sweet et al., 2012). NMs at present are being used as broad-spectrum antimicrobial agents, such as silver nanoparticles attached to clothing for surgery and medical care. Also, the use of nanomodified surfaces has been effective in inhibiting bacterial growth and biofilm formation, decreasing the incidence of infections associated with medical procedures (Machado et al., 2010). Equally, NMs can be incorporated into polymer matrices to make antibacterial nanocomposites, which selectively lyse microbial membranes and are very useful as macromolecular antimicrobial polymers (MAPs) in biosensors and biomedical devices, fibers for wound dressing, membranes for water purification, and dispersion formulations (Santos et al., 2012; Aruguete et al., 2013). The antimicrobial activity of nanoparticles can be justified by several mechanisms of direct toxicity such as disruption cell wall, interruption of electron transport, alteration of the membrane potential, escape of the cell contents by physical damage, and generation of reactive oxygen species (ROS) (Figure 2.2) (Hajipour et al., 2012; Pagnout et al., 2012). In antibiotic development against MDR bacteria, there is a strong interest in the use of NMs for drug delivery to improve the pharmacokinetics, therapeutic index, and intracellular drug therapy (Zhang et al., 2010; Armstead and Li, 2011).

TABLE 2.1 Antimicrobial nanomedicines (Yacoby and Benhar, 2008; Matthews et al., 2010; Hajipour et al., 2012)

Nanomedicine	Mechanism of action	Antimicrobial activity
SYNTHETIC ANTIBACTERIAL NANOMEDICINES		
Carbon nanotubes and fullerenes	Cell membrane damage, fullerenes can produce radical-oxygen species	*E. coli* DH5α, *Vibrio fischeri*, and *Bacillus subtilis*
Silver nanoparticles	Cell membrane damage by accumulation	*E. coli*, *P. aeruginosa*, and *S. aureus*
Gold nanoparticles	Disruption of cell membrane	*E. coli*, *Salmonella tiphymurium*
Bioactive glasses	Production of alkaline species	*E. faecalis*
Metal oxide nanoparticles	Electrostatic interaction between nanoparticle and bacteria	*E. coli*, *B. subtilis*, and *S. aureus*
Magnesium oxide nanoparticles	Formation of superoxide anions	*Bacillus subtilis*, *S. aureus*
Zinc oxide nanoparticles	Induction of oxidative stress	*C. jejuni*, *Salmonella enterica* serovar Enteritidis, and *E. coli* O157:H7
Silicon dioxide nanoparticles	Release of nitric oxide	*S. aureus*
BIOLOGICAL-BASED ANTIBACTERIAL NANOMEDICINES		
Chitosan nanofiber	Loss of membrane permeability	*E. coli*, *S aureus*
Targeted drug-carrying phage medicines	Delivery of antimicrobial in the target pathogen	*S. aureus*, *Streptococcus pyogenes*, *E. coli*
Poly-L-lactide nanoparticles	Release of antimicrobial protein nisin	*Lactobacillus delbrueckeii*

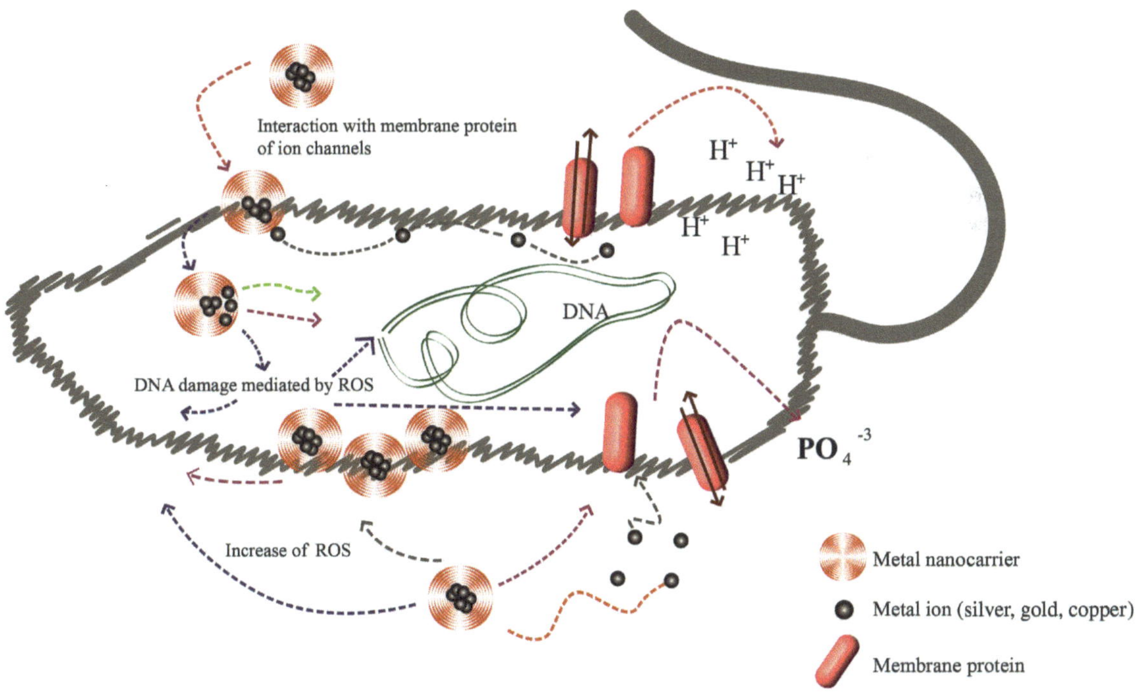

FIGURE 2.2 Antimicrobial activity of nanoparticles.

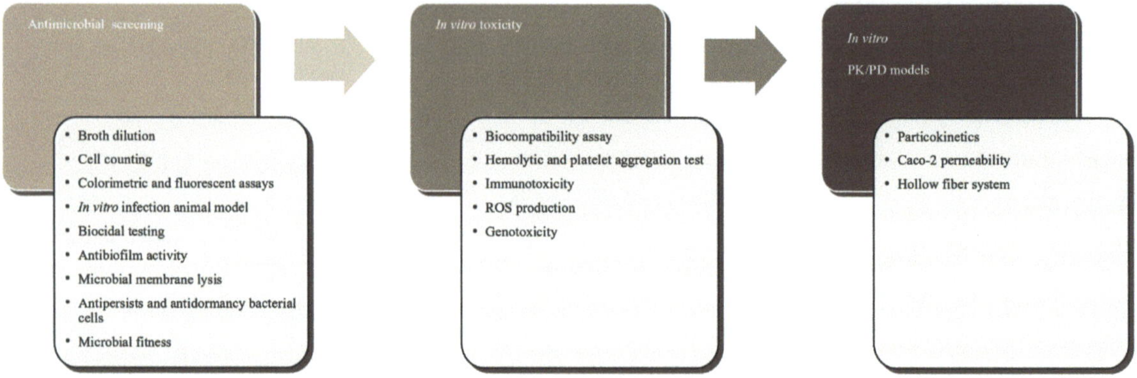

FIGURE 2.3 Antimicrobial drug discovery model in nanotechnology. *In vitro* testing from activity identification until absorption.

Determining the pharmacological activity of a nanoparticle for use as an antimicrobial agent requires experimental techniques that measure microorganism viability after exposure and can determine minimum inhibitory concentrations (MICs) or minimum bactericidal concentrations (MBCs) (Seil and Webster, 2012). Although numerous techniques have been developed to determine the antibacterial activity of NMs, the optimization of a rationale protocol for developing a nanotechnological program of antimicrobial drug discovery is necessary and there is a need for *in vitro* screening platforms with the ability to predict activity, toxicity, and clinical outcome with great accuracy (Bueno, 2012).

The aim of this chapter is to provide an interdisciplinary approach combining activity detection, toxicological evaluation, and pharmacokinetics *in vitro* in a useful model for antimicrobial drug discovery and product development using nanotechnology (Figure 2.3).

2.2 ANTIMICROBIAL SUSCEPTIBILITY TESTING METHODS OF NMs

Before developing antimicrobial test for nanoparticles, it is very important to evaluate the dependence between nanoparticle shape and antimicrobial activity (Pal et al., 2007). It was found that the antibacterial activity of the nanoparticles vary when their size diminishes (Martínez-Castañón et al., 2008) and by strain specificity (Ruparelia et al., 2008). Determination of zeta potential, which shows the degree of repulsion depending on the particle surface charge, affects the stability of the suspension by electrostatic repulsion (a high zeta potential means a more stable suspension). This charge also influences the biological activity of nanoparticles in bacterial cell membranes, which are negatively charged (Qi et al., 2004). For that reason, paper disk diffusion bioassays as well as agar diffusion assays are not described because of the difficulties of diffusion of compounds through agar medium and the disadvantages with these platforms being adapted to a system of high-throughput screening (Burygin et al., 2009; Sánchez and Kouznetsov, 2010; Allahverdiyev et al., 2011; Elkhair, 2014).

2.2.1 Broth Dilution Test

Conventional broth microdilution methods based on the Clinical and Laboratory Standards Institute (CLSI) have been used to evaluate the antimicrobial activity of nanoparticles (Hwang et al., 2012), especially Ag nanoparticles (Kim et al., 2007), for both bacteria and fungi (Lee et al., 2010). This is a test developed in a 96-multiwell plate that expresses the results in micrograms per milliliter (μg/mL) to identify the MIC that is the lowest concentration by visual observation of an antimicrobial agent that inhibits the growth (Figure 2.4) of American Type Culture Collection (ATCC) quality control (QC) microorganisms and clinical strains. Using QC microorganisms in antimicrobial susceptibility testing, the laboratory screening can retain accuracy and reproducibility in the procedures and compare them with the MICs obtained with strains from clinical samples. Depending on the type of microorganism, the protocols are classified as M7 and M11 (bacteria), M27 and M38 (fungi), M24 (mycobacteria), M33 (virus), and M26 for microbicidal activity evaluation by performing time kill curves (Qaiyumi, 2007). These methods have the limitation that the MIC alone does not provide the mechanism of action of the antimicrobial; for that motive, each antimicrobial protocol should be complemented using the M26 method for identifying static or cidal activity. Similar pharmacodynamic and pharmacokinetic parameters of antimicrobial agents should be included for predicting the efficacy *in vivo* (Wiegand et al., 2008).

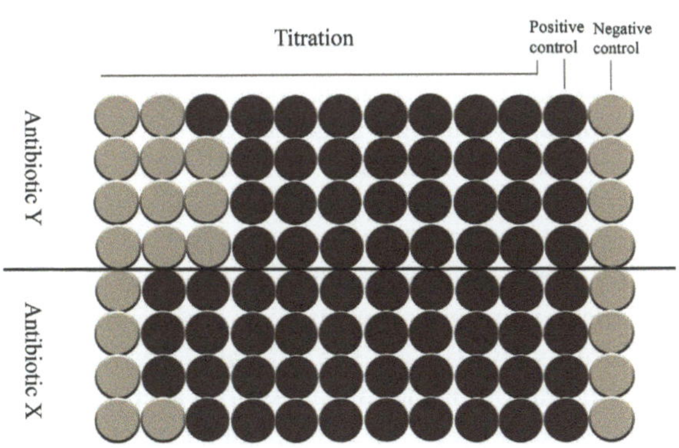

FIGURE 2.4 Antimicrobial susceptibility testing. MIC standard assay.

2.2.2 Spectrophotometric Measurement

As an adaptation of CLSI methods for automation, a spectrophotometric analysis of multiwell plates can be developed for establishment of the number of colony-forming units (CFUs) by turbidity based on a measurement of optical density (OD) 0.1 at 600 nm (OD of 0.1 corresponds to a concentration of 10^8 CFU/mL of culture medium) (Shrivastava et al., 2007). With this microplate proliferation assay, microbiostatic and microbicicidal activities of biomaterials after cell exposure both in bacteria and yeast can be determined in comparison with an untreated control (Alt et al., 2004; Montigny and Wagener, 2007; Pfaller et al., 2011). The precision with which cell density and MICs can be measured with this technique at low cell densities is questionable (Seil and Webster, 2012) and is very difficult in *Mycobacterium* spp. strains with inhomogeneous growth (Penuelas-Urquides et al., 2013).

2.2.3 Cell Counting

Spectrophotometric reading as a qualitative measure of cell growth is useful to evaluate cell viability and calculate antimicrobial activity of surfaces and materials after the inoculation of 10^5–10^6 CFU/mL (by survival percentage) using the following formula (Maneerung et al., 2008; Jones et al., 2008):

$$\left(\frac{\text{Viable count at 0 h} - \text{Viable count at 24 h}}{\text{Viable count at 0 h}}\right) \times 100\%$$

For improving the cell counts, it could be necessary to apply sonication treatment of liquid culture for disruption of bacterial clusters. It is equally important to take into consideration that the antimicrobial particles in liquid medium can interact with destroyed cells, producing coagulation and altering microbicidal activity and cell counts; therefore, these measurements should be performed at various concentrations and time periods (Sondi and Salopek-Sondi, 2004).

2.2.4 Colorimetric and Fluorescent Assays

For developing antimicrobial screening platforms, robust, reproducible, and automatable use of colorimetric and fluorescent dyes can be very useful in a multiwell plate format for cell viability assays. However, it is very important to take into account that NMs cannot be treated in the same manner as chemical compounds, and the interactions with vital dyes that can alter the results are possible; for example, carbon-based NMs and carbon nanotubes interact with 3-(4,5-dimethylthiazole-2-yl)-2,5-biphenyl tetrazolium bromide (MTT), neutral red, alamar blue, 2-(4-iodophenyl)-3-(4-nitrophenyl)-5-(2,4-disulfophenyl)-2H-tetrazolium, monosodium salt (WST-1), and coomassie blue, producing uncertain results because carbon NMs have the ability to adsorb dyes, altering fluorescence or absorbance properties (Doak et al., 2009). Similarly, silver nanoparticles present the phenomenon known as metal-enhanced fluorescence (MEF) that causes spectral alterations when interacting with fluorescent dyes (Aslan et al., 2006). Also, some NMs present the same absorbance used in MTT assays (525 nm), interfering with the results obtained in colorimetric tests (Wörle-Knirsch et al., 2006; Díaz et al., 2008; Kong et al., 2011). Unlike other colorimetric tests, certain classes of NMs have not been shown to interfere with the Alamar blue assay, thus making them an alternative as long as they are not being evaluated with carbon NMs (Fahmy and Cormier, 2009). Another interesting alternative is the use of the LIVE/DEAD® *Bac*Light™ Bacterial Viability kit that uses SYTO 9 and propidium iodide (PI) to differentiate cells with intact (live organisms stained in green) and compromised (dead organisms stained in red) membranes. Exposed cells can be examined using an epifluorescence microscope or measured by flow cytometry with the appropriate filter cube (Jung et al., 2008; Pagnout et al., 2012). Also, it is useful for determination of antimicrobial activity of carbon nanotubes and silver nanoparticles (Simon-Deckers et al., 2009; Flores et al., 2010). In addition, gold nanoparticles are used in phototherapy (Mocan et al., 2014). However, it is recommended to control the final concentration of nanoparticles and that each formulation should be evaluated regarding its compatibility with the tests to be performed (Ong et al., 2014).

2.2.5 *In Vitro* Infection Animal Model

Nonmammalian models of infectious disease are currently a valuable approach for the discovery and evaluation of new antimicrobial agents as well as their toxicity (Desbois and Coote, 2012). In infectious disease research, nematode models have been implemented and have the highest number of reports of use and reproducibility. These models have the advantage of providing results at low cost and do not need ethical approval (Desbois and Coote, 2012). Although *Caenorhabditis elegans* have been used as a classic model in whole-animal antimicrobial screening, the sensitivity of this nematode to oxidative stress does not make it ideal for measuring survival rates of infection with NMs. Instead, *Galleria mellonella*, have been observed that can survive high doses of the silver [Ag(I)] complexes administered *in vivo*, so it is more favorable for evaluating antimicrobial activity of nanoparticles (McCann et al., 2012).

An infectious disease model using *G. mellonella* is a very reproducible tool for measuring the survival of the disease because the administration of inoculum is more accurate than other models such as *Drosophila melanogaster* or *C. elegans* and requires minimal training. Also, it is recommended that cells of microbial inoculum should be washed to prevent the injection of various virulence factors that affect the assay (Desbois and Coote, 2012). So, this wax moth larva model is very useful for performing *in vitro* antimicrobial screening against human pathogens because of its ability to mimic the infectious process in mammalian cells (Rowan et al., 2009; Desbois and Coote, 2011).

2.2.6 Biocidal Testing

Although currently NMs are being used as disinfectants for destroying pathogens in water treatment plants (Godara et al., 2013), there is an increased interest in the use of NMs as antiseptic agents because environmental contamination is a concern of health care. It is necessary to perform proper surface decontamination procedures to reduce and control pathogen transmission as well as develop biocide agents with low toxicity and high efficacy (Gebel et al., 2013). For that reason, the European standard EN 14885 gives guidance for standards of disinfectants and antiseptics evaluation. EN 14885 includes suspension tests and application tests (carrier tests or surface tests) that determine the appropriate use and concentration of the biocide agent, and defines a group of test organisms in which the product must be active to determine their spectrum of action (Meyer and Cookson, 2010; Humphreys, 2011). Other tests such as handwash, handrub, and surface tests simulating practical conditions are also used (Humphreys, 2011).

2.2.7 Antibiofilm Activity

Bacteria and fungi growing as biofilms can cause persistent infections that are highly refractory to antimicrobial therapy; surface chemical agents may be used to inhibit biofilm growth (Epstein et al., 2011). Nanotechnology has been considered an important strategy for avoiding biofilm formation and persistent-related infections (Chen et al., 2013). Drug delivery systems, such as liposomes and polymer carriers, can have great utility for biofilm infections treatment because of their penetration ability and the number of drugs that NMs can carry (Zhang et al., 2010). Also, the use of bacteriophages is another biostrategy studied for the prevention and control of clinical biofilms; these bacteriophages have strong bactericidal activity and specificity. Other interesting approaches are nanotextured surfaces that can prevent biofilm formation during early stages by using their antimicrobial properties (Taylor and Webster, 2011).

Among the *in vitro* assays performed for antibiofilm screening of NMs, the static biofilm assay using violet crystal staining is the most conventional method; it has been used for determination of antibiofilm activity of nanoparticles and can be measured by spectrophotometry at 590 nm (Palanisamy et al., 2014). However, MBEC™ (Minimum Biofilm Eradication Concentration) assay is a high-throughput screening assay with the capacity to determine the activity of antimicrobials against biofilms produced by various microorganisms; it consists of a plastic lid with 96 pegs and a corresponding culture base (Figure 2.5) (Jardeleza et al., 2014). Also, using a platform with plastic 12-well plate, the Biofilm Eradication Surface Test (BEST) Assay™ is designed for testing specialized coatings and surfaces for antibiofilm activity and can be an interesting assay for product testing of NMs (Harding et al., 2011). For studying biofilm structure and biological changes during NM exposure, the use of a confocal laser scanning microscopy (CLSM) can offer more data about the mechanism of action of the compound and formulation with antibiofilm activity (He et al., 2012). CLSM can also be used to study antibiofilm activity in NMs deposited in antimicrobial coatings or surfaces (Applerot et al., 2012). Recently, a cellular microarray system consisting of nanobiofilms of *Candida albicans* encapsulated in an alginate matrix was developed. This

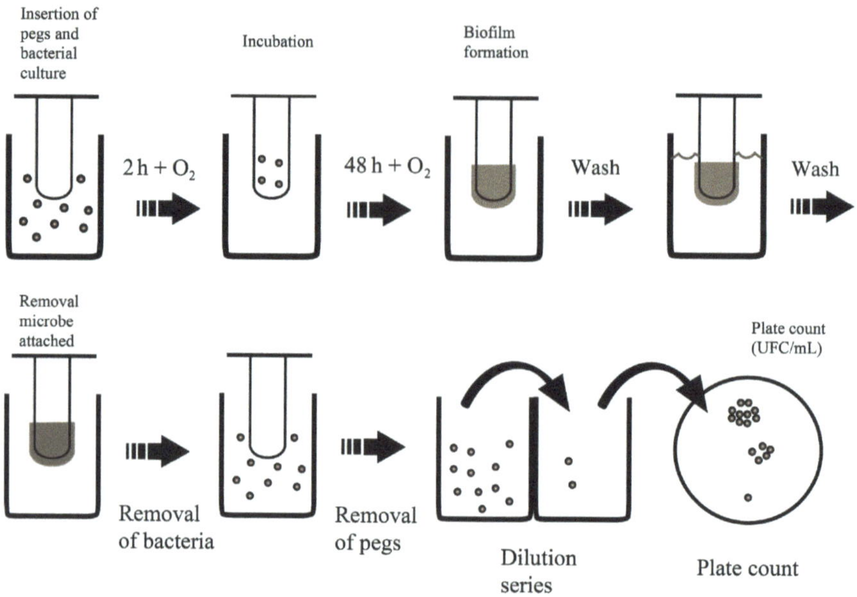

FIGURE 2.5 MBEC™ assay.

nanobiofilm platform can be used for antifungal drug screening of chemical repositories and new formulations of selective novel drug candidates (Srinivasan et al., 2011, 2013).

2.2.8 Quorum-Sensing Inhibitors

Bacteria communicate with each other by the production, distribution, and detection of *N*-acyl-homoserine lactones (AHL), which diffuse outside the cell and are detected by the LuxR receptor protein; this chemical communication is known as quorum sensing and it is an important target in the inhibition of virulence factors production (Figure 2.6) (Tillotson and Theriault, 2013). In nanomedicine, the use of NMs in liposomal formulations as quorum-sensing inhibitors has been determined by measuring AHL secretion in a model of interaction between *P. aeruginosa* isolates and *Agrobacterium tumefaciens* reporter strain (Halwani et al., 2009; Alhariri and Omri, 2013). Also, the antagonistic activity of NMs toward AHL effects has been determined in violacein production or inhibition of quorum sensing in the bacterial pathogen *Chromobacterium violaceum*. Because *C. violaceum* uses LuxR receptor quorum-sensing that is related to virulence factor production, whereby antagonist molecules can replace AHL ligands and develop inhibition, it is an interesting model for discovering quorum-sensing inhibitors using NMs (Martinelli et al., 2004; Stauff and Bassler, 2011; Arunkumar et al., 2014).

2.2.9 Microbial Membrane Lysis

Because the most important mechanism of action of NMs is the disruption of the microbial cell membrane, which is an indicator of microbicidal activity, it is essential to determinate if the NMs create this attribute. The effects of nano-Ag or $AgNO_3$ on the bacterial cell membrane were studied after detergent-mediated bacteriolysis (Lok et al., 2006). A scanning electron microscope (SEM) has

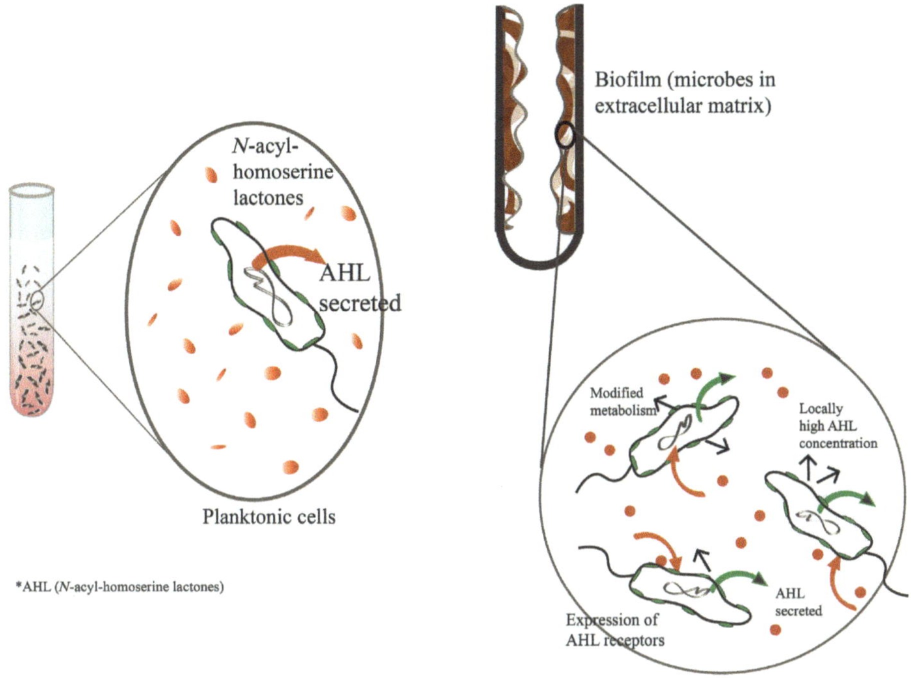

FIGURE 2.6 Quorum-sensing role of AHL.

traditionally been used to study the interactions between graphene-based materials and *E. coli* cell wall. However, other techniques using an atomic force microscope (AFM) have been shown to be extremely useful for analyzing the three-dimensional (3D) structure of this phenomenon (Braga and Ricci, 2011). Because AFM offers data acquisition in 3D, the analysis results of the images obtained are greater and permit investigation of the shape and the surface of bacteria under the action of antibiotics, as well as internal biochemical action (Braga and Ricci, 2011). AFM has changed the research of the microbial cell surface because, different from other techniques, 3D images are acquired without conventional sample preparation (staining, labeling, or fixation) (Alsteens et al., 2013).

Another method for evaluation of the impact of NMs on bacterial membrane integrity is using the LIVE/DEAD® *Bac*Light™ Bacterial Viability kit; following the manufacturer's instructions, it is possible to calculate the proportion of dead cells with SYTO 9/PI fluorescence with the time kill curve of the antimicrobial agent that produces cell membrane disruption in the exposed bacteria (Simon-Deckers et al., 2009).

2.2.10 Microbial Oxidative Stress

The production of ROS in microbial cells by NMs is other antimicrobial mechanism of action; at the same time, it is an important inducer of toxicity processes and adaptive responses such as the hormesis (Forman et al., 2010). *In vitro* induction of ROS production can be evaluated by colorimetric and fluorometric methods. For example, superoxide anion radicals can be measured using the vital dye XTT

[2,3-bis-(2-methoxy-4-nitro-5-sulfophenyl)-2H-tetrazolium-5-carboxanilide], which is reduced in presence of this free radical oxygen (Gil-Lamaignere et al., 2002; Liu et al., 2011). However, before developing the experimental procedures, it is necessary to investigate the possible interactions between NMs and XTT (Doak et al., 2009). A fluorometric method of determination of the intracellular level of ROS is using the fluorescence dye 5-(and-6)-chloromethyl-2′,7′-dichlorodihydrofluorescein diacetate acetyl ester (CMH_2DCF). This indicator changes the state of fluorescence when cleaved by intracellular esterases and oxidized (Pulskamp et al., 2007). However, it is slowly oxidized in the air and is also susceptible to photo-oxidation by laser light during exposure to fluorescence microscopy. Thus, detection of intracellular fluorescence is prone to false-positive results and the laboratory conditions should be regulated (Doak et al., 2009; Kalyanaraman et al., 2012).

2.2.11 Antipersister and Antidormancy Bacterial Cells

Dormant microbes are the cause of bacterial persistence; a subpopulation of dormant cells (persisters) tolerates antibiotic treatment and metabolically inactive cells cannot be destroyed when they are exposed to bactericidal concentrations of antibiotics (Fattorini et al., 2013; Lewis, 2013). For this reason, new antimicrobial agents need to be developed and a screening platform needs to be taken into account. Resazurin microplate assays (REMAs) have correlated very well with CFU counts in models of antimicrobial activity against dormant cells in hypoxic conditions (Taneja and Tyagi, 2007), and the fluorescence of the resazurin metabolite resorufin can be determined in a microplate reader to excitation at 560 nm and emission at 590 nm (Sala et al., 2010). Nevertheless, it is important to consider that nanoparticles such as CdSe and TiO_2 can affect the results obtained with resazurin, so possible interactions should be evaluated before developing experiments (Ong et al., 2014).

2.2.12 Microbial Fitness

Because sublethal concentrations of NMs may intensify microbial fitness, it is necessary to include hormesis effects in antimicrobial experiments (Xiu et al., 2012). Hormesis is a cellular process of adaptation in which lower doses of a given substance produce stimulation, whereas higher doses of the same substance inhibit this function in a complex adaptive system and increase antimicrobial resistance (Yim et al., 2007; Bell et al., 2013). It is very useful to explore dose–response relations and hormetic effects with nonlinear growth models in response to several concentrations of antimicrobial agents and to measure the ratio of the OD in multiwell plates (Calabrese et al., 2006; Drage et al., 2012). The results of nonlinear growth can be shown in a U-shaped or inverted U-shaped dose–response, in which it is possible to determine the level of response (Cook and Calabrese, 2007).

2.3 NANOTOXICOLOGY

To develop antimicrobial products with a broad spectrum of activity and safety, it is mandatory to evaluate the toxicity effects of NMs and the possible risks with human exposure to establish a suitable selectivity criterion (Kroll et al., 2011). Nanoparticles can have cytotoxic, genotoxic, and inflammatory effects in mammalian cells, and can induce oxidative stress response, which should be considered before selection of a nanotechnological approach (Sharma et al., 2012). The toxic effects of nanoparticles are determined by the combination of various factors; among these factors, the most important are the high surface area and the intrinsic toxicity of the surface

(Figure 2.7) (Lacerda et al., 2006). Routes of entry include inhalation, skin absorption, and ingestion routes of entry, and also voluntary injection or drug-delivery systems (Crosera et al., 2009).

2.3.1 Nano-Genotoxicology

NMs produce genotoxicity by indirect DNA damage in the excessive generation of ROS and by induction of chronic inflammatory responses in the exposure site (Doak et al., 2012). Generally, *in vitro* assays to identify potential DNA damage are focused on single-strand and double-strand breaks, mutations, deletions, and chromosomal aberrations, as well as impairment in DNA repair and the cell cycle (Ng et al., 2010). The comet assay (cell gel electrophoresis), *Salmonella* reversion mutation assay (Ames test), fibroblast assay, CHO cell proliferation assay, micronucleus (MN) test, and hydroxy-deoxyguanosine (8-OHdG) analysis (Moreira et al., 2009; Ng et al., 2010) have been used to detect this damage. However, real-time NM exposure is still controversial, and some studies, such as the Ames test, do not appear to be suitable for the assessment of NMs because of uptake by the bacterial cells is less than that in human cells because of absence of the endocytosis process in prokaryotics and the barrier formed by the cell wall; therefore, this lack of uptake could potentially lead to false-negative results, and *in vitro* mammalian cell tests for comet assays and MN assays are recommended because the comet assays more sensitive in detecting early DNA breakage (Stone et al., 2009; Landsiedel et al., 2009; Kim et al., 2013).

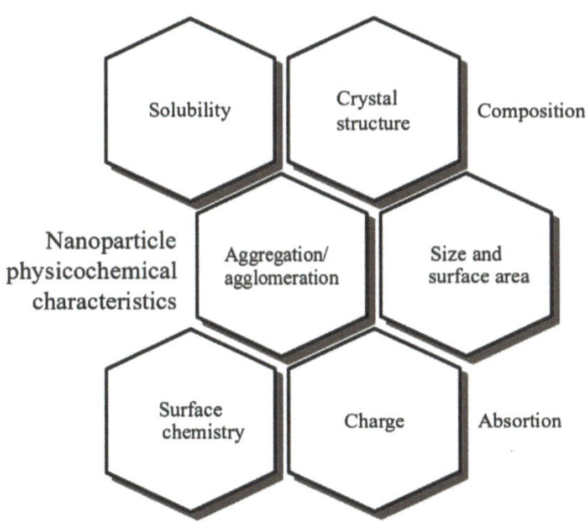

FIGURE 2.7 Physicochemical features of NMs that influence their toxicity.

2.3.2 Cytotoxicity

It has been demonstrated that conventional cytotoxic assays may not be appropriate to assess NM toxicity because it can interfere with assay reagents or detection systems, generating false-positive/false-negative results (Kroll et al., 2011). The authors have evaluated and standardized common *in vitro* assays measuring three different cytotoxic endpoints, oxidative stress (DCF assay), cell death (LDH assay), and cellular metabolic activity (MTT assay), and have discovered that all NM dispersions interfered with the original protocol for the detection of DCF fluorescence at concentrations of 10 $\mu g/cm^2$; equally light absorption of MTT was strongly altered in the presence of all NM dispersions at concentrations of 10 $\mu g/cm^2$, and removing NM suspensions before incubation with DCF or MTT is recommended to eliminate interference (Kroll et al., 2011, 2012). Recently, Ong et al. (2014) have recommended controlling the final concentration of nanoparticles and their compatibility with the tests to be performed.

2.3.3 Immunotoxicity

NMs can act on the immune system and produce various responses such as immunosuppression or immunostimulation. Nanoparticles can be potentially immunotoxic; however, specific immunotoxicity of NMs has not been reported, so it must be evaluated using the respective reference methods and control compounds (Dobrovolskaia and McNeil, 2012). It is necessary to control the factors that can affect the *in vitro* assays used for analyzing the effects of NPs on immune cells and immune responses because variable levels of endotoxin are common contaminants of glassware, culture media, and additives and can activate monocytes/macrophages, altering the results (Peer, 2012). The choice of the biological assay is also of central importance; representative *in vitro* cellular assays should be based on human blood leukocytes (such as monocytes), which are the first cells to come in contact with the injected NMs and have biological endpoints representing the early innate/inflammatory-type responses (Lankveld et al., 2010). Primary cells should be the first choice in this kind of assay, because the primary response to stimulation is more sensitive than continuous cell lines and is qualitatively different. Also, an important requirement is the selection of endpoint, which should be representative of primary cell activation (activation of IL-8 expression for example) (Oostingh et al., 2011). Another interesting model is the use of coelomocytes from the earthworm *Lumbricus rubellus*, with results that are comparable with immunotoxicity assays using NR8383 rat macrophage cells (van der Ploeg et al., 2014).

2.3.4 *In Vitro* Skin Irritation

NMs, such as quantum dots (QDs) and fullerenes, are able to penetrate into the stratum corneum. A model of *in vitro* skin toxicity using an assay of cytotoxicity of nanoparticles in human epidermal keratinocytes (HEKs) is being needed, but it should follow the recommendations expressed previously regarding the use of dyes for cytotoxic tests (Johnstone et al., 2010; Samberg et al., 2010).

2.3.5 *Caenorhabditis elegans* Toxicity Model

C. elegans is a globally distributed nematode species encountered in nutrient-rich and/or bacteria-rich soil-associated substrates. It has been used extensively in genetic, developmental biology, and toxicological studies, where it is typically cultured either on agar plates or in liquid medium and fed bacteria. It has recently been used for assessment of NM toxicity because *C. elegans* is sensitive to ROS induction, genotoxins, and metals (Meyer et al., 2010; Xiu et al., 2012).

2.3.6 Nanotoxicity in Embryonic and Adult Zebrafish

Although zebrafish (*Danio rerio*) have been used as an *in vivo* infectious model, especially in the discovery of antituberculosis drugs (Makarov et al., 2014), it has been explored for nanotoxicity testing because the size, short life cycle, and adaptability to platforms of high-throughput screening, which makes it suitable for acute and chronic toxicity studies (Figure 2.8) (Wang et al., 2012).

2.3.7 Bioluminescence-Based Nanotoxicity Test

Bioluminescence bacteria have been used as a biosensor agent to quantify the toxicity of nanoparticles and nanotubes, based on their bioluminescence inhibition; in these organisms, the bioluminescence intensity is directly proportional to the metabolic activity of the bacterial population. This bacterial toxicity test is faster (15 min by assay) and offers results comparable with reference methods (Wani et al., 2011).

2.4 *IN VITRO* PHARMACOKINETICS/ PHARMACODYNAMIC MODELS

Pharmacokinetics/pharmacodynamic (PK/PD) models in antimicrobial nanotechnology

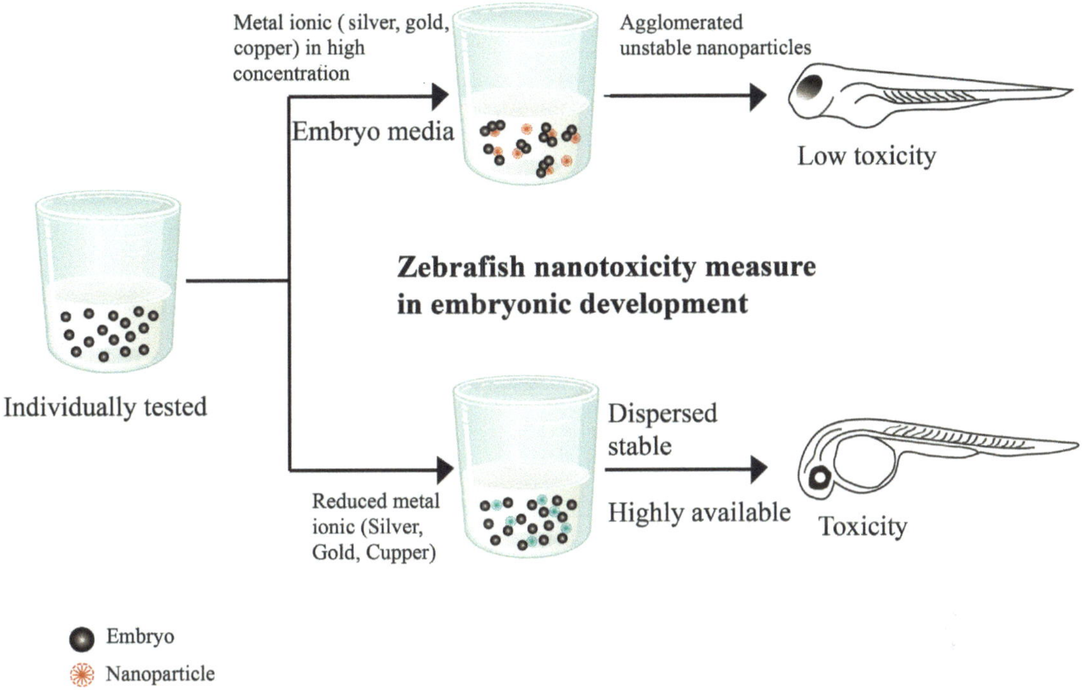

FIGURE 2.8 Nanotoxicity model using zebrafish (*D. rerio*).

can estimate the optimal effective dose and predict the effect of new drugs in a specific microorganism (Mouton et al., 2011; Bhatta et al., 2012). In addition, PK/PD modeling provides a platform for extrapolating data from *in vitro* to *in vivo* studies (Toutain and Lees, 2004).

2.4.1 Particokinetics

Unlike soluble compounds, particles can settle, diffuse, and aggregate in relation to their size, density, and surface. These features can significantly affect the cellular dose. For that reason, the definition of proper dose for nanoparticles using an *in vitro* system should be more dynamic and less comparable than other types of formulations (Teeguarden et al., 2007). It is necessary to calculate the correct dose/exposure interaction depending on material characteristics and experimental conditions using the following equations:

$$\text{Surface area concentration} = \frac{\text{Mass concentration}}{\text{Particle density}} \cdot \frac{6}{d}$$

$$= \#\text{Concentration} \cdot \pi d^2$$

$$\#\text{Concentration} = \frac{\text{Mass concentration}}{\text{Particle density}} \cdot \frac{6}{\pi d^3}$$

$$= \frac{\text{Surface area concentration}}{\pi d^2}$$

where spherical shape of particles is assumed as d (their diameter) in cm, surface area concentration is expressed in cm^2/mL of culture medium, mass concentration is expressed in g/mL of culture medium, and # is the amount of particles, with particle density expressed in g/cm^3 (Figure 2.9) (Teeguarden et al., 2007). It is valid to use an *in vitro* sedimentation and diffusion and

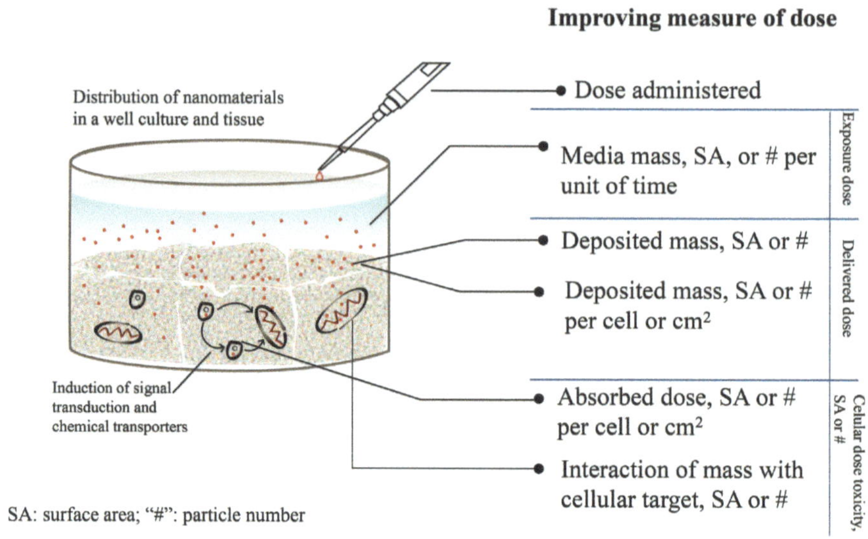

FIGURE 2.9 Particokinetics. Dynamical model of dose exposure.

dosimetry model (ISDD) to obtain particle sedimentation and diffusion data in cell culture media, and it is essential to perform direct interpretation of results from *in vitro* particle–cell interaction studies (Hinderliter et al., 2010; Ahmad Khanbeigi et al., 2012; Cohen et al., 2014).

2.4.2 Caco-2 Permeability

The Caco-2 cells isolated from human colorectal adenocarcinoma are widely used to screen for absorption rate of new compounds in the initial stage of drug discovery. This cell line in culture forms monolayers with an enterocyte-like phenotype and exhibits properties associated with the physical and metabolic barrier of the intestinal epithelium. The permeability measure across a monolayer of Caco-2 cells can predict the compound absorption in the human intestinal epithelium across tight junctions between epithelial cells (Vila et al., 2004; Sevin et al., 2013). Likewise, nanoparticles can cross the epithelial barriers by two possible routes: transcellular pathway via passive diffusion or transcytosis involving both endocytosis and exocytosis (Prokop and Davidson, 2008). Because of its similarity to the small intestinal epithelium, Caco-2 cells provide an interesting *in vitro* model to evaluate the passing of nanoparticles across epithelial barriers that can be compared by cellular transport in the body (Lin et al., 2011).

2.4.3 Hollow Fiber System

To study the most important parameters that influence an optimal nanoparticle suspension, the use of a hollow fiber system is a useful tool (Laouini et al., 2011). Also, this system can be an *in vitro* PK/PD model for infectious diseases treatment (Vidaillac et al., 2009), with the ability to become a two-compartment PK/PD model (Werth et al., 2013). It also has the ability to imitate modifications in drug concentrations over time, as would occur in humans, and to provide useful data for development of antimicrobial agents and their dosages (Cadwell, 2012).

2.5 CONCLUSIONS

Antimicrobial drug resistance is a global public health problem that requires innovation and development in pursuit of the next generation of control measures to reduce mortality and morbidity of infectious disease. To overcome this new biomedical challenge, it is necessary to implement multidisciplinary, interdisciplinary, and transdisciplinary research projects where several disciplines can provide clinicians with novel methods, theoretical approaches, and technologies. In that way, nanotechnology will be an important source of potent tools for design and manufacture of a new generation of compounds and formulations with specific antimicrobial properties for avoiding emergence and spread of drug-resistant strains, thus providing and testing new drugs, medical devices, and sanitizers for control of microbial surface contamination. Likewise, nanotechnology is opening the possibility of obtaining multipurpose materials by mixing several mechanisms of action and pharmacological properties, but at the same time it is necessary to develop toxicity screening methods that can predict potential environmental damage and risk that this advancement may cause. A multidisciplinary, interdisciplinary, and transdisciplinary approach for design strategies to control antibiotic resistance requires integrity among antimicrobial drug screening, toxicity testing, and drug metabolism and pharmacokinetics. Future efforts should be directed to the development of nanocomposites with the ability to comply with the following three simple characteristics: potent antimicrobial activity and specificity; safety and excellent tissue distribution and metabolism; and applicable to medicine, agriculture, and industry.

Acknowledgment

The author acknowledges Claudia Marcela Montes for designing the figures in this chapter.

References

Ahmad Khanbeigi, R., Kumar, A., Sadouki, F., Lorenz, C., Forbes, B., Dailey, L.A., et al., 2012. The delivered dose: applying particokinetics to *in vitro* investigations of nanoparticle internalization by macrophages. J. Control. Release. 162, 259–266.

Alhariri, M., Omri, A., 2013. Efficacy of liposomal bismuth–ethanedithiol-loaded tobramycin after intratracheal administration in rats with pulmonary *Pseudomonas aeruginosa* infection. Antimicrob. Agents Chemother. 57, 569–578.

Allahverdiyev, A.M., Kon, K.V., Abamor, E.S., Bagirova, M., Rafailovich, M., 2011. Coping with antibiotic resistance: combining nanoparticles with antibiotics and other antimicrobial agents. Expert Rev. Anti Infect. Ther. 9, 1035–1052.

Alsteens, D., Beaussart, A., El-Kirat-Chatel, S., Sullan, R.M.A., Dufrêne, Y.F., 2013. Atomic force microscopy: a new look at pathogens. PLoS Pathog. 9, e1003516.

Alt, V., Bechert, T., Steinrücke, P., Wagener, M., Seidel, P., Dingeldein, E., et al., 2004. *In vitro* testing of antimicrobial activity of bone cement. Antimicrob. Agents Chemother. 48, 4084–4088.

Applerot, G., Lellouche, J., Perkas, N., Nitzan, Y., Gedanken, A., Banin, E., 2012. ZnO nanoparticle-coated surfaces inhibit bacterial biofilm formation and increase antibiotic susceptibility. RSC Adv. 2, 2314–2321.

Armstead, A.L., Li, B., 2011. Nanomedicine as an emerging approach against intracellular pathogens. Int. J. Nanomed. 6, 3281–3293.

Aruguete, D.M., Kim, B., Hochella, M.F., Ma, Y., Cheng, Y., Hoegh, A., et al., 2013. Antimicrobial nanotechnology: its potential for the effective management of microbial drug resistance and implications for research needs in microbial nanotoxicology. Environ. Sci. Process. Impacts. 15, 93–102.

Arunkumar, M., Suhashini, K., Mahesh, N., Ravikumar, R., 2014. Quorum quenching and antibacterial activity of silver nanoparticles synthesized from *Sargassum polyphyllum*. Bangladesh J. Pharmacol. 9, 54–59.

Aslan, K., Holley, P., Geddes, C.D., 2006. Metal-enhanced fluorescence from silver nanoparticle-deposited polycarbonate substrates. J. Mater. Chem. 16, 2846–2852.

Bell, I.R., Schwartz, G.E., Boyer, N.N., Koithan, M., Brooks, A.J., 2013. Advances in integrative nanomedicine for improving infectious disease treatment in public health. Eur. J. Integr. Med. 5, 126–140.

Bhatta, R., Chandasana, H., Chhonker, Y., Rathi, C., Kumar, D., Mitra, K., et al., 2012. Mucoadhesive nanoparticles for prolonged ocular delivery of natamycin: *in vitro* and pharmacokinetics studies. Int. J. Pharm. 432, 105–112.

Braga, P.C., Ricci, D., 2011. Imaging bacterial shape, surface and appendages before and after treatment with antibiotics. Methods Mol. Biol. 736, 391–399.

Bueno, J., 2012. *In vitro* antimicrobial activity of natural products using minimum inhibitory concentrations, looking for new chemical entities or predicting clinical response. Med. Aromat. Plants 7, 113–114.

Burygin, G., Khlebtsov, B., Shantrokha, A., Dykman, L., Bogatyrev, V., Khlebtsov, N., 2009. On the enhanced antibacterial activity of antibiotics mixed with gold nanoparticles. Nanoscale Res. Lett. 4, 794–801.

Cadwell, J., 2012. The hollow fiber infection model for antimicrobial pharmacodynamics and pharmacokinetics. Adv. Pharmacoepidem. Drug Safety S 1, 2167–1052.

Calabrese, E.J., Staudenmayer, J.W., Stanek III, E.J., Hoffmann, G.R., 2006. Hormesis outperforms threshold model in National Cancer Institute antitumor drug screening database. Toxicol. Sci. 94, 368–378.

Cars, O., Hedin, A., Heddini, A., 2011. The global need for effective antibiotics-moving towards concerted action. Drug Resist. Updat. 14, 68–69.

Chen, M., Yu, Q., Sun, H., 2013. Novel strategies for the prevention and treatment of biofilm related infections. Int. J. Mol. Sci. 14, 18488–18501.

Cohen, J.M., Teeguarden, J.G., Demokritou, P., 2014. An integrated approach for the *in vitro* dosimetry of engineered nanomaterials. Part. Fibre Toxicol. 11, 20

Cook, R., Calabrese, E.J., 2007. The importance of hormesis to public health. Cien. Saude Colet. 12, 955–963.

Crosera, M., Bovenzi, M., Maina, G., Adami, G., Zanette, C., Florio, C., et al., 2009. Nanoparticle dermal absorption and toxicity: a review of the literature. Int. Arch. Occup. Environ. Health 82, 1043–1055.

Davies, J., Davies, D., 2010. Origins and evolution of antibiotic resistance. Microbiol. Mol. Biol. Rev. 74, 417–433.

Desbois, A.P., Coote, P.J., 2011. Wax moth larva (*Galleria mellonella*): an *in vivo* model for assessing the efficacy of antistaphylococcal agents. J. Antimicrob. Chemother. 66, 1785–1790.

Desbois, A.P., Coote, P.J., 2012. Utility of greater wax moth larva (*Galleria mellonella*) for evaluating the toxicity and efficacy of new antimicrobial agents. Adv. Appl. Microbiol. 78, 25–53.

Díaz, B., Sánchez-Espinel, C., Arruebo, M., Faro, J., de Miguel, E., Magadánn, S., et al., 2008. Assessing methods for blood cell cytotoxic responses to inorganic nanoparticles and nanoparticle aggregates. Small 4, 2025–2034.

Doak, S., Griffiths, S., Manshian, B., Singh, N., Williams, P., Brown, A., et al., 2009. Confounding experimental considerations in nanogenotoxicology. Mutagenesis 24, 285–293.

Doak, S., Manshian, B., Jenkins, G., Singh, N., 2012. *In vitro* genotoxicity testing strategy for nanomaterials and the adaptation of current OECD guidelines. Mutat. Res. 745, 104–111.

Dobrovolskaia, M.A., McNeil, S.E., 2012. Immunological properties of engineered nanomaterials: an introduction. In: Yarmush, M.L., Shi, D. (Eds.), Frontiers in Nanobiomedical Research. World Scientific Publishing, Singapore, pp. 1–23.

Drage, S., Engelmeier, D., Bachmann, G., Sessitsch, A., Mitter, B., Hadacek, F., 2012. Combining microdilution with MicroResp: microbial substrate utilization, antimicrobial susceptibility and respiration. J. Microbiol. Methods 88, 399–412.

Elkhair, E.K.A., 2014. Antidermatophytic activity of essential oils against locally isolated *Microsporum canis*—Gaza strip. Nat. Sci. 6, 676–684.

Epstein, A., Hochbaum, A., Kim, P., Aizenberg, J., 2011. Control of bacterial biofilm growth on surfaces by nanostructural mechanics and geometry. Nanotechnology 22, 494007.

Fahmy, B., Cormier, S.A., 2009. Copper oxide nanoparticles induce oxidative stress and cytotoxicity in airway epithelial cells. Toxicol. In Vitro 23, 1365–1371.

Fattorini, L., Piccaro, G., Mustazzolu, A., Giannoni, F., 2013. Targeting dormant bacilli to fight tuberculosis. Mediterr. J. Hematol. Infect. Dis. 5, e2013072.

Fauci, A.S., Marston, H.D., 2014. The perpetual challenge of antimicrobial resistance. JAMA. Available from: <http://dx.doi.org/10.1001/jama.2014.2465>.

Flores, C., Diaz, C., Rubert, A., Benítez, G., Moreno, M., Fernández Lorenzo de Mele, M., et al., 2010. Spontaneous adsorption of silver nanoparticles on Ti/TiO_2 surfaces. Antibacterial effect on *Pseudomonas aeruginosa*. J. Colloid. Interface Sci. 350, 402–408.

Forman, H.J., Maiorino, M., Ursini, F., 2010. Signaling functions of reactive oxygen species. Biochemistry 49, 835–842.

Gebel, J., Exner, M., French, G., Chartier, Y., Christiansen, B., Gemein, S., et al., 2013. The role of surface disinfection in infection prevention. GMS. Hyg. Infect. Control. 8. Available from: <http://dx.doi.org/10.3205/dgkh000210>.

Gil-Lamaignere, C., Roilides, E., Maloukou, A., Georgopoulou, I., Petrikkos, G., Walsh, T.J., 2002. Amphotericin B lipid complex exerts additive antifungal activity in combination with polymorphonuclear leucocytes against *Scedosporium prolificans* and *Scedosporium apiospermum*. J. Antimicrob. Chemother. 50, 1027–1030.

Godara, A., Srivastava, S., Anandan, A., Dayal, B., Kumar, A., 2013. Engineered nanomaterials as disinfectants: benefits and health concerns. Int. J. Biotechnol. Bioeng. Res. 3, 191–196.

Hajipour, M.J., Fromm, K.M., Ashkarran, A.A., Jimenez de Aberasturi, D., de Larramendi, I.R., Rojo, T., et al., 2012. Antibacterial properties of nanoparticles. Trends Biotechnol. 30, 499–511.

Halwani, M., Hebert, S., Suntres, Z.E., Lafrenie, R.M., Azghani, A.O., Omri, A., 2009. Bismuth–thiol incorporation enhances biological activities of liposomal tobramycin against bacterial biofilm and quorum sensing molecules production by *Pseudomonas aeruginosa*. Int. J. Pharm. 373, 141–146.

Harding, M., Howard, R., Daniels, G., Mobbs, S., Lisowski, S., Allan, N., et al., 2011. A multi-well plate method for rapid growth, characterization and biocide sensitivity testing of microbial biofilms on various surface materials. Formatex. 3, 872–877.

He, W., Wang, D., Ye, Z., Qian, W., Tao, Y., Shi, X., et al., 2012. Application of a nanotechnology antimicrobial spray to prevent lower urinary tract infection: a multicenter urology trial. J. Transl. Med. 10 (Suppl. 1), S14.

Hecker, A., Uhle, F., Schwandner, T., Padberg, W., Weigand, M., 2014. Diagnostics, therapy and outcome prediction in abdominal sepsis: current standards and future perspectives. Langenbecks Arch. Surg. 399, 11–22.

Hinderliter, P.M., Minard, K.R., Orr, G., Chrisler, W.B., Thrall, B.D., Pounds, J.G., et al., 2010. ISDD: a computational model of particle sedimentation, diffusion and target cell dosimetry for *in vitro* toxicity studies. Part. Fibre Toxicol. 7, 36.

Humphreys, P.N., 2011. Testing standards for sporicides. J. Hosp. Infect. 77, 193–198.

Hwang, I., Hwang, J.H., Choi, H., Kim, K., Lee, D.G., 2012. Synergistic effects between silver nanoparticles and antibiotics and the mechanisms involved. J. Med. Microbiol. 61, 1719–1726.

Jardeleza, C., Rao, S., Thierry, B., Gajjar, P., Vreugde, S., Prestidge, C.A., et al., 2014. Liposome-encapsulated ISMN: a novel nitric oxide-based therapeutic agent against staphylococcus aureus biofilms. PLoS One 9, e92117.

Johnston, H.J., Hutchison, G.R., Christensen, F.M., Aschberger, K., Stone, V., 2010. The biological mechanisms and physicochemical characteristics responsible for driving fullerene toxicity. Toxicol. Sci. 114, 162–182.

Jones, N., Ray, B., Ranjit, K.T., Manna, A.C., 2008. Antibacterial activity of ZnO nanoparticle suspensions on a broad spectrum of microorganisms. FEMS Microbiol. Lett. 279, 71–76.

Jung, W.K., Koo, H.C., Kim, K.W., Shin, S., Kim, S.H., Park, Y.H., 2008. Antibacterial activity and mechanism of action of the silver ion in *Staphylococcus aureus* and *Escherichia coli*. Appl. Environ. Microbiol. 74, 2171–2178.

Kalyanaraman, B., Darley-Usmar, V., Davies, K.J., Dennery, P.A., Forman, H.J., Grisham, M.B., et al., 2012. Measuring reactive oxygen and nitrogen species with fluorescent probes: challenges and limitations. Free Radic. Biol. Med. 52, 1–6.

Kim, H.R., Park, Y.J., Shin, D.Y., Oh, S.M., Chung, K.H., 2013. Appropriate *in vitro* methods for genotoxicity testing of silver nanoparticles. Environ. Health Toxicol. 28, e2013003.

Kim, J.S., Kuk, E., Yu, K.N., Kim, J., Park, S.J., Lee, H.J., et al., 2007. Antimicrobial effects of silver nanoparticles. Nanomedicine 3, 95–101.

Kong, B., Seog, J.H., Graham, L.M., Lee, S.B., 2011. Experimental considerations on the cytotoxicity of nanoparticles. Nanomedicine 6, 929–941.

Kroll, A., Dierker, C., Rommel, C., Hahn, D., Wohlleben, W., Schulze-Isfort, C., et al., 2011. Cytotoxicity screening of 23 engineered nanomaterials using a test matrix of ten cell lines and three different assays. Part. Fibre Toxicol. 8. Available from: <http://dx.doi.org/10.1186/1743-8977-8-9>.

Kroll, A., Pillukat, M.H., Hahn, D., Schnekenburger, J., 2012. Interference of engineered nanoparticles with *in vitro* toxicity assays. Arch. Toxicol. 86, 1123–1136.

Lacerda, L., Bianco, A., Prato, M., Kostarelos, K., 2006. Carbon nanotubes as nanomedicines: from toxicology to pharmacology. Adv. Drug Deliv. Rev. 58, 1460–1470.

Landsiedel, R., Kapp, M.D., Schulz, M., Wiench, K., Oesch, F., 2009. Genotoxicity investigations on nanomaterials: methods, preparation and characterization of test material, potential artifacts and limitations-many questions, some answers. Mutat. Res. 681, 241–258.

Lankveld, D.P.K., Van Loveren, H., Baken, K.A., Vandebriel, R.J., 2010. *In vitro* testing for direct immunotoxicity: state of the art. Methods Mol. Biol. 598, 401–423.

Laouini, A., Jaafar-Maalej, C., Sfar, S., Charcosset, C., Fessi, H., 2011. Liposome preparation using a hollow fiber membrane contactor—application to spironolactone encapsulation. Int. J. Pharm. 415, 53–61.

Lee, J., Kim, K., Sung, W.S., Kim, J.G., Lee, D.G., 2010. The silver nanoparticle (nano-Ag): a new model for antifungal agents. In: Pozo, D. (Ed.), Silver Nanoparticles. InTech publishing, Croatia, pp. 295–308.

Lewis, K., 2013. Platforms for antibiotic discovery. Nat. Rev. Drug Discov. 12, 371–387.

Lin, I., Liang, M., Liu, T., Ziora, Z.M., Monteiro, M.J., Toth, I., 2011. Interaction of densely polymer-coated gold nanoparticles with epithelial Caco-2 monolayers. Biomacromolecules 12, 1339–1348.

Liu, S., Zeng, T.H., Hofmann, M., Burcombe, E., Wei, J., Jiang, R., et al., 2011. Antibacterial activity of graphite, graphite oxide, graphene oxide and reduced graphene oxide: membrane and oxidative stress. ACS Nano. 5, 6971–6980.

Lok, C., Ho, C., Chen, R., He, Q., Yu, W., Sun, H., et al., 2006. Proteomic analysis of the mode of antibacterial action of silver nanoparticles. J. Proteome Res. 5, 916–924.

Machado, M.C., Cheng, D., Tarquinio, K.M., Webster, T.J., 2010. Nanotechnology: pediatric applications. Pediatr. Res. 67, 500–504.

Makarov, V., Lechartier, B., Zhang, M., Neres, J., van der Sar, A.M., Raadsen, S.A., et al., 2014. Towards a new combination therapy for tuberculosis with next generation benzothiazinones. EMBO Mol. Med. 6, 372–383.

Maneerung, T., Tokura, S., Rujiravanit, R., 2008. Impregnation of silver nanoparticles into bacterial cellulose for antimicrobial wound dressing. Carbohydr. Polym. 72, 43–51.

Martinelli, D., Grossmann, G., Sequin, U., Brandl, H., Bachofen, R., 2004. Effects of natural and chemically synthesized furanones on quorum sensing in *Chromobacterium violaceum*. BMC Microbiol. 4, 25.

Martínez-Castañón, G., Niño-Martínez, N., Martínez-Gutiérrez, F., Martínez-Mendoza, J., Ruiz, F., 2008. Synthesis and antibacterial activity of silver nanoparticles with different sizes. J. Nanopart. Res. 10, 1343–1348.

Matthews, L., Kanwar, R.K., Zhou, S., Punj, V., Kanwar, J. R., 2010. Applications of nanomedicine in antibacterial medical therapeutics and diagnostics. Open Trop. Med. J. 3, 1–9.

McCann, M., Curran, R., Ben-Shoshan, M., McKee, V., Tahir, A.A., Devereux, M., et al., 2012. Silver (I) complexes of 9-anthracenecarboxylic acid and imidazoles: synthesis, structure and antimicrobial activity. Dalton Trans. 41, 6516–6527.

Meyer, B., Cookson, B., 2010. Does microbial resistance or adaptation to biocides create a hazard in infection prevention and control? J. Hosp. Infect. 76, 200–205.

Meyer, J.N., Lord, C.A., Yang, X.Y., Turner, E.A., Badireddy, A.R., Marinakos, S.M., et al., 2010. Intracellular uptake and associated toxicity of silver nanoparticles in *Caenorhabditis elegans*. Aquat. Toxicol. 100, 140–150.

Mishra, R.P., Oviedo-Orta, E., Prachi, P., Rappuoli, R., Bagnoli, F., 2012. Vaccines and antibiotic resistance. Curr. Opin. Microbiol. 15, 596–602.

Mocan, L., Ilie, I., Matea, C., Tabaran, F., Kalman, E., Iancu, C., et al., 2014. Surface plasmon resonance-induced photoactivation of gold nanoparticles as bactericidal agents against methicillin-resistant *Staphylococcus aureus*. Int. J. Nanomed. 9, 1453.

Montigny, R., Wagener, M., 2007. A new antimicrobial ALCERU® fibre with silver nanoparticles. FIBRES TEXT. East. Eur. 15, 64–65.

Moreira, S., Silva, N.B., Almeida-Lima, J., Rocha, H.A.O., Medeiros, S.R.B., Alves Jr., C., et al., 2009. BC nanofibres: *in vitro* study of genotoxicity and cell proliferation. Toxicol. Lett. 189, 235–241.

Mouton, J.W., Ambrose, P.G., Canton, R., Drusano, G.L., Harbarth, S., MacGowan, A., et al., 2011. Conserving antibiotics for the future: new ways to use old and new drugs from a pharmacokinetic and pharmacodynamic perspective. Drug Resist. Update 14, 107–117.

Ng, C., Li, J.J., Bay, B., Yung, L.L., 2010. Current studies into the genotoxic effects of nanomaterials. J. Nucleic Acids 2010.

Ong, K.J., MacCormack, T.J., Clark, R.J., Ede, J.D., Ortega, V.A., Felix, L.C., et al., 2014. Widespread nanoparticle-assay interference: implications for nanotoxicity testing. PLoS One 9, e90650.

Oostingh, G.J., Casals, E., Italiani, P., Colognato, R., Stritzinger, R., Ponti, J., et al., 2011. Problems and challenges in the development and validation of human cell-based assays to determine nanoparticle-induced immunomodulatory effects. Part. Fibre Toxicol. 8, 8.

Pagnout, C., Jomini, S., Dadhwal, M., Caillet, C., Thomas, F., Bauda, P., 2012. Role of electrostatic interactions in the toxicity of titanium dioxide nanoparticles toward *Escherichia coli*. Colloids Surf. B. Biointerfaces 92, 315–321.

Pal, S., Tak, Y.K., Song, J.M., 2007. Does the antibacterial activity of silver nanoparticles depend on the shape of the nanoparticle? A study of the Gram-negative bacterium *Escherichia coli*. Appl. Environ. Microbiol. 73, 1712–1720.

Palanisamy, N.K., Ferina, N., Amirulhusni, A.N., Mohd-Zain, Z., Hussaini, J., Ping, L.J., et al., 2014. Antibiofilm properties of chemically synthesized silver nanoparticles found against *Pseudomonas aeruginosa*. J. Nanobiotechnol. 12, 2.

Peer, D., 2012. Immunotoxicity derived from manipulating leukocytes with lipid-based nanoparticles. Adv. Drug Deliv. Rev. 64, 1738–1748.

Penuelas-Urquides, K., Villarreal-Trevino, L., Silva-Ramirez, B., Rivadeneyra-Espinoza, L., Said-Fernandez, S., de Leon, M.B., 2013. Measuring of *Mycobacterium tuberculosis* growth. A correlation of the optical measurements with colony forming units. Braz. J. Microbiol. 44, 287–289.

Pfaller, M., Espinel-Ingroff, A., Boyken, L., Hollis, R., Kroeger, J., Messer, S., et al., 2011. Comparison of the broth microdilution (BMD) method of the European Committee on Antimicrobial Susceptibility Testing with the 24-hour CLSI BMD method for testing susceptibility of *Candida* species to fluconazole, posaconazole, and voriconazole by use of epidemiological cutoff values. J. Clin. Microbiol. 49, 845–850.

Prokop, A., Davidson, J.M., 2008. Nanovehicular intracellular delivery systems. J. Pharm. Sci. 97, 3518–3590.

Pulskamp, K., Diabaté, S., Krug, H.F., 2007. Carbon nanotubes show no sign of acute toxicity but induce intracellular reactive oxygen species in dependence on contaminants. Toxicol. Lett. 168, 58–74.

Qaiyumi, S., 2007. Macro- and microdilution methods of antimicrobial susceptibility testing. In: Schwalbe, R., Steele-Moore, L., Goodwin, A. (Eds.), Antimicrobial Susceptibility Testing Protocols. CRC Press, Boca Raton, FL, pp. 75–79.

Qi, L., Xu, Z., Jiang, X., Hu, C., Zou, X., 2004. Preparation and antibacterial activity of chitosan nanoparticles. Carbohydr. Res. 339, 2693–2700.

Rowan, R., Moran, C., McCann, M., Kavanagh, K., 2009. Use of *Galleria mellonella* larvae to evaluate the in vivo anti-fungal activity of [Ag2 (mal)(phen) 3]. Biometals 22, 461–467.

Ruparelia, J.P., Chatterjee, A.K., Duttagupta, S.P., Mukherji, S., 2008. Strain specificity in antimicrobial activity of silver and copper nanoparticles. Acta Biomater. 4, 707–716.

Sala, C., Dhar, N., Hartkoorn, R.C., Zhang, M., Ha, Y.H., Schneider, P., et al., 2010. Simple model for testing drugs against nonreplicating *Mycobacterium tuberculosis*. Antimicrob. Agents Chemother. 54, 4150–4158.

Samberg, M.E., Oldenburg, S.J., Monteiro-Riviere, N.A., 2010. Evaluation of silver nanoparticle toxicity in skin *in vivo* and keratinocytes in vitro. Environ. Health Perspect. 118, 407.

Sánchez, J.G.B., Kouznetsov, V.V., 2010. Antimycobacterial susceptibility testing methods for natural products research. Braz. J. Microbiol. 41, 270–277.

Santos, C.M., Mangadlao, J., Ahmed, F., Leon, A., Advincula, R.C., Rodrigues, D.F., 2012. Graphene nanocomposite for biomedical applications: fabrication, antimicrobial and cytotoxic investigations. Nanotechnology 23, 395101.

Seil, J.T., Webster, T.J., 2012. Antimicrobial applications of nanotechnology: methods and literature. Int. J. Nanomed. 7, 2767–2781.

Sevin, E., Dehouck, L., Fabulas-da Costa, A., Cecchelli, R., Dehouck, M., Lundquist, S., et al., 2013. Accelerated Caco-2 cell permeability model for drug discovery. J. Pharmacol. Toxicol. Methods 68, 334–339.

Sharma, A., Madhunapantula, S.V., Robertson, G.P., 2012. Toxicological considerations when creating nanoparticle-based drugs and drug delivery systems. Expert Opin. Drug Metab. Toxicol. 8, 47–69.

Shrivastava, S., Bera, T., Roy, A., Singh, G., Ramachandrarao, P., Dash, D., 2007. Characterization of enhanced antibacterial effects of novel silver nanoparticles. Nanotechnology 18, 225103.

Simon-Deckers, A., Loo, S., Mayne-L'hermite, M., Herlin-Boime, N., Menguy, N., Reynaud, C., et al., 2009. Size-, composition-and shape-dependent toxicological impact of metal oxide nanoparticles and carbon nanotubes toward bacteria. Environ. Sci. Technol. 43, 8423–8429.

Sondi, I., Salopek-Sondi, B., 2004. Silver nanoparticles as antimicrobial agent: a case study on *E. coli* as a model for Gram-negative bacteria. J. Colloid Interface Sci. 275, 177–182.

Srinivasan, A., Uppuluri, P., Lopez-Ribot, J., Ramasubramanian, A.K., 2011. Development of a high-throughput *Candida albicans* biofilm chip. PLoS One 6, e19036.

Srinivasan, A., Leung, K.P., Lopez-Ribot, J.L., Ramasubramanian, A.K., 2013. High-throughput nano-biofilm microarray for antifungal drug discovery. Mbio. 4, e00331–13.

Stauff, D.L., Bassler, B.L., 2011. Quorum sensing in *Chromobacterium violaceum*: DNA recognition and gene regulation by the CviR receptor. J. Bacteriol. 193, 3871–3878.

Stone, V., Johnston, H., Schins, R.P., 2009. Development of *in vitro* systems for nanotoxicology: methodological considerations. Crit. Rev. Toxicol. 39, 613–626.

Sweet, M.J., Chesser, A., Singleton, I., 2012. Review: metal-based nanoparticles; size, function, and areas for advancement in applied microbiology. Adv. Appl. Microbiol. 80, 113–142.

Talbot, G.H., Bradley, J., Edwards, J.E.J., Gilbert, D., Scheld, M., Bartlett, J.G., 2006. Bad bugs need drugs: an update on the development pipeline from the Antimicrobial Availability Task Force of the Infectious Diseases Society of America. Clin. Infect. Dis. 42, 657–668.

Taneja, N.K., Tyagi, J.S., 2007. Resazurin reduction assays for screening of anti-tubercular compounds against dormant and actively growing *Mycobacterium tuberculosis*, *Mycobacterium bovis* BCG and *Mycobacterium smegmatis*. J. Antimicrob. Chemother. 60, 288–293.

Taylor, E., Webster, T.J., 2011. Reducing infections through nanotechnology and nanoparticles. Int. J. Nanomed. 6, 1463–1473.

Teeguarden, J.G., Hinderliter, P.M., Orr, G., Thrall, B.D., Pounds, J.G., 2007. Particokinetics *in vitro*: dosimetry considerations for *in vitro* nanoparticle toxicity assessments. Toxicol. Sci. 95, 300–312.

Tillotson, G.S., Theriault, N., 2013. New and alternative approaches to tackling antibiotic resistance. F1000Prime Rep. 5, 51.

Toutain, P., Lees, P., 2004. Integration and modelling of pharmacokinetic and pharmacodynamic data to optimize dosage regimens in veterinary medicine. J. Vet. Pharmacol. Therap. 27, 467–477.

van der Ploeg, M.J., van den Berg, J.H., Bhattacharjee, S., de Haan, L.H., Ershov, D.S., Fokkink, R.G., et al., 2014. *In vitro* nanoparticle toxicity to rat alveolar cells and coelomocytes from the earthworm *Lumbricus rubellus*. Nanotoxicology 8, 28–37.

Veerapandian, M., Yun, K., 2011. Functionalization of biomolecules on nanoparticles: specialized for antibacterial applications. Appl. Microbiol. Biotechnol. 90, 1655–1667.

Vidaillac, C., Leonard, S.N., Rybak, M.J., 2009. *In vitro* activity of ceftaroline against methicillin-resistant *Staphylococcus aureus* and heterogeneous vancomycin-intermediate *S. aureus* in a hollow fiber model. Antimicrob. Agents Chemother. 53, 4712–4717.

Vila, A., Sánchez, A., Janes, K., Behrens, I., Kissel, T., Jato, J. L.V., et al., 2004. Low molecular weight chitosan nanoparticles as new carriers for nasal vaccine delivery in mice. Eur. J. Pharm. Biopharm. 57, 123–131.

Wang, J., Zhu, X., Chen, Y., Chang, Y., 2012. Application of embryonic and adult zebrafish for nanotoxicity assessment. Methods Mol. Biol. 926, 317–329.

Wani, M.Y., Hashim, M.A., Nabi, F., Malik, M.A., 2011. Nanotoxicity: dimensional and morphological concerns. Adv. Phys. Chem. 2011, 450912.

Werth, B.J., Steed, M.E., Kaatz, G.W., Rybak, M.J., 2013. Evaluation of ceftaroline activity against heteroresistant vancomycin-intermediate *Staphylococcus aureus* and vancomycin-intermediate methicillin-resistant *S. aureus* strains in an in vitro pharmacokinetic/pharmacodynamic model: exploring the "seesaw effect". Antimicrob. Agents Chemother. 57, 2664–2668.

Wiegand, I., Hilpert, K., Hancock, R.E., 2008. Agar and broth dilution methods to determine the minimal inhibitory concentration (MIC) of antimicrobial substances. Nat. Protoc 3, 163–175.

Wijnhoven, S.W., Peijnenburg, W.J., Herberts, C.A., Hagens, W.I., Oomen, A.G., Heugens, E.H., et al., 2009. Nano-silver-a review of available data and knowledge gaps in human and environmental risk assessment. Nanotoxicology 3, 109–138.

Wörle-Knirsch, J., Pulskamp, K., Krug, H., 2006. Oops they did it again! Carbon nanotubes hoax scientists in viability assays. Nano Lett. 6, 1261–1268.

Xiu, Z., Zhang, Q., Puppala, H.L., Colvin, V.L., Alvarez, P. J., 2012. Negligible particle-specific antibacterial activity of silver nanoparticles. Nano Lett. 12, 4271–4275.

Yacoby, I., Benhar, I., 2008. Antibacterial nanomedicine. Nanomedicine (Lond). 3, 329–341.

Yim, G., Wang, H.H., Davies, J., 2007. Antibiotics as signalling molecules. Philos. Trans. R. Soc. Lond. B. Biol. Sci. 362, 1195–1200.

Zhang, L., Pornpattananangkul, D., Hu, C., Huang, C., 2010. Development of nanoparticles for antimicrobial drug delivery. Curr. Med. Chem. 17, 585–594.

CHAPTER

3

Silver Nanoparticles for the Control of Vector-Borne Infections

Kateryna Kon[1] *and Mahendra Rai*[2]

[1]Department of Microbiology, Virology and Immunology, Kharkiv National Medical University, Kharkiv, Ukraine [2]Biotechnology Department, SGB Amravati University, Amravati, Maharashtra, India

3.1 INTRODUCTION

Arthropods play a role in the transmission of various bacterial, viral, and protozoal diseases, such as plague, tularemia, yellow fever, Japanese encephalitis, malaria, leishmaniasis, and many others. The tendency of recent time is that many vector-borne pathogens are appearing in new geographical regions, and in endemic diseases there is also a tendency for an increase of incidence (Kilpatrick and Randolph, 2012; Medlock et al., 2012; El-Bahnasawy et al., 2013). Moreover, growing rates of insecticide resistance among arthropods transmitting infections are a serious problem influencing control of vector-borne diseases (Bridges et al., 2012; Smith and Goldman, 2012; Aïkpon et al., 2013; Yang and Liu, 2013; Coetzee and Koekemoer, 2013).

Resistance was described among all types of vectors (Raghavendra et al., 2011; Durand et al., 2012; Faza et al., 2013; Blayneh and Mohammed-Awel, 2014), and in many studies it was shown to reach extremely high levels. In a study on tick resistance, Faza et al. (2013) reported resistance levels among larvae of *Rhipicephalus microplus* (Acari: Ixodidae) to the organophosphates and pyrethroids as high as 75.49% and 97.44%, respectively. Among *Aedes aegypti* mosquitoes, resistance was observed in 100% of localities in which mosquitoes were exposed to insecticides DDT, bendiocarb, and temephos (Ocampo et al., 2011). In head lice, *Pediculus humanus capitis* De Geer (Phthiraptera: Pediculidae), frequencies of pyrethroid insecticide resistance kdr alleles were also found to be high (67–100%) (Toloza et al., 2014).

At the same time, presently available insecticidal substances possess many serious drawbacks, including drug residue contamination of the environment and, as a consequence, also of milk and meat products (Jayaseelan and Rahuman, 2012). Because of this, the interest of researchers has been directed to the

Nanotechnology in Diagnosis, Treatment and Prophylaxis of Infectious Diseases.
DOI: http://dx.doi.org/10.1016/B978-0-12-801317-5.00003-7

anti-arthropod potential of different types of other bioactive agents, including silver nanoparticles (AgNPs) (Rai and Ingle, 2012; Adhikaria et al., 2013; Suganya et al., 2014).

Different beneficial properties of silver have been known since ancient times, but development of technologies making it possible to easily produce silver in nanoparticle form has opened a new era in the biomedical application of silver. AgNPs have physical properties that are different from both silver ions and bulk material; their much higher biological activity caused by high surface area-to-volume ratio makes them promising agents in fighting against all possible types of pathogenic microorganisms—bacteria, fungi, protozoa, and viruses (Rai et al., 2009, 2014)—and also against vectors transmitting these microorganisms. Activity of AgNPs against microorganisms has been proven by hundreds of studies; however, investigations of anti-arthropod properties of AgNPs have been attracted attention recently. The aim of the present review is to summarize studies on activity of AgNPs against arthropod vectors of infectious diseases, such as lice, mosquito, ticks, and flies, and to formulate tasks for the studies that can be completed in the near feature.

3.2 LOUSE-BORNE INFECTIONS AND ACTIVITY OF AgNPs AGAINST LICE

There are more than 3,000 species of lice (order Phthiraptera), and among them only three are classified as human pathogens. They belong to sucking lice (Arthropoda: Insecta: order Phthiraptera: suborder Anoplura) and placed into two families—family Pediculidae, genus *Pediculus* (head louse and body louse) and family Pthiridae, genus *Pthirus* (pubic louse—*Pthirus pubis*). Although head and body lice have distinct ecology niches and minor variations in morphology and biology, recent genetic studies suggest that they can be considered more as different phenotypes of the same species, with differences mainly in gene expression and not in gene content (Veracx and Raoult, 2012). The traditional name for head louse is *P. humanus capitis*, and for body louse it is *Pediculus humanus humanus*, but because both lice belong to the same species, calling both of them *Pediculus humanus* (Smith, 2009) has been proposed; however, traditionally used names are still present in many studies.

The role of vectors transmitting infections is carried by both head and body lice. They can transmit life-threatening infections such as epidemic typhus (caused by *Rickettsia prowazekii*), relapsing fever (*Borrelia recurrentis*), and trench fever (*Bartonella quintana*) (Raoult and Roux, 1999; Fournier et al., 2002; El-Bahnsawy et al., 2012). These infections are of high concern in developing and developed countries; even if outbreaks of epidemic typhus and epidemic relapsing fever were fixed only in developing countries, significantly high seroprevalence against both bacteria among the homeless population of developed countries remind that these diseases have a high risk of outbreak throughout the world (Badiaga and Brouqui, 2012).

Despite the great importance of lice in transmission of life-threatening infections, there are only few studies on lousicidal activity of AgNPs (Jayaseelan et al., 2011; Marimuthu et al., 2012). In both studies AgNPs were biosynthesized using aqueous leaf extracts of *Tinospora cordifolia* and *Lawsonia inermis*, respectively, and they demonstrated significant activity against not only human head louse but also sheep body louse *Bovicola ovis* Schrank (Marimuthu et al., 2012). However, although the sizes of nanoparticles in both studies are equal, the activity of AgNPs produced using *L. inermis* was approximately 10-times higher, which is demonstrated by LC_{50} (Table 3.1).

TABLE 3.1 Studies on Pediculocidal, Larvicidal, and Acaricidal Activity of AgNPs

Aqueous leaf extract utilized for synthesis of AgNPs	Size and shape of AgNPs	Arthropod	LC_{50} (mg/L)	Ref.
LOUSICIDAL ACTIVITY				
T. cordifolia Miers (Menispermaceae)	55–80 nm	*P. humanus capitis* De Geer	12.46	Jayaseelan et al. (2011)
L. inermis	Average size of 60 nm, spherical	*P. humanus capitis* De Geer	1.33	Marimuthu et al. (2012)
LARVICIDAL ACTIVITY: AEDES MOSQUITO LARVAE[a]				
Fungus *Cochliobolus lunatus*	3–21 nm, spherical	*Ae. aegypti*	1.29, 1.48, and 1.58 against second, third, and fourth instar larvae, respectively	Salunkhe et al. (2011)
R. mucronata	60–95 nm, spherical	*Ae. aegypti*	0.585	Gnanadesigan et al. (2011)
Pergularia daemia	44–255 nm, spherical	*Ae. aegypti*	4.39, 5.12, 5.66, 6.18 against first to fourth instar larvae, respectively	Patil et al. (2012a)
Plumeria rubra plant latex	32–200 nm, spherical	*Ae. aegypti*	1.49, 1.82 against second and fourth instar larvae, respectively	Patil et al. (2012b)
Fungus *Chrysosporium tropicum*	20–50 nm, spherical	*Ae. aegypti*	3.47, 4, and 2 for the first, third and fourth larvae, respectively	Soni and Prakash (2012)
LARVICIDAL ACTIVITY: ANOPHELES MOSQUITO LARVAE				
T. cordifolia Miers (Menispermaceae)	55–80 nm	*Anopheles subpictus* Grassi	6.43	Jayaseelan et al. (2011)
Mimosa pudica Gaertn (Mimosaceae)	25–60 nm, spherical	*A. subpictus* Grassi	13.90	Marimuthu et al. (2011)
Nelumbo nucifera Gaertn. (Nymphaeaceae)	25–80 nm, spherical, triangle, truncated triangles, and decahedral-shaped	*A. subpictus* Grassi	0.69	Santhoshkumar et al. (2011)
Eclipta prostrata (L.) L. (Asteraceae)	35–60 nm, spherical	*A. subpictus* Grassi	5.14	Rajakumar and Rahuman (2011)
Fungus *Cochliobolus lunatus*	3–21 nm, spherical	*Anopheles stephensi* Liston	1.17, 1.30, and 1.41 against second, third, and fourth instar larvae, respectively	Salunkhe et al. (2011)
M. paradisiaca L. (Musaceae)	60–150 nm, rod-shaped	*A. stephensi* Liston	1.39	Jayaseelan et al. (2012)
P. daemia (Forssk.) Chiov. (Apocynaceae) plant latex	44–255 nm, spherical	*A. stephensi* Liston	4.41, 5.35, 5.91, 6.47 against first to fourth instar larvae, respectively	Patil et al. (2012a)

(Continued)

TABLE 3.1 (Continued)

Aqueous leaf extract utilized for synthesis of AgNPs	Size and shape of AgNPs	Arthropod	LC_{50} (mg/L)	Ref.
P. rubra L. (Apocynaceae) plant latex	32–200 nm, spherical	*A. stephensi* Liston	1.10, 1.74 against second and fourth instar larvae, respectively	Patil et al. (2012b)
Euphorbia hirta L. (Euphorbiaceae)	30–60 nm, spherical and with cubic structures	*A. stephensi* Liston	10.14, 16.82, 21.51, and 27.89 against first to fourth instar larvae, respectively	Priyadarshini et al. (2012)
Cassia occidentalis L. (Calsalpiniaceae)	450 nm	*A. stephensi* Liston	0.30 ppm for 3 mg/L, 0.41 ppm for 1.50 mg/L, and 2.12 ppm for 0.75 mg/L	Murugan et al. (2012)
Vinca rosea (L.) (Apocynaceae)	25–47 nm, spherical	*A. stephensi* Liston	12.47	Subarani et al. (2013)
Nerium oleander L. (Apocynaceae)	20–35 nm, spherical- and cubic-shaped	*A. stephensi* Liston	20.60, 24.90, 28.22, and 33.99 against first to fourth instar larvae, respectively	Roni et al. (2013)
Dried green fruits of *Drypetes roxburghii* (Wall.) Hur. (Euphorbiaceae)	10–35 nm, polyhedral	*A. stephensi* Liston	0.795, 0.964, and 1.134 against second, third, and fourth instar larvae, respectively	Haldar et al. (2013)
LARVICIDAL ACTIVITY: CULEX MOSQUITO LARVAE				
T. cordifolia Miers (Menispermaceae)	55–80 nm	*C. quinquefasciatus* say	6.96	Jayaseelan et al. (2011)
M. pudica L. (Fabaceae)	25–60 nm, spherical	*C. quinquefasciatus* say	11.73	Marimuthu et al. (2011)
Nelumbo nucifera Gaertn. (Nymphaeaceae)	25–80 nm, spherical, triangle, truncated triangles, and decahedral-shaped	*C. quinquefasciatus* say	1.10	Santhoshkumar et al. (2011)
R. mucronata L. (Rhizophoraceae)	60–95 nm, spherical	*C. quinquefasciatus* say	0.891	Gnanadesigan et al. (2011)
Eclipta prostrata (L.) L. (Asteraceae)	35–60 nm, spherical	*C. quinquefasciatus* say	4.56	Rajakumar and Rahuman (2011)
Pithecellobium dulce Roxb. (Benth.) (Fabaceae)	50–100 nm, spherical	*C. quinquefasciatus* say	21.56	Raman et al. (2012)
M. paradisiaca L. (Musaceae)	60–150 nm, rod-shaped	*Culex tritaeniorhynchus* Giles	1.63	Jayaseelan et al. (2012)

(*Continued*)

TABLE 3.1 (Continued)

Aqueous leaf extract utilized for synthesis of AgNPs	Size and shape of AgNPs	Arthropod	LC_{50} (mg/L)	Ref.
V. rosea (L.) (Apocynaceae)	25–47 nm, spherical	*C. quinquefasciatus* say	43.80	Subarani et al. (2013)
Dried green fruits of *D. roxburghii* (Wall.) Hur. (Euphorbiaceae)	10–35 nm, polyhedral	*C. quinquefasciatus* say	0.92, 1.27, and 1.40 against second, third, and fourth instar larvae, respectively	Haldar et al. (2013)
Bark extract of *Ficus racemosa* L. (Moraceae)	Average size of 251 nm, cylindrical, uniform, and rod-shaped	*C. quinquefasciatus* Say	12.00	Velayutham et al. (2013)
Bark extract of *F. racemosa* L. (Moraceae)	Average size of 251 nm, cylindrical, uniform, and rod-shaped	*Culex gelidus*	11.21	Velayutham et al. (2013)
ACTIVITY AGAINST TICKS				
M. pudica L. (Fabaceae)	25–60 nm, spherical	*R. (B.) microplus* Canestrini	8.98 against larvae	Marimuthu et al. (2011)
Manilkara zapota (L.) P. Royen (Sapotaceae)	70–140 nm, spherical and oval	*R. (B.) microplus*	3.44 against larvae	Rajakumar and Rahuman (2012)
Cissus quadrangularis L. (Vitaceae)	Average size of 42 nm, spherical and oval	*R. (B.) microplus*	7.61 against larvae	Santhoshkumar et al. (2012)
O. canum Sims (Labiatae)	25–110 nm, rods and cylindrical	*H. anatolicum* (a.) *anatolicum* Koch, 1844	0.78 against larvae	Jayaseelan and Rahuman (2012)
O. canum Sims (Labiatae)	25–110 nm, rods and cylindrical	*H. marginatum* (m.) *isaaci* Sharif, 1928	1.51 against larvae	Jayaseelan and Rahuman (2012)
M. paradisiaca L. (Musaceae)	60–150 nm, rod-shaped	*H. bispinosa* Neumann	1.87 against larvae	Jayaseelan et al. (2012)
Euphorbia prostrata Ait. (Euphorbiaceae)	25–80, rod-shaped	*H. bispinosa* Neumann	2.30 against adult ticks	Zahir and Rahuman (2012)
ACTIVITY AGAINST FLIES				
M. paradisiaca L. (Musaceae)	60–150 nm, rod-shaped	*H. maculata* Leach	2.02 against larvae	Jayaseelan et al. (2012)
E. prostrata Ait. (Euphorbiaceae)	25–80 nm, rod-shaped	*H. maculata* Leach	2.55 against adult flies	Zahir and Rahuman (2012)
C. quadrangularis L. (Vitaceae)	Average size of 42 nm, spherical and oval	*H. maculata* Leach	18.14 against adult flies	Santhoshkumar et al. (2012)
M. zapota (L.) P. Royen (Sapotaceae)	70–140 nm, spherical and oval	*M. domestica*	3.64 against adult flies	Kamaraj et al. (2012)

[a] *If not indicated, fourth instar larvae of mosquito were used.*

3.3 MOSQUITO-BORNE INFECTIONS AND ACTIVITY OF AgNPs AGAINST MOSQUITOES

Mosquitoes are involved in the transmission of a number of life-threatening diseases that have a great impact on worldwide morbidity and mortality (Tolle, 2009; Kamareddine, 2012), including malaria caused by Plasmodium, filariasis caused by worms, and viral-generated yellow fever, dengue infection, chikungunya virus infection, Rift Valley fever, Japanese encephalitis, West Nile encephalitis, and others (Reiner et al., 2013).

Mosquitoes transmitting human infections mainly belong to *Aedes*, *Anopheles*, and *Culex* genera. *A. aegypti* is the common vector of yellow fever and dengue; dengue is also often transmitted by another *Aedes* species—*Ae. albopictus*; *Ae. triseriatus* transmits La Crosse encephalitis, *Ae. japonicus* is a vector of Japanese encephalitis;, *Anopheles* spp. transmit plasmodia parasites causing malaria, and *Culex* spp. transmit several types of arboviral encephalitis (Eastern and Western equine encephalites, St. Louis encephalitis) and West Nile virus (Tolle, 2009). All these genera can transmit filariasis depending on geographical location; *Anopheles* is the most common vector in Africa, *Culex quinquefasciatus* is the most common in America, and *Aedes* is most common in the Pacific and in Asia (CDC, http://www.cdc.gov/parasites/lymphaticfilariasis/gen_info/vectors.html).

Most studies exploring activity of AgNPs against insects are directed at anti-mosquito activity (Table 3.1). Published studies used various sources for the biosynthesis of AgNPs, particularly different species of plants and fungi, and various stages of larvae life cycle as experimental models; nevertheless, all types of AgNPs demonstrated significant activity against all studied larvae of the *Ae. aegypti* mosquito, with ranges of LC_{50} from 0.59 mg/L (Gnanadesigan et al., 2011) to 6.18 mg/L (Patil et al., 2012a).

Activity of AgNPs against larvae of *Anopheles* mosquito was a little lower with LC_{50} reaching more than 20 mg/L in some studies (Priyadarshini et al., 2012; Roni et al., 2013); however, in other studies AgNPs also demonstrated good larvicidal effect with LC_{50} of less than 1.5 mg/L (Santhoshkumar et al., 2011; Salunkhe et al., 2011; Jayaseelan et al., 2012; Patil et al., 2012b; Haldar et al., 2013). Interesting results are being demonstrated in the ongoing study by Murugan et al. (2012); the authors evaluated not only larvicidal toxicity of AgNPs expressed in LC_{50} but also adult longevity in mosquitoes treated as larvae with an AgNP concentration of 0.1 ppm and number of eggs laid by female mosquitoes exposed as larvae to AgNPs exposed to the same concentration. Adult longevity (measured in days) in male and female mosquitoes was reduced by 29% and the number of eggs decreased by 32%; both results were significantly different ($P < 0.05$) from results of nonexposed mosquitoes.

In larvae of *Culex* mosquitoes, activity of AgNPs changed in wide ranges, starting from LC_{50} of 0.89 mg/L in AgNPs produced using aqueous extract of *Rhizophora mucronata* (Gnanadesigan et al., 2011) to 43.8 mg/L in AgNPs produced with aqueous extract of *Catharanthus roseus* (Subarani et al., 2013).

There are many studies that proved the anti-arthropod effects of AgNPs, but there is still no comprehensible scientific explanation for it. Some studies stated that larvicidal activity of AgNPs is concentration-dependent and is supposed to be caused by penetration of nanoparticles through the membrane of larvae (Sap-Iam et al., 2010; Salunkhe et al., 2011); however, more studies should be conducted to reveal all specific mechanisms of anti-arthropod activity of AgNPs.

3.4 TICK-BORNE INFECTIONS AND ACTIVITY OF AgNPs AGAINST TICKS

Ticks can carry viruses, bacteria, and protozoans, and they are vectors of many life-threatening infections. Tick-borne viruses ("tiboviruses") cause febrile illnesses with a rapid onset, fever, sweating, headache, nausea, weakness, myalgia, arthralgia, and sometimes polyarthritis and rash, infections that affect the central nervous system, such as meningitis, meningoencephalitis, or encephalomyelitis with paresis, paralysis, and other sequelae, and hemorrhagic diseases (Hubálek and Rudolf, 2012). Among bacterial diseases transmitted by ticks are Lyme disease, rickettsioses, and tularemia (Parola et al., 2005; Foley and Nieto, 2010; Kung et al., 2013); furthermore, ticks transmit the protozoan infection babesiosis (Schnittger et al., 2012). A bite from one tick may transmit several infections simultaneously (Pujalte and Chua, 2013).

Ticks most commonly transmitting infections belong to genera *Ixodes* and *Dermacentor*. *Ixodes scapularis* transmits Lyme disease, babesiosis, and anaplasmosis; *Dermacentor andersoni* and *Dermacentor variabilis* transmit Rocky Mountain spotted fever and tularemia (CDC, http://www.cdc.gov/ticks/diseases/).

Some ticks produce toxins that can cause tick paralysis, for example, *D. andersoni* (the Rocky Mountain wood tick), *D. variabilis* (the American dog tick), and *Ixodes holocyclus* (the marsupial tick) (Diaz, 2010; Pecina, 2012).

During evaluation of AgNP effects on ticks, most scientists studied the activity against larvae of ticks (Table 3.1), but only Zahir and Rahuman (2012) used adult ticks in the experiment. Another limitation of the present data is that only a few tick species were studied—*R. (Boophilus) microplus*, *Hyalomma anatolicum anatolicum*, *Hyalomma marginatum isaaci*, and *Haemaphysalis bispinosa*. However, all performed studies showed promising activity of AgNPs against both larvae and adult ticks with LC_{50} ranging from 0.79 mg/L (Jayaseelan and Rahuman, 2012) to 8.98 mg/L (Marimuthu et al., 2011). The best results were obtained with AgNPs produced using extract of *Ocimum canum* against ticks of *Hyalomma* spp. (Jayaseelan and Rahuman, 2012).

3.5 FLIES, THEIR ROLE IN TRANSMISSION AND SPREAD OF INFECTIONS, AND ACTIVITY OF AgNPs AGAINST FLIES

Flies play a double role in the transmission of infectious diseases. They can transmit infections through biting or can be a mechanical factor contributing to the spread of infections. Biting or hematophagous flies are involved in the transmission of bacterial infections, such as tularemia (deer fly, *Chrysops* spp.), protozoan infections such as leishmaniasis (sand fly, Diptera: Psychodidae) and African sleeping sickness (tsetse fly, *Glossina* spp.), and worm invasions such as onchocerciasis (blackfly, *Simulium* spp.) (Petersen et al., 2008; Traore et al., 2012; Holmes et al., 2013; Cruz et al., 2013).

Nonbiting synanthropic flies (some species are in the families Sarcophagidae, such as flesh flies, Muscidae, such as house flies and latrine flies, and Calliphoridae, such as blow flies and bottle flies) can contribute to the spread of infections by mechanical carrying of bacteria causing gastrointestinal infections (cholera, typhoid fever, salmonellosis) or contact infections (e.g., trachoma) (Graczyk et al., 2005). Likewise, flies can transmit oocysts of *Toxoplasma gondii* and of diarrhea-producing protozoan *Cryptosporidium parvum*, which recently has contributed significantly to the mortality of immunocompromised or immunosuppressed patients (Graczyk et al., 2005).

Transmission of microorganisms by nonbiting flies occurs after their feeding on some infected sources and by mechanical dislodgement from the exoskeleton of flies or from their feces and vomit (Graczyk et al., 2004).

Studies of the activity of AgNPs against flies are very scarce. Several authors demonstrated activity of plant-synthesized AgNPs against hematophagous fly *Hippobosca maculata* Leach, including effects on fly larvae (Jayaseelan et al., 2012) and adult flies (Zahir and Rahuman, 2012; Santhoshkumar et al., 2012). Kamaraj et al. (2012) reported activity of AgNPs against the synanthropic fly *Musca domestica*. In all these studies AgNPs were biosynthetically produced using plant extracts with a wide range of obtained sizes of nanoparticles and their activity. The best results were demonstrated in AgNPs produced by using aqueous leaf extract of *Musa paradisiaca* with LC_{50} 2.02 mg/L (Jayaseelan et al., 2012).

3.6 CONCLUSIONS AND FUTURE PROSPECTS

Vector-borne infections are very important among infectious diseases with high morbidity and mortality worldwide. One approach to fighting against such infections is by controlling vectors transmitting them. The control of populations of mosquitoes, lice, flies, and ticks may help to reduce the prevalence of vector-borne infections, and production of substances that have high anti-arthropod activity and simultaneously are environmentally safe is an important challenge in modern science.

Published studies demonstrated promising activity of AgNPs against all types of vectors of infectious diseases. They showed broad-spectrum insecticidal activity that was especially well-studied against larvae of mosquitoes; insecticidal activity was also examined in a few studies that investigated the activity of AgNPs against flies, lice, and ticks. Interestingly, in all the published studies AgNPs were produced in a biosynthetical manner utilizing plant extracts or, in a few studies, fungi. In all studies AgNPs had significantly higher activity than that of corresponding plant extracts and higher than that of 1 mM $AgNO_3$ solution used for nanoparticle synthesis.

Despite high anti-arthropod activity, biosynthetically produced AgNPs and solvent extracts used for their production did not show any notable toxicity on environmental organisms, such as water fleas *Daphnia magna* and *Ceriodaphnia dubia*, and they did not show any undesirable effects in the animal model against cattle *Bos indicus* (Zahir and Rahuman, 2012); likewise, no toxicity was detected against fish *Poecilia reticulate* (Salunkhe et al., 2011; Patil et al., 2012a,b; Subarani et al., 2013). Thus, the control of arthropods with biosynthetically produced AgNPs is environmentally friendly.

At the same time, many questions in this area are still not clear and require future investigations. There is no complete understanding of the mechanisms of the anti-arthropod effects of AgNPs. Only a few studies hypothesized membrane-damaging larvicidal activity, but more efforts should be directed to formulating theories of mechanisms of larvicidal, lousicidal, and acaricidal effects. A broader species spectrum of flies and ticks should be evaluated. Furthermore, concentrations of AgNPs with pediculicidal effects should be studied regarding toxicity in mammal organisms. A better understanding of all these questions will help control arthropod-borne diseases.

References

Adhikaria, U., Ghoshb, A., Chandra, G., 2013. Nano particles of herbal origin: a recent eco-friend trend in mosquito control. Asian Pac. J. Trop. Dis. 3 (2), 167–168.

Aïkpon, R., Agossa, F., Ossè, R., Oussou, O., Aïzoun, N., Oké-Agbo, F., et al., 2013. Bendiocarb resistance in *Anopheles gambiae* s.l. populations from Atacora department in Benin, West Africa: a threat for malaria vector control. Parasit. Vectors 6 (1), 192.

Badiaga, S., Brouqui, P., 2012. Human louse-transmitted infectious diseases. Clin. Microbiol. Infect. 18 (4), 332–337.

Blayneh, K.W., Mohammed-Awel, J., 2014. Insecticide-resistant mosquitoes and malaria control. Math. Biosci. 252C, 14–26.

Bridges, D.J., Winters, A.M., Hamer, D.H., 2012. Malaria elimination: surveillance and response. Pathog. Glob. Health 106 (4), 224–231.

Coetzee, M., Koekemoer, L.L., 2013. Molecular systematics and insecticide resistance in the major African malaria vector *Anopheles funestus*. Annu. Rev. Entomol. 58, 393–412.

Cruz, C.F., Cruz, M.F., Galati, E.A., 2013. Sandflies (Diptera: Psychodidae) in rural and urban environments in an endemic area of cutaneous leishmaniasis in southern Brazil. Mem. Inst. Oswaldo Cruz. 108 (3), pii: S0074-02762013000300303.

Diaz, J.H., 2010. A 60-year meta-analysis of tick paralysis in the United States: a predictable, preventable, and often misdiagnosed poisoning. J. Med. Toxicol. 6 (1), 15–21.

Durand, R., Bouvresse, S., Berdjane, Z., Izri, A., Chosidow, O., Clark, J.M., 2012. Insecticide resistance in head lice: clinical, parasitological and genetic aspects. Clin. Microbiol. Infect. 18 (4), 338–344.

El-Bahnasawy, M.M., Khater, M.K., Morsy, T.A., 2013. The mosquito borne West Nile virus infection: is it threating to Egypt or a neglected endemic disease? J. Egypt. Soc. Parasitol. 43 (1), 87–102.

El-Bahnsawy, M.M., Labib, N.A., Abdel-Fattah, M.A., Ibrahim, A.M., Morsy, T.A., 2012. Louse and tick borne relapsing fevers. J. Egypt. Soc. Parasitol. 42 (3), 625–638.

Faza, A.P., Pinto, I.S., Fonseca, I., Antunes, G.R., Monteiro, C. M., Daemon, E., et al., 2013. A new approach to characterization of the resistance of populations of *Rhipicephalus microplus* (Acari: Ixodidae) to organophosphate and pyrethroid in the state of Minas Gerais, Brazil. Exp. Parasitol. pii: S0014-4894(13)00127-6. 10.1016/j.exppara.2013.04.006.

Foley, J.E., Nieto, N.C., 2010. Tularemia. Vet. Microbiol. 140 (3–4), 332–338.

Fournier, P.E., Ndihokubwayo, J.B., Guidran, J., Kelly, P.J., Raoult, D., 2002. Human pathogens in body and head lice. Emerg. Infect. Dis. 8 (12), 1515–1518.

Gnanadesigan, M., Anand, M., Ravikumar, S., Maruthupandy, M., Vijayakumar, V., Selvam, S., et al., 2011. Biosynthesis of silver nanoparticles by using mangrove plant extract and their potential mosquito larvicidal property. Asian Pac. J. Trop. Med. 4 (10), 799–803.

Graczyk, T.K., Grimes, B.H., Knight, R., Szostakowska, B., Kruminis-Lozowska, W., Racewicz, M., et al., 2004. Mechanical transmission of *Cryptosporidium parvum* oocysts by flies. Wiad. Parazytol. 50 (2), 243–247.

Graczyk, T.K., Knight, R., Tamang, L., 2005. Mechanical transmission of human protozoan parasites by insects. Clin. Microbiol. Rev. 18 (1), 128–132.

Haldar, K.M., Haldar, B., Chandra, G., 2013. Fabrication, characterization and mosquito larvicidal bioassay of silver nanoparticles synthesized from aqueous fruit extract of putranjiva, *Drypetes roxburghii* (Wall.). Parasitol. Res. 112 (4), 1451–1459.

Holmes, P., 2013. Tsetse-transmitted trypanosomes—their biology, disease impact and control. J. Invertebr. Pathol. 112 (Suppl), S11–S14.

Hubálek, Z., Rudolf, I., 2012. Tick-borne viruses in Europe. Parasitol. Res. 111 (1), 9–36.

Jayaseelan, C., Rahuman, A.A., 2012. Acaricidal efficacy of synthesized silver nanoparticles using aqueous leaf extract of *Ocimum canum* against *Hyalomma anatolicum anatolicum* and *Hyalomma marginatum isaaci* (Acari: Ixodidae). Parasitol. Res. 111 (3), 1369–1378.

Jayaseelan, C., Rahuman, A.A., Rajakumar, G., Vishnu Kirthi, A., Santhoshkumar, T., Marimuthu, S., et al., 2011. Synthesis of pediculocidal and larvicidal silver nanoparticles by leaf extract from heartleaf moonseed plant, *Tinospora cordifolia* Miers. Parasitol. Res. 109 (1), 185–194.

Jayaseelan, C., Rahuman, A.A., Rajakumar, G., Santhoshkumar, T., Kirthi, A.V., Marimuthu, S., et al., 2012. Efficacy of plant-mediated synthesized silver nanoparticles against hematophagous parasites. Parasitol. Res. 111 (2), 921–933.

Kamaraj, C., Rajakumar, G., Rahuman, A.A., Velayutham, K., Bagavan, A., Zahir, A.A., et al., 2012. Feeding deterrent activity of synthesized silver nanoparticles using *Manilkara zapota* leaf extract against the house fly, *Musca domestica* (Diptera: Muscidae). Parasitol. Res. 111 (6), 2439–2448.

Kamareddine, L., 2012. The biological control of the malaria vector. Toxins (Basel). 4 (9), 748–767.

Kilpatrick, A.M., Randolph, S.E., 2012. Drivers, dynamics, and control of emerging vector-borne zoonotic diseases. Lancet 380 (9857), 1946–1955.

Kung, F., Anguita, J., Pal, U., 2013. *Borrelia burgdorferi* and tick proteins supporting pathogen persistence in the vector. Future Microbiol. 8 (1), 41–56.

Marimuthu, S., Rahuman, A.A., Rajakumar, G., Santhoshkumar, T., Kirthi, A.V., Jayaseelan, C., et al., 2011. Evaluation of green synthesized silver nanoparticles against parasites. Parasitol. Res. 108 (6), 1541–1549.

Marimuthu, S., Rahuman, A.A., Santhoshkumar, T., Jayaseelan, C., Kirthi, A.V., Bagavan, A., et al., 2012. Lousicidal activity of synthesized silver nanoparticles using *Lawsonia inermis* leaf aqueous extract against *Pediculus humanus capitis* and *Bovicola ovis*. Parasitol. Res. 111 (5), 2023–2033.

Medlock, J.M., Hansford, K.M., Schaffner, F., Versteirt, V., Hendrickx, G., Zeller, H., et al., 2012. A review of the invasive mosquitoes in Europe: ecology, public health risks, and control options. Vector Borne Zoonotic Dis. 12 (6), 435–447.

Murugan, K., Shri, K.P., Barnard, D., 2012. Green synthesis of silver nanoparticles from botanical sources and their use for control of medical insects and malaria parasites. <http://www.ars.usda.gov/research/publications/publications.htm?seq_no_115=281989>.

Ocampo, C.B., Salazar-Terreros, M.J., Mina, N.J., McAllister, J., Brogdon, W., 2011. Insecticide resistance status of *Aedes aegypti* in 10 localities in Colombia. Acta Trop. 118 (1), 37–44.

Parola, P., Paddock, C.D., Raoult, D., 2005. Tick-borne rickettsioses around the world: emerging diseases challenging old concepts. Clin. Microbiol. Rev. 18 (4), 719–756.

Patil, C.D., Borase, H.P., Patil, S.V., Salunkhe, R.B., Salunke, B.K., 2012a. Larvicidal activity of silver nanoparticles synthesized using *Pergularia daemia* plant latex against *Aedes aegypti* and *Anopheles stephensi* and nontarget fish *Poecillia reticulata*. Parasitol. Res. 111 (2), 555–562.

Patil, C.D., Patil, S.V., Borase, H.P., Salunke, B.K., Salunkhe, R.B., 2012b. Larvicidal activity of silver nanoparticles synthesized using *Plumeria rubra* plant latex against *Aedes aegypti* and *Anopheles stephensi*. Parasitol. Res. 110 (5), 1815–1822.

Pecina, C.A., 2012. Tick paralysis. Semin. Neurol. 32 (5), 531–532.

Petersen, J.M., Carlson, J.K., Dietrich, G., Eisen, R.J., Coombs, J., Janusz, A.M., et al., 2008. Multiple *Francisella tularensis* subspecies and clades, tularemia outbreak, Utah. Emerg. Infect. Dis. 14 (12), 1928–1930.

Priyadarshini, K.A., Murugan, K., Panneerselvam, C., Ponarulselvam, S., Hwang, J.S., Nicoletti, M., 2012. Biolarvicidal and pupicidal potential of silver nanoparticles synthesized using *Euphorbia hirta* against *Anopheles stephensi* Liston (Diptera: Culicidae). Parasitol. Res. 111 (3), 997–1006.

Pujalte, G.G., Chua, J.V., 2013. Tick-borne infections in the United States. Prim. Care. 40 (3), 619–635.

Raghavendra, K., Barik, T.K., Reddy, B.P., Sharma, P., Dash, A.P., 2011. Malaria vector control: from past to future. Parasitol. Res. 108 (4), 757–779.

Rai, M., Ingle, A., 2012. Role of nanotechnology in agriculture with special reference to management of insect pests. Appl. Microbiol. Biotechnol. 94 (2), 287–293.

Rai, M., Yadav, A., Gade, A., 2009. Silver nanoparticles as a new generation of antimicrobials. Biotechnol. Adv. 27 (1), 76–83.

Rai, M., Kon, K., Ingle, A., Duran, N., Galdiero, S., Galdiero, M., 2014. Broad-spectrum bioactivities of silver nanoparticles: the emerging trends and future prospects. Appl. Microbiol. Biotechnol. Available from: <http://dx.doi.org/10.1007/s00253-013-5473-x>.

Rajakumar, G., Rahuman, A.A., 2011. Larvicidal activity of synthesized silver nanoparticles using *Eclipta prostrata* leaf extract against filariasis and malaria vectors. Acta Trop. 118 (3), 196–203.

Rajakumar, G., Rahuman, A.A., 2012. Acaricidal activity of aqueous extract and synthesized silver nanoparticles from *Manilkara zapota* against *Rhipicephalus (Boophilus) microplus*. Res. Vet. Sci. 93 (1), 303–309.

Raman, N., Sudharsan, S., Veerakumar, V., Pravin, N., Vithiya, K., 2012. *Pithecellobium dulce* mediated extracellular green synthesis of larvicidal silver nanoparticles. Spectrochim. Acta A Mol. Biomol. Spectrosc. 96, 1031–1037.

Raoult, D., Roux, V., 1999. The body louse as a vector of reemerging human diseases. Clin. Infect. Dis. 29 (4), 888–911.

Reiner Jr., R.C., Perkins, T.A., Barker, C.M., Niu, T., Chaves, L.F., Ellis, A.M., et al., 2013. A systematic review of mathematical models of mosquito-borne pathogen transmission: 1970-2010. J. R. Soc. Interface 10 (81), 20120921.

Roni, M., Murugan, K., Panneerselvam, C., Subramaniam, J., Hwang, J.S., 2013. Evaluation of leaf aqueous extract and synthesized silver nanoparticles using *Nerium oleander* against *Anopheles stephensi* (Diptera: Culicidae). Parasitol. Res. 112 (3), 981–990.

Salunkhe, R.B., Patil, S.V., Patil, C.D., Salunke, B.K., 2011. Larvicidal potential of silver nanoparticles synthesized using fungus *Cochliobolus lunatus* against *Aedes aegypti* (Linnaeus, 1762) and *Anopheles stephensi* Liston (Diptera; Culicidae). Parasitol. Res. 109 (3), 823–831.

Santhoshkumar, T., Rahuman, A.A., Rajakumar, G., Marimuthu, S., Bagavan, A., Jayaseelan, C., et al., 2011. Synthesis of silver nanoparticles using *Nelumbo nucifera* leaf extract and its larvicidal activity against malaria and filariasis vectors. Parasitol. Res. 108 (3), 693–702.

Santhoshkumar, T., Rahuman, A.A., Bagavan, A., Marimuthu, S., Jayaseelan, C., Kirthi, A.V., et al., 2012. Evaluation of stem aqueous extract and synthesized silver nanoparticles using *Cissus quadrangularis* against *Hippobosca maculata* and *Rhipicephalus (Boophilus) microplus*. Exp. Parasitol. 132 (2), 156–165.

Sap-Iam, N., Homklincha, C., Larpudomle, R., Warisnoich, W., Sereemaspu, A., Dubas, S.T., 2010. UV irradiation-induced silver nanoparticles as mosquito larvicides. J. App. Sci. 10 (23), 3132–3136.

Schnittger, L., Rodriguez, A.E., Florin-Christensen, M., Morrison, D.A., 2012. Babesia: a world emerging. Infect. Genet. Evol. 12 (8), 1788–1809.

Smith, C.H., Goldman, R.D., 2012. An incurable itch: head lice. Can. Fam. Physician 58 (8), 839–841.

Smith, V., 2009. Taxonomy of human lice. <http://phthiraptera.info/content/taxonomy-human-lice>.

Soni, N., Prakash, S., 2012. Efficacy of fungus mediated silver and gold nanoparticles against *Aedes aegypti* larvae. Parasitol. Res. 110 (1), 175–184.

Subarani, S., Sabhanayakam, S., Kamaraj, C., 2013. Studies on the impact of biosynthesized silver nanoparticles (AgNPs) in relation to malaria and filariasis vector control against *Anopheles stephensi* Liston and *Culex quinquefasciatus* Say (Diptera: Culicidae). Parasitol. Res. 112 (2), 487–499.

Suganya, G., Karthi, S., Shivakumar, M.S., 2014. Larvicidal potential of silver nanoparticles synthesized from Leucas aspera leaf extracts against dengue vector *Aedes aegypti*. Parasitol. Res. 2014 [Epub ahead of print].

Tolle, M.A., 2009. Mosquito-borne diseases. Curr. Probl. Pediatr. Adolesc. Health Care 39 (4), 97–140.

Toloza, A.C., Ascunce, M.S., Reed, D., Picollo, M.I., 2014. Geographical distribution of pyrethroid resistance allele frequency in head lice (Phthiraptera: Pediculidae) from Argentina. J. Med. Entomol. 51 (1), 139–144.

Traore, M.O., Sarr, M.D., Badji, A., Bissan, Y., Diawara, L., Doumbia, K., et al., 2012. Proof-of-principle of onchocerciasis elimination with ivermectin treatment in endemic foci in Africa: final results of a study in Mali and Senegal. PLoS. Negl. Trop. Dis. 6 (9), e1825.

Velayutham, K., Rahuman, A.A., Rajakumar, G., Roopan, S. M., Elango, G., Kamaraj, C., et al., 2013. Larvicidal activity of green synthesized silver nanoparticles using bark aqueous extract of *Ficus racemosa* against *Culex quinquefasciatus* and *Culex gelidus*. Asian Pac. J. Trop. Med. 6 (2), 95–101.

Veracx, A., Raoult, D., 2012. Biology and genetics of human head and body lice. Trends Parasitol. 28 (12), 563–571.

Yang, T., Liu, N., 2013. Permethrin resistance profiles in a field population of mosquitoes, *Culex quinquefasciatus* (Diptera: Culicidae). J. Med. Entomol. 50 (3), 585–593.

Zahir, A.A., Rahuman, A.A., 2012. Evaluation of different extracts and synthesised silver nanoparticles from leaves of *Euphorbia prostrata* against *Haemaphysalis bispinosa* and *Hippobosca maculata*. Vet. Parasitol. 187 (3–4), 511–520.

CHAPTER

4

Magnetite Nanostructures: Trends in Anti-Infectious Therapy

Alina Maria Holban[1,2,3], *Alexandru Mihai Grumezescu*[2,3] *and Florin Iordache*[4]

[1]Microbiology Immunology Department, Faculty of Biology, University of Bucharest, Bucharest, Romania [2]AMG Transcend, Bucharest, Romania [3]Department of Science and Engineering of Oxide Materials and Nanomaterials, Faculty of Applied Chemistry and Materials Science, University Politehnica of Bucharest, Bucharest, Romania [4]Institute of Cellular Biology and Pathology of Romanian Academy, "Nicolae Simionescu," Department of Fetal and Adult Stem Cell Therapy, Bucharest, Romania

4.1 INTRODUCTION

The alarming rates of antimicrobial resistance and the potential risks that this issue brings for health, society, and economy highlight the need for developing novel, efficient antimicrobials. Unfortunately, the rate of pathogens to adapt and obtain resistance overwhelms our capacity of developing novel therapeutic compounds with antimicrobial properties (Moy et al., 2006). The most investigated factors that lead to a low rate of newly developed antimicrobials are the following: (i) the high percent of noncultivable or fastidious microorganisms (Osburne et al., 2000); (ii) a high number of toxic or poorly pharmacokinetic compounds found in the synthetic compound libraries (Lipinski and Hopkins, 2004); (iii) the incapacity of most therapeutic compounds to nullify the multidrug-resistant (MDR) barrier of bacteria cells (especially in the Gram-negative MDRs) (Li and Nikaido, 2004); and (iv) the small resemblance of the results obtained *in vitro* and *in vivo* in the whole organisms where the tested compound should actually function (Lipinski and Hopkins, 2004).

Trying to overcome these technical or natural issues, researchers have developed new or uncommon strategies with the same aim: to efficiently fight infections. Moy and collaborators suggest that the putative antimicrobial compounds should be tested on *in vivo* models to prove their true efficiency. Their work resembles survival tests of the nematode *Caenorhabditis elegans* persistently infected with the human

Nanotechnology in Diagnosis, Treatment and Prophylaxis of Infectious Diseases.
DOI: http://dx.doi.org/10.1016/B978-0-12-801317-5.00004-9

opportunistic bacterial pathogen *Enterococcus faecalis*, an infection that usually leads to nematode death. Their screening of a synthetic compound library and also of natural compounds, which seems to cure *C. elegans* of *E. faecalis* persistent infections, suggests that, in contrast to the traditional antibiotics, the proposed assay identifies compounds that block pathogen multiplication *in vitro* and also helps to identify compounds that may be used as prodrugs, affect the virulence of the pathogen, suppress survival, and enhance the immune response of the host (Moy et al., 2006).

Because the synthetic compounds are usually more prone to select microbial resistance, researchers focus more on identifying and using natural compounds with antimicrobial activity (Holban et al., 2013a,b).

Natural compounds derived from microorganisms and their hosts have also been proven to be effective in infection control (Holban et al., 2014a). Vegetal extracts and essential oils have been extensively used in traditional medicine for treating infections (WHO, 2002). Recent studies revealed the great impact of mixture extracts and pure compounds obtained from Angiosperms and Gymnosperms to impair bacteria growth and also to modulate their virulence and resistance (Saviuc et al., 2013; Anghel and Grumezescu, 2013; Anghel et al., 2013a). Despite their proven role, studies revealed that natural compounds usually have low stability, and the technical procedures necessary for their testing and use significantly reduce their antimicrobial effect, especially when used in low doses (Holban et al., 2014b). Therefore, stabilizing agents that are able to preserve their action, deliver them to the intended site, and ensure an adequate release are desirable.

4.2 NANOPARTICLES WITH BIOMEDICAL APPLICATIONS

Nanotechnology has significantly evolved to improve the efficiency of therapeutic drugs and offer new opportunities for both diagnosis and therapy. Nanoparticles usually range in dimension from 1 to 100 nm and have unique properties, many of which being different from their bulk equivalent. These unique physicochemical, optical, and biological properties possessed by nanoparticles can be easily manipulated for desired applications (Rai and Bai, 2011).

Many types of nanostructures have been reported as useful in the biomedical field. Among the most used nanostructures are: liposomes; polymer nanoparticles; magnetic nanoparticles; and dendrimers. They allowed the development of fast and more efficient biosensors and ensured targeted drug delivery to specific cells and organs, novel cancer therapy, and hyperthermia treatments (Hofmann-Amtenbrink et al., 2009).

Iron oxide nanoparticles are particles with diameters between approximately 1 and 100 nm. These nanoparticles were found in a large variety of organisms, and they have attracted extensive interest because of their superparamagnetic properties and their potential applications in many fields (Teja and Koh, 2009).

Synthetic magnetic nanoparticles and, in particular, iron oxide nanoparticles are currently the subject of wide basic and applied research because of their useful properties. Among current applications we can list magnetic resonance imaging (MRI), magnetic detection of bimolecular interactions, biocompatible films, anticancer agents, drug and gene delivery systems, DNA analysis, and cell labeling (Giersig and Khomutov, 2008). Because of their biocompatibility and good magnetic properties, iron oxide nanoparticles are used in the development of nanostructured materials (Giersig and Khomutov, 2008).

In the microbiology field, nanoparticles have been used as novel antimicrobials with intrinsic therapeutic effects (Prabhu and Poulose, 2012), and also as efficient shuttles for the delivery and controlled release of antimicrobial compounds. Studies revealed that biomimetic nanoparticles also represent a promising approach in handling infections because it has been proved that they have the potential to act as targets for several toxins produced by bacteria and are considered true molecular decoys for microbial compounds with detrimental effects within the host (Hu et al., 2013).

4.2.1 Design of Tailored Magnetic Nanoparticles with Applications in Microbiology

In recent years, many methods aiming to obtain nanoparticles with biomedical applications have been optimized. Choosing the most appropriate method of synthesis of magnetite nanoparticles, depending on the final purpose, represents a very important step that influences the particle dimensions and shape, size distribution, surface chemistry, and also magnetic properties.

The most intensively used and optimized synthesis methods are: (i) the co-precipitation method; (ii) the hydrothermal method; (iii) solvothermal method; (iv) microwave-assisted method; and (v) the ultrasound-assisted method (Mauricio et al., 2013; Meng et al., 2013). Figure 4.1 shows the step-by-step design of

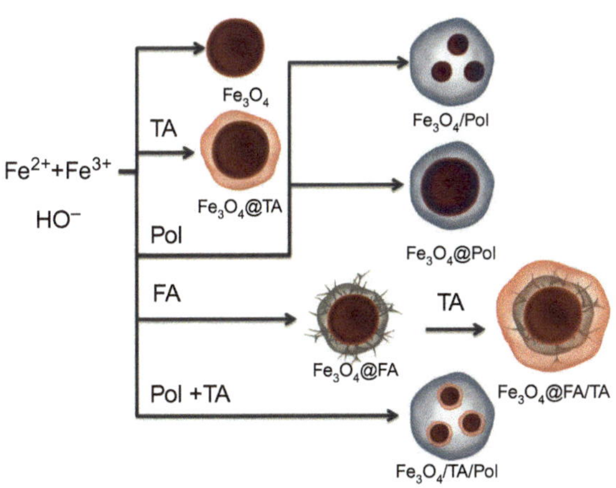

FIGURE 4.1 The design of magnetite nanoparticles by co-precipitation (TA, therapeutic agents; FA, fatty acids; Pol, polymers).

functionalized magnetite nanostructures used for biomedical applications.

Buteică et al. (2010) reported the synthesis of a magnetite-based nanofluid by adapting the Massart method using Fe^{2+} and Fe^{3+} salts with oleic acid as the surfactant under microwave conditions. The synthesis of Fe_3O_4 and $CoFe_2O_4$ by the Massart method using Fe^{3+}, Fe^{2+} salts, and HO^- under microwave conditions was confirmed by Grumezescu et al. (2010) and by Mihaiescu et al. (2011).

To orientate the function of a magnetite nanomaterial through a specific biological effect, researchers usually functionalize the nanoparticles to dictate their activity and reach the proposed aim. To improve the antimicrobial activity of several antimicrobial compounds, a series of studies reported the synthesis of antimicrobial magnetite nanoparticles that contain specific antibiotics. Other studies used natural compounds, such as essential oils and major vegetal extracted fractions with proven antimicrobial effects for several biomedical applications, that have the potential to be used further in the clinical field (Saviuc et al., 2011a). Because many nanoparticles have unsatisfactory biocompatibility, with many of them being toxic to human and animal cells, a good strategy for improving their biocompatibility proved to be the use of natural polymers in the synthesis of bioactive magnetite nanoparticles. The most utilized polymers are chitosan and cellulose, and the preferred method of synthesis of magnetite polymeric nanoparticles is the co-precipitation method (Grumezescu et al., 2011a,b).

Nanoparticles based on magnetite and polyethylene glycol have been synthesized under microwave conditions; to improve their antimicrobial activity, essential oil of *Citrus maxima* was adsorbed on the surface of nanostructured material as an extra-shell (Saviuc et al., 2011a). Nanoparticles functionalized with different natural and synthetic antimicrobials significantly enhance the effect of the bioactive drug by targeted delivery and controlled release

(Figure 4.2). A diethylaminoethyl-cellulose/Fe_3O_4/cephalosporin hybrid material utilized as a magnetic drug delivery system in microbiological applications have been also obtained (Grumezescu et al., 2011b).

Water-soluble magnetite nanoparticles significantly improved the activity of currently used antibiotics, representing great potential to be used as a nanocarrier for these antimicrobial substances and to achieve extracellular and intracellular targets. An adapted diffusion method was developed to assess the influence of the water-soluble nanovehicle on the antimicrobial activity of vancomycin, clindamycin, azithromycin, oxacyllin, trimethoprim/sulfamethoxazole, rifampicin, ofloxacin, tetracycline, penicillin, ciprofloxacin, gentamicin, piperacillin/tazobactam, cefepime, aztreonam, ceftazidim, and piperacillin against *Staphylococcus aureus* and *Pseudomonas aeruginosa* strains. The results demonstrate that

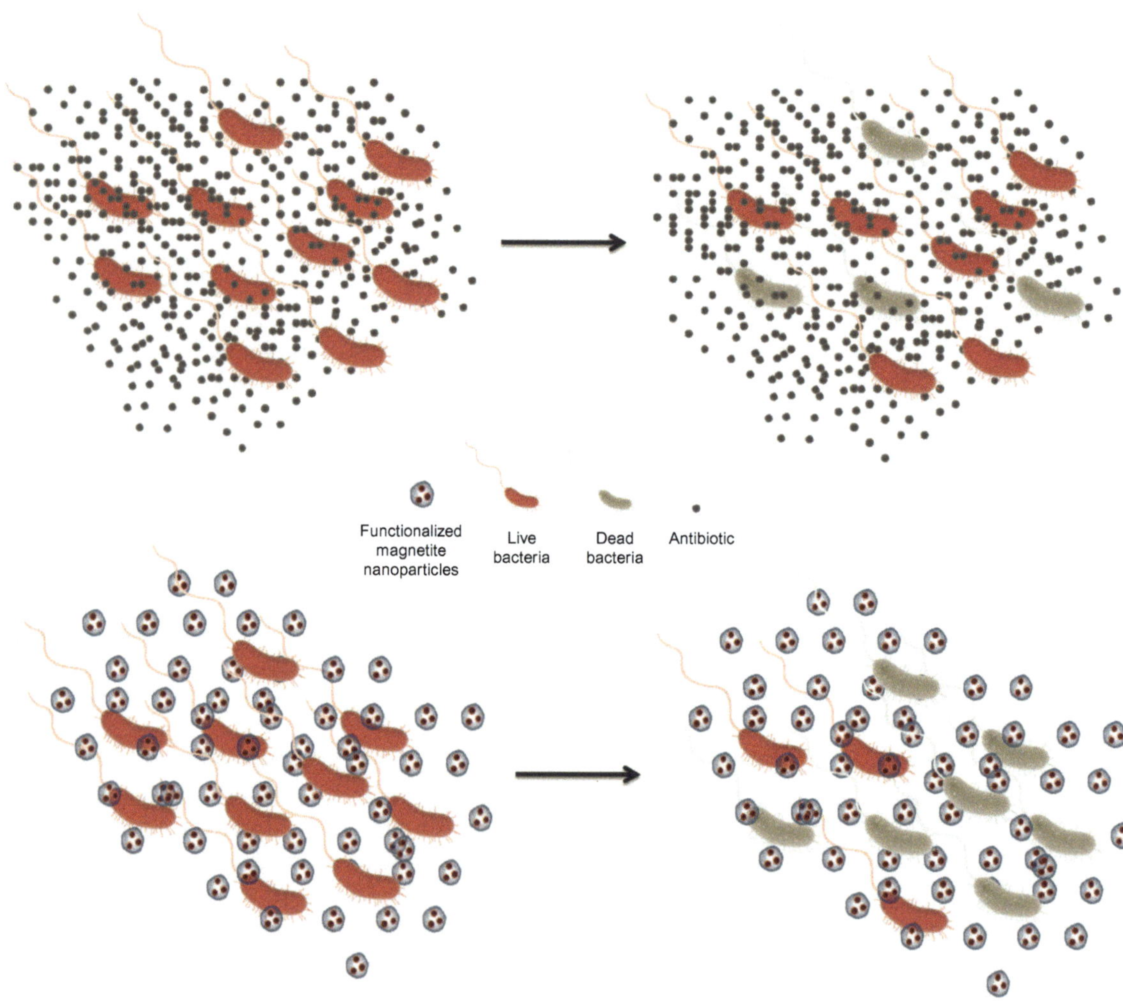

FIGURE 4.2 The enhanced antimicrobial effect of functionalized magnetite nanoparticles as compared with the antimicrobial drug.

the tested nanosystems have the ability to carry and deliver antibiotics in active forms. In the case of *P. aeruginosa*, authors demonstrate that the incorporation of the anti-pesudomonal antibiotics in the nanoparticles led, in all tested cases except piperacillin plus tazobactam, to an increase in the bacterial growth inhibition diameters. In the case of the *S. aureus* strain, the potentiation of the antimicrobial activity of different antibiotics was limited to four of the tested spectrum, namely penicillin, oxacillin, aztreonam, and doxycycline. However, the results reveal that for other antibiotics, incorporation into the developed magnetite nanosystem did not affect antibiotic efficiency, which remained the same as in the case of antibiotic controls (Mihaiescu et al., 2012). These studies reveal that further optimizations in the synthesis and functionalization methods have to be performed to obtain the desired effect for a particular drug, depending on its physicochemical properties.

Other studies used magnetite nanoparticles functionalized with cefotaxime and polymyxin B to test their effect on *Escherichia coli* and *P. aeruginosa*, and also amoxicillin, kanamycin, and streptomycin to test *E. coli* only. The results showed a decrease in the minimum inhibitory concentration (MIC) value of magnetite nanoparticles functionalized with the antibiotics amoxicillin and kanamycin against *E. coli* that was comparable with the MIC values of antibiotics solutions. For *P. aeruginosa*, the MIC values of Fe_3O_4/cefotaxime decreased at 5.85 μg/mL, which is comparable with the MIC value of 8 μg/mL of the antibiotic solution against *P. aeruginosa* reference strain (Cotar et al., 2013).

Another recent study demonstrates that magnetites are able to enhance and insure controlled drug release and to significantly improve the efficacy of antimicrobial agents against *E. faecalis*, which is one of the most resistant opportunistic pathogens.

The results showed that the magnetic nanoparticles clearly improved the activity of the tested antibiotics, as revealed by the increase of the growth zone inhibition diameter as compared with the antibiotic disks charged with the same antibiotic concentration. Furthermore, these newly developed magnetite nanoparticles slightly improved the antibiotics effect against *E. faecalis* biofilms development (Chifiriuc et al., 2013a).

Results demonstrate that different bioactive magnetite nanoparticles have a significant effect on bacteria virulence modulation, not just on their viability. Magnetite nanoparticles functionalized with *Eugenia carryophyllata* essential oil exhibited an early microbicidal effect on *P. aeruginosa* and *S. aureus* stains, with significantly lower MIC values for *S. aureus* strains, but the essential oil modulated the expression of soluble virulence factors by increasing the production of lipase in both species and of DN-ase, gelatinase, hemolysins, and lecithinase, particularly in *P. aeruginosa* strains (Saviuc et al., 2011a,b,c).

To improve their biocompatibility, polymers have been added to obtain efficient magnetite nanoparticles. A hybrid material consisting of magnetite, chitosan, and second-generation cephalosporins, cefuroxime had significantly higher antimicrobial activity as compared with the respective antibiotic alone against *E. coli* and *S. aureus* (Grumezescu et al., 2011a).

The effect of hybrid magnetic materials based on traditional antibiotic formulations against bacterial strains is probably due to the morphological and/or physiological changes induced in the plasma membrane by the nanoparticles and better adsorption of the small size nanoparticles, which is facilitated by the polymeric shell (Grumezescu et al., 2011b).

Other studies demonstrated that antibiotics incorporated into magnetic chitosan microspheres improve the delivery of these antibiotics in active forms (Andronescu et al., 2012). Chitosan was evaluated as a potential drug delivery system, and it was demonstrated that it can elute antibiotics in an active form that would be efficacious in inhibiting *S. aureus* and *E. coli* growth. The incorporation of the tested cephalosporins in chitosan demonstrates a drastic decrease of the minimal inhibitory concentration from 2- to 7.8-times, both on *E. coli* as well as on *S. aureus* strains (Chifiriuc et al., 2012a). It is well-known that the size and the electric charge of the active drugs influence the specific interactions between the drug carrier and the active substance. Magnetic dextran microspheres could be used as macromolecular carriers for large-spectrum antibiotics, particularly for those with small, polar molecules belonging to penicillins, aminoglycosides, rifampicines, and quinolones classes. The magnetic dextran microspheres slightly improved the antimicrobial activity of several anti-staphylococcal drugs, which was the most significant improvement of the antimicrobial activity, highlighted by the enlargement of the growth inhibition zones, with 20% being obtained for clindamycin and rifampicin. These results demonstrate the specific interaction of the proposed delivery system with different antibiotics and the necessity to develop "personalized" drug carriers for different therapeutic substances (Andronescu et al., 2012).

The antimicrobial effects of gentamycin and piperacillin against *S. aureus* and *P. aeruginosa* tested strains have been significantly potentiated in the presence of the polymeric magnetic silica drug loader. The antimicrobial activity of the tested antibiotics proved to be differently influenced by the drug loader system, depending on either the tested antibiotic or the tested microbial strains, indicating that there are probably specific interactions between the tested antibiotic and the carrier system that interfere with the diffusion rate of the antibiotic from the polymeric magnetic silica microspheres. The authors also state that their results demonstrate that the composite particles charged with antibiotics probably penetrate the bacterial cell wall and deliver the antibiotic in active forms to the intracellular targets (Grumezescu et al., 2012a).

Another study reports the fabrication of water-soluble magnetite nanoparticles, protected by chitosan and polyvinyl alcohol to increase the bioaccommodation of the system (Grumezescu et al., 2012b). The results demonstrate that the nanobiocomposite has the ability to modify and improve the antimicrobial activity of gentamicin, ciprofloxacin, and cefotaxime against *S. aureus* and *P. aeruginosa*. It has been revealed that loading kanamycin sulfate into the water-dispersible metal oxide nanobiocomposite improves the delivery of this drug in its active form, reducing minimum inhibitory concentration of kanamycin by two-fold (when tested on *S. aureus*) to four-fold (when tested on *E. coli*) as compared with the kanamycin control. Furthermore, cytotoxicity tests performed by the same authors revealed that the nanobiocomposite has a very low toxic effect on eukaryotic cells (Grumezescu et al., 2012b).

Balaure and coworkers reported the fabrication of a new drug delivery system based on polyanionic matrix (e.g., sodium alginate), polycationic matrix (e.g., chitosan), and silica network (Balaure et al., 2013). Their results demonstrated the biocompatibility and the ability of the fabricated biocomposite to maintain or improve the efficacy of the following antibiotics: piperacillin–tazobactam, cefepime, piperacillin, imipenem, gentamicin, and ceftazidime against *P. aeruginosa*; and cefazolin, cefaclor, cefuroxime, ceftriaxone, cefoxitin, and trimethoprim/sulfamethoxazole against *E. coli* reference strains (Balaure et al., 2013).

Further research led to the development and bioevaluation of a novel biocompatible, resorbable, and bioactive wound dressing prototype based on anionic polymers (sodium alginate [AlgNa], carboximethylcellulose [CMC]) and magnetic nanoparticles loaded with usnic acid (Fe_3O_4/UA) (Grumezescu et al., 2013a). Because of their structural and functional properties, CMC/Fe_3O_4/UA and AlgNa/Fe_3O_4/UA are considered suitable candidates for further applications on the biomedical field, especially in regenerative medicine (Grumezescu et al., 2013a).

Vlad et al. (2014) published a paper revealing the synthesis of carboxymethyl-cellulose/magnetite (CMC/Fe_3O_4). Their results showed that CMC/Fe_3O_4/ATB might be a promising candidate for the development of efficient and cheap antimicrobial drug carriers under the magnetic field. The most evident improvement of the antimicrobial activity was observed for the CEF loaded onto the magnetic scaffold tested on the *E. coli* reference strain, although they could not observe any change in the efficiency of the two tested ATBs in the presence of the magnetic scaffold in case of *S. aureus* (Vlad et al., 2014).

4.2.2 Magnetite Nanoparticles Used to Control Microorganisms Attachment and Biofilm Formation

Biofilm-associated infections have the highest rates of antibiotic resistance; therefore, the treatment options of such infections are very limited. Because recent findings demonstrate that nanomaterials may be efficiently used in medicine, and particularly in anti-infectious approaches, researchers aimed to develop nanobioactive magnetite systems to eradicate and control biofilms formation.

Recently, a new water-dispersible nanostructure based on magnetite (Fe_3O_4) and usnic acid (UA) was prepared in a well-shaped spherical form using a precipitation method (Grumezescu et al., 2013b). It was shown that the nanoparticles were well-individualized and homogeneous in size. The UA was entrapped in the magnetic nanoparticles during preparation. These nanostructures were tested on planktonic cell growth and biofilm development on Gram-positive *S. aureus* and *E. faecalis* and on Gram-negative *E. coli* and *P. aeruginosa* reference strains. UA-functionalized Fe_3O_4 nanoparticles displayed improved anti-pathogenic activity against *E. faecalis* and *E. coli* tested strains, as compared with the magnetite control, without UA. This magnetic nanosystem also revealed a significant effect on the *E. faecalis* and *E. coli* biofilm inhibition in a concentration-dependent manner. Despite the great antibiofilm activity observed for *E. faecalis* and *E. coli*, in case of *P. aeruginosa* biofilms, no or very low inhibitory effects were observed (Grumezescu et al., 2013b). Functionalized magnetite nanoparticles also proved to be efficient against fungal biofilm development.

Recently developed modified wound dressings coated with magnetite nanoparticles exhibit enhanced anti-adherence and anti-biofilm effects against the versatile *Candida tropicalis.* The nanomodified wound dressing surfaces do not allow *C. tropicalis* biofilm formation. Both early and mature biofilm formation phases were significantly impaired when the modified wound dressings were used. Furthermore, the effect of nanomodified bioactive wound care materials seems to be highly stable during time, because its activity is maintained for at least 3 days (Holban et al., 2013a,b).

A hybrid nanomaterial composed from Fe_3O_4, PEG_{600}, and *C. maxima* exhibited antibacterial activity on *E. coli, S. aureus,* and *E. faecalis* strains. The *in vitro* assay also demonstrates the influence of the magnetic biomaterial on bacterial adherence to cellular and inert substrata and biofilm development, suggesting that this material could be used for the development of novel antimicrobial materials with anti-adherent properties (Saviuc et al., 2011a).

Anghel et al. (2012a) reported the successful fabrication of functionalized magnetite (Fe_3O_4/C_{18}), with an average size not exceeding 20 nm, that has been synthesized by precipitation of ferric and ferrous salts in aqueous solution of oleic acid (C_{18}) and NaOH. The functionalized magnetite was further used for coating textile wound dressings and designed to treat different wounds. The results of this study demonstrated that the nano-modified textile dressings are more resistant to *Candida albicans* attachment and biofilm formation, as compared with the uncoated dressings. These data suggest that magnetite-based nanocoated surfaces may be useful for the prevention of wound contamination, and also for the treatment of infected wounds (Anghel et al., 2012a).

Magnetic nanoparticles of Fe_3O_4/C_{18} also interfered with adherence to cellular and inert substrata of clinical isolated *Candida* species. A recent study proved that Fe_3O_4/C_{18} nanoparticles have the ability to differently impair yeast biofilms in a species-dependent manner. The results revealed that the biofilm inhibition decreases in the following order: *C. albicans* > *C. tropicalis* > *C. glabrata* > *S. cerevisiae* > *C. krusei* > *C. famata*, and this phenotype is correlated with early germ tube production. The *in vitro* fungal attachment to the cellular substratum was also reduced in the presence of magnetite nanoparticles, suggesting that they may be used for obtaining improved surface materials with anti-adherence properties (Chifiriuc et al., 2013b).

In 2011, Saviuc and collaborators obtained Fe_3O_4/C_{18} nanostructures by using the Massart method adapted for microwave conditions (Saviuc et al., 2011b). To achieve improved anti-adherent activity, the extra-shell, consisting of diluted *A. graveolens*, was applied by adsorption in a secondary covering treatment (Saviuc et al., 2011b). Their dynamic study of development of fungal biofilms on the glass cover slips and, respectively, on essential oil-based nanosystems revealed that at 48 h of incubation, the development of fungal biofilms on coated coverslips was much more reduced and exhibited simplified architecture of CLSM images of the uncoated coverslips inoculated with *C. tropicalis*–tested strains and showed mature and compact biofilm with pseudohyphae and pluristratified zones and rare adherent cells to the coated coverslip, whereas the *C. famata*–tested strain showed an adherence pattern "in patches" for the macrocolonies formed by adherent fungal cells on uncoated coverslips and rare isolated yeast cells adhered to nanosystem-coated surface (Saviuc et al., 2011b).

Similar magnetite nanostructures functionalized with UA evidenced a reduced number of adherent bacteria and a simplified structure of the biofilm as compared with the biofilm developed on the coverslips coated with iron oxide nanoparticles, which revealed a more complex architecture and nonhomogenous distribution on the coverslip, with dense pluristratified macrocolonies separated by free bacterial cell areas, probably functioning as water channels (Grumezescu et al., 2011c).

The *Ralvia officinalis*–coated nanoparticles strongly inhibited the adherence ability and biofilm development on the catheter surface of the *C. albicans* and *C. tropicalis* strains. These materials-based approaches to controlling fungal adherence could provide new tools for studying mechanisms of fungal virulence and biofilm formation, and also new approaches for the design of film-coated surfaces or for treating the surfaces of solid and fiber-based materials that prevent or disrupt the formation of fungal biofilms (Chifiriuc et al., 2012b).

Recent studies aimed to evaluate a novel nanobiosystem based on magnetic nanofluid and *S. officinalis* for coating the Provox voice section prostheses surfaces with antibiofilm properties. The synthesized nanobiosystem composed of Fe_3O_4/C_{18}/essential oils demonstrated a great fungicidal effect and also altered fungal adherence and biofilm development, as revealed by the CLSM examination of the catheter sections colonized with the tested *C. albicans* strains. The analysis showed that the tested nanobiosystem exhibited an intensive antibiofilm effect, as demonstrated by a low number of yeast cells that adhered to the coated surface (Anghel et al., 2012b).

Limban et al. (2012) reported the synthesis of new 2-((4-ethylphenoxy)methyl)-*N*-(substituted-phenylcarbamothioyl)-benzamides and their usage as coatings for a core/shell nanostructure. The purpose of this study was to design a new nanosystem for catheter surface functionalization with improved resistance to *S. aureus* and *P. aeruginosa* colonization and subsequent biofilm development. Viable cell counts and SEM examination demonstrated that the nano-coated surfaces inhibited both the initial attachment and biofilm development of *S. aureus* and *P. aeruginosa* on the functionalized catheter surfaces (Limban et al., 2012).

Natural *Mentha piperita* essential oil combined with a 5-nm core/shell nanosystem-improved surface also proved anti-adherence and antibiofilm properties (Anghel and Grumezescu, 2013). This nanosystem acts as a controlled release machine for the essential oil and is very efficient for inhibiting biofilm formation; it is a good candidate for the design of novel material surfaces used for prosthetic devices (Anghel and Grumezescu, 2013; Anghel et al., 2013b).

As the laser techniques continue to mount, they have been adopted in chemistry, biotechnology, and biomedical fields, where they are mainly used for the deposition of thin bioactive films (Figures 4.3 and 4.4).

Mihaiescu et al. (2013) reported the fabrication and deposition of magnetite/salicylic acid/silica-shell/antibiotics (Fe_3O_4/SA/SiO_2/ATB) thin films by matrix-assisted pulsed laser evaporation (MAPLE) to inert substrates. Fe_3O_4/SA/SiO_2/ATB thin films inhibited the ability of microbial strains to initiate and develop mature biofilms in a strain-dependent and antibiotic-dependent manner (Mihaiescu et al., 2013).

In another study, Grumezescu et al. (2014) reported that magnetic polylactic-co-glycolic acid-polyvinyl alcohol (PLGA-PVA) microspheres were loaded with UA, and these nanostructures were used to obtain coatings by MAPLE deposition of the thin films. The bio-nano-active–modified surface inhibited the initial attachment of *S. aureus* to coated surfaces and also their control of biofilm formation and their development to mature biofilms.

4.2.3 The Biocompatibility of Magnetite Nanoparticles

Preparation and characterization of magnetic composite scaffolds including the morphology, crystallinity, and the *in vitro* efficacy as drug delivery vehicles as well as their influence on the eukaryotic cells and the whole host body represent the most important parameters that are important in the establishment of the biomedical application of the system.

4.2.3.1 Biocompatibility Evaluation of Magnetite Nanoparticles at the Cellular Level

The first step in the evaluation of magnetite nanoparticles effect on cells is to look at cellular

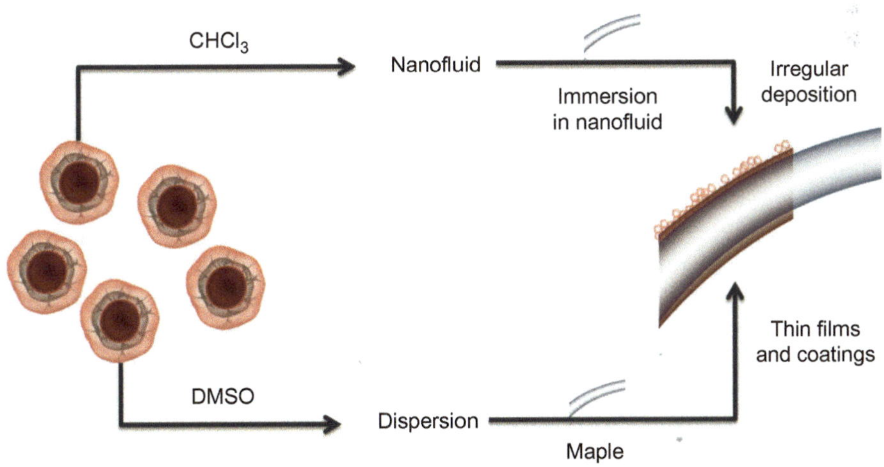

FIGURE 4.3 Different coating techniques of medical devices. Immersion in nanofluid versus MAPLE technique.

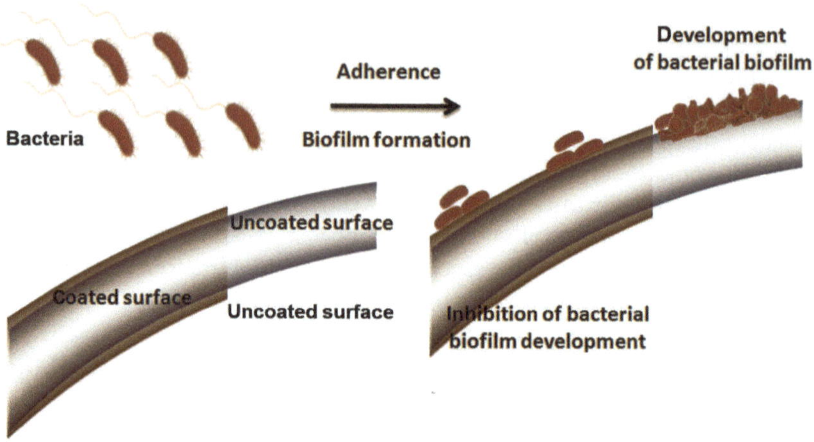

FIGURE 4.4 The microbial growth on a surface of a regular catheter and on modified nanobioactive surfaces, deposited as thin films.

morphology and viability. There are many methods that can reveal the condition of cells in the presence of nanoparticles. First, the easiest methods are by optic or fluorescent microscopy. These methods provide information regarding the modification of cellular shape, attachment to substrate, integrity of cellular membrane, and the number of vacuoles. For the evaluation of cellular viability, fluorescent microscopy is used widely. The cells are marked with fluorescent dyes that enter the cells; by the action of cellular enzymes these dyes become fluorescent, suggesting that cells are viable and metabolic active.

For detection of apoptosis, cells can be stained with annexin V, Hoechst, propidium iodide, DAPI, and trypan blue, and analyzed both by fluorescent microscopy and flow cytometry (Cimpean, 2006). Fe_3O_4/Au nanoparticles synthesized by reducing metal salts in the presence of oleylamine and oleic acid could offer the possibility of functionalization of these nanoparticles with different molecules. The interaction with cells and the cytotoxicity of the Fe_3O_4/Au were determined on incubation with the HeLa cell line. These nanoparticles showed no cytotoxicity when evaluated by the MTT assay and a higher level of accumulation in the cells for glucose-conjugated nanoparticles (Salado et al., 2012). For better biocompatibility, magnetic nanoparticles require surface modification to promote dispersibility in aqueous solutions. *In vitro* studies confirmed that the silica layer significantly reduced cellular toxicity as assessed by MTT assay. The results showed an increase in cell viability and reduction in reactive oxygen species (ROS) production during 48 h of culture (Singh et al., 2012). Other studies that tested several functionalized magnetite nanoparticles, including those prepared with silane and silica, show that not only nanoparticles could determine cytotoxicity but also the nutrient medium and the time of suspension before exposure to cells also contribute to nanoparticles cytotoxicity (Mbeh et al., 2012).

Another step in the investigation of cellular biocompatibility with magnetite nanoparticles is evaluating the cellular metabolism. The assays that are most frequently used are MTT assay, LDH assay, BrdU assay, cell cycle assays, and TUNEL assay. MTT and LDH assays are based on measuring the activity of mitochondrial and cytosolic enzymes that transform the tetrazolium compound MTT in formazan, or on measuring the level of LDH that is rapidly released on damage of the cell membrane. Another popular test is the BrdU (5-bromo-2-deoxyuridine) assay, which permits the evaluation of cell proliferation (Cimpean, 2006). An important parameter in the evaluation of toxicity of magnetite nanoparticles is the level of oxidative stress in the cells that is thought to be the main cause of cell toxicity. Oxidative stress appeared when damaging oxidants also known as ROS, such as hydrogen peroxide and hydroxyl radicals, cannot be reduced by protective antioxidants such as vitamin C, glutathione, catalase, superoxide dismutase, and various peroxidases. The accumulation of oxidants eventually leads to destruction of cellular proteins, enzymes, lipids, and nucleic acids, and finally to cell apoptosis and necrosis. ROS can be generated from the surface of magnetite nanoparticles, the leaching of metal ions from the core, or release of oxidants by enzymatic degradation of the magnetite nanoparticles. ROS production can be measured using dichlorofluorescein diacetate fluorescent probe (Markides et al., 2012). Evaluation of the *in vitro* biocompatibility of 33 nm Fe_3O_4/FITC-MSNs/PEG nanoparticles using the MTT assay showed that in all the cell lines tested (HeLa, PC-12, and HCT-116 cell lines), the viability was not affected after 12 h of exposure to up to 200 μg/mL nanoparticles (Lin and Haynes, 2009). The concentration of Fe_3O_4 nanoparticles is another parameter that is very important for the viability of the cells. After 24 h of incubation with 10 mM of Fe_3O_4 nanoparticles coated by sodium oleate or PEG more than 90% and 70%, respectively, of cell viability was achieved. However, when the concentration of Fe_3O_4 nanoparticles was 20 mM, cell viability was lower than 60% even though the incubation time was only 4 h. Furthermore, the results showed that the toxicity of Fe_3O_4 nanoparticles coated by sodium oleate was lower than that of Fe_3O_4 nanoparticles coated by PEG. Prussian blue staining of fibroblast cell line 3T3 incubated with 2 mM Fe_3O_4 nanoparticles coated by sodium oleate for 4 h showed no changes in the quantity and shape of 3T3 cells, suggesting that Fe_3O_4 nanoparticles can be considered to be biocompatible (Sun et al., 2007). Because of their potential *in vivo* biomedical applications such as targeted drug delivery, cancer cell diagnostics, and therapeutics, the biocompatibility of Fe_3O_4 nanoparticles should be evaluated not only in normal cell lines but also in cancer cell lines. After incubation of normal fibroblast WI-38 cell lines, normal glia cell lines (SVGp12), glia cancer cell lines (D54MG, G9T/VGH, SF126, U87, U251, and U373), normal breast epithelial cell lines (H184B5F5/M10), and breast cancer cell lines (MB157, SKBR3, and T47D) with Fe_3O_4 nanoparticles at maximum exposure dosage (100 μg/mL) for 72 h, the cells responded differently. Fe_3O_4 nanoparticles are nontoxic for all cell lines tested in the range of 0.1–10 μg/mL. In SKBR3 and T47D cell lines, cell injuries were clearly visible. However, in the rest of the cell lines, nanoparticles entered into the cell membrane and nucleus, suggesting that there are attractive forces between the cell membrane and nanoparticles (Ankamwar et al., 2010). Cellular interactions of lauric acid and dextran-coated magnetite nanoparticles evaluated with two different cell lines (mouse fibroblast and human cervical carcinoma) showed that lauric acid–coated magnetite nanoparticles were less cytocompatible than dextran-coated magnetite nanoparticles and cellular uptake of lauric acid–coated magnetic nanoparticles was much higher than that of dextran-coated magnetite nanoparticles. Therefore, coating plays an important role in modulation of biocompatibility and cellular interaction of magnetite nanoparticles (Pradhan et al., 2007). Because some magnetite nanoparticles are designed for tracking cells to their destination *in vivo* and are administrated intravenously, it is important to test the compatibility of these nanoparticles with blood cells. The *in vitro* biocompatibility with human red blood cells of apatite-coated magnetite nanoparticles showed no hemolytic effects at concentrations lower than 3 mg/mL (Múzquiz-Ramos et al., 2013). The *in vitro* interaction of animal hemoglobin with biocompatible magnetite nanoparticles showed a hemolytic effect in many cases. There are two possible ways that the magnetite nanoparticles influence the heme energetic levels—either electronic or vibrational—by interacting with heme from released hemoglobin molecules or by exerting chronic magnetic exposure on the heme iron after the addition to the red blood cell membranes, but further investigations should be addressed (Creangă et al., 2009).

4.2.3.2 Biocompatibility Evaluation of Magnetite Nanoparticles at Biochemical and Molecular Level

The evaluation of proinflammatory potential of magnetite nanoparticles can be assessed by enzyme-linked immunosorbent assay (ELISA) for interleukin (IL)-1, IL-3, IL-6, IL-12, and IFNγ. Superparamagnetic iron oxide nanoparticles (SPION) exposure to human fibroblasts caused the increase in expression of genes involved in cell signaling, including integrin subunits, tyrosine kinases, protein kinase C family ($\alpha/\delta/\theta/\zeta$), and actin filaments, suggesting that SPION can have an impact on signaling transduction pathways (Soenen et al., 2010). A study conducted on pancreatic islet cells labeled with Resovist (carboxydextran-coated SPION, MRI agent), revealed that insulin expression in labeled islets was significantly elevated (Kim et al., 2009). The proliferation of cells can be determined by molecular biology techniques following genes involved in cell-cycle progression. Studies on mesenchymal stem cells showed that in the presence of SPION, the expressions of hyperphosphorylated retinoblastoma tumor suppressor protein pRb, cyclins, and cyclin-dependent kinases, such as cyclins B, D1, E, CDK2, and CDK4, were increased, demonstrating a stimulation of cell proliferation (Huang et al., 2009). The pH and Ca^{2+} are also parameters that are very important for maintaining cellular homeostasis. The interactions of red blood cells and Caco-2 cells with 11-nm magnetic iron oxide nanoparticles showed that the nanoparticles do not have a significant influence on the pH and Ca^{2+} content of Caco-2 cells, whereas in red blood cells the intracellular pH was slightly reduced (Moersdorf et al., 2010). *In vivo* studies showed that when mice were treated with different concentrations of magnetite nanoparticles by a single intratracheal instillation, they presented a decreased level of intracellular reduced glutathione (GSH) in the cells of bronchoalveolar lavage fluid. Also, the concentrations of IL-1, TNF-α, and IL-6 pro-inflammatory cytokines were dose-dependently increased in the alveolar cells and in the blood. The same authors demonstrated that the expressions of many genes related to inflammation and tissue damage are significantly induced (Park et al., 2010).

Another interesting study evaluated the cytotoxic, genotoxic, and inflammatory responses of nanoparticles from photocopiers. In epithelial cells, the level of cytokines IL-8, VEGF, EGF, IL-1α, TNF-α, IL-6, and GM-CSF was significantly elevated and apoptosis was also induced, consistent with the upregulation of key apoptosis-regulating genes p53 and caspase 8 (Khatri et al., 2013). Macrophages and normal hepatocytes treated with SPIONs revealed that at a dose of more than 25 μg/mL, nanoparticles become cytotoxic. However, the macrophage cells and hepatocytes responded differently to the SPIONs. The liver cells present protein binding to iron while macrophages take-up foreign particles via phagocitosis (Shubayev et al., 2009). Lipid peroxidation evaluated by TBARS assay indicated oxidative stress as a self-defense antioxidant response (Priprem et al., 2010).

4.2.3.3 Biocompatibility Evaluation of Magnetite Nanoparticles Using Animal Models

Data revealing no toxicity via *in vitro* tests can then be moved to *in vivo* studies. For *in vivo* studies, mice are the most used animal models. Validation tests include histology analysis on major metabolic sites (liver, pancreas, kidney, and brain) to look for signs of magnetite nanoparticles spreading and accumulation. Sections could be stained for iron by Prussian blue stain and caspase 3 as an indicator for apoptosis within these areas. The results obtained *in vitro* may not be the same with *in vivo* results. This may be attributed to the facts that in organisms there are homeostasis processes that try to eliminate or arrest the unknown particles. Studies revealed that Balb/c mice injected intravenously or intraperitoneally with apatite-coated magnetite nanoparticles in doses ranging from 100 to 2,500 mg/kg showed normal kidney and liver function. No significant changes were found in body weight or organ weight, and iron levels in liver were unchanged (Gomaa et al., 2012). For successful application of nonmaterial in bioscience, it is essential to evaluate mutagenicity and genotoxicity potential using short-term and long-term experimental models. *Salmonella typhimurium* strains were cultivated in the presence of magnetite nanoparticles at concentration ranging from 10 to 70 ppm. The mutagenic potential of nanoparticles was evaluated using the Salmonella Ames test in the presence and the absence of metabolic activation with S9-liver extract. The results showed significant mutagenic activity at 70 ppm and higher concentrations (Szalay et al., 2012). In a rat model after iron oxide nanoparticles exposure, the results based on pathological examination indicated no abnormalities in the exposed rat organs except in lungs. In the treated rats, the lungs revealed focal, interstitial inflammation and weak pulmonary fibrosis developed by the end of the first month (Szalay et al., 2008). Recent results revealed that iron oxide nanoparticles may induce inflammatory responses via oxidative stress in the lungs and also may lead to the formation of microgranuloma (Park et al., 2010). All these data show that the behavior of magnetite nanoparticles is influenced by the type of cells. Liver cells that present iron binding protein are more resistant to magnetite nanoparticles than immune cells or endothelial cells (Khatri et al., 2013). The magnetite nanoparticles should be tested on specific cell types according to their destination for a better determination of the cytotoxicity effect (Dousset et al., 2008). Once the magnetite nanoparticles have entered the body, they come into direct contact with all biological macromolecules and tissues found nearby. Studies revealed that some biomolecules may bind to the surface of the magnetite nanoparticles and are able to form a multilayered complex of molecules around nanoparticles. Physicochemical properties of magnetite dictate the binding pattern of the biomolecules and affect the interaction with various cells (Markides et al., 2012). SPIONs nanoparticles have been used in diagnostics as a contrast agent in MRI and magnetic resonance angiography (MRA). The transport of most of the contrast agents is by intravenous administration, and successful delivery depends on the size of the particle that has to pass through the vascular capillary wall. Depending on their size, charge, and the configuration of the coating (Stella et al., 2000; Pankhurst et al., 2003), these nanoparticles are metabolized by the reticuloendothelial system (RES) consisting of monocytes and macrophages. These cells accumulate in lymph nodes and the spleen as well as in the liver, and are favored for the uptake of SPIONs and influence the delivery time and the diffusion to certain tissues (Hofmann-Amtenbrink et al., 2009).

4.3 CONCLUSIONS

In conclusion, nanoparticles have to be highly specific and efficient, and they should be rapidly internalized by the target cells. These characteristics are limited by several factors: (i) nanoparticle aggregation; (ii) the short half-life of the nanoparticles in blood circulation (when nanoparticles agglomerate, or adsorb plasma proteins, they are quickly eliminated from the bloodstream by macrophages of the mononuclear phagocyte system before they can reach the target cells); (iii) the low efficiency of the intracellular uptake of nanoparticles; and (iv) nonspecific targeting. The coating and functionalization of nanoparticles could improve their specific application and their biocompatibility in the human body. The ultimate challenge after *in vivo* animal tests is the

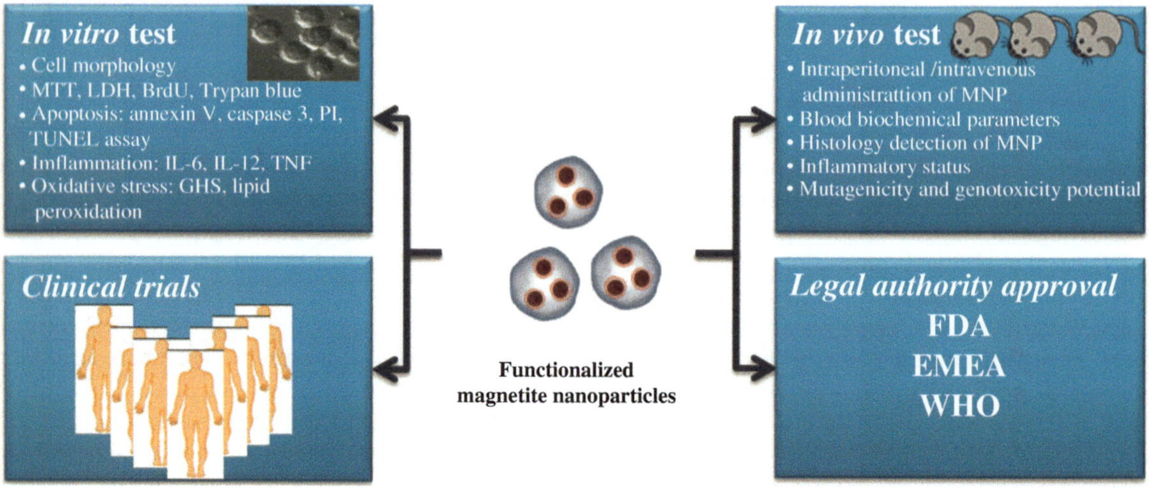

FIGURE 4.5 Overview of biocompatibility assays and the steps for moving from bench to bed side using functionalized magnetite nanoparticles (PI, propidium iodide; GHS, glutathione reductase; FDA, Food and Drug Administration; EMEA, European Medicines Agency; WHO, World Health Organization).

use of nanoparticles in human clinical trials (Figure 4.5) that are controlled by regulatory bodies such as the Food and Drug Agency (FDA). For this reason, extensive safety assessment of these particles must be performed to satisfy not only the regulators but also the patient.

In recent years, the number of types of magnetite nanoparticles has increased very rapidly. There is an enormous variety of particles available both commercially and in research (Figure 4.5), and this makes it very difficult to provide a certain answer regarding the questions of whether magnetite nanoparticles are toxic. A large number of *in vitro* toxicity investigations have shown no adverse side effects; however, long-term *in vivo* studies should be addressed to improve the applications of magnetite nanoparticles.

Acknowledgments

This work is supported by the Sectoral Operational Programme Human Resources Development and is financed by the European Social Fund and by the Romanian Government under the contract number ID 132397 (ExcelDOC).

References

Andronescu, E., Grumezescu, A.M., Ficai, A., Gheorghe, I., Chifiriuc, M.C., Mihaiescu, D.E., et al., 2012. *In vitro* efficacy of antibiotic magnetic dextran microspheres complexes against *Staphylococcus aures* and *Pseudomonas aeruginosa* strains. Biointerface Res. Appl. Chem. 2, 332–338.

Anghel, I., Grumezescu, A.M., 2013. Hybrid nano-structured coating for increased resistance of prosthetic devices to staphylococcal colonization. Nanoscale Res. Lett. 8, 6.

Anghel, I., Grumezescu, A.M., Andronescu, E., Anghel, A. G., Ficai, A., Saviuc, C., et al., 2012a. Magnetite nanoparticles for functionalized textile dressing to prevent fungal biofilms development. Nanoscale Res. Lett. 7, 501.

Anghel, I., Grumezescu, V., Andronescu, E., Anghel, G.A., Grumezescu, A.M., Mihaiescu, D.E., et al., 2012b. Protective effect of magnetite nanoparticle/*Salvia officinalis* essential oil hybrid nanobiosystem against fungal colonization on the Provox® voice section prosthesis. Digest J. Nanomat. Biostruct. 7, 1205–1212.

Anghel, I., Holban, A.M., Andronescu, E., Grumezescu, A. M., Chifiriuc, M.C., 2013a. Efficient surface functionalization of wound dressings by a phytoactive nanocoating refractory to *Candida albicans* biofilm development. Biointerphases 8, 12.

Anghel, I., Grumezescu, A.M., Holban, A.M., Ficai, A., Anghel, A.G., Chifiriuc, M.C., 2013b. Biohybrid nanostructured iron oxide nanoparticles and *Satureja hortensis* to prevent fungal biofilm development. Int. J. Mol. Sci. 14, 18110–18123.

Ankamwar, B., Lai, T.C., Huang, J.H., Liu, R.S., Hsiao, M., Chen, C.H., et al., 2010. Biocompatibility of Fe_3O_4 nanoparticles evaluated by *in vitro* cytotoxicity assays using normal, glia and breast cancer cells. Nanotechnology 21, 1–9.

Balaure, P.C., Andronescu, E., Grumezescu, A.M., Ficai, A., Huang, K.S., Yang, C.H., et al., 2013. Fabrication, characterization and *in vitro* profile based interaction with eukaryotic and prokaryotic cells of alginate–chitosan–silica biocomposite. Int. J. Pharm. 441, 555–561.

Buteică, A.S., Mihaiescu, D.E., Grumezescu, A.M., Vasile, B.S., Popescu, A., Mihaiescu, A.M., et al., 2010. The antibacterial activity of magnetic nanofluid: Fe_3O_4/oleic acid/cephalosporins—core/shell/adsorbtion-shell proved on *S. aureus* and *E. coli* and possible applications as drug delivery systems. Digest J. Nanomat. Biostruct. 5, 927–932.

Chifiriuc, M.C., Grumezescu, A.M., Saviuc, C., Croitoru, C., Mihaiescu, D.E., Lazar, V., 2012a. Improved antibacterial activity of cephalosporins loaded in magnetic chitosan microspheres. Int. J. Pharm. 436, 201–205.

Chifiriuc, M.C., Grumezescu, V., Grumezescu, A.M., Saviuc, C.M., Lazar, V., Andronescu, E., 2012b. Hybrid magnetite nanoparticles/*Rosmarinus officinalis* essential oil nanobiosystem with antibiofilm activity. Nanoscale Res. Lett. 7, 209.

Chifiriuc, M.C., Grumezescu, A.M., Andronescu, E., Ficai, A., Cotar, A.I., Grumezescu, V., et al., 2013a. Water dispersible magnetite nanoparticles influence the efficacy of antibiotics against planktonic and biofilm embedded *Enterococcus faecalis* cells. Anaerobe 22, 14–19.

Chifiriuc, M.C., Grumezescu, A.M., Saviuc, C., Hristu, R., Grumezescu, V., Bleotu, C., et al., 2013b. Magnetic nanoparticles for controlling *in vitro* fungal biofilms. Curr. Org. Chem. 17, 1023–1028.

Cimpean, A., 2006. Animal Cell Culture. Ars Docendi Publishing, Bucharest, Romania.

Cotar, A.I., Grumezescu, A.M., Andronescu, E., Voicu, G., Ficai, A., Ou, K.L., et al., 2013. Magnetic nanoparticle effects on the red blood cells. Lett. Appl. NanoBioSci. 2, 97–104.

Creangă, D.E., Culea, M., Nădejde, C., Oancea, S., Curecheriu, L., Racuciu, M., 2009. Magnetic nanoparticle effects on the red blood cells. J. Phys. Conf. S. 170, 1–6.

Dousset, V., Tourdias, T., Brochet, B., Boiziau, C., Petry, K. G., 2008. How to trace stem cells for MRI evaluation? J. Neuro. Sci. 265, 122–126.

Giersig, M., Khomutov, G.B., 2008. Nanomaterials for Application in Medicine and Biology. Springer, Bonn, Germany.

Gomaa, I.O., Abdel Kader, M.H., Salah Eldin, T.A., Heikal, O. A., 2012. Evaluation of *in vitro* mutagenicity and genotoxicity of magnetite nanoparticles. Drug Discov. Ther. 7, 116–123.

Grumezescu, A.M., Mihaiescu, D.E., Mogoşanu, D.E., Chifiriuc, M.C., Lazăr, V., Călugărescu, I., et al., 2010. *In vitro* assay of the antimicrobial activity of Fe_3O_4 and $CoFe_2O_4$/oleic acid—core/shell on clinical isolates of bacterial and fungal strains. Optoelect. Adv. Mat. R. Comm. 4, 1798–1801.

Grumezescu, A.M., Saviuc, C., Holban, A., Hristu, R., Stanciu, G., Chifiriuc, C., et al., 2011a. Magnetic chitosan for drug targeting and *in vitro* drug delivery response. Biointerface Res. Appl. Chem. 1, 160.

Grumezescu, A.M., Andronescu, E., Ficai, A., Saviuc, C., Mihaiescu, D., Chifiriuc, M.C., 2011b. Deae-cellulose/Fe_3O_4/cephalosporins hybrid materials for targeted drug delivery. Rom. J. Mat. 41, 383–387.

Grumezescu, A.M., Saviuc, C., Chifiriuc, M.C., Hristu, R., Mihaiescu, D.E., Balaure, P., et al., 2011c. Inhibitory activity of Fe_3O_4/oleic acid/usnic acid—core/shell/extra-shell nanofluid on *S. aureus* biofilm development. IEEE T. NanoBioSci. 10, 269–274.

Grumezescu, A.M., Ficai, A., Ficai, D., Prdean, G., Chifiriuc, M.C., 2012a. Polymeric magnetic silica microspheres as a drug loader for antimicrobial delivery substances. Digest J. Nanomat. Biosctruct. 7, 1891–1896.

Grumezescu, A.M., Holban, A.M., Andronescu, E., Ficai, A., Bleotu, C., Chifiriuc, M.C., 2012b. Water dispersible metal oxide nanobiocomposite as a potentiator of the antimicrobial activity of kanamycin. Lett. Appl. NanoBioSci. 1, 77–82.

Grumezescu, A.M., Holban, A.M., Andronescu, E., Mogosanu, G.D., Vasile, B.S., Chifiriuc, M.C., et al., 2013a. Anionic polymers and 10 nm Fe_3O_4@UA wound dressings support human fetal stem cells normal development and exhibit great antimicrobial properties. Int. J. Pharm. 463, 146–154.

Grumezescu, A.M., Cotar, A.I., Andronescu, E., Ficai, A., Ghitulica, C.D., Grumezescu, V., et al., 2013b. *In vitro* activity of the new water dispersible Fe_3O_4@usnic acid nanostructure against planktonic and sessile bacterial cells. J. Nano Res. 15, 1766.

Grumezescu, V., Holban, A.M., Grumezescu, A.M., Socol, G., Ficai, A., Vasile, B.S., et al., 2014. Usnic acid loaded biocompatible magnetic PLGA-PVA microspheres thin films fabricated by MAPLE with increased resistance to staphylococcal colonization. Biofabrication. 6, 035002.

Hofmann-Amtenbrink, M., von Rechenberg, B., Hofmann, H., 2009. Superparamagnetic nanoparticles for biomedical applications. In: Tan, M.C. (Ed.), Nanostructured Materials for Biomedical Applications. Transworld Research Network, 37/661 (2), Fort P.O., Trivandrum-695 023, Kerala, India.

Holban, A.M., Bleotu, C., Chifiriuc, M.C., Lazar, V., 2013a. Control of bacterial virulence by cell-to-cell signalling molecules. In: Méndez-Vilas, A. (Ed.), Microbial Pathogens and Strategies for Combating them: Science, Technology and Education. Formatex Research Center, Spain, pp. 311–321.

Holban, A.M., Grumezescu, A.M., Ficai, A., Chifiriuc, M.C., Lazar, V., Radulescu, R., 2013b. Fe3O4@C18-carvone to prevent *Candida tropicalis* biofilm development. Rom. J. Mat. 43 (3), 300–305.

Holban, A.M., Gestal, M.C., Grumezescu, A.M., 2014a. New molecular strategies for reducing implantable medical devices associated infections. Curr. Med. Chem. accepted.

Holban, A.M., Grumezescu, A.M., Gestal, M.C., Mogoanta, L., Mogosanu, G.D., 2014b. Novel drug delivery magnetite nano-systems used in antimicrobial therapy. Curr. Org. Chem. 18, 185–191.

Hu, C.M.J., Fang, R.H., Copp, J., Luk, B.T., Zhang, L., 2013. A biomimetic nano sponge that absorbs pore-forming toxins. Nat. Nanotechnol. 8, 336–340.

Huang, D.M., Hsiao, J.K., Chen, Y.C., Chien, L.Y., Yao, M., Chen, Y.K., 2009. The promotion of human mesenchymal stem cell proliferation by superparamagnetic iron oxide nanoparticles. Biomaterials 30, 3645–3651.

Khatri, M., Bello, D., Pal, A.K., Cohen, J.M., Woskie, S., Gassert, T., et al., 2013. Evaluation of cytotoxic, genotoxic and inflammatory responses of nanoparticles from photocopiers in three human cell lines. Part. Fibre Toxicol. 10 (42), 1–22.

Kim, H.S., Choi, Y., Song, I.C., Moon, W.K., 2009. Magnetic resonance imaging and biological properties of pancreatic islets labeled with iron oxide nanoparticles. NMR Biomed. 22, 852–856.

Li, X.Z., Nikaido, H., 2004. Efflux-mediated drug resistance in bacteria. Drugs 64, 159–204.

Limban, C., Grumezescu, A.M., Saviuc, C., Voicu, G., Chifiriuc, C., 2012. Optimized anti-pathogenic agents based on core/shell nanostructures and 2-((4-ethylphenoxy)methyl)-*N*-(substituted-phenylcarbamothioyl)-benzamides. Int. J. Mol. Sci. 13, 12584–12597.

Lin, Y.S., Haynes, C.L., 2009. Synthesis and characterization of biocompatible and size-tunable multifunctional porous silica nanoparticles. Chem. Mater. 21, 3979–3986.

Lipinski, C., Hopkins, A., 2004. Navigating chemical space for biology and medicine. Nature. 432, 855–861.

Markides, H., Rotherham, M., El Haj, A.J., 2012. Biocompatibility and toxicity of magnetic nanoparticles in regenerative medicine. J. Nanomat. 12, 1–11.

Mauricio, M., Ribeiro de Barros, H., Guilherme, M.R., Radovanovic, E., Rubira, A., De Carvalho, G., 2013. Synthesis of highly hydrophilic magnetic nanoparticles of Fe_3O_4 for potential use in biologic systems. J. Colloids Surf. A Physicochem. Eng. Aspects 417, 224–229.

Mbeh, D.A., França, R., Merhi, Y., Zhang, X.F., Veres, T., Sacher, E., et al., 2012. *In vitro* biocompatibility assessment of functionalized magnetite nanoparticles: biological and cytotoxicological effects. J. Biomed. Mater. Res. A. 100, 1637–1646.

Meng, H., Zhang, Z., Zhao, F., Qiu, T., Yang, J., 2013. Orthogonal optimization design for preparation of Fe_3O_4 nanoparticles via chemical coprecipitation. Appl. Surf. Sci. 280, 679–685.

Mihaiescu, D.E., Grumezescu, A.M., Mogosanu, D.E., Traistaru, V., Balaure, P.C., Buteica, A., 2011. Hybrid organic/inorganic nanomaterial for controlled cephalosporins release. Biointerface Res. Appl. Chem. 1, 41.

Mihaiescu, D.E., Horja, H., Gheorghe, I., Ficai, A., Grumezescu, A.M., Bleotu, C., et al., 2012. Water soluble magnetite nanoparticles for antimicrobial drugs delivery. Lett. Appl. NanoBioSci. 1, 45–49.

Mihaiescu, D.E., Cristescu, R., Dorcioman, G., Popescu, C., Nita, C., Socol, G., et al., 2013. Functionalized magnetite silica thin films fabricated by MAPLE with antibiofilm properties. Biofabrication 5, 015007.

Moersdorf, D., Hugounenq, P., Truonc Phuoc, L., Mamlouk Chaouachi, H., Felder-Flesch, D., Begin-Colin, S., et al., 2010. Influence of magnetic iron oxide nanoparticles on red blood cells and Caco-2 cells. Adv. Biosci. Biotechnol. 1, 439–443.

Moy, T., Ball, A.R., Anklesaria, Z., Casadei, G., Lewis, K., Ausubel, F.M., 2006. Identification of novel antimicrobials using a live-animal infection model. Proc. Natl. Acad. Sci. U.S.A. 103, 10414–10419.

Múzquiz-Ramos, E.M., Cortés-Hernández, D.A., Escobedo-Bocardo, J.C., Zugasti-Cruz, A., Ramírez-Gómez, X.S., Osuna-Alarcón, J.G., 2013. *In vitro* and *in vivo* biocompatibility of apatite-coated magnetite nanoparticles for cancer therapy. J. Mater. Sci. Mater. Med. 24, 1035–1041.

Osburne, M.S., Grossman, T.H., August, P.R., MacNeil, I. A., 2000. Tapping into microbial diversity for natural products drug discovery. ASM News. 66, 411–417.

Pankhurst, Q.A., Connolly, J., Jones, S.K., Dobson, J., 2003. Applications of magnetic nanoparticles in biomedicine. J. Phys. D: Appl. Phys. 36, R167–R181.

Park, E.J., Kim, H., Kim, Y., Yi, J., Choi, K., Park, K., 2010. Inflammatory responses may be induced by a single intratracheal instillation of iron nanoparticles in mice. Toxicology 275, 65–71.

Prabhu, S., Poulose, E.K., 2012. Silver nanoparticles: mechanism of antimicrobial action, synthesis, medical applications, and toxicity effects. Int. Nano Lett. 2, 32.

Pradhan, P., Giri, J., Banerjee, R., Bellare, J., Bahadur, D., 2007. Cellular interactions of lauric acid and dextran-coated magnetite nanoparticles. J. Magnet. Magnet. Mat. 311, 282–287.

Priprem, A., Mahakunakorn, P., Thomas, C., Thomas, I., 2010. Cytotoxicity studies of superparamagnetic iron oxide nanoparticles in macrophage and liver cells. Am. J. Nanotechnol. 1, 78–85.

Rai, R.V., Bai, J.A., 2011. Nanoparticles and their potential application as antimicrobials. Science against microbial pathogens: communicating current research and technological advances. In: Méndez-Vilas, A. (Ed.), Microbial Pathogens and Strategies for Combating them: Science, Technology and Education, vol. 197–209. Formatex Research Center, Spain.

Salado, J., Insausti, M., Lezama, L., Gil de Muro, I., Moros, M., Pelaz, B., et al., 2012. Functionalized Fe_3O_4@Au superparamagnetic nanoparticles: *in vitro* bioactivity. Nanotechnology 23, 1–14.

Saviuc, C., Grumezescu, A.M., Bleotu, C., Holban, A., Chifiriuc, M.C., Balaure, P., et al., 2011a. Phenotipical studies for raw and nanosystem embedded *Eugenia carryophyllata* buds essential oil effect on *Pseudomonas aeruginosa* and *Staphylococcus aureus* strains. Biointerface Res. Appl. Chem. 1, 111.

Saviuc, C., Grumezescu, A.M., Chifiriuc, C.M., Mihaiescu, D.E., Hristu, R., Stanciu, G., et al., 2011b. Hybrid nanosystem for stabilizing essential oils in biomedical applications. Digest J. Nanomat. Biostruct. 6, 1657–1666.

Saviuc, C., Grumezescu, A.M., Holban, A., Chifiriuc, C., Mihaiescu, D., Lazar, V., 2011c. Hybrid nanostructured material for biomedical applications. Biointerface Res. Appl. Chem. 1, 64.

Saviuc, C., Cotar, A.I., Holban, A.M., Banu, O., Grumezescu, A.M., Chifiriuc, M.C., 2013. Phenotypic and molecular evaluation of *Pseudomonas aeruginosa* and *Staphylococcus aureus* virulence patterns in the presence of some essential oils and their major compounds. Lett. Appl. NanoBioSci. 2, 91–96.

Shubayev, V.I., Pisanic, T.R., Jin, S., 2009. Magnetic nanoparticles for theragnostics. Adv. Drug Deliv. Rev. 61, 467–477.

Singh, R.K., Kim, T.H., Patel, K.D., Knowles, J.C., Kim, H. W., 2012. Biocompatible magnetite nanoparticles with varying silica-coating layer for use in biomedicine: physicochemical and magnetic properties, and cellular compatibility. J. Biomed. Mater. Res. A. 100, 1734–1742.

Soenen, S.J., Nuytten, N., De Meyer, S.F., De Smedt, S.C., De Cuyper, M., 2010. High intracellular iron oxide nanoparticle concentrations affect cellular cytoskeleton and focal adhesion kinase-mediated signaling. Small 6, 832–842.

Stella, B., Arpicco, S., Peracchia, M.T., Desmaële, D., Hoebeke, J., Renoir, M., et al., 2000. Design of folic acid-conjugated nanoparticles for drug targeting. J. Pharm. Sci. 89, 1452–1464.

Sun, J., Zhou, S., Hou, P., Yang, Y., Weng, J., Li, X., et al., 2007. Synthesis and characterization of biocompatible Fe_3O_4 nanoparticles. J. Biomed. Mater. Res. A. 80A, 333–341.

Szalay, B., Kováčikova, Z., Brózik, M., Pándics, T., Tátrai, E., 2008. Effects of iron oxide nanoparticles on pulmonary morphology, redox system, production of immunoglobulins and chemokins in rats: *in vivo* and *in vitro* studies. Central E. J. Occup. Environ. Med. 14, 149–164.

Szalay, B., Tátrai, E., Nyírő, G., Vezér, T., Dura, G., 2012. Potential toxic effects of iron oxide nanoparticles in *in vivo* and *in vitro* experiments. J. Appl. Toxicol. 32, 446–453.

Teja, A.S., Koh, P.Y., 2009. Synthesis, properties, and applications of magnetic iron oxide nanoparticles. Prog. Crys. Grow. Charact. Mat. 55, 22–45.

Vlad, M., Andronescu, E., Grumezescu, A.M., Ficai, A., Voicu, G., Bleotu, C., et al., 2014. Carboxymethyl-cellulose/Fe_3O_4 nanostructures for antimicrobial substances delivery. Biomed. Mater. Eng. 24, 1639–1646.

Who, 2002. Report on Infectious Diseases. World Health Organization, Geneva, Switzerland.

CHAPTER

5

Photodynamic Therapy of Infectious Disease Mediated by Functionalized Fullerenes

Zeyd Issa[1,2] *and Michael R Hamblin*[2,3,4]

[1]University of Exeter Medical School, Exeter, Devon, UK [2]Wellman Center for Photomedicine, Massachusetts General Hospital, Boston, MA, USA [3]Department of Dermatology, Harvard Medical School, Boston, MA, USA [4]Harvard-MIT Division of Health Sciences and Technology, Cambridge, MA, USA

5.1 INTRODUCTION

Photodynamic therapy (PDT) is a new treatment for infections that involves the killing of bacteria and other microorganisms through the use of a combination of harmless light and a nontoxic but light-activated dye known as a photosensitizer (PS). The PS, when activated, produces radical oxygen species (ROS) by one of two different pathways known as type 1 and type 2. These ROS, once produced, have been proven to be highly effective at killing cancer and bacterial cells, but they have also demonstrated similar effects against fungi and viruses as well (Zeina et al., 2001; Huang et al., 2010; Kessel, 2014).

As far as is known, PDT was first mentioned as a treatment for leprosy by an ancient Egyptian text, the Ebers Papyrus, and a Hindu text, the *"Arthava veda,"* dating to approximately 1550 BC. The practice continued to be mentioned, appearing in the Buddhist Chinese literature in approximately 200 AD, and in the Sung period in approximately the 10th century; the Islamic Caliphate scholar Ibn El Bitar also described treatments involving honey and powdered seeds being exposed to sunlight during the Islamic "golden age" of the 13th century (Wyss et al., 2000; Magicray.ru, http://www.magicray.ru/ENG/lecture/L2/2.html).

However, despite this long history, PDT as it is currently used was only first explored in detail in the final years of the 19th century by a German Scientist named Oskar Raab. Raab accidently discovered that the combination of acridine orange solution, a dye, with sunlight resulted in the killing of *Paramecia*, an effect that was not perceived when light was absent.

Nanotechnology in Diagnosis, Treatment and Prophylaxis of Infectious Diseases.
DOI: http://dx.doi.org/10.1016/B978-0-12-801317-5.00005-0

As a result, with the assistance of his mentor, Professor H. von Tappeiner, he hypothesized that the fluorescent dyes converted the sunlight into chemical energy, resulting in the killing of the *Paramecia*. von Tappeiner, along with A. Jodlbauer, would later go on to test the use of the technique against facial carcinoma; while doing so, they coined the phrase "*Photodynamische wirkung*" or "photodynamic action" to describe the effect by which the photoactivated dyes killed cells and required the presence of oxygen (Krasnovsky, 2007).

Current studies involving PDT have concentrated on a wide variety of diseases, but they are mostly divided into antimicrobial and anticancer treatments. This chapter focuses on the usage of PDT as a treatment for microbial infections, specifically PDT that uses the closed cage carbon allotrope, fullerenes, as a PS, and how these fullerenes are modified and adapted in multiple different ways to function as a PS against microbial infections.

5.2 ANTIBIOTIC RESISTANCE AND THE NEED FOR PDT

Antibiotic resistance is one of the most challenging issues facing modern medicine to date. There have been numerous efforts to counter the adaption of bacteria to antibiotics, such as education of the public and of health care personnel regarding the importance of finishing courses of antibiotics, as well as thorough and frequent cleansing of surfaces and surgical equipment; however, despite this, the WHO describes current rates of infection by antimicrobial-resistant (AMR) microbes as "alarming" in its 2014 surveillance report on AMR (WHO, http://www.who.int/drugresistance/documents/surveillancereport/en/). These concerns are not unfounded. The increase in MRSA in the developed world, for example, has been widespread, with the CDC estimating that there may have been as many as 75,309 invasive cases of MRSA in the United States in 2012, or approximately 23.99 cases per 100,000 US citizens (CDC, http://www.cdc.gov/abcs/reports-findings/survreports/mrsa12.pdf).

XDR TB (otherwise known as extensive drug-resistant TB) is another emerging danger in the field of multidrug-resistant infections with an alarmingly poor prognosis. The CDC estimates that it can be cured in only 30–50% of cases (CDC, http://www.cdc.gov/tb/publications/factsheets/drtb/xdrtb.htm). This growing body of evidence has raised serious concerns about a future where modern medicine has to make do without antibiotics. So long a staple of our surgical techniques, medical prescriptions, and even livestock farming, these drugs are rapidly becoming obsolete. The ultimate outcome of this situation has been widely speculated. The WHO, for instance, goes so far as to say that "A post-antibiotic era—in which common infections and minor injuries can kill—far from being an apocalyptic fantasy, is instead a very real possibility for the 21st century" (WHO, http://www.who.int/drugresistance/documents/surveillancereport/en/).

There have been several calls for research into newer types of antibiotics to combat these infections, but this effort still requires significant investments because pharmaceutical companies remain rather reluctant to continue investigation into new forms of antibiotics. To many companies, the field appears to be one of diminishing returns based on cost and timing alone. For example, in the United Kingdom it can take up to 12 years for new drugs to be introduced as treatments, and expenditure during this process may be as high as £1 billion (approximately $1.68 billion at the time of writing). Although governments are looking for ways to encourage pharmaceutical efforts to find solutions, there are many who believe that there most likely needs to be a more permanent replacement for antibiotics (Stephens,

2014) or at least a supplement to standard treatment that can kill drug-resistant species outright, and many competing treatments exist that could be considered for this purpose (*New York Times*, 2014).

Recently, PDT has entered the spotlight because of its possibilities as such a treatment. The broad-spectrum activity of PDT, including the destruction of antibiotic-resistant bacteria, viruses, fungi, and even parasites, has made it of considerable interest. Additionally, though certain species of the microbes may resist some doses/varieties of PDT, there have not been any reports of adaptive resistance being developed toward PDT in the same way as toward antibiotics, and *in vitro* studies reveal that adding higher doses of PS to these hardier species will result in their destruction (Dai et al., 2009). Of course, although this does not rule out that resistance may be possible, it does bode well for the treatment as a possible replacement for antibiotics in the future. So far, PDT has been successfully used for treating infectious conditions as diverse as surgical wounds, papilloma, acne, and biliary, and ongoing research is investigating the use of PDT for many others. Outside of the laboratory, PDT is currently performed using 5-aminolevulanic acid and phenothiazinium dyes in the fields of dermatology and dentistry, and PDT has also been demonstrated as a method of purification of blood through the inactivation of viruses. As such, PDT is an antibiotic alternative that is growing steadily in popularity and it is thought to be one that may prove to be of great importance in the future (Hamblin and Mroz, 2008; Kharkwal et al., 2011).

5.3 PDT MECHANISM OF ACTION

5.3.1 Use as an Antimicrobial Treatment

To understand how PDT effectively functions requires a detailed explanation, because the mechanism by which the treatment affects a microbial infection involves different effects working in tandem. Although there are two different photochemical pathways of PDT, type 1 and type 2, both lead to the same overall product: reactive oxygen species and radical chain reactions. Figure 5.1 shows graphically how these two different photochemical pathways operate in photoexcited fullerenes. In both cases, the PS absorbs photons from the light source, which results in it being converted to an excited state. During this excited state, in type 1 reactions, a substrate (this being the pathogenic cell in the case of PDT) donates an electron forming the PS radical anion, which then reacts with oxygen, resulting in production of several different ROS such as superoxide and hydroxyl radicals; in type 2, the PS reaches a long-lived excited triplet state with which oxygen (under standard conditions, a

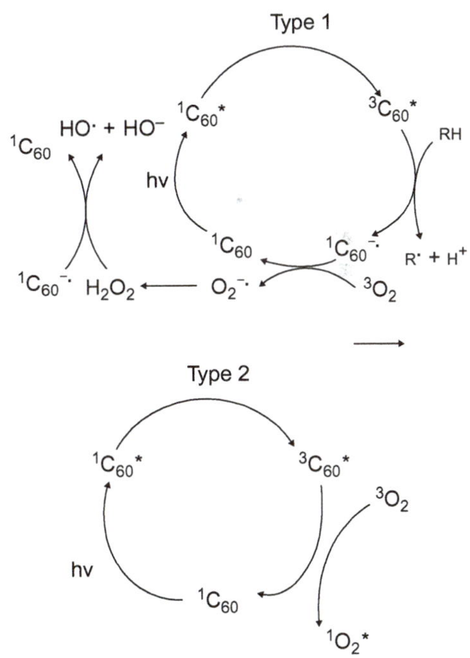

FIGURE 5.1 Diagram demonstrating type 1 and type 2 photosensitization of C_{60} fullerenes.

triplet itself) can interact. This reaction results in the energy transfer forming highly cytotoxic singlet state oxygen, but this reactive species only lasts for approximately 3 μs.

Thus far, PDT has been demonstrated to be destructive to any biological tissue at high enough doses, but it has proven most effective against Gram-positive bacteria cells, in which the PS can penetrate the peptidoglycan structures of the cell membranes and thus do major damage inside these cells rather easily.

Gram-negative bacteria, like *Escherichia coli*, are more difficult to treat with PDT because of their unique double-layer membrane structure that is difficult to effectively penetrate with either PS or antibiotics. The most difficult of the Gram-negative species to treat with PDT are *Pseudomonas*, whose cell membrane has a more efficient permeability barrier than the majority of other Gram-negative species. Gram-negative bacteria are susceptible to only a few types of PS; however, when well-optimized, this results in high levels of killing (Caminos et al., 2006; Sperandio et al., 2013).

Fungal cells, such as *Candida*, are more difficult to kill than Gram-negative cells because of their mannan layer composed of lipoteichoic acid or beta-glucan, which makes penetration even more difficult. Although fungal species have been proven to be susceptible to a wider range of PS than Gram-negative bacteria, they require higher doses to achieve the same amount of killing (Mroz et al., 2007a; Oleinick, 2011).

5.4 APPLICATIONS

5.4.1 Antiviral

There are a few different antiviral applications that have been demonstrated thus far by PDT, namely the purification of blood through the use of methylene blue as a PS and the use of PDT for treatment of papillomatosis.

5.4.1.1 *Papillomatosis*

Recurrent respiratory papillomatosis (RRP) is a condition associated with human papillomavirus (HPV) that is characterized by papillomatous lesions, which are benign growths in the mucosal layer of the upper respiratory tract (Shikowitz et al., 1998). Recurrence of these symptoms can be indicative of life-threatening conditions (Lieder et al., 2014). Lee et al. (2010) examined the use of PDT with a phthalocyanine PS for treating mice grafted with cottontail rabbit skin and then infected with cottontail rabbit papillomavirus; 0.6 or 1.0 mg/kg of the PS were applied, along with 100 or 150 J/cm^2 of 675 nm light. At lower doses, this treatment proved to be ineffective with little regression of the papilloma; however, at 1.0 mg/kg and 150 J/cm^2 of light, there was "complete regression" of the papilloma at the site of infection with a significant difference in the controls (P value <0.001). Follow-up analysis showed no recurrence during the follow-up period, as well as an absence of residual tumors by histological analysis. The conclusion of this was that PDT "warrants further study as a treatment for HPV-induced papillomas."

There have been additional studies involving the use of PDT to treat other papilloma conditions, such as papillomatosis of the bile duct or biliary papillomatosis (BP), a rare (Bechmann et al., 2008) condition (fewer than 100 cases have been reported in the cited literature) with high chances of progressing to carcinoma. Bechmann et al. (2008) studied the use of PDT (using 2 mg/kg of Photofrin II® as the PS and 633 nm light) to treat the condition in a 72-year-old patient; the patient demonstrating limited recurrence of the papilloma after the treatment and was treated with PDT again until he died 4 years later of multiple organ failure stemming from secondary biliary cirrhosis, not from cholestasis or cholangitis. This is a significantly longer period of survival than others with BP, who have an average survival time of 28 months. Quality of life using the treatment was also apparently markedly improved with the use of PDT. The conclusion was that this treatment modality might be suitable as a therapeutic option after resection (the current treatment is Whipple treatment), as well as for those not eligible for surgery.

Likewise, nasal inverted papilloma (NIP) treatment with PDT has been examined by Zhang et al. (2013), who applied ALA treatment (5-aminolevulinic acid as a PS) using 635 nm of light and a fluence of 100–120 J/cm^2 (applied using an endoscope) to three patients with continued follow-up medical examinations for 6 months. The result of the treatment was complete regression of the NIP, with no recurrence in the next 6–8 months. Despite side effects of some pain, erosion, and exudation, the treatment was "well-tolerated" by the patients. The authors concluded that PDT "appears to be an effective treatment of [NIP]."

5.4.1.2 *Blood Purification*

One notable area suggested as a possible application for PDT is that of blood/blood product purification. The use of PDT as a method of cleansing viruses from blood has already been performed with methylene blue, where the ROS produced as a result of the photosensitization leads to the deactivation of viruses within the blood. Methylene blue is also nontoxic in the blood, so this method of filtration is highly recommended. This use of PDT was first approved in Germany by the Red Cross in 1992 to deactivate the hepatitis B and hepatitis C viruses, as well as HIV, which had previously been transmitted to patients during blood transfusion. However, there are still risks associated with transfusion of blood products that have been purified in this manner because serological tests cannot accurately determine if infection has occurred during the window of opportunity that follows infection. Nonetheless, despite extensive testing *in vivo* to examine possible side effects, there appear to have been none discovered as of yet. For instance, in rabbits (Yin et al., 2014a, b), there was no significant damage to organs, and the treatment is believed to be a safe and novel method for inactivating blood-borne pathogens.

5.4.2 Dentistry

The use of PDT to treat bacterial biofilms on teeth has been tested on *ex vivo*–extracted teeth, specifically *Enterococcus faecalis*, the pathogen most associated with endodontic infections (Rocas et al., 2004). A study by Soukos et al. (2006) claimed to have observed 97% killing of *E. faecalis* by using methylene blue and exposure to 665 nm of light, provided that the dosage was increased from 30 to 222 J/cm^2, whereas a later experiment by Fonseca et al. (2008) demonstrating the use of the PS toluidine blue and 660 nm of light showed a decrease in colony-forming units (CFU) of the same species by 99.9% in the treated group. Similarly, encouraging results were found by Garcez et al. (2007) against Gram-negative species (*Pseudomonas aeurginosa* and *Proteus mirabilis*) using a conjugate of polyethylenimine and chlorin(e6) as a PS, where 90% of the bacteria were killed with a standard endodontic treatment, 95% were killed with PDT, and 98% were killed with a combined approach.

Despite these favorable results, however, skepticism remains. According to Muhammad et al. (2014), who compared ultrasound irrigation with PDT using toluidine blue as the PS and a 650-nm laser, the PDT treatment was ineffectual over the period of time that it was used. This point appears to vindicate the earlier concerns of Siddigui et al. (2013), who stated that use of PDT for removal of *E. faecalis* was still "questionable."

5.4.3 Dermatology

Apart from dentistry, another field that has begun to use PDT widely has been that of dermatology, where the possible use of ALA-PDT to combat acne (among other conditions) has proven popular.

5.4.3.1 Acne

Treatment of acne by PDT has been widely explored by dermatologists in recent years. For example, Hongcharu et al. (2000), effectively demonstrated that use of ALA-PDT leads to significantly decreased acne lesions after 20 weeks and four applications of PDT. The authors reported that the effects of the PDT inhibited multiple pathogenic factors of acne and suggested that it could prove an interesting treatment for the future. Akaraphanth et al. (2007) also compared the use of ALA-PDT with blue light alone for the treatment of acne, demonstrating 71.1% reduction of inflamed lesion counts seen on the patient, although at high cost and with several side effects. More recently, Mei et al. (2013) demonstrated rates of reductions in inflammatory lesions as high as 83.6% using the technique, with no severe side effects. Yin et al. (2014a, b) noted that when treating cases of severe acne using ALA-PDT and an ablative fractional laser, in addition to "good to excellent" improvement in scarring in 85% of patients tested, none of the 32 male and 8 female patients demonstrated recurrent acne lesions, which is a highly promising result for the future use of PDT for acne treatment.

5.4.4 Wounds

Multiple studies have explored the use of PDT on surface wounds for antimicrobial effects. Dai et al. (2009) reviewed how localized infections can be treated with PDT and stated that wounds are easily infected by microbes, with surgical wounds accounting for "25% of nosocomial infections." In two studies by Hamblin et al. (2002, 2003), mice were given excisional wounds that were then infected with bioluminescent *E. coli* and *P. aeruginosa*, and the infection was monitored in real time through the use of a charge-coupled camera. This infection was treated with polylysine (pL)–ce6 conjugate as a PS and 665 nm of light at a fluence of 240 J/cm^2. All three groups of control mice (control, those treated with PS only, and those treated with light only) died within 5 days, whereas 90% of the PDT-treated mice survived for the full 20-day duration of study.

Likewise, Zolfagari et al. (2009) demonstrated that MRSA in wounds could also be effectively killed using PDT; in their study, 100 μg/mL MB was used as PS, with a 360-J/cm^2 dose of a 670-nm laser light applied. The result was a 25-fold decrease in the number of viable MRSA in the wound. Wong et al. (2005) examined a similar effect with *Vibrio vulnificus*, where septicemia was well-established in the wound with a bacterial inoculation nearly 100-times more than the 50% lethal dose. When treated with 100 μg/mL of TBO and exposed to 150 J/cm^2 of broad-spectrum red light, 10 out of 19 mice (53%) survived. Given the ability of PDT in these studies to prevent death in otherwise fatal infections of wounds, PDT appears to be a promising alternative treatment for surgical wounds, especially in the case of MDR bacteria such as MRSA.

5.4.5 MRSA

As previously mentioned, PDT has proven formidable against drug-resistant bacteria such as MRSA, and considerable interest has been expressed in both *in vitro* and *in vivo* studies that can demonstrate the viability of PDT as a treatment. One example of such a study is that by Fu et al. (2013), who reviewed multiple works on the subject and reached numerous conclusions. It is unclear whether any method by which MRSA can develop resistance toward ROS exists, and use of PDT did not indicate major damage to the patients' healthy tissues, meaning that the treatment should be safe for use in patients. Additionally, although some PS demonstrate little bactericidal effects toward MRSA, efflux mechanisms in MRSA might not influence certain PS. As a result, although the use of PDT topically against MRSA during clinical practice may happen in the future, whether it is an appropriate replacement for topical antibiotics will be determined by the development of specific cellular delivery systems.

5.5 THE IDEAL PS

Many of the fundamental issues with PDT mentioned in Section 1.3 stem from failures of the PS; therefore, it can be considered important to examine just what would make an "ideal photosensitizer," something that many researchers, such as Pushpan et al. (2002), Allison et al. (2004), and Kudinova and Berenov (2009), have tried to analyze. First and foremost, the potential PS must demonstrate a high molar extinction coefficient, meaning that it must be able to absorb photons efficiently. Second, it must have a long-lived triplet state with a high quantum yield, and also produce large amounts of singlet oxygen. The PS should be highly stable and also demonstrate negligible or no dark toxicity, meaning that unless it is directly exposed to the wavelength required for treatment, it will not cause damage to tissues. This wavelength, too, should ideally be comparatively long (between 700 and 850 nm) to ensure that the treatment occurs with the best possible light penetration into tissue, and there should be as little quenching of the PS as possible, because that results in the reduction of the efficacy of the treatment. Taken together, these attributes would result in increased selectivity and comparatively high levels of bacterial killing, something that would go a long way to offset many of the perceived downsides of PDT as a treatment.

5.5.1 Limitations

5.5.1.1 Penetration

Although PDT has been proven to be highly effective in the local areas to which PS are applied, it does have some limitations. One of the most important of these limitations is low penetration. When applied to a target area at low doses, PDT PS have a tendency to have limited uptake into the infectious organisms, particularly in Gram-negative bacteria such as *E. coli*, where the double membrane structure of these bacteria, in addition to efflux pumps, results in minimal uptake of drugs in general, including both PS and antibiotics.

Because of this efflux and lack of penetration of the double membrane, the PS can sometimes be activated outside of the target cells, which results in greatly reduced amounts of killing because the oxidative species are unable to effectively attack the interior structure of the pathogen. That said, these issues are somewhat offset by certain PS. Because most microbial cells have an overall negative charge, it is thought that positively charged PS can bind to these charges on the outer layers of target cells and diffuse inward to the cellular interior, where they can do damage under illumination. Even in the case of Gram-negative bacteria, cationic compounds are able to displace divalent magnesium and calcium cations that play an important role in the structural integrity of the double membrane, allowing the fullerene to penetrate the cell, in a method referred to as "self-promoted uptake." Nonetheless, despite this advantage, PDT has thus far been able to kill any biological tissue in a large enough dose *in vitro*; *in vivo*, the PS needs to be applied in comparatively large doses to achieve the same amount of killing.

5.5.1.2 Nonspecificity

The nonspecificity of PDT in the antimicrobial role gives it a great advantage in dealing with bacteria and other pathogens that might otherwise adapt to it and allows the therapy to act as a broad-spectrum therapy against viruses, fungi, and both Gram-negative and Gram-positive bacteria. As previously stated, PDT has been used to kill most forms of microbial infection *in vitro*. However, this major boon for the therapy is something of a double-edged sword. Although the same PS can be applied for many different pathogens, it might not differentiate between the host tissue and the pathogen at higher incubation times, and there can be side effects as a result of poorly targeted or use of an overly long incubation time for the PS. Therefore, a great deal of time and effort have been expended by research laboratories the world over trying to analyze the best manner in which to reduce uptake of various PS in healthy tissue, but at the same time promoting their uptake in the pathogens. In the words of Fu et al. (2014), the successful implementation of PDT will rely heavily on being able to produce "high efficient inactivation of specific bacteria but minimal side effects." Therefore, the focus of much research is on finding a PS as close to the "ideal" as possible.

5.5.1.3 *Side Effects*

5.5.1.3.1 DAMAGE TO SURROUNDING TISSUE

PDT does not demonstrate specificity in terms of targeting cells with the release of ROS. As a result, it is very possible for oxidative stress to occur in patients' cells where the use of the PS has not been restricted to the area of the microbial infection, which undermines the case for PDT as a safe treatment. There are methods that can be used to reduce this damage and still ensure the elimination of the pathogen.

Apart from variation of the dose of PS applied and the choice of both a suitable wavelength and dosimetry of light for the treatment, application of the PS to the precisely correct area is also vital for preventing photo damage to healthy cells, even when applied topically. The most effective optimization of the treatment, however, is to vary the amount of time between the application of the PS and the use of light on the afflicted area. Because the PS is generally taken up by the pathogenic cells much quicker than by host cells, the result of this, given the correct time delay, is that the PS is given enough time to be taken up into a significant number of the targeted cells, but not enough for it to be taken up into the non-target host cells before the light is applied.

However, although these simple methods do reduce damage of healthy tissue, they are not necessarily the single most effective optimization of the treatment. Methods of modifying the PS to be more selective toward microbial vectors would in all likelihood be a considerably more effective approach (some of which are detailed in Section 1.7). Together with choosing appropriate delivery vehicles, finding the most effective photosensitive compounds and modifying them appropriately allow us to come closer to developing the "ideal PS."

There are, of course, additional challenges posed by PDT to the body. Certain PS demonstrate the so-called dark toxicity, and can damage cells without being exposed to light. Although unmodified fullerenes show "low toxicity," according to Aschberger et al. (2010) there are concerns. No short-term harmful effects of functionalized fullerenes have yet to be found, but no long-term effects can be or have been extrapolated. With such uncertainties, they go on to add that "it seems relevant to clarify whether certain fullerene types may potentially induce genotoxic and/or carcinogenic effects via physiologically relevant routes" (Ferreira et al., 2014).

5.5.1.3.2 RESISTANT PATHOGENS AND THE DANGERS OF DEVELOPING A RESISTANCE

One of the most important features of PDT is its ability to kill pathogens that antibiotics simply cannot affect *in vitro*. Although there has not yet been a case of a reported resistance to PDT, certain cells can be highly resistant to

PDT treatment, except at doses that would be unacceptable during *in vivo* treatment, especially Gram-negative bacteria. This issue is of course compounded by the poor uptake of the PS in some cases, and without a large enough dose applied to the afflicted area there is a significant risk that the pathogen can survive the damage. This condition, continued over an extended period, was in essence how antibiotic resistance was first allowed to emerge, which is a deeply concerning prospect if PDT is to be a permanent replacement of traditional treatment.

As such, although PDT has demonstrated clear benefits in both antimicrobial and anticancer fields, without overcoming the challenge of resistant pathogens adequately, the treatment could become defunct. It is therefore of the utmost importance for the future of this treatment to discover new PS that are capable of eliminating these resistant bacteria without allowing the development of resistance.

5.5.2 Optimization

As previously stated, there have been many different methods explored for the purpose of PDT optimization. These can be broadly divided into two different categories: optimizations that focus on getting the PS into the target cells and those that focus on getting light into the cells to activate that PS.

5.5.2.1 Getting the PS to the Cell

One example of how current research has been exploring how to promote PS uptake is variation of drug delivery systems to ensure increased drug uptake into the infectious organism. Many PS are hydrophobic, as stated by Rijcken et al. (2007) and as demonstrated by Ikeda et al. (2009), whereby a fullerene is encapsulated within a surface cross-linked liposome, known as a cerasome. The fullerene did not need to be released from the liposome to be photosensitized, and this can increase the retention time of the fullerene in the blood while maintaining a high morphological stability, which increases cellular uptake; this has also been discussed as a possibility for a "nano-delivery vehicle" for PS.

There are many methods by which these so-called conventional liposomes can be modified for their purposes, as reviewed by Sadasivam et al. (2013). One example is that of light-sensitive liposomes, such as those containing photopolymerizable phospholipid bis-SorbPC that was able to deliver a hydrophobic PS and, on activation with light of wavelength 550 nm, release the PS. Another is that of enzyme-triggered liposomes, which only release their contained PS in the presence of a specific enzyme. In the case of cancer, this has proven highly effective because tumors overexpress certain enzymes, although the approach is rather difficult. This chapter does not mention antimicrobial use of these liposomes, but they remain a possibility for future research.

Fusogenic liposomes, inspired by fusogenic proteins found in many viruses, are a type of liposome that fuses to the membrane of a target cell and thus delivers its contents into the cytosol of the cell. Although no experiments appear to have been performed thus far using this system for release of PS, the possibility of releasing the PS molecule directly into the microbial cell, given the difficulties of uptake in Gram-negative and fungal cells, is highly appealing. pH-sensitive liposomes are mediated by the presence of lower or higher pH conditions, resulting in the release of the contained PS. In the case of cancers and microbial infections, which can have different pH conditions, this can result in more targeted release of the PS. Thermosensitive liposomes are liposomes that are sensitive to heat and are capable of being modified in such a way that different applications of heat result in different phases of release of the PS, which allows them more permeability and higher selectivity for target cells.

Antibody- and ligand-targeted liposomes are liposomes that target specific unique molecules that are expressed by their target and have antibodies (or other moieties that mediate specific binding) on their surface that recognize and bind to specific targets. Given the presence of several antigens on microbial cells, it is very possible for such liposomes to bind to and release their PS inside or in close proximity to microbial cells, greatly increasing the specificity of the PS for these targets. These varied types of liposomes, in coordination with newer, more effective PS, appear highly promising for overcoming some of the traditional issues with PDT and making it a more effective and selective treatment modality.

5.5.2.2 *Getting the Light into the Tissue*

Getting light into tissues underneath the skin, such as in the digestive or respiratory tracts, can be another serious issue with PDT, given that many wavelengths are absorbed by melanin in the skin, making treatment of deep tissue infections difficult. Even with the right wavelength, light can still be scattered, which affects the distribution of light to the target site.

5.5.2.2.1 INTERSTITIAL FIBERS

One of the suggested methods for improving light penetration of the skin has been the use of interstitial fibers to transmit the light. One example of this is the use by Liang et al. (2013) of existing cylindrical diffusing fibers (currently used for diffuse optical tomography) as a light source for PDT inside a phantom prostate. The study does mostly focus on the use of PDT for anticancer purposes, but the use of such techniques to deliver light to otherwise difficult to illuminate tissues is intriguing.

5.5.2.2.2 CHANGING TISSUE OPTICAL PROPERTIES

Another suggested method has been that of changing the optical properties of the tissues that are being illuminated by one of two main methods, optical clearing or photobleaching chromophores. Optical clearing is a technique by which light penetration through a tissue can be improved by modifying the tissue itself to be less scattering and more transparent to certain forms of light. As reviewed by Zhu et al. (2013), there are several different methods through which this can be achieved, including optical clearing agents (OCAs). OCAs can "match the refractive indices of tissue components with extracellular fluid," which reduces scattering of light and can make the tissue more transparent. There are also mechanical improvements that can facilitate penetration of OCAs, such as intensive pulsed light, ultrasound, and mechanical compression, as well as chemical improvements such as chemical penetration enhancers.

Bonnett and Martinez (2001) described photobleaching as the "loss of absorption or emission intensity as caused by exposure to light." There have also been studies to examine possibilities in the field of PDT, with the use of photobleaching being recognized as beneficial by reducing skin photosensitivity, with a group of mice suffering only 1 week of photosensitivity instead of the usual 5 weeks associated with use of Photofrin®, suggesting that photobleaching might be useful to reduce oxidative damage done to healthy cells by high doses of PDT.

5.5.2.2.3 EXTENDING ABSORPTION SPECTRA

While not strictly a modification of the light, extending the absorption spectra of the PS so that they are excited by a wider range of light wavelengths is a promising method of increasing input of ROS for the same dose of PS. This can be achieved mainly by addition of antennae to the PS.

5.5.2.2.4 NONLINEAR EXCITATION

Two-photon excitation, as explained by Probodh and Cramb (2012), is a method using a femtosecond pulsed laser with high peak power at the focus, whereby the targeted molecule (the PS) is hit by two photons at the same time as half the energy (equivalent to twice the wavelength) of one photon (standard) PDT. The equivalence of tissue-penetrating near-infrared (NIR) to this lower wavelength allows for a great deal more penetration, whereas the activation occurs at the focus of the laser beam, resulting in a great deal of selectivity of the area photosensitized.

Upconverting nanoparticles are particles composed of specific rare earth compounds such as ytterbium and ytrium, which have a long-lived metastable state and, when excited with one NIR photon, will remain stable long enough to absorb a second photon and emit fluorescence at just over half the excitation wavelength (upconversion), allowing the use of a standard CW light source; however, in this case, the nanoparticle needs to be decorated with PS that will absorb the emitted fluorescent light. Wang et al. (2014) studied this method for the purposes of deep cancer therapy, although the use of such nanoparticles can be extended to antimicrobial PDT as well.

5.6 PDT USING FULLERENES

5.6.1 Introduction to Fullerenes

Fullerenes are molecules formed from pure carbon atoms only in the form of a closed cage and exist in multiple forms. For the purposes of this chapter, however, we refer exclusively to the spherical and elliptical fullerenes that are referred to as "buckyballs" because of their historical discovery by a team that named them "Buckminsterfullerenes" as a tribute to the famous architect, Buckminster Fuller (Kasermannand and Kempf, 1997).

5.6.1.1 Photochemistry

Research in the field of PDT has begun to explore the possibilities of using fullerenes as PS only recently, and their use as such for cancer treatment is far more definitively explored than their possibilities as an antibacterial treatment. Fullerenes represent an intriguing class of PS molecules prospect because of their very particular photochemistry. First, they are much more photostable than other PS, such as tetrapyrroles and dyes, and are able to accept up to six electrons; with excited states, they become even more effective electron acceptors. The fullerene can also undergo both forms of photochemistry, allowing production of both free radicals from the radical fullerene anion and singlet oxygen from the triplet state. This means that fullerenes can often inflict a greater variety of photodamage and possibly a greater quantity (higher quantum yield) than other PS. Fullerenes can also be photoexcited with visible light, whereas the so-called pristine fullerenes (fullerenes that are completely unmodified) have been found to be nontoxic and during treatment have been significantly more selective for microbial cells over mammalian ones. Although fullerenes do demonstrate some flaws as a proposed PS, these have been overcome with functionalization and formulation (Tegos et al., 2005; Milanesio et al., 2013).

5.6.1.2 Functionalization

As a result of this photochemistry, pristine fullerenes might at first appear to be a most likely candidate for a PS. They are, however, unsuitable for use because of their extremely hydrophobic nature, meaning that they are not soluble in biologically compatible solvents. Therefore, these pristine fullerene molecules must be prepared for any use in a biological system by being "functionalized" through the addition to their structure of chemical groups that provide water solubility and biological targeting, like hydrophilic or amphiphilic side chains such as quaternary ammonium groups. These groups not only allow fullerenes to be effectively administered to a site of infection *in vivo* but also have demonstrated a notable increase in production of singlet oxygen species, hydroxyl radicals, and superoxide anion, making these particular fullerenes more effective PS. Another result of the addition of these groups is the functionalized fullerene being given multiple overall positive charges, which, as previously mentioned, can result in higher antimicrobial selectivity as well as improved uptake in Gram-negative cells (Wang et al., 2013; Li et al., 2014).

5.6.1.3 *Formulation*

Fullerenes are rather stable molecules, but that does not mean they are completely inert. In fact, some specialized buckyballs have been created that have attached antennae to the exterior of the carbon shell for various roles. An example of this is the addition of light-harvesting antennae to functionalized fullerenes, which extend the absorption spectrum of the PS and also result in overall increased quantum yield and ROS production. Fullerenes can also be modified with the trapping of additional metal atoms within their cage-like structure; these particular compounds are known as "endohedral fullerenes." One of the major advantages of this is that it allows for significant modification of the fullerene molecule, overcoming some of its major disadvantages and allowing tailor-engineering of the fullerene (Ikeda et al., 2009; Mizuno et al., 2011).

5.7 IN VITRO STUDIES

5.7.1 Viruses

Kässerman and Kempf (1997) and Hirayama et al. (1999) are among several groups that studied the use of photoactivated fullerenes against viruses *in vitro*, and both found that fullerenes mainly inactivated viruses through their production of singlet oxygen, which is virucidal. The result of this is that the main antiviral action of fullerenes is largely oxygen-dependent, and the use of fullerenes requires a higher concentration of PS than would be required with methylene blue, despite the high yield of singlet oxygen. However, the studies also pointed out that the fullerenes could be removed from aqueous solutions with relative ease, allowing for their use in purification of blood because they could be immobilized and then immersed into the blood for the PDT.

As mentioned previously, blood purification remains an approved application for PDT, with examples of it being used with methylene blue as a PS rather effectively. Given the high performance of many fullerenes, it has also been suggested that they could be used as a PS for the same purpose, although they would need to be immobilized and immersed in the blood for the effect because fullerenes uptake might not be beneficial for the bloodstream, whereas methylene blue is harmless when taken up in the same way. Although such studies have yet to be released in detail, studies of antiviral blood purification remain a tantalizing area of interest for the PDT research community.

One recent study by Banerjee et al. (2012), has examined the use of nanotube mesh, a different form of fullerene, for possible use as an antiviral agent. This particular mesh, an acid-functionalized multiwalled carbon nanotubule, to which protoporphyrin IX (PPIX) was attached, significantly reduced the ability of influenza A virus to infect mammalian cells. In addition, this nanotube can be used against influenza viruses with "little to no danger" of the development of viral resistance and can be easily recovered by filtration, which allows for excellent reuse of the active porphyrin. Although nanotubules may not be buckyballs, the fact that this successful demonstration of their utility, combined with the similar general purpose that they share with possible immobilized buckyballs, is something that offers hope for the future.

5.7.2 Bacteria

Recently there have been numerous studies of the use of fullerenes as an antibacterial treatment that have explored its use against both Gram-negative and Gram-positive bacteria.

Tegos et al. (2005) looked at the use of six different functionalized fullerenes, three (BF-1, BF-2, BF-3) noncationic and three (BF-4, BF-5, BF-6) cationic, and performed a comparison with toluidine blue O, a common PS, and used all seven of these different PS against both *E. coli* and *Staphylococcus aureus*. The synthetic strategy and chemical structures of BF4-6 are shown in Figure 5.2. BF-5 and BF-6 demonstrated extremely strong PDT toxicity, giving a killing rate as high as 99.9999% in *S. aureus* regardless of light applied. When diluted to a concentration of 1 μM, they managed to kill 4 to 5 logs of bacteria (up to 99.999%) with 2 J/cm^2 for BF-5 and 1 J/cm^2 for BF-6. These PS were then placed under conditions similar to those inside a mammalian organism using a concentration of 10 μM of PS and 10-min incubation in 10% fetal bovine serum (FBS) and then a wash. Under these conditions, TBO was almost entirely ineffective at photodynamic inactivation (PDI), whereas BF-5 and BF-6 suffered the loss of approximately 1 log of killing. Tegos et al. (2005) also conducted an experiment to determine the amount of selectivity of the fullerenes as compared with toluidine blue by incubating the PS in murine L929 fibroblasts and examining the outcomes of the therapy. The fullerenes did kill the L929 cells with both dark toxicity (20–60% of cells) and

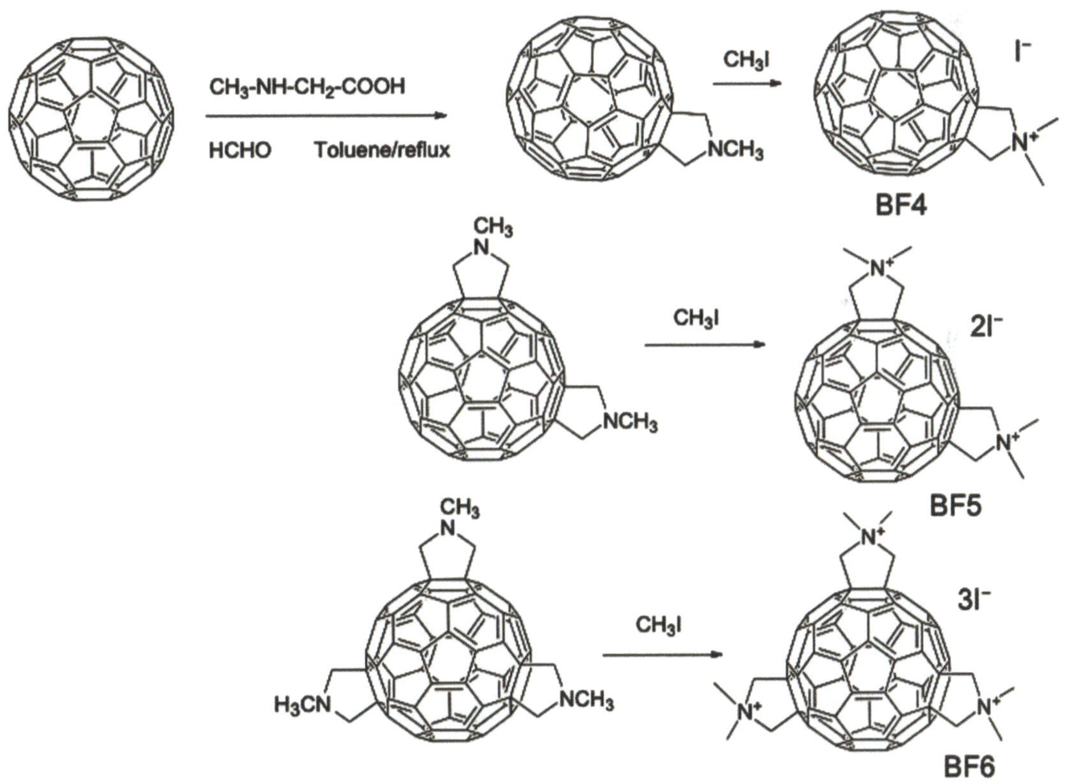

FIGURE 5.2 Formation of the BF-4, BF-5, and BF-6 buckyballs from pristine C_{60} fullerene.

during the treatment (an additional 20% of cells), but they did not display the "pronounced light toxicity" of toluidine blue.

Mroz et al. (2007b) reviewed the use of this same series of functionalized fullerenes in treatment of both *S. aureus* and *E. coli*. Although the noncationic fullerenes (BF-1, BF-2, and BF-3) did not demonstrate dark toxicity, they required very high doses of both the PS (100 μM) and light (120 J/cm^2) to achieve a "significant killing" rate of approximately 99.9% in Gram-positive cells, whereas against *E. coli* it achieved a 90% killing rate at best. By comparison, BF-5 and BF-6 of the polycationic fullerenes of the series (BF-4, BF-5, and BF-6) gave high levels of dark toxicity but achieved a 4-log (99.99%) killing rate against *S. aureus* with only 10 μM concentration and 1–2 J/cm^2 of light. When tested against *E. coli* at a concentration of 10 μM, these same two polycationic fullerenes resulted in notably high levels of killing, as high as 6 logs (up to 99.9999%).

Regarding the treatment of *E. coli* with fullerenes, Spesia et al. (2008) explored the possibility while using DTC_{60}^{2+}, a dicationic fullerene composed of a hydrophobic carbon sphere, and two attached cationic groups that form a single monoadduct to the sphere (structure shown in Figure 5.3). It was shown that this particular fullerene resulted in a 99.97% (3.5 log) killing rate of bacteria at approximately 2 μM concentration after 30 min of irradiation, and that a similar PDI effect of 3.2 log was demonstrated with just 1 μM concentration of DTC_{60}^{2+} as 5 μM of cis-dicationic porphyrin. Combined with the evidence brought forward by Mroz et al. (2007a), this represents an effective demonstration of the capabilities of fullerenes *in vitro* against Gram-negative species, especially the excellent selectivity that some fullerenes can demonstrate toward pathogens, even at lower doses, something that is far more likely to be safer for the patient.

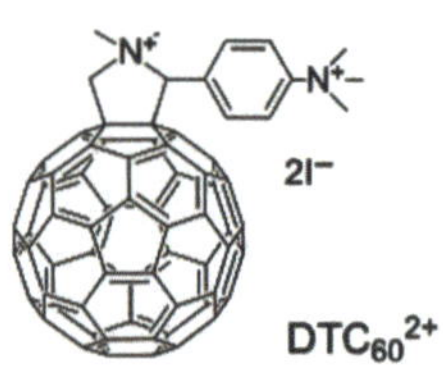

FIGURE 5.3 *N,N*-dimethyl-2-(4′-*N,N,N*-trimethylaminophenyl)fulleropyrrolidinium iodide (DTC(60)(2 +)) synthesized by 1,3-dipolar cycloaddition using 4-(*N,N*-dimethylamino) benzaldehyde, *N*-methylglycine, and fullerene C(60).

5.7.3 Fungi

Candida albicans is a particularly common fungal skin infection in both humans and other mammals. One group at particular risk for the infection are HIV-positive patients and cancer patients undergoing chemotherapy, and the recent emergence of drug-resistant fungal species is of great concern because without effective treatment, these patients have no effective defense against these pathogens. Again, PDT is being suggested as a possible treatment for these conditions, one that may be able to damage pathogens.

Recent analysis has identified that Candida is effectively killed by PDT utilizing either methylene blue or toluidine blue, although such rates of killing are lower when compared with other infectious organisms, such as those of *Staphylococcus* or *Streptococcus* species. A notable benefit of the treatment is that these forms of PDT have proven highly selective to the fungi, with much lower rates of keratinocyte damage resulting from its use than might be expected against other targets (Calzavara-Pinton et al., 2005).

Use of dicationic C_{60} fullerenes has also been explored as a PS for treatment of *C. albicans*, with encouraging results. According to Milanesio et al. (2013), use of the DTC_{60}^{2+} cationic fullerene against Candida species *in vitro* resulted in a 5-log decrease in cellular survival after 30 min of irradiation, representing "approximately 99.999%" of cellular inactivation from the treatment, with the Candida demonstrating growth delay and inactivation during the illumination.

5.8 IN VIVO STUDIES

As Tegos et al. (2005) stated, cationic fullerenes have demonstrated specificity for bacterial cells over mammalian cells during *in vitro* testing, although they have demonstrated some dark toxicity toward mammalian cells as well. Although an *in vitro* experiment, it nonetheless demonstrates the sound nature of *in vivo* PDT using fullerenes against bacteria and, as Hamblin et al. (2003) makes clear, the use of PDT to treat Gram-negative infections such as *P. aeruginosa* is clinically relevant at present, comprising 8% of all surgical wound infections and 10% of all bloodstream infections.

Use of a nonfullerene compound (pL−ce6) as a PS for this traditionally difficult to treat species was effectively reported *in vivo* with mice by Hamblin et al. (2003). During this study, a wound was made and infected with a possibly fatal strain of *Pseudomonas*. This proved fatal to the control group, as well as to the mice treated with only the PS, and those treated using light alone, after 24 to 60 h. By comparison, 90% of mice treated with the PS and light survived the treatment. Although these surviving mice did end up suffering the symptoms of a bacterial infection, they appeared to be fully recovered after 5 days.

In the case of *Pseudomonas*, as already mentioned, the use of the positive charge of cationic fullerenes can result in damage to the double cellular membrane through "self-promoted uptake." Sharma et al. (2011) mentioned that Lu et al. (2010) effectively demonstrated a similar treatment using fullerenes as the PS against bioluminescent Gram-negative bacteria, specifically, *P. aeruginosa* and *P. mirabilis*. Mice were infected with both of these species through an excisional wound before being treated with the tris-cationic fullerene buckyball BF-6 as a PS and 180 J/cm^2 of white broadband light (400−700 nm). In both cases, bioluminescence from the bacteria in the wounds was reduced when exposed to PDT; in mice infected with *P. mirabilis*, there was an 82% survival rate, as compared with a mere 8% without treatment. In the case of *Pseudomonas*, although the PDT alone did not result in an increase in survival, when combined with a "suboptimal" dose of tobramycin (6 mg/kg for 1 day) there was a 60% survival rate compared with 20% using the tobramycin alone. Therefore, it is indicative of the possibilities of fullerenes in the future treatment *in vivo* of *Pseudomonas*. However, compared with previous treatments using nonfullerene PS, there is still work to be done to optimize buckyballs for the purpose.

5.9 CONCLUSIONS

In conclusion, although the field of PDT has demonstrated a great deal of innovation and promise, substantial challenges remain, particularly lack of specificity and the concerning developments of resistant bacteria. To be an effective *in vivo* treatment, there are multiple fields to be effectively examined, such as optimization of the treatment itself, perfection of new methods of localizing the PS, and illuminating it once it is in position, as well as the ongoing search for the ideal PS. Buckyballs may pose some issues and are still a long way from being perfected, but their excellent properties *in vivo* do demonstrate that they remain one of the best hopes for such a PS, and the constant modification and remodification of their structures as well as the methods by which they can be localized to the site of infection are some of the most important uses of nanotechnology in the field of PDT.

Acknowledgments

Research in the Hamblin laboratory is supported by the US NIH grant R01Ai050875. We are grateful to Pinar Avci, MD, for editorial and scientific assistance.

References

Akaraphanth, R., Kanjanawanitchkul, W., Gritiyarangsan, P., 2007. Efficacy of ALA-PDT vs blue light in the treatment of acne. Photodermatol. Photoimmunol. Photomed. 23, 186–190.

Allison, R.R., Downie, G.H., Cuenca, R., Hu, X.H., Childs, C.J., Sibata, C.H., 2004. Photosensitizers in clinical PDT. Photodiagn. Photodyn. Ther.27–42.

Aschberger, K., Johnston, H.J., Stone, V., Aitken, R.J., Tran, C.L., Hankin, S.M., et al., 2010. Review of fullerene toxicity and exposure—appraisal of a human health risk assessment, based on open literature. Regul. Toxicol. Pharmacol. 58, 455–473.

Banerjee, I., Douaisi, M.P., Mondal, D., Kane, R.S., 2012. Light-activated nanotube-porphyrin conjugates as effective antiviral agents. Nanotechnology 23, 105101.

Bechmann, L.P., Hilgard, P., Frilling, A., Schumacher, B., Baba, H.A., Gerken, G., et al., 2008. Successful photodynamic therapy for biliary papillomatosis: a case report. World J. Gastroenterol. 14, 4234–4237.

Bonnett, R., Martinez, G., 2001. Photobleaching of sensitizers used in photodynamic therapy. Tetrahedron 57, 9513–9547.

Calzavara-Pinton, P.G., Venturini, M., Sala, R., 2005. A comprehensive overview of photodynamic therapy in the treatment of superficial fungal infections of the skin. J. Photochem. Photobiol. B. 78, 1–6.

Caminos, D.A., Spesia, M.B., Durantini, E.N., 2006. Photodynamic inactivation of *Escherichia coli* by novel meso-substituted porphyrins by 4-(3-*N,N,N*-trimethylammoniumpropoxy)phenyl and 4-(trifluoromethyl) phenyl groups. Photochem. Photobiol. Sci. 5, 56–65.

Dai, T., Huang, Y.Y., Hamblin, M.R., 2009. Photodynamic therapy for localized infections—state of the art. Photodiagn. Photodyn. Ther. 6, 170–188.

Ferreira, J.L., Lonne, M.N., Franca, T.A., Maximilla, N.R., Lugokenski, T.H., Costa, P.G., et al., 2014. Co-exposure of the organic nanomaterial fullerene C(6)(0) with benzo[a]pyrene in *Danio rerio* (zebrafish) hepatocytes: evidence of toxicological interactions. Aquat. Toxicol. 147, 76–83.

Fonseca, M.B., Junior, P.O., Pallota, R.C., Filho, H.F., Denardin, O.V., Rapoport, A., et al., 2008. Photodynamic therapy for root canals infected with *Enterococcus faecalis*. Photomed. Laser Surg. 26, 209–213.

Fu, X.J., Fang, Y., Yao, M., 2013. Antimicrobial photodynamic therapy for methicillin-resistant *Staphylococcus aureus* infection. Biomed. Res. Int. 159157.

Fu, X.J., Zhu, Y.Q., Peng, Y.B., Chen, Y.S., Hu, Y.P., Lu, H. X., et al., 2014. Enzyme activated photodynamic therapy for methicillin-resistant *Staphylococcus aureus* infection both in-vitro and *in vivo*. J. Photochem. Photobiol. B. 136, 72–80.

Garcez, A.S., Ribeiro, M.S., Tegos, G.P., Nunez, S.C., Jorge, A.O., Hamblin, M.R., 2007. Antimicrobial photodynamic therapy combined with conventional endodontic treatment to eliminate root canal biofilm infection. Lasers Surg. Med. 39, 59–66.

Hamblin, M.R., Mroz, P., 2008. Advances in Photodynamic Therapy: Basic, Translational, and Clinical. Artech House, Norwood, MA.

Hamblin, M.R., O'donnell, D.A., Murthy, N., Contag, C.H., Hasan, T., 2002. Rapid control of wound infections by targeted photodynamic therapy monitored by *in vivo* bioluminescence imaging. Photochem. Photobiol. 75, 51–57.

Hamblin, M.R., Zahra, T., Contag, C.H., Mcmanus, A.T., Hasan, T., 2003. Optical monitoring and treatment of potentially lethal wound infections *in vivo*. J. Infect. Dis. 187, 1717–1725.

Hirayama, J., Abe, H., Kamo, N., Shinbo, T., Ohnishi-Yamada, Y., Kurosawa, S., et al., 1999. Photoinactivation of vesicular stomatitis virus with fullerene conjugated with methoxy polyethylene glycol amine. Biol. Pharm. Bull. 22, 1106–1109.

Hongcharu, W., Taylor, C.R., Chang, Y., Aghassi, D., Suthamjariya, K., Anderson, R.R., 2000. Topical ALA-photodynamic therapy for the treatment of acne vulgaris. J. Invest. Dermatol. 115, 183–192.

Huang, L., Dai, T., Hamblin, M.R., 2010. Antimicrobial photodynamic inactivation and photodynamic therapy for infections. Methods Mol. Biol. 635, 155–173.

Ikeda, A., Nagano, M., Akiyama, M., Matsumoto, M., Ito, S., Mukai, M., et al., 2009. Photodynamic activity of C70 caged within surface-cross-linked liposomes. Chem. Asian. J. 4, 199–205.

Kasermann, F., Kempf, C., 1997. Photodynamic inactivation of enveloped viruses by buckminsterfullerene. Antiviral Res. 34, 65–70.

Kessel, D., 2014. Introduction to photodynamic therapy [online]. Detroit. Available from: <http://www.photobiology.info/Kessel.html>.

Kharkwal, G.B., Sharma, S.K., Huang, T.Y., Dai, T., Hamblin, M.R., 2011. Photodynamic therapy for infections: clinical applications. Lasers Surg. Med. 43, 755–767.

Krasnovsky Jr., AA, 2007. Primary mechanisms of photoactivation of molecular oxygen. History of development and the modern status of research. Biochemistry (Mosc). 72, 1065–1180.

Kudinova, N.V., Berezov, T.T., 2009. Photodynamic therapy: search for ideal photosensitizer. Biomed. Khim. 55, 558–569.

Lee, R.G., Vecchiotti, M.A., Heaphy, J., Panneerselvam, A., Schluchter, M.D., Oleinick, N.L., et al., 2010. Photodynamic therapy of cottontail rabbit papillomavirus-induced papillomas in a severe combined immunodeficient mouse xenograft system. Laryngoscope 120, 618–624.

Li, Z., Pan, L.L., Zhang, F.L., Wang, Z., Shen, Y.Y., Zhang, Z.Z., 2014. Preparation and characterization of fullerene (C60) amino acid nanoparticles for liver cancer cell treatment. J. Nanosci. Nanotechnol. 14, 4513–4518.

Liang, X., Wang, K.K., Zhu, T.C., 2013. Feasibility of interstitial diffuse optical tomography using cylindrical diffusing fibers for prostate PDT. Phys. Med. Biol. 58, 3461–3480.

Lieder, A., Khan, M.K., Lippert, B.M., 2014. Photodynamic therapy for recurrent respiratory papillomatosis. Cochrane Database Syst. Rev. 6, CD009810.

Lu, Z., Dai, T., Huang, L., Kurup, D.B., Tegos, G.P., Jahnke, A., et al., 2010. Photodynamic therapy with a cationic functionalized fullerene rescues mice from fatal wound infections. Nanomedicine (Lond). 5, 1525–1533.

Mei, X., Shi, W., Piao, Y., 2013. Effectiveness of photodynamic therapy with topical 5-aminolevulinic acid and intense pulsed light in Chinese acne vulgaris patients. Photodermatol. Photoimmunol. Photomed. 29, 90–96.

Milanesio, M.E., Spesia, M.B., Cormick, M.P., Durantini, E. N., 2013. Mechanistic studies on the photodynamic effect induced by a dicationic fullerene C60 derivative on *Escherichia coli* and *Candida albicans* cells. Photodiagn. Photodyn. Ther. 10, 320–327.

Mizuno, K., Zhiyentayev, T., Huang, L., Khalil, S., Nasim, F., Tegos, G.P., et al., 2011. Antimicrobial photodynamic therapy with functionalized fullerenes: quantitative structure–activity relationships. J. Nanomed. Nanotechnol. 2, 1–9.

Mroz, P., Pawlak, A., Satti, M., Lee, H., Wharton, T., Gali, H., et al., 2007a. Functionalized fullerenes mediate photodynamic killing of cancer cells: Type I versus Type II photochemical mechanism. Free Radic. Biol. Med. 43, 711–719.

Mroz, P., Tegos, G.P., Gali, H., Wharton, T., Sarna, T., Hamblin, M.R., 2007b. Photodynamic therapy with fullerenes. Photochem. Photobiol. Sci. 6, 1139–1149.

Muhammad, O.H., Chevalier, M., Rocca, J.P., Brulat-Bouchard, N., Medioni, E., 2014. Photodynamic therapy versus ultrasonic irrigation: Interaction with endodontic microbial biofilm, an *ex vivo* study. Photodiagn. Photodyn. Ther. 11, 171–181.

New York Times, 2014. Editorial: The Rise of Antibiotic Resistance. *New York Times*, p. 12. <http://www.nytimes.com/2014/05/11/opinion/sunday/the-rise-of-antibiotic-resistance.html?_r50>.

Oleinick, N.L., 2011. Basic Photosensitization [Online]. Cleveland. <http://www.photobiology.info/Oleinick.html>.

Probodh, I., Cramb, D.T., 2012. Two-photon excitation photodynamic therapy: working toward a new treatment for wet age-related macular degeneration. In: Ying, G.-S. (Ed.), Age Related Macular Degeneration The Recent Advances in Basic Research and Clinical Care. Intech Open.

Pushpan, S.K., Venkatraman, S., Anand, V.G., Sankar, J., Parmeswaran, D., Ganesan, S., et al., 2002. Porphyrins in photodynamic therapy—a search for ideal photosensitizers. Curr. Med. Chem. Anticancer Agents 2, 187–207.

Rijcken, C.J., Hofman, J.W., Van Zeeland, F., Hennink, W. E., Van Nostrum, C.F., 2007. Photosensitiser-loaded biodegradable polymeric micelles: preparation, characterization and *in vitro* PDT efficacy. J. Control. Release 124, 144–153.

Rocas, I.N., Siqueira Jr., J.F., Santos, K.R., 2004. Association of *Enterococcus faecalis* with different forms of periradicular diseases. J. Endod. 30, 315–320.

Sadasivam, M., Avci, P., Gupta, G.K., Lakshmanan, S., Chandran, R., Huang, Y.Y., et al., 2013. Self-assembled liposomal nanoparticles in photodynamic therapy. Eur. J. Nanomed. 5. Available from: <http://dx.doi.org/10.1515/ejnm-2013-0010>.

Sharma, S.K., Chiang, L.Y., Hamblin, M.R., 2011. Photodynamic therapy with fullerenes *in vivo*: reality or a dream?. Nanomedicine (Lond). 6, 1813–1825.

Shikowitz, M.J., Abramson, A.L., Freeman, K., Steinberg, B. M., Nouri, M., 1998. Efficacy of DHE photodynamic therapy for respiratory papillomatosis: immediate and long-term results. Laryngoscope. 108, 962–967.

Siddiqui, S.H., Awan, K.H., Javed, F., 2013. Bactericidal efficacy of photodynamic therapy against *Enterococcus faecalis* in infected root canals: a systematic literature review. Photodiagn. Photodyn. Ther. 10, 632–643.

Soukos, N.S., Chen, P.S.Y., Morris, J.T., Ruggiero, K., Abernethy, A.D., Som, S., et al., 2006. Photodynamic therapy for endodontic disinfection. J. Endod. 32, 979–984.

Sperandio, F.F., Huang, Y.Y., Hamblin, M.R., 2013. Antimicrobial photodynamic therapy to kill Gram-negative bacteria. Recent Pat. Antiinfect. Drug Discov. 8, 108–120.

Spesia, M.B., Milanesio, M.E., Durantini, E.N., 2008. Synthesis, properties and photodynamic inactivation of *Escherichia coli* by novel cationic fullerene C60 derivatives. Eur. J. Med. Chem. 43, 853–861.

Stephens, P., 2014. Call to discover new antibiotics to stop global crisis. <http://www.bbc.com/news/health-27323474>.

Tegos, G.P., Demidova, T.N., Arcila-Lopez, D., Lee, H., Wharton, T., Gali, H., et al., 2005. Cationic fullerenes are effective and selective antimicrobial photosensitizers. Chem. Biol. 12, 1127–1135.

Wang, H., Dong, C., Zhao, P., Wang, S., Liu, Z., Chang, J., 2014. Lipid coated upconverting nanoparticles as NIR remote controlled transducer for simultaneous photodynamic therapy and cell imaging. Int. J. Pharm. 466, 307–313.

Wang, M., Maragani, S., Huang, L., Jeon, S., Canteenwala, T., Hamblin, M.R., et al., 2013. Synthesis of decacationic [60] fullerene decaiodides giving photoinduced production of superoxide radicals and effective PDT-mediation on antimicrobial photoinactivation. Eur. J. Med. Chem. 63, 170–184.

Wong, T.W., Wang, Y.Y., Sheu, H.M., Chuang, Y.C., 2005. Bactericidal effects of toluidine blue-mediated photodynamic action on *Vibrio vulnificus*. Antimicrob. Agents Chemother. 49, 895–902.

Wyss, P., Tadir, Y., Tromberg, B.J., Haller, U. (Eds.), 2000. Photomedicine in Gynecology and Reproduction. Karger, Basel. <http://dx.doi.org/10.1159/000062800>.

Yin, H., Zhang, G., Chen, H., Wang, W., Kong, D., Li, Y., 2014a. Preliminary safety evaluation of photodynamic therapy for blood purification: an animal study. Artif. Organs 38, 510–515.

Yin, R., Lin, L., Xiao, Y., Hao, F., Hamblin, M.R., 2014b. Combination ALA-PDT and ablative fractional Er:YAG laser (2,940 nm) on the treatment of severe acne. Lasers Surg. Med. 46, 165–172.

Zeina, B., Greenman, J., Purcell, W.M., Das, B., 2001. Killing of cutaneous microbial species by photodynamic therapy. Br. J. Dermatol. 144, 274–278.

Zhang, Y., Yang, Y., Zou, X., 2013. Efficacy of 5-aminolevulinic acid photodynamic therapy in treatment of nasal inverted papilloma. Photodiagn. Photodyn. Ther. 10, 549–551.

Zhu, D., Larin, K.V., Luo, Q., Tuchin, V.V., 2013. Recent progress in tissue optical clearing. Laser Photon. Rev. 7, 732–757.

Zolfaghari, P.S., Packer, S., Singer, M., Nair, S.P., Bennett, J., Street, C., et al., 2009. *In vivo* killing of *Staphylococcus aureus* using a light-activated antimicrobial agent. BMC Microbiol. 9, 27. <http://dx.doi.org/10.1186/1471-2180-9-27>.

CHAPTER

6

Nonconventional Routes to Silver Nanoantimicrobials: Technological Issues, Bioactivity, and Applications*

Mauro Pollini[1], *Federica Paladini*[1], *Alessandro Sannino*[1], *Rosaria Anna Picca*[2], *Maria Chiara Sportelli*[2], *Nicola Cioffi*[2], *Maria Angela Nitti*[3], *Marco Valentini*[3] *and Antonio Valentini*[3]

[1]Department of Engineering for Innovation, University of Salento, Lecce, Italy [2]Department of Chemistry, University of Bari "Aldo Moro," Bari, Italy [3]Department of Physics "M. Merlin", University of Bari "Aldo Moro," Bari, Italy

6.1 INTRODUCTION

Silver-based antimicrobial nanomaterials have a significant impact on different fields and applications, such as modifiers for medical devices (e.g., implants, catheters, and wound dressings), additives in food packaging, air/water purification, and textiles, just to cite a few. There is huge academic interest and industrial interest toward these nanoantimicrobials derived from the high bioactivity because they generally exert against a broad spectrum of pathogens (see other chapters of the book for more details on this point) and show limited phenomena of bacterial resistance (Rai et al., 2012). Antimicrobial action of nanosized silver is not only due to the bioactivity of the metal itself, which has been known for centuries, but also due to the peculiar properties of nanostructures (NSs) (e.g., large surface area-to-volume ratio, different electronic and crystallographic structures, and surface defects, enhanced chemical reactivity), which greatly enhance nanomaterial action against dangerous microorganisms (Rai et al., 2009; Cioffi and Rai, 2012). Consequently, the number of

* This chapter was prepared by three research groups that equally contributed to the work. The first author listed for each institution represents the first author for his/her own research group. Mauro Pollini and Federica Paladini contributed equally to this chapter.

Nanotechnology in Diagnosis, Treatment and Prophylaxis of Infectious Diseases.
DOI: http://dx.doi.org/10.1016/B978-0-12-801317-5.00006-2

research papers related to this topic has been constantly growing, as demonstrated by the histogram reported in Figure 6.1, which shows the trend of publications in the time span from 2000 to 2013 ("Scifinder © Database," 2014). Among these reports, Ag nanoparticles (AgNPs) are generally synthesized by wet chemical reduction of a silver precursor in the presence of stabilizers or micelles (Marambio-Jones and Hoek, 2010), and nanostructured coatings/composites are often fabricated by plasma or sputtering physical approaches (Yuranova et al., 2003; Chen et al., 2008). Besides these common strategies that have been reviewed elsewhere (Xirouchaki and Palmer, 2004; Wiley et al., 2007; Krutyakov et al., 2008; Kumar, 2009; Cheruthazhekatt et al., 2010), this chapter presents a bird's eye view of alternative routes for the preparation of Ag nanoantimicrobials, which have been successfully developed in our research groups. In particular, electrochemical synthesis of AgNP colloids, ion beam sputtering, and photo-assisted depositions of AgNP-based coatings are reviewed. These strategies offer a high degree of versatility, ease of implementation, and industrial scalability. A brief overview of the most accepted bioactivity mechanisms and selected applications of silver nanoantimicrobials is also provided.

6.2 ION BEAM SPUTTERING DEPOSITION OF AgNP-BASED COATINGS

Innovative nanocoatings that confer desired characteristics to a treated surface are receiving increasing attention from the scientific community for a wide variety of real-life applications. In particular, in recent years research and technology have been constantly involved to develop antimicrobial coatings because of their high demand in medical applications as well as in health care and hygiene. In all these contexts it is crucial to exert strong inhibitory action against the growth of undesirable microorganisms through the use of bioactive substances, whose release properties can be modulated to achieve a concentration that is toxic to target organisms without being harmful to humans.

The recent increase of new resistant strains of bacteria to the most potent antibiotics has promoted research in the activity of silver-based compounds, including AgNPs, because of the well-known extraordinary antimicrobial properties of this metal (Olson et al., 2000; Balogh et al., 2001; Aymonier et al., 2002; Alt et al., 2004; Baker et al., 2005; Lee et al., 2005; Melaiye et al., 2005; Podsiadlo et al., 2005; Sun et al., 2005). AgNPs represent an interesting

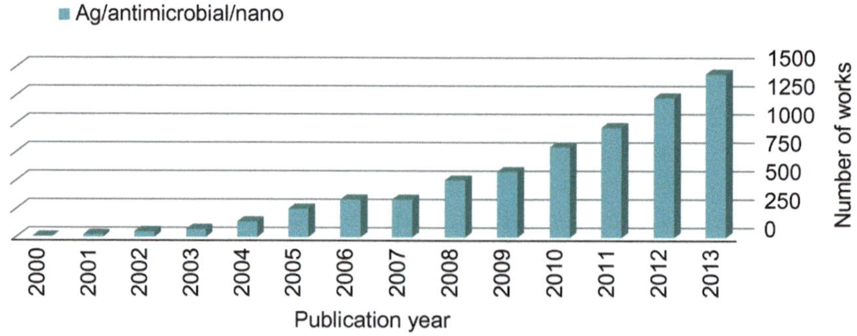

FIGURE 6.1 Number of works published during 2000–2013 using the reported key words in Scifinder © Database, 2014.

candidate for research as microbicides because of their effectiviness in small doses, minimal toxicity, and side effects (Lara et al., 2010, 2011).

As already mentioned, AgNP-based coatings with antibacterial properties have been successfully produced using surface modification techniques such as plasma deposition (Yuranova et al., 2003), RF magnetron sputtering (Asanithi et al., 2012; Herrera et al., 2013), and ion beam co-sputtering (IBS). Among these technologies, IBS is a low-cost deposition method that proves to be successful in the controlled deposition of thin nanoantimicrobial coatings in all those contexts in which wet impregnation of manufactured goods with chemical solvents are not feasible (Pollini et al., 2009; Giannossa et al., 2013).

Until now, the IBS of an inorganic target of metal (Au, Pd, Cu) or metal oxide (ZnO) and a poly-tetrafluoroethylene (PTFE) target have been successfully used in our research group for the production of new multifunctional coatings composed of inorganic NPs finely dispersed in a polymer matrix, thus combining the antimicrobial properties of NPs with the antistain ones of the PTFE matrix (Cioffi et al., 2002, 2003; Convertino et al., 2002, 2003; Farella et al., 2005; Sportelli et al., 2014). After our previous research on metal-fluoropolymer nanomaterials, our research group optimized the development of antimicrobial AgNP-based coatings on different substrates. Typically, AgNPs are dispersed in PTFE by simultaneously sputtering Ag and PTFE targets by Ar^+ ion beams at room temperature at a pressure of 10^{-2} Pa. Before any deposition run, the growth rates (r) of each component of the nanocomposite material are separately measured by means of a quartz microbalance sensor placed close to the substrate. The individual sputtering rates of each target, PTFE or Ag, therefore can be separately changed to modify the volume fraction (φ) of the silver phase into the organic film. The φ can be effectively set to any value between 0 and 1 by a proper combination of the sputtering conditions (ion-beam energy and current). The silver volume fraction can be fixed using the relation:

$$\varphi = \frac{V_{Ag}}{V_{PTFE} + V_{Ag}} = \frac{r_{Ag}}{r_{PTFE} + r_{Ag}} = \frac{r_{Ag}/r_{PTFE}}{1 + (r_{Ag}/r_{PTFE})} \tag{6.1}$$

where V_{PTFE} and V_{Ag} are the deposited volumes of PTFE and Ag, and r_{PTFE} and r_{Ag} are the deposition rates of the two materials experimentally measured for any deposition run and expressed as film thickness per second.

Another important parameter necessary to control the resulting composite is its density. The knowledge of this parameter is essential for the measurement in real time of the thickness of the deposited film using a quartz crystal microbalance placed in proximity of the substrate. Assuming the rate of deposition r uniform over the entire surface of the substrate, one obtains the relationship existing between the density of the composite ρ_{comp} and the volume fraction φ of silver included in it (Cioffi et al., 2005b):

$$\rho_{comp} = \rho_{PTFE} + \varphi\,(\rho_{Ag} - \rho_{PTFE}) \tag{6.2}$$

Typical IBS parameters for the controlled growth of Ag-fluorocarbon composites with the silver phase volume fraction φ in the interval $0 < \varphi < 0.15$ are summarized in Table 6.1. E_{Ag} and I_{Ag} are the energy and the current intensity of the Ar^+ ion beam bombarding the Ag target, respectively. An analogous notation is used to indicate the energy and current of the Ar^+ beam for the sputtering of the PTFE target. The ρ represents the density of the Ag-PTFE nanocomposite.

The definition of these characteristic parameters of IBS constitutes a real functionalization protocol applicable to any type of substrate materials to be treated with PTFE-silver at a higher value of silver loading φ. As a result, the IBS allows control of the

TABLE 6.1 IBS Parameters for Deposition of Ag-PTFE Nanocomposites, as a Function of the Silver Volume Fraction φ

Sample	φ	E_{PTFE} (V)	I_{PTFE} (mA)	E_{Ag} (V)	I_{Ag} (mA)	ρ_{comp} (g/cm^3)
PTFE	0	1000	40	/	/	2.20
PTFE-Ag (5%)	0.05	1000	40	350	20	2.52
PTFE-Ag (10%)	0.10	1000	40	400	25	2.94
PTFE-Ag (15%)	0.15	1000	40	500	30	3.36

percentage volume of AgNPs embedded in a fluoropolymer matrix, enabling the overcoating of each type of materials in an operating regime that prevents toxicity for humans. The combination of silver-polymer has the advantage that only the polymeric component is directly in contact with the skin, therefore simplifying the problems related to allergies or intolerances from contact with the metal that constitutes the nanoparticles. Moreover, the entrapment of NSs in the polymeric matrix prevents the direct release of the whole NP in the environment or in contact media, such as sweat or other biological fluids, thus avoiding problems related to NP environmental disseminaton and toxicity.

The morphological characterization by means of transmission electron microscopy (TEM) of the AgNP-based coatings produced by IBS at different silver loading φ highlights the uniform in-plane distribution of AgNPs in the composite and their excellent dimensional homogeneity.

6.3 PHOTO-ASSISTED DEPOSITION OF AgNP-BASED COATINGS

Antibacterial treatments based on the photo-assisted deposition of silver nanoparticles have been developed and patented in 2005 at the University of Salento in Italy (Pollini et al., 2005). In 2008, a spin-off company named Silvertech S.r.l. started the technology transfer and the industrial scale-up of the research and also explored many different fields of application. Some of the most interesting features of this technology are represented by the ease and the versatility of the process, thus allowing the treatment of a wide range of natural and synthetic materials. Over the course of the years, the process has been optimized regarding function of the specific nature and application of the materials. Advanced textiles, antibacterial biomedical devices such as catheters and wound dressings, polymeric air/water filters, and natural and synthetic leather for domestic and automotive applications are just some examples of application fields explored by the authors. For the specific application of the material and the function of the required antimicrobial properties, the technology can be adopted for silver deposition on different substrates through the definition of the most appropriate process parameters. In general, the process is based on the following procedure. The first step is the preparation of the silver solution containing silver nitrate as the precursor for metallic silver and methanol as both a solvent and a reducing agent. The percentage of the silver precursor is selected for each specific treatment according to the expected antibacterial capability and by evaluating the most convenient cost/effectiveness ratio. As a reducing agent, the presence of methanol in the silver solution is necessary to allow the reaction of photo-reduction; however, the percentage of methanol adopted can be selected proportionally to the silver content and can be reduced to low percentages for low silver contents. The presence of water can be suggested to reduce the cost of the silver solution when the percentage of silver solution is low enough to be reduced by low percentages of methanol. Then, the silver solution is deposited on the surface of the material through dip-coating or spray-coating according to the nature of the material. The deposition method is defined for the specific surface by evaluating the quality of the coating in terms of size and distribution of AgNPs. Immediately after the impregnation, the wet substrates are exposed to ultraviolet (UV) irradiation to induce *in situ* synthesis and deposition of the silver nanoparticles on the surface. Because of the presence of methanol and the UV radiation, the silver precursor is converted to elemental silver through the following photochemical reaction:

$$\begin{aligned} Ag_n + 2Ag^+ + CH_3OH + h\nu \rightarrow \\ Ag_{n+2} + 2H^+ + CH_2O \end{aligned} \tag{6.3}$$

The UV exposure time is defined by evaluating the degree of conversion from silver salt to silver coating. Thus, silver solutions containing higher amounts of silver nitrate require the presence of higher percentages of methanol and longer UV exposure time. The success of the photo-reduction is evaluated through thermo-gravimetric analysis (TGA) by calculating the amount of silver initially deposited and amount of silver still present on the substrate after repeated washings.

Once defined the process parameters for each substrate, such as chemical composition of the silver solution, amount and deposition method of the silver solution, and UV exposure time, the silver coating exhibits strong adhesion to the substrate and long-term antimicrobial properties.

6.4 ELECTROCHEMICAL METHODS FOR NANOMATERIAL SYNTHESIS

6.4.1 Fundamentals of the Electrocrystallization Process

Electrochemical procedures for the synthesis of Ag nanomaterials, although less diffused than chemical and physical routes, represent a powerful tool for preparing differently shaped NPs over a wide range of metals and compounds (Fedlheim and Foss, 2001; Djokić, 2012; Magagnin and Cojocaru, 2012). One of the advantages of these methods reside in the strong theoretical background that exists regarding the fundamentals of (nano)-metals electrosynthesis. A concise description of nanoparticles nucleation and growth in the electrocrystallization process is provided.

Any electrode process that leads to the formation of a solid phase is generically defined as "electrocrystallization" (Plieth, 1985; Pletcher et al., 2001). In the case of the electrochemical reduction of metallic ions, forming a metal layer on the electrode surface, the process can be divided into seven distinct steps (Plieth,1985; Pletcher et al., 2001):

1. Metal ions diffuse from the solution to the electrode surface
2. Electron transfer
3. Loss of the solvation shell, with consequent formation of metal adsorbed atoms (ad-atoms)
4. Diffusion of ad-atoms on the electrode surface
5. Assembling of ad-atoms into clusters as critical nuclei
6. Integration of ad-atoms at lattice sites
7. Growth of crystalline deposits.

The heterogeneous process that occurs at the electrode/solution interface depends on the nature and properties of the substrate. In the following, the deposited cluster nuclei are assumed spherical (or quasi-spherical) and, consequently, their contact area with the substrate becomes negligible. In the electrochemical production of a solid phase, the equilibrium condition is reached by electron transfer (Pletcher et al., 2001). For the general process (Eq. (6.4)):

$$\mathrm{Me}^{n+} + ne^- \rightarrow \mathrm{Me} \quad (6.4)$$

the equilibrium potential, defined as $E(\mathrm{eq})$, is calculated from the Nernst equation, and the equilibrium surface activity of the ad-atoms is defined as $a_{\mathrm{Me}_{\mathrm{ads(eq)}}}$. If the electrode potential is changed by a quantity defined as the *overpotential* (η), the new electrode potential is $(E_{(\mathrm{eq})} + \eta)$ and the surface ad-atoms activity is $a_{\mathrm{Me}_{\mathrm{ads}}}$, whose value can be determined by the Nernst equation (Pletcher et al., 2001) (Eq. (6.5)):

$$\frac{a_{\mathrm{Me}_{\mathrm{ads}}}}{a_{\mathrm{Me}_{\mathrm{ads}}(\mathrm{eq})}} = \exp(-nF\eta/RT) \quad (6.5)$$

Hence, the free energy per volume (V) unit, related to the solid phase formation, is (Eq. (6.6)):

$$\Delta G_{\mathrm{V}} = nF\eta/V \quad (6.6)$$

where F is the Faraday constant, V is the molar volume of the solid phase, and η is (for a cathodic process) a negative number. V can be replaced by the M/ρ ratio, where M is the metal atomic mass and ρ is the metal density (Eq. (6.7)):

$$\Delta G_{\mathrm{V}} = nF\rho/M \quad (6.7)$$

The total free energy of formation relevant to a spherical nucleus (Eq. (6.8)) is given by two contributors, a bulk term and a surface term (Pletcher et al., 2001):

$$\Delta G_{\mathrm{tot}} = \Delta G_{\mathrm{bulk}} + \Delta G_{\mathrm{surf}} \quad (6.8)$$

Assuming the formation of spherical particles and defining γ as the surface free energy, Eq. (6.8) becomes:

$$\Delta G_{\mathrm{tot}} = 4\pi r^3 nF\eta\rho/3M + 4\pi r^2\gamma \quad (6.9)$$

For metal electro-reduction, η assumes negative values, and this makes it evident that *bulk* and *surface* terms provide opposite contributions to the total nucleation free energy ΔG_{tot}, which reaches a maximum at specific values of r and η. ΔG_{tot} and r values relevant to this maximum are defined as *critical* parameters of the nucleation process (ΔG_c and r_c). r_c represents the minimum radius that a nucleus must have to evolve into a stable entity; ΔG_c can be considered as the free activation energy of the nucleation process.

Nevertheless, this discussion has intrinsic limitations because of the simplifications involved. First, nucleation process is assumed to not be influenced by the nucleus/electrode interface; this ideal situation does not correspond to most real cases in which nuclei growth is strongly influenced by the substrate chemistry. Furthermore, in this simplified model, the thermodynamic properties of nanoparticles are supposed to be equal to those of bulk materials. Finally, Eq. (6.8), which divides the nucleus free energy of formation into two terms, is based on the assumption that discontinuity exists in the atomic environment at the interface and that the two terms are totally independent. However, the thickness of the surface layer involved in the phase formation (that is influenced by both surface and bulk parameters) could be large enough to sensibly alter calculations above all those relevant to clusters composed of less than a few hundred atoms. Despite these limitations, this simple model provides a useful qualitative description of the nucleation process and several renowned groups have frequently used it as a basis to interpret important experimental evidence about electro-produced nanoparticles.

6.4.2 Electrochemical Synthesis of Colloidal AgNPs

Electrochemical techniques can be proficiently used to produce metal nanoparticles with high purity and crystallinity, controlling particle size by means of a few experimental parameters (Khaydarov et al., 2009). The main advantages of the electrochemical methods lie in the high purity of particles and the possibility of controlling NP size by simple tuning of current density (or potential) without the need for expensive equipment or vacuum methods (Khaydarov et al., 2009). Reetz and Helbig (1994) were the first to develop an electrochemical process known as sacrificial anode electrolysis (SAE), for the synthesis of transition metal NPs in a colloidal form. In SAE, a bulk metal sheet is anodically dissolved in an electrolytic solution (typically a mixture of acetonitrile [ACN] and tetrahydrofuran [THF]) in the presence of tetraalkylammonium salts able to stabilize in solution the metal clusters initially formed at the cathode by reduction of metal ions. In 2000, this strategy was successfully adopted for the electrochemical synthesis of AgNPs by Rodríguez-Sánchez and co-workers (2000) and by Cioffi et al. (2000). Rodríguez-Sánchez and co-workers (2000) were the first who demonstrated that ACN/THF mixtures were unsuitable for Ag synthesis, because the presence of THF induced particles aggregation. Hence, pure ACN containing tetraalkylammonium bromides was adopted as an electrolytic bath for this specific process. Following a similar approach, AgNPs were also obtained in ethanol (Starowicz et al., 2006; Chulovskaya et al., 2009), hexane (Jian et al., 2005), thioglycolic acid (Rabinal et al., 2013), *N,N*-dimethylformamide (Rabinal et al., 2013), and dimethyl sulfoxide (Wadkar et al., 2006). Interestingly, none of these works reported the potential application of such NPs as antimicrobials. In our research group, based on the successful development of Cu nanocolloids obtained using a similar SAE strategy (Cioffi et al., 2004), we optimized the synthesis of AgNPs for the preparation of spin-deposited low-cost bioactive nanocomposites (Cioffi et al., 2005a). We showed not only that the octyl-chain of the used quaternary ammonium salt (tetraoctylammonium bromide [TOAB]) was particularly effective for NP stabilization and size control (Figure 6.2) but also that the capping agent might provide a beneficial synergistic bioactivity, supporting the effects of the Ag nanophase.

Moreover, the presence of a stabilizer layer allowed a controlled release of ions into culture broths, as demonstrated in previous studies (Cioffi et al., 2004), without direct leaching of NPs and consequent occurrence of nanotoxicology phenomenon. The influence of stabilizers on NP morphology was also explored in

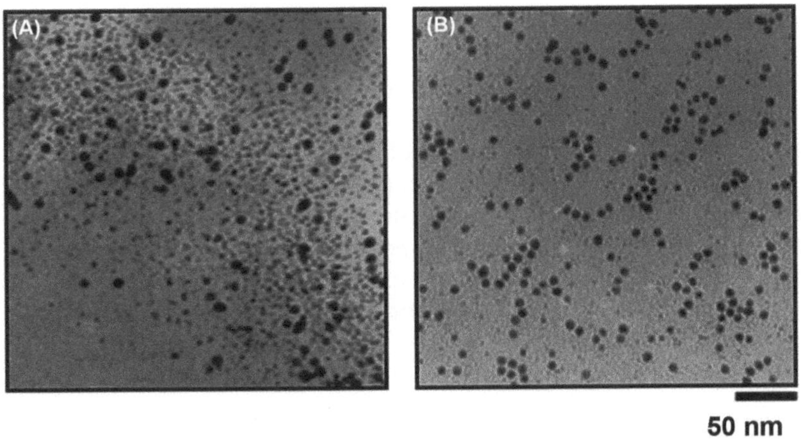

FIGURE 6.2 TEM images of AgNPs electrochemically prepared in the presence of TOAB as a stabilizer at +0.5 V (A) and at +1.5 V (B). *Adapted from Cioffi et al. (2005a), Figure 6.1. Copyright (2005), with kind permission from Springer Science and Business Media.*

subsequent work in which the use of a zwitterionic surfactant in ACN caused the growth of branched hierarchical nanostructures (nanofractals) (Cioffi et al., 2009). Despite many advantages, such as size tuning and cheap experimental set-up, this method has its main drawback in the use of organic solvents, somehow limiting real-life applications of the as-prepared colloids (Lee and Syu, 2010; Huang et al., 2012). As a result, several authors have proposed the preparation of aqueous and long-lived colloids (Yin et al., 2003; Khaydarov et al., 2009; Lee and Syu, 2010) to reduce the impact of toxic chemicals and solvents. A very elegant example was presented by Penner's group, who prepared silver (I) oxide colloids with tunable NP morphology by anodizing a sacrificial silver wire in sulfate aqueous solutions at pH of 12 (Murray et al., 2005). Ten Kortenaar et al. (1999) found that long-lived, subnanometer-sized silver clusters could be prepared by anodic dispersion of a silver electrode in basic aqueous solutions (pH 10.5–12) free of stabilizing polymers on the application of a high-DC voltage (65 V) between two silver electrodes. Unfortunately, this process is energy-consuming and has some limits in terms of morphological control on the produced particles.

Regarding the use of surfactants and other capping agents in the electrochemical reduction of silver ions, it should be noted that this electrolysis involves two competitive processes: the formation of AgNPs and the deposition of a silver film onto the cathode surface (see paragraph 6.4.1). Thermodynamics show that the second occurrence should dominate the first one, hence the need for good stabilizers (Ma et al., 2004). Typical stabilizing agents are both synthetic and natural substances, such as poly(*N*-vinylpyrrolidone) (Yin et al., 2003; Ma et al., 2004; Jovanović et al., 2013), and in combination with co-stabilizers (Petica et al., 2008), poly(amide-hydroxyurethane) (Marius et al., 2011), polyphenylpyrrole (Johans, 2002), polyethylene-glycol (Zhu et al., 2001; Roldán et al., 2013), Pluronic f127® (Blandon et al., 2012), saccharides (Panáček et al., 2006), and chitosan (CS) (Reicha et al., 2012). Supported metal NSs, (i.e., electrochemically deposited onto different substrates) are the most developed nanomaterials in electrochemistry. One of the first examples of supported Ag NS was provided by Zoval et al. (1996), who used a potentiostatic pulsed method to deposit ordered silver NPs onto atomically flat graphite. The intrinsic flexibility of this technique is responsible for its rapid technological development: many different substrates, like metal sheets (Tang et al., 2009; Yuan et al., 2011; Guan and Wang, 2012) (Figure 6.3) or graphene and graphite (Tang et al., 2001; Mazur, 2004; Fang et al., 2010; Zhong et al., 2013), were used to deposit variously shaped NS. Direct embedding of NPs in polymeric matrices can be also performed in a one-pot approach using this technique. Recently, Zhitomirsky and Hashambhoy (2007) proposed a deposition of inorganic AgNPs contextually with a CS film, paving the way for a fast and controllable preparation of cheap and antimicrobial films for biomedical applications.

6.5 OVERVIEW OF THE MOST WIDELY ACCEPTED BIOACTIVITY MECHANISMS

A concise overview of the recent advances in the understanding of the biocidal action mechanisms of silver nanoparticles is provided in this paragraph, whereas for more specialistic information, the reader can refer to other sections of this book. Antimicrobial properties of AgNPs make them useful in the treatment and prophylaxis of infections (because of bacteria, fungi, viruses, and protozoa), although their anti-arthropod properties can be used in controlling the spread of infections by affecting arthropod vectors (Rai et al., 2009, 2014).

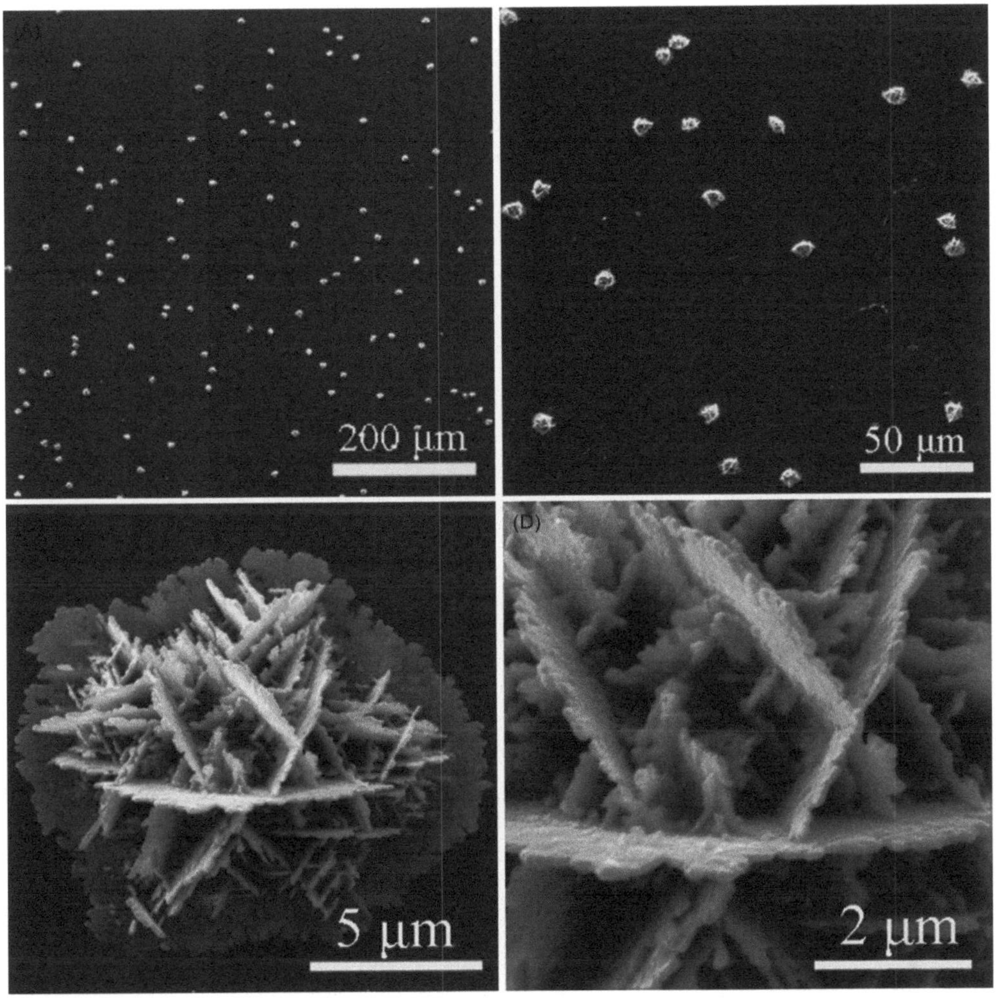

FIGURE 6.3 Typical SEM images at increasing magnifications of flower-like silver microparticles with two-dimensional nanoflakes as building blocks electrochemically deposited on a platinum film electrode. *Reproduced from Tang et al. (2009). Copyright (2009), with permission from Elsevier.*

The antibacterial properties of AgNPs, exploited in many practical applications, are not completely well-known. For a long time, the release of silver ions (Ag^{+}) was thought to be the only bioactive mechanism of silver-based coatings, without any direct "particle-specific" effects exerted by AgNPs (Xiu et al., 2012). The Ag^{+} release inhibits respiratory enzymes and also generates reactive oxygen species (ROS), which are extremely deleterious to cells and cause damage to lipids or DNA (Pal et al., 2007). Currently, at least other two main mechanisms have been proposed to describe their antibacterial activity: (i) direct interaction of AgNPs with the bacterial cell membrane, causing subsequent membrane damage and complexation with components located inside the cells (Sondi and Salopek-Sondi, 2004) and (ii) interaction with thiol ($-SH$)

groups and production of ROS (Banerjee et al., 2010). Many authors report on the size-dependent activity of AgNPs, with smaller NPs having higher activity, probably because of a relative increase in contact surface area and surface chemical reactivity (Zaporojtchenko et al., 2010; Shameli et al., 2012).

The antifungal activity of AgNPs has been less studied compared with the antibacterial activity (70 with respect to 590 published articles, respectively), so a precise bioactivity mechanism of silver nanoparticles against fungal pathogens has not been indicated. Some authors report on significant antifungal activity of AgNPs against *Penicillium citrinum*, *Aspergillus niger*, *Trichophyton mentagrophytes*, and *Candida albicans* (Zhang, 2008; Kim et al., 2008, 2009). One possible explanation is the destruction of fungal membrane and inhibition of the normal budding process in yeasts, with subsequent other effects similar to the mechanisms depicted for bacteria. AgNPs have also proven to be active against a broad range of viruses, including human imunodeficiency virus (HIV), hepatitis B virus, herpes simplex virus, respiratory syncytial virus, and monkey pox virus, with a lower possibility of developing resistance compared with conventional antivirals (Galdiero et al., 2011). However, a precise mechanism of AgNP antiviral activity and an exact stage of infection at which AgNPs exert their antiviral properties have yet to be determined. Moreover, further studies are necessary to clarify the exact site of interaction and how to modify the nanoparticle surface characteristics for a broader and more effective use. In the case of antiviral action of AgNPs against HIV, direct interaction with viral surface glycoproteins and interactions with the viral genome (DNA or RNA) into the cell could be hypothesized as mechanisms. In particular, size-dependent interaction of silver nanoparticles with HIV-1 was found, resulting in NPs within the range of 1–10 nm only being able to bind to the virus (Elechiguerra et al., 2005). More details regarding this type of antivirial activity of AgNPs can be found in the dedicated chapter. Similarly, influenza virus can be efficiently inhibited by AgNPs with an average size of 10 nm (Xiang et al., 2011).

Despite the fact that many research works report anti-anthropod activity of AgNPs, further studies are necessary to reveal their precise mechanism of action as acaricide, larvicide, and lousicide.

Broad-spectrum bioactivities of AgNPs make them promising agents for tumors, with particular research interest in the treatment of different types of drug-resistant cancer cells.

6.6 OVERVIEW OF THE MOST PROMISING APPLICATIONS

The emergence of bacterial resistance to multiple classes of antibiotics has led to a growing interest in the antimicrobial properties of silver nanoparticles in a wide range of domestic and clinical products (Sotiriou and Pratsinis, 2011; Mohammadzadeh and Ahmadiyan, 2013; Mulley et al., 2014). The research on medical applications of nanosilver has been extremely active and great attention has been recently addressed regarding hospital-acquired infections (Chaloupka et al., 2010). Despite advances in preventive care and antimicrobial therapy, nosocomial infections associated with indwelling devices still represent a serious concern in hospitals (Sampathkumar et al., 2011; Sousa et al., 2011; Chaudhury et al., 2012; Parker and Doebbeling, 2012; Esposito et al., 2013; Nayeemuddin et al., 2013; Frykholm et al., 2014; Lombardi et al., 2014; Paredes et al., 2014; Silva et al., 2014; Stickler, 2014).

Nanotechnology approaches are considered promising for preventing clinical infection (Monzillo et al., 2012; Mansouri et al., 2013; Zhang et al., 2013). Silver-based antimicrobials and AgNPs have been proposed for catheter impregnation and resulted in effective prevention of early bacterial colonization (Beattie and Taylor, 2011; Agarwala et al., 2014). Since 1999, numerous studies of the incorporation of AgNPs in the catheter matrix have been published, wherein the substantial reduction of bacterial colonization has been assessed (Böswald et al., 1999; Hentschel and Münstedt, 1999; Joyce-Wöhrmann and Münstedt, 1999; Martus et al., 1999; Rösch and Lugauer, 1999; Schoerner et al., 1999; Samuel and Guggenbichler, 2004). More recently, the research performed at the University of Salento (Lecce, Italy) demonstrated the possibility to deposit nanosilver on both the luminal and the outer surface of catheters by adopting the photo-assisted deposition of silver nanoparticles. The obtained results indicated the impressive reduction in bacteria biofilm proliferation, no significant cytotoxic effect on fibroblasts, and very low values of silver ion release even in contact with biological fluids for extended periods (Pollini et al., 2011; Paladini et al., 2012, 2013b).

Silver coatings have also been considered attractive for deposition on other medical device surfaces such as implants and wound dressings (Taheri et al., 2014). Implant-associated bacterial infection is still one of the most serious complications in orthopedic surgery. The incorporation and deposition of silver nanoparticles on titanium implants and screws were demonstrated effective in preventing bacteria biofilm adhesion and infections (Secinti et al., 2011; Cheng et al., 2013). The use of silver nanoparticles has also been explored for the production of antimicrobial bone graft (Marsich et al., 2013) and in orthodontia, where antifungal acrylic resins containing AgNPs were developed as antifungal compounds for denture bases (Acosta-Torres et al., 2012; Nam et al., 2012; Sodagar et al., 2012). A wide range of silver dressings with broad-spectrum antimicrobial activity is commercially available for topical application and new wound dressing biomaterials have also been proposed in the literature for the management of exuding venous leg ulcers for bioburden and infection reduction (White, 2013; Forlee et al., 2014). The application of dressings based on silver nanoparticles was demonstrated effective on the restoration of the normal skin (Rigo et al., 2013). Nanobiocomposites prepared by embedding silver nanoparticles in a polymeric matrix were demonstrated effective in promoting the wound healing by modulation of collagen deposition (Ghosh Auddy et al., 2013). Cotton fabrics treated with AgNPs also demonstrated wound-contracting ability that was significantly higher than that for untreated cotton fabrics and can be considered promising for the production of smart textiles for medical purposes and other biological fields (Hebeish et al., 2014). The silver deposition treatments described in this chapter have been extensively applied to textile substrates for biomedical application and for the development of advanced textiles with improved comfort. Natural and synthetic fibers deposited with silver nanoparticles demonstrated long-lasting antifungal and antibacterial capabilities on many micro-organisms, no skin irritation effect *in vivo*, excellent adhesion of the coating to the fibers, and durability in laundry (Paladini et al., 2014c). Silver-coated cotton and flax substrates have been developed as effective wound dressing biomaterials for the prevention of wound infections (Paladini et al., 2013a, 2014b); silver-treated textiles have also been proposed for other biomedical applications such as the production of antimicrobial bed linens for the prevention of cross-transmission among patients (Paladini et al., 2014c). The transmission of infectious diseases can also be associated with public places such as waiting rooms, schools, or public transportation, where hand-touch surfaces can easily get colonized by bacteria and fungi. For this purpose, silver-treated leather with durable antibacterial properties has been proposed by the authors for application in the public transport systems (Pollini et al., 2013), and silver-treated polymeric foams also demonstrated strong efficacy against pathogen strains responsible for respiratory diseases associated with the contamination of air filtration units in domestic and public places (Paladini et al. 2014a). Filters able to remove potential pathogens also have significant medical applications (Islam et al., 2013), so filters containing silver have been proposed for tap water purification, wastewater and groundwater treatment, and other bio-related applications (Vonberg et al., 2008; Dankovich and Gray, 2011; Nowack et al., 2011; Mpenyana-Monyatsi et al., 2012; Pathak and Gopal, 2012; Seo et al., 2012; Yuan et al., 2013). Other promising applications of AgNPs described in recent research works include those of the food industry (as packing material) and those of the marine industry (as antibiofouling material) (Gottesman et al., 2011; Song et al., 2011; Echegoyen and Nerín, 2013; Inbakandan et al., 2013; Liu et al., 2014).

6.7 CONCLUSIONS AND FUTURE PERSPECTIVES

Nonconventional procedures for the size-controlled preparation of silver nanoantimicrobials, such as electrolysis, ion beam sputtering, photo-deposition, have been reviewed in this chapter. Fundamentals and experimental details of these uncommon procedures have been explicated for the material scientist interested in reproducing the process. Several case studies and selected applications have been reported, as well as a very concise overview of the most diffused bioactivity mechanisms of Ag nanoantimicrobials.

The reviewed techniques offer several advantages (i.e., industrial scaling-up) that make them appealing for the reader interested in the reproducible and cost-effective development of Ag nanomaterials for real-life applications. The nonconventional methods reviewed here could be adopted to develop other efficient nanoantimicrobials based on (or in combination with) alternative bioactive materials to widen the range of syntheses available to obtain nanomaterials with enhanced activity against pathogens.

However, toxicology studies should be considered in the future to correlate the different physicochemical and biological properties of the proposed silver-based antimicrobials with the exerted bioactivity and potential toxicity to humans and the environment. The development of hybrid nanostructures based on supported and well-confined Ag nanophases, in place of barely dispersed AgNPs, could reduce exposure risks.

References

Acosta-Torres, L.S., Mendieta, I., Nuñez-Anita, R.E., Cajero-Juárez, M., Castaño, V.M., 2012. Cytocompatible antifungal acrylic resin containing silver nanoparticles for dentures. Int. J. Nanomed. 7, 4777–4786. Available from: http://dx.doi.org/10.2147/IJN.S32391.

Agarwala, M., Barman, T., Gogoi, D., Choudhury, B., Pal, A.R., Yadav, R.N.S., 2014. Highly effective antibiofilm coating of silver-polymer nanocomposite on polymeric medical devices deposited by one step plasma process. J. Biomed. Mater. Res. B Appl. Biomater. 102, 1223–1235. Available from: http://dx.doi.org/10.1002/jbm.b.33106.

Alt, V., Bechert, T., Steinrücke, P., Wagener, M., Seidel, P., Dingeldein, E., et al., 2004. An *in vitro* assessment of the antibacterial properties and cytotoxicity of nanoparticulate silver bone cement. Biomaterials 25, 4383–4391. Available from: http://dx.doi.org/10.1016/j.biomaterials.2003.10.078.

Asanithi, P., Chaiyakun, S., Limsuwan, P., 2012. Growth of silver nanoparticles by DC magnetron sputtering. J. Nanomater. 2012, e963609. Available from: http://dx.doi.org/10.1155/2012/963609.

Aymonier, C., Schlotterbeck, U., Antonietti, L., Zacharias, P., Thomann, R., Tiller, J.C., et al., 2002. Hybrids of silver nanoparticles with amphiphilic hyperbranched macromolecules exhibiting antimicrobial properties. Chem. Commun. 3018–3019. Available from: http://dx.doi.org/10.1039/B208575E.

Baker, C., Pradhan, A., Pakstis, L., Pochan, D.J., Shah, S.I., 2005. Synthesis and antibacterial properties of silver nanoparticles. J. Nanosci. Nanotechnol. 5, 244–249. Available from: http://dx.doi.org/10.1166/jnn.2005.034.

Balogh, L., Swanson, D.R., Tomalia, D.A., Hagnauer, G.L., McManus, A.T., 2001. Dendrimer – silver complexes and nanocomposites as antimicrobial agents. Nano Lett. 1, 18–21. Available from: http://dx.doi.org/10.1021/nl005502p.

Banerjee, M., Mallick, S., Paul, A., Chattopadhyay, A., Ghosh, S.S., 2010. Heightened reactive oxygen species generation in the antimicrobial activity of a three component iodinated chitosan – silver nanoparticle composite. Langmuir 26, 5901–5908. Available from: http://dx.doi.org/10.1021/la9038528.

Beattie, M., Taylor, J., 2011. Silver alloy vs. uncoated urinary catheters: a systematic review of the literature. J. Clin. Nurs. 20, 2098–2108. Available from: http://dx.doi.org/10.1111/j.1365-2702.2010.03561.x.

Blandon, L., Vazquez, M.V., Benjumea, D.M., Ciro, G., 2012. Electrochemical synthesis of silver nanoparticles and their potential use as antimicrobial agent: a case study on *Escherichia coli*: port. Electrochim. Acta 30, 135–144. Available from: http://dx.doi.org/10.4152/pea.201202135.

Böswald, M., Lugauer, S., Regenfus, A., Braun, G.G., Martus, P., Geis, C., et al., 1999. Reduced rates of catheter-associated infection by use of a new silver-impregnated central venous catheter. Infection 27, S56–S60. Available from: http://dx.doi.org/10.1007/BF02561616.

Chaloupka, K., Malam, Y., Seifalian, A.M., 2010. Nanosilver as a new generation of nanoproduct in biomedical applications. Trends Biotechnol. 28, 580–588. Available from: http://dx.doi.org/10.1016/j.tibtech.2010.07.006.

Chaudhury, A., Rangineni, J., Venkatramana, B., 2012. Catheter lock technique: *in vitro* efficacy of ethanol for eradication of methicillin-resistant staphylococcal biofilm compared with other agents. FEMS Immunol. Med. Microbiol. 65, 305–308. Available from: http://dx.doi.org/10.1111/j.1574-695X.2012.00950.x.

Chen, Q., Yue, L., Xie, F., Zhou, M., Fu, Y., Zhang, Y., et al., 2008. Preferential facet of nanocrystalline silver embedded in polyethylene oxide nanocomposite and its antibiotic behaviors. J. Phys. Chem. C. 112, 10004–10007. Available from: http://dx.doi.org/10.1021/jp800306c.

Cheng, H., Li, Y., Huo, K., Gao, B., Xiong, W., 2013. Long-lasting *in vivo* and *in vitro* antibacterial ability of nanostructured titania coating incorporated with silver nanoparticles. J. Biomed. Mater. Res. A. Available from: http://dx.doi.org/10.1002/jbm.a.35019.

Cheruthazhekatt, S., Černák, M., Slavíček, P., Havel, J., 2010. Gas plasmas and plasma modified materials in medicine. J. Appl. Biomed. 8, 55–66. Available from: http://dx.doi.org/10.2478/v10136-009-0013-9.

Chulovskaya, S.A., Garas'ko, E.V., Parfenyuk, V.I., 2009. Electrochemical preparation and properties of ultradisperce silver powder. Russ. J. Appl. Chem. 82, 1396–1400. Available from: http://dx.doi.org/10.1134/S107042720908014X.

Cioffi, N., Rai, M., 2012. Nano-antimicrobials: Progress and Prospects. Springer-Verlag, Berlin Heidelberg.

Cioffi, N., Torsi, L., Sabbatini, L., Zambonin, P.G., Bleve-Zacheo, T., 2000. Electrosynthesis and characterisation of nanostructured palladium–polypyrrole composites. J. Electroanal. Chem. 488, 42–47.

Cioffi, N., Farella, I., Torsi, L., Valentini, A., Tafuri, A., 2002. Correlation between surface chemical composition and vapor sensing properties of gold-fluorocarbon nanocomposites. Sens. Actuators B Chem. 84, 49–54. Available from: http://dx.doi.org/10.1016/S0925-4005(01)01073-5.

Cioffi, N., Farella, I., Torsi, L., Valentini, A., Sabbatini, L., Zambonin, P.G., 2003. Ion-beam sputtered palladium-fluoropolymer nano-composites as active layers for organic vapours sensors. Sens. Actuators B Chem. 93, 181–186. Available from: http://dx.doi.org/10.1016/S0925-4005(03)00182-5.

Cioffi, N., Torsi, L., Ditaranto, N., Sabbatini, L., Zambonin, P.G., Tantillo, G., et al., 2004. Antifungal activity of polymer-based copper nanocomposite coatings. Appl. Phys. Lett. 85, 2417–2419.

Cioffi, N., Ditaranto, N., Torsi, L., Picca, R.A., De Giglio, E., Sabbatini, L., et al., 2005a. Synthesis, analytical characterization and bioactivity of Ag and Cu nanoparticles embedded in poly-vinyl-methyl-ketone films. Anal. Bioanal. Chem. 382, 1912–1918. Available from: http://dx.doi.org/10.1007/s00216-005-3334-x.

Cioffi, N., Ditaranto, N., Torsi, L., Picca, R.A., Sabbatini, L., Valentini, A., et al., 2005b. Analytical characterization of bioactive fluoropolymer ultra-thin coatings modified by copper nanoparticles. Anal. Bioanal. Chem. 381, 607–616. Available from: http://dx.doi.org/10.1007/s00216-004-2761-4.

Cioffi, N., Colaianni, L., Pilolli, R., Calvano, C., Palmisano, F., Zambonin, P., 2009. Silver nanofractals: electrochemical synthesis, XPS characterization and application in LDI-MS. Anal. Bioanal. Chem. 394, 1375–1383. Available from: http://dx.doi.org/10.1007/s00216-009-2820-y.

Convertino, A., Valentini, A., Bassi, A., Cioffi, N., Torsi, L., Cirillo, E.N.M., 2002. Effect of metal clusters on the swelling of gold–fluorocarbon–polymer composite films. Appl. Phys. Lett. 80, 1565–1567. Available from: http://dx.doi.org/10.1063/1.1448858.

Convertino, A., Capobianchi, A., Valentini, A., Cirillo, E.N. M., 2003. A new approach to organic solvent detection: high-reflectivity bragg reflectors based on a gold nanoparticle/teflon-like composite material. Adv. Mater. 15, 1103–1105. Available from: http://dx.doi.org/10.1002/adma.200304777.

Dankovich, T.A., Gray, D.G., 2011. Bactericidal paper impregnated with silver nanoparticles for point-of-use water treatment. Environ. Sci. Technol. 45, 1992–1998. Available from: http://dx.doi.org/10.1021/es103302t.

Djokić, S.S. (Ed.), 2012. Electrochemical Production of Metal Powders, Modern Aspects of Electrochemistry. Springer US, Boston, MA.

Echegoyen, Y., Nerín, C., 2013. Nanoparticle release from nano-silver antimicrobial food containers. Food Chem. Toxicol. 62, 16–22. Available from: http://dx.doi.org/10.1016/j.fct.2013.08.014.

Elechiguerra, J.L., Burt, J.L., Morones, J.R., Camacho-Bragado, A., Gao, X., Lara, H.H., et al., 2005. Interaction of silver nanoparticles with HIV-1. J. Nanobiotechnol. 3, 1–10. Available from: http://dx.doi.org/10.1186/1477-3155-3-6.

Esposito, S., Purrello, S.M., Bonnet, E., Novelli, A., Tripodi, F., Pascale, R., et al., 2013. Central venous catheter-related biofilm infections: an up-to-date focus on meticillin-resistant *Staphylococcus aureus*. J. Glob. Antimicrob. Resist. 1, 71–78. Available from: http://dx.doi.org/10.1016/j.jgar.2013.03.002.

Fang, Y.-M., Lin, Z.-B., Zeng, Y.-M., Chen, W.-K., Chen, G.-N., Sun, J.-J., et al., 2010. Facile electrochemical preparation of Ag nanothorns and their growth mechanism. Chem. Eur. J. 16, 6766–6770. Available from: http://dx.doi.org/10.1002/chem.201000068.

Farella, I., Valentini, A., Cioffi, N., Torsi, L., 2005. Dual ion-beam sputtering deposition of palladium-fluoropolymer nano-composites. Appl. Phys. A. 80, 791–795. Available from: http://dx.doi.org/10.1007/s00339-003-2444-6.

Fedlheim, D.L., Foss, C.A., 2001. Metal Nanoparticles: Synthesis, Characterization, and Applications. CRC Press, New York.

Forlee, M., Rossington, A., Searle, R., 2014. A prospective, open, multicentre study to evaluate a new gelling fibre dressing containing silver in the management of venous leg ulcers. Int. Wound J. Available from: http://dx.doi.org/10.1111/iwj.12239.

Frykholm, P., Pikwer, A., Hammarskjöld, F., Larsson, A.T., Lindgren, S., Lindwall, R., et al., 2014. Clinical guidelines on central venous catheterisation. Acta Anaesthesiol. Scand. 58, 508–524. Available from: http://dx.doi.org/10.1111/aas.12295.

Galdiero, S., Falanga, A., Vitiello, M., Cantisani, M., Marra, V., Galdiero, M., 2011. Silver nanoparticles as potential antiviral agents. Molecules. 16, 8894–8918. Available from: http://dx.doi.org/10.3390/molecules16108894.

Ghosh Auddy, R., Abdullah, M.F., Das, S., Roy, P., Datta, S., Mukherjee, A., 2013. New guar biopolymer silver nanocomposites for wound healing applications. BioMed. Res. Int. 2013, e912458. Available from: http://dx.doi.org/10.1155/2013/912458.

Giannossa, L.C., Longano, D., Ditaranto, N., Nitti, M.A., Paladini, F., Pollini, M., et al., 2013. Metal nanoantimicrobials for textile applications. Nanotechnol. Rev. 2, 307–331. Available from: http://dx.doi.org/10.1515/ntrev-2013-0004.

Gottesman, R., Shukla, S., Perkas, N., Solovyov, L.A., Nitzan, Y., Gedanken, A., 2011. Sonochemical coating of paper by microbiocidal silver nanoparticles. Langmuir ACS J. Surf. Colloids. 27, 720–726. Available from: http://dx.doi.org/10.1021/la103401z.

Guan, D., Wang, Y., 2012. Electrodeposition of Ag nanoparticles onto bamboo-type TiO_2 nanotube arrays to improve their lithium-ion intercalation performance. Ionics. 19, 879–885. Available from: http://dx.doi.org/10.1007/s11581-012-0814-9.

Hebeish, A., El-Rafie, M.H., EL-Sheikh, M.A., Seleem, A.A., El-Naggar, M.E., 2014. Antimicrobial wound dressing and anti-inflammatory efficacy of silver nanoparticles. Int. J. Biol. Macromol. 65, 509–515. Available from: http://dx.doi.org/10.1016/j.ijbiomac.2014.01.071.

Hentschel, T., Münstedt, H., 1999. Thermoplastic polyurethane—the material used for the Erlanger silver catheter. Infection 27, S43–S45. Available from: http://dx.doi.org/10.1007/BF02561617.

Herrera, B., Bruna, T., Guerra, D., Yutronic, N., Kogan, M. J., Jara, P., 2013. Silver nanoparticles produced by magnetron sputtering and selective nanodecoration onto α-cyclodextrin/carboxylic acid inclusion compounds crystals. Adv. Nanoparticles 02, 112–119. Available from: http://dx.doi.org/10.4236/anp.2013.22019.

Huang, R.-H., Chao, W.-K., Yu, R.-S., Huang, R.-T., Hsueh, K.-L., Shieu, F.-S., 2012. Facile synthesis of silver nanoparticles by electrochemical method in the presence of sodium montmorillonite. J. Electrochem. Soc. 159, E122–E126. Available from: http://dx.doi.org/10.1149/2.037206jes.

Inbakandan, D., Kumar, C., Abraham, L.S., Kirubagaran, R., Venkatesan, R., Khan, S.A., 2013. Silver nanoparticles with anti microfouling effect: a study against marine biofilm forming bacteria. Colloids Surf. B Biointerfaces 111C, 636–643. Available from: http://dx.doi.org/10.1016/j.colsurfb.2013.06.048.

Islam, M.S., Larimer, C., Ojha, A., Nettleship, I., 2013. Antimycobacterial efficacy of silver nanoparticles as deposited on porous membrane filters. Mater. Sci. Eng. C Mater. Biol. Appl. 33, 4575–4581. Available from: http://dx.doi.org/10.1016/j.msec.2013.07.013.

Jian, Z., Xiang, Z., Yongchang, W., 2005. Electrochemical synthesis and fluorescence spectrum properties of silver nanospheres. Microelectron. Eng. 77, 58–62. Available from: http://dx.doi.org/10.1016/j.mee.2004.08.005.

Johans, C., 2002. Electrosynthesis of polyphenylpyrrole coated silver particles at a liquid–liquid interface. Electrochem. Commun. 4, 227–230. Available from: http://dx.doi.org/10.1016/S1388-2481(02)00256-4.

Jovanović, Ž., Radosavljević, A., Stojkovska, J., Nikolić, B., Obradovic, B., Kačarević-Popović, Z., et al., 2013. Silver/poly(*N*-vinyl-2-pyrrolidone) hydrogel nanocomposites obtained by electrochemical synthesis of silver nanoparticles inside the polymer hydrogel aimed for biomedical applications. Polym. Compos. Available from: http://dx.doi.org/10.1002/pc.22653.

Joyce-Wöhrmann, R.M., Münstedt, H., 1999. Determination of the silver ion release from polyurethanes enriched with silver. Infection. 27, S46–S48. Available from: http://dx.doi.org/10.1007/BF02561618.

Khaydarov, R.A., Khaydarov, R.R., Gapurova, O., Estrin, Y., Scheper, T., 2009. Electrochemical method for the synthesis of silver nanoparticles. J. Nanoparticle Res. 11, 1193–1200. Available from: http://dx.doi.org/10.1007/s11051-008-9513-x.

Kim, K.-J., Sung, W.S., Moon, S.-K., Choi, J.-S., Kim, J.G., Lee, D.G., 2008. Antifungal effect of silver nanoparticles on dermatophytes. J. Microbiol. Biotechnol. 18, 1482–1484.

Kim, K.-J., Sung, W.S., Suh, B.K., Moon, S.-K., Choi, J.-S., Kim, J.G., et al., 2009. Antifungal activity and mode of action of silver nano-particles on *Candida albicans*. Biometals Int. J. Role Met. Ions Biol. Biochem. Med. 22, 235–242. Available from: http://dx.doi.org/10.1007/s10534-008-9159-2.

Krutyakov, Y.A., Kudrinskiy, A.A., Olenin, A.Y., Lisichkin, G.V., 2008. Synthesis and properties of silver nanoparticles: advances and prospects. Russ. Chem. Rev. 77, 233–257. Available from: http://dx.doi.org/10.1070/RC2008v077n03ABEH003751.

Kumar, C.S.S.R., 2009. Metallic Nanomaterials, Nanomaterials for the Life Sciences. Wiley-VCH Verlag GmbH & Co. KGaA, Weinheim, Germany.

Lara, H.H., Ayala-Nuñez, N.V., Ixtepan-Turrent, L., Rodriguez-Padilla, C., 2010. Mode of antiviral action of silver nanoparticles against HIV-1. J. Nanobiotechnol. 8, 1. Available from: http://dx.doi.org/10.1186/1477-3155-8-1.

Lara, H.H., Garza-Treviño, E.N., Ixtepan-Turrent, L., Singh, D.K., 2011. Silver nanoparticles are broad-spectrum bactericidal and virucidal compounds. J. Nanobiotechnol. 9, 30. Available from: http://dx.doi.org/10.1186/1477-3155-9-30.

Lee, C.-L., Syu, C.-M., 2010. Electrochemical synthesis of hexadecyltrimethylammonium-coated Ag nanopeanuts and their self-assembly to nanonets. Colloids Surf. Physicochem. Eng. Asp. 358, 158–162. Available from: http://dx.doi.org/10.1016/j.colsurfa.2010.01.045.

Lee, D., Cohen, R.E., Rubner, M.F., 2005. Antibacterial properties of Ag nanoparticle loaded multilayers and formation of magnetically directed antibacterial microparticles. Langmuir. 21, 9651–9659. Available from: http://dx.doi.org/10.1021/la0513306.

Liu, T., Song, X., Guo, Z., Dong, Y., Guo, N., Chang, X., 2014. Prolonged antibacterial effect of silver nanocomposites with different structures. Colloids Surf. B Biointerfaces. 116, 793–796. Available from: http://dx.doi.org/10.1016/j.colsurfb.2014.01.010.

Lombardi, S., Scutell, M., Felice, V., Di Campli, E., Di Giulio, M., Cellini, L., 2014. Central vascular catheter infections in a hospital of central Italy. New Microbiol. 37, 41–50.

Ma, H., Yin, B., Wang, S., Jiao, Y., Pan, W., Huang, S., et al., 2004. Synthesis of silver and gold nanoparticles by a novel electrochemical method. ChemPhysChem. 5, 68–75. Available from: http://dx.doi.org/10.1002/cphc.200300900.

Magagnin, L., Cojocaru, P., 2012. Electrochemical synthesis of dispersed metallic nanoparticles. In: Djokić, S.S. (Ed.), Electrochemical Production of Metal Powders, Modern Aspects of Electrochemistry. Springer US, Boston, MA, pp. 345–368.

Mansouri, M.D., Hull, R.A., Stager, C.E., Cadle, R.M., Darouiche, R.O., 2013. *In vitro* activity and durability of a combination of an antibiofilm and an antibiotic against vascular catheter colonization. Antimicrob. Agents Chemother. 57, 621–625. Available from: http://dx.doi.org/10.1128/AAC.01646-12.

Marambio-Jones, C., Hoek, E.M.V., 2010. A review of the antibacterial effects of silver nanomaterials and potential implications for human health and the environment. J. Nanopart Res. 12, 1531–1551. Available from: http://dx.doi.org/10.1007/s11051-010-9900-y.

Marius, S., Lucian, H., Marius, M., Daniela, P., Irina, G., Romeo-Iulian, O., et al., 2011. Enhanced antibacterial effect of silver nanoparticles obtained by electrochemical synthesis in poly(amide-hydroxyurethane) media. J. Mater. Sci. Mater. Med. 22, 789–796. Available from: http://dx.doi.org/10.1007/s10856-011-4281-z.

Marsich, E., Bellomo, F., Turco, G., Travan, A., Donati, I., Paoletti, S., 2013. Nano-composite scaffolds for bone tissue engineering containing silver nanoparticles: preparation, characterization and biological properties. J. Mater. Sci. Mater. Med. 24, 1799–1807. Available from: http://dx.doi.org/10.1007/s10856-013-4923-4.

Martus, P., Geis, C., Lugauer, S., Böswald, M., Guggenbichler, J.P., 1999. Clinical study of the Erlanger silver catheter—data management and biometry. Infection. 27, S61–S68. Available from: http://dx.doi.org/10.1007/BF02561622.

Mazur, M., 2004. Electrochemically prepared silver nanoflakes and nanowires. Electrochem. Commun. 6, 400–403. Available from: http://dx.doi.org/10.1016/j.elecom.2004.02.011.

Melaiye, A., Sun, Z., Hindi, K., Milsted, A., Ely, D., Reneker, D.H., et al., 2005. Silver(I) – imidazole cyclophane gem-diol complexes encapsulated by electrospun tecophilic nanofibers: formation of nanosilver particles and antimicrobial activity. J. Am. Chem. Soc. 127, 2285–2291. Available from: http://dx.doi.org/10.1021/ja040226s.

Mohammadzadeh, R., Ahmadiyan, N., 2013. Skin infection management using novel antibacterial agents. Adv. Pharm. Bull. 3, 247–248. Available from: http://dx.doi.org/10.5681/apb.2013.040.

Monzillo, V., Corona, S., Lanzarini, P., Dalla Valle, C., Marone, P., 2012. Chlorhexidine-silver sulfadiazine-impregnated central venous catheters: *in vitro* antibacterial activity and impact on bacterial adhesion. New Microbiol. 35, 175–182.

Mpenyana-Monyatsi, L., Mthombeni, N.H., Onyango, M.S., Momba, M.N.B., 2012. Cost-effective filter materials coated with silver nanoparticles for the removal of pathogenic bacteria in groundwater. Int. J. Environ. Res. Public. Health. 9, 244–271. Available from: http://dx.doi.org/10.3390/ijerph9010244.

Mulley, G., Jenkins, A.T.A., Waterfield, N.R., 2014. Inactivation of the antibacterial and cytotoxic properties of silver ions by biologically relevant compounds. PLoS One 9, e94409. Available from: http://dx.doi.org/10.1371/journal.pone.0094409.

Murray, B.J., Li, Q., Newberg, J.T., Menke, E.J., Hemminger, J.C., Penner, R.M., 2005. Shape- and Size-selective electrochemical synthesis of dispersed silver(I) oxide colloids. Nano Lett. 5, 2319–2324. Available from: http://dx.doi.org/10.1021/nl051834o.

Nam, K.-Y., Lee, C.-H., Lee, C.-J., 2012. Antifungal and physical characteristics of modified denture base acrylic incorporated with silver nanoparticles. Gerodontology 29, e413–e419. Available from: http://dx.doi.org/10.1111/j.1741-2358.2011.00489.x.

Nayeemuddin, M., Pherwani, A.D., Asquith, J.R., 2013. Imaging and management of complications of central venous catheters. Clin. Radiol. 68, 529–544. Available from: http://dx.doi.org/10.1016/j.crad.2012.10.013.

Nowack, B., Krug, H.F., Height, M., 2011. 120 Years of nanosilver history: implications for policy makers. Environ. Sci. Technol. 45, 1177–1183. Available from: http://dx.doi.org/10.1021/es103316q.

Olson, M.E., Wright, J.B., Lam, K., Burrell, R.E., 2000. Healing of porcine donor sites covered with silver-coated dressings. Eur. J. Surg. 166, 486–489. Available from: http://dx.doi.org/10.1080/110241500750008817.

Pal, S., Tak, Y.K., Song, J.M., 2007. Does the antibacterial activity of silver nanoparticles depend on the shape of the nanoparticle? A study of the gram-negative bacterium *Escherichia coli*. Appl. Environ. Microbiol. 73, 1712–1720. Available from: http://dx.doi.org/10.1128/AEM.02218-06.

Paladini, F., Pollini, M., Talà, A., Alifano, P., Sannino, A., 2012. Efficacy of silver treated catheters for haemodialysis in preventing bacterial adhesion. J. Mater. Sci. Mater. Med. 23, 1983–1990. Available from: http://dx.doi.org/10.1007/s10856-012-4674-7.

Paladini, F., Meikle, S.T., Cooper, I.R., Lacey, J., Perugini, V., Santin, M., 2013a. Silver-doped self-assembling diphenylalanine hydrogels as wound dressing biomaterials. J. Mater. Sci. Mater. Med. 24, 2461–2472. Available from: http://dx.doi.org/10.1007/s10856-013-4986-2.

Paladini, F., Pollini, M., Deponti, D., Di Giancamillo, A., Peretti, G., Sannino, A., 2013b. Effect of silver nanocoatings on catheters for haemodialysis in terms of cell viability, proliferation, morphology and antibacterial activity. J. Mater. Sci. Mater. Med. 24, 1105–1112. Available from: http://dx.doi.org/10.1007/s10856-013-4870-0.

Paladini, F., Cooper, I.R., Pollini, M., 2014a. Development of antibacterial and antifungal silver coated polyurethane foams as air filtration units for the prevention of respiratory diseases. J. Appl. Microbiol. 116, 710–717. Available from: http://dx.doi.org/10.1111/jam.12402.

Paladini, F., De Simone, S., Sannino, A., Pollini, M., 2014b. Antibacterial and antifungal dressings obtained by photochemical deposition of silver nanoparticles. J. Appl. Polym. Sci. 131, 40326(8 pp). Available from: http://dx.doi.org/10.1002/app.40326.

Paladini, F., Sannino, A., Pollini, M., 2014c. *In vivo* testing of silver treated fibers for the evaluation of skin irritation effect and hypoallergenicity. J. Biomed. Mater. Res. B Appl. Biomater. 102, 1031–1037. Available from: http://dx.doi.org/10.1002/jbm.b.33085.

Panáček, A., Kvítek, L., Prucek, R., Kolář, M., Večeřová, R., Pizúrová, N., et al., 2006. Silver colloid nanoparticles: synthesis, characterization, and their antibacterial activity. J. Phys. Chem. B. 110, 16248–16253. Available from: http://dx.doi.org/10.1021/jp063826h.

Paredes, J., Alonso-Arce, M., Schmidt, C., Valderas, D., Sedano, B., Legarda, J., et al., 2014. Smart central venous port for early detection of bacterial biofilm related infections. Biomed. Microdevices 16, 365–374. Available from: http://dx.doi.org/10.1007/s10544-014-9839-3.

Parker, M.G., Doebbeling, B.N., 2012. The challenge of methicillin-resistant *Staphylococcus aureus* prevention in hemodialysis therapy. Semin. Dial. 25, 42–49. Available from: http://dx.doi.org/10.1111/j.1525-139X.2011.00999.x.

Pathak, S.P., Gopal, K., 2012. Evaluation of bactericidal efficacy of silver ions on *Escherichia coli* for drinking water disinfection. Environ. Sci. Pollut. Res. Int. 19, 2285–2290. Available from: http://dx.doi.org/10.1007/s11356-011-0735-6.

Petica, A., Gavriliu, S., Lungu, M., Buruntea, N., Panzaru, C., 2008. Colloidal silver solutions with antimicrobial properties. Mater. Sci. Eng. B. 152, 22–27. Available from: http://dx.doi.org/10.1016/j.mseb.2008.06.021.

Pletcher, D., Greff, R., Peat, R., Peter, L.M., Robinson, J., 2001. —Electrocrystallisation. Instrumental Methods in Electrochemistry. Horwood Publishing Limited, Cambridge, pp. 283–316 (Chapter 9).

Plieth, W.J., 1985. The work function of small metal particles and its relation to electrochemical properties. Surf. Sci. 156 (Part 1), 530–535. Available from: http://dx.doi.org/10.1016/0039-6028(85)90615-6.

Podsiadlo, P., Paternel, S., Rouillard, J.-M., Zhang, Z., Lee, J., Lee, J.-W., et al., 2005. Layer-by-layer assembly of nacre-like nanostructured composites with antimicrobial properties. Langmuir ACS J. Surf. Colloids. 21, 11915–11921. Available from: http://dx.doi.org/10.1021/la051284.

Pollini, M., Sannino, A., Maffezzoli, A., Licciulli, A., 2005. Antibacterial surface treatments based on silver clusters deposition. WO2007074484A2.

Pollini, M., Russo, M., Licciulli, A., Sannino, A., Maffezzoli, A., 2009. Characterization of antibacterial silver coated yarns. J. Mater. Sci. Mater. Med. 20, 2361–2366. Available from: http://dx.doi.org/10.1007/s10856-009-3796-z.

Pollini, M., Paladini, F., Catalano, M., Taurino, A., Licciulli, A., Maffezzoli, A., et al., 2011. Antibacterial coatings on haemodialysis catheters by photochemical deposition of silver nanoparticles. J. Mater. Sci. Mater. Med. 22, 2005–2012. Available from: http://dx.doi.org/10.1007/s10856-011-4380-x.

Pollini, M., Paladini, F., Licciulli, A., Maffezzoli, A., Sannino, A., Nicolais, L., 2013. Antibacterial natural leather for application in the public transport system. J. Coat. Technol. Res. 10, 239–245. Available from: http://dx.doi.org/10.1007/s11998-012-9439-1.

Rabinal, M.K., Kalasad, M.N., Praveenkumar, K., Bharadi, V.R., Bhikshavartimath, A.M., 2013. Electrochemical synthesis and optical properties of organically capped silver nanoparticles. J. Alloys Compd. 562, 43–47. Available from: http://dx.doi.org/10.1016/j.jallcom.2013.01.043.

Rai, M., Yadav, A., Gade, A., 2009. Silver nanoparticles as a new generation of antimicrobials. Biotechnol. Adv. 27, 76–83. Available from: http://dx.doi.org/10.1016/j.biotechadv.2008.09.002.

Rai, M., Kon, K., Ingle, A., Duran, N., Galdiero, S., Galdiero, M., 2014. Broad-spectrum bioactivities of silver nanoparticles: the emerging trends and future prospects. Appl. Microbiol. Biotechnol. 98, 1951–1961. Available from: http://dx.doi.org/10.1007/s00253-013-5473-x.

Rai, M.K., Deshmukh, S.D., Ingle, A.P., Gade, A.K., 2012. Silver nanoparticles: the powerful nanoweapon against multidrug-resistant bacteria. J. Appl. Microbiol. 112, 841–852. Available from: http://dx.doi.org/10.1111/j.1365-2672.2012.05253.x.

Reetz, M.T., Helbig, W., 1994. Size-selective synthesis of nanostructured transition metal clusters. J. Am. Chem. Soc. 116, 7401–7402. Available from: http://dx.doi.org/10.1021/ja00095a051.

Reicha, F.M., Sarhan, A., Abdel-Hamid, M.I., El-Sherbiny, I. M., 2012. Preparation of silver nanoparticles in the presence of chitosan by electrochemical method. Carbohydr. Polym. 89, 236–244. Available from: http://dx.doi.org/10.1016/j.carbpol.2012.03.002.

Rigo, C., Ferroni, L., Tocco, I., Roman, M., Munivrana, I., Gardin, C., et al., 2013. Active silver nanoparticles for wound healing. Int. J. Mol. Sci. 14, 4817–4840. Available from: http://dx.doi.org/10.3390/ijms14034817.

Rodríguez-Sánchez, L., Blanco, M.C., López-Quintela, M.A., 2000. Electrochemical synthesis of silver nanoparticles. J. Phys. Chem. B. 104, 9683–9688. Available from: http://dx.doi.org/10.1021/jp001761r.

Roldán, M.V., Pellegri, N., de Sanctis, O., 2013. Electrochemical method for Ag-PEG nanoparticles synthesis. J. Nanoparticles 2013, 1–7. Available from: http://dx.doi.org/10.1155/2013/524150.

Rösch, W., Lugauer, S., 1999. Catheter-associated infections in urology: possible use of silver-impregnated catheters and the Erlanger silver catheter. Infection 27, S74–S77. Available from: http://dx.doi.org/10.1007/BF02561624.

Sampathkumar, K., Ramakrishnan, M., Sah, A.K., Sooraj, Y., Mahaldhar, A., Ajeshkumar, R., 2011. Tunneled central venous catheters: experience from a single center. Indian J. Nephrol. 21, 107–111. Available from: http://dx.doi.org/10.4103/0971-4065.82133.

Samuel, U., Guggenbichler, J.P., 2004. Prevention of catheter-related infections: the potential of a new nano-silver impregnated catheter. Int. J. Antimicrob. Agents 23 (Suppl. 1), S75–S78. Available from: http://dx.doi.org/10.1016/j.ijantimicag.2003.12.004.

Schoerner, C., Guggenbichler, J.P., Lugauer, S., Regenfus, A., 1999. Silver catheter study: methods and results of microbiological investigations. Infection 27, S54–S55. Available from: http://dx.doi.org/10.1007/BF02561620.

Scifinder © Database, American Chemical Society, 2014.

Secinti, K.D., Özalp, H., Attar, A., Sargon, M.F., 2011. Nanoparticle silver ion coatings inhibit biofilm formation on titanium implants. J. Clin. Neurosci. Off. J. Neurosurg. Soc. Australas. 18, 391–395. Available from: http://dx.doi.org/10.1016/j.jocn.2010.06.022.

Seo, Y.I., Hong, K.H., Kim, S.H., Chang, D., Lee, K.H., Kim, Y.D., 2012. Removal of bacterial pathogen from wastewater using Al filter with Ag-containing nanocomposite film by *in situ* dispersion involving polyol process. J. Hazard. Mater. 227–228, 469–473. Available from: http://dx.doi.org/10.1016/j.jhazmat.2012.05.026.

Shameli, K., Ahmad, M.B., Jazayeri, S.D., Shabanzadeh, P., Sangpour, P., Jahangirian, H., et al., 2012. Investigation of antibacterial properties silver nanoparticles prepared via green method. Chem. Cent. J. 6, 73. Available from: http://dx.doi.org/10.1186/1752-153X-6-73.

Silva, T.N.V., de Marchi, D., Mendes, M.L., Barretti, P., Ponce, D., 2014. Approach to prophylactic measures for central venous catheter-related infections in hemodialysis: a critical review. Hemodial. Int. Int. Symp. Home Hemodial. 18, 15–23. Available from: http://dx.doi.org/10.1111/hdi.12071.

Sodagar, A., Kassaee, M.Z., Akhavan, A., Javadi, N., Arab, S., Kharazifard, M.J., 2012. Effect of silver nano particles on flexural strength of acrylic resins. J. Prosthodont. Res. 56, 120–124. Available from: http://dx.doi.org/10.1016/j.jpor.2011.06.002.

Sondi, I., Salopek-Sondi, B., 2004. Silver nanoparticles as antimicrobial agent: a case study on *E. coli* as a model for Gram-negative bacteria. J. Colloid Interface Sci. 275, 177–182. Available from: http://dx.doi.org/10.1016/j.jcis.2004.02.012.

Song, H., Li, B., Lin, Q.-B., Wu, H.-J., Chen, Y., 2011. Migration of silver from nanosilver-polyethylene composite packaging into food simulants. Food Addit. Contam. Part Chem. Anal. Control Expo. Risk Assess. 28, 1758–1762. Available from: http://dx.doi.org/10.1080/19440049.2011.603705.

Sotiriou, G.A., Pratsinis, S.E., 2011. Engineering nanosilver as an antibacterial, biosensor and bioimaging material. Curr. Opin. Chem. Eng. 1, 3–10. Available from: http://dx.doi.org/10.1016/j.coche.2011.07.001.

Sousa, C., Henriques, M., Oliveira, R., 2011. Mini-review: antimicrobial central venous catheters—recent advances and strategies. Biofouling 27, 609–620. Available from: http://dx.doi.org/10.1080/08927014.2011.593261.

Sportelli, M.C., Nitti, M.A., Valentini, M., Picca, R.A., Bonerba, E., Sabbatini, L., et al., 2014. Ion beam sputtering deposition and characterization of ZnO-fluoropolymer nano-antimicrobials. Sci. Adv. Mater. 6, 1019–1025. Available from: http://dx.doi.org/10.1166/sam.2014.1852.

Starowicz, M., Stypuła, B., Banaś, J., 2006. Electrochemical synthesis of silver nanoparticles. Electrochem. Commun. 8, 227–230. Available from: http://dx.doi.org/10.1016/j.elecom.2005.11.018.

Stickler, D.J., 2014. Clinical complications of urinary catheters caused by crystalline biofilms: something needs to be done. J. Intern. Med. Available from: http://dx.doi.org/10.1111/joim.12220.

Sun, Y., Fuge, G.M., Fox, N.A., Riley, D.J., Ashfold, M.N.R., 2005. Synthesis of aligned arrays of ultrathin ZnO nanotubes on a Si wafer coated with a thin ZnO film. Adv. Mater. 17, 2477–2481. Available from: http://dx.doi.org/10.1002/adma.200500726.

Taheri, S., Cavallaro, A., Christo, S.N., Smith, L.E., Majewski, P., Barton, M., et al., 2014. Substrate independent silver nanoparticle based antibacterial coatings. Biomaterials. 35, 4601–4609. Available from: http://dx.doi.org/10.1016/j.biomaterials.2014.02.033.

Tang, S., Meng, X., Wang, C., Cao, Z., 2009. Flowerlike Ag microparticles with novel nanostructure synthesized by an electrochemical approach. Mater. Chem. Phys. 114, 842–847. Available from: http://dx.doi.org/10.1016/j.matchemphys.2008.10.048.

Tang, Z., Liu, S., Dong, S., Wang, E., 2001. Electrochemical synthesis of Ag nanoparticles on functional carbon surfaces. J. Electroanal. Chem. 502, 146–151. Available from: http://dx.doi.org/10.1016/S0022-0728(01)00344-8.

Ten Kortenaar, M.V., Kolar, Z.I., Tichelaar, F.D., 1999. Formation of long-lived silver clusters in aqueous solution by anodic dispersion. J. Phys. Chem. B. 103, 2054–2060. Available from: http://dx.doi.org/10.1021/jp983621b.

Vonberg, R.-P., Sohr, D., Bruderek, J., Gastmeier, P., 2008. Impact of a silver layer on the membrane of tap water filters on the microbiological quality of filtered water. BMC Infect. Dis. 8, 133. Available from: http://dx.doi.org/10.1186/1471-2334-8-133.

Wadkar, M.M., Chaudhari, V.R., Haram, S.K., 2006. Synthesis and characterization of stable organosols of silver nanoparticles by electrochemical dissolution of silver in DMSO. J. Phys. Chem. B. 110, 20889–20894. Available from: http://dx.doi.org/10.1021/jp063422p.

White, R., 2013. An open response to the technology scoping report examining the clinical and cost effectiveness of silver. J. Wound Care 22, 514–519.

Wiley, B., Sun, Y., Xia, Y., 2007. Synthesis of silver nanostructures with controlled shapes and properties. Acc. Chem. Res. 40, 1067–1076. Available from: http://dx.doi.org/10.1021/ar7000974.

Xiang, D., Chen, Q., Pang, L., Zheng, C., 2011. Inhibitory effects of silver nanoparticles on H1N1 influenza A virus *in vitro*. J. Virol. Methods 178, 137–142. Available from: http://dx.doi.org/10.1016/j.jviromet.2011.09.003.

Xirouchaki, C., Palmer, R.E., 2004. Deposition of size-selected metal clusters generated by magnetron sputtering and gas condensation: a progress review. Philos. Trans. R. Soc. Lond. Ser. Math. Phys. Eng. Sci. 362, 117–124. Available from: http://dx.doi.org/10.1098/rsta.2003.1306.

Xiu, Z., Zhang, Q., Puppala, H.L., Colvin, V.L., Alvarez, P. J.J., 2012. Negligible particle-specific antibacterial activity of silver nanoparticles. Nano Lett. 12, 4271–4275. Available from: http://dx.doi.org/10.1021/nl301934w.

Yin, B., Ma, H., Wang, S., Chen, S., 2003. Electrochemical synthesis of silver nanoparticles under protection of poly(*N*-vinylpyrrolidone). J. Phys. Chem. B. 107, 8898–8904. Available from: http://dx.doi.org/10.1021/jp0349031.

Yuan, G., Chang, X., Zhu, G., 2011. Electrosynthesis and catalytic properties of silver nano/microparticles with different morphologies. Particuology 9, 644–649. Available from: http://dx.doi.org/10.1016/j.partic.2011.03.011.

Yuan, Z., Chen, Y., Li, T., Yu, C.-P., 2013. Reaction of silver nanoparticles in the disinfection process. Chemosphere 93, 619–625. Available from: http://dx.doi.org/10.1016/j.chemosphere.2013.06.010.

Yuranova, T., Rincon, A.G., Bozzi, A., Parra, S., Pulgarin, C., Albers, P., et al., 2003. Antibacterial textiles prepared by RF-plasma and vacuum-UV mediated deposition of silver. J. Photochem. Photobiol. Chem. 161, 27–34. Available from: http://dx.doi.org/10.1016/S1010-6030(03)00204-1.

Zaporojtchenko, V., Chakravadhanula, V.S.K., Faupel, F., Tamulevičius, S., Andrulevičius, M., Tamulevičienė, A., et al., 2010. Residual stress in polytetrafluoroethylene-metal nanocomposite films prepared by magnetron sputtering. Thin Solid Films 518, 5944–5949. Available from: http://dx.doi.org/10.1016/j.tsf.2010.05.097.

Zhang, L., Keogh, S., Rickard, C.M., 2013. Reducing the risk of infection associated with vascular access devices through nanotechnology: a perspective. Int. J. Nanomed. 8, 4453–4466. Available from: http://dx.doi.org/10.2147/IJN.S50312.

Zhang, Y.H., 2008. Composite fluorocarbon/ZnO films prepared by RF magnetron sputtering of Zn and PTFE. Surf. Coat. Technol. 202, 2612–2615. Available from: http://dx.doi.org/10.1016/j.surfcoat.2007.09.029.

Zhitomirsky, I., Hashambhoy, A., 2007. Chitosan-mediated electrosynthesis of organic–inorganic nanocomposites. J. Mater. Process. Technol. 191, 68–72. Available from: http://dx.doi.org/10.1016/j.jmatprotec.2007.03.043.

Zhong, L., Gan, S., Fu, X., Li, F., Han, D., Guo, L., et al., 2013. Electrochemically controlled growth of silver nanocrystals on graphene thin film and applications for efficient nonenzymatic H_2O_2 biosensor. Electrochim. Acta. 89, 222–228. Available from: http://dx.doi.org/10.1016/j.electacta.2012.10.161.

Zhu, J.-J., Liao, X.-H., Zhao, X.-N., Chen, H.-Y., 2001. Preparation of silver nanorods by electrochemical methods. Mater. Lett. 49, 91–95. Available from: http://dx.doi.org/10.1016/S0167-577X(00)00349-9.

Zoval, J.V., Stiger, R.M., Biernacki, P.R., Penner, R.M., 1996. Electrochemical deposition of silver nanocrystallites on the atomically smooth graphite basal plane. J. Phys. Chem. 100, 837–844. Available from: http://dx.doi.org/10.1021/jp952291h.

CHAPTER

7

Application of Nanomaterials in Prevention of Bone and Joint Infections

Nusret Kose[1] *and Aydan Ayse Kose*[2]

[1]Department of Orthopedics and Traumatology, Eskisehir Osmangazi University, Eskisehir, Turkey
[2]Department of Plastic and Reconstructive Surgery, Eskisehir Osmangazi University, Eskisehir, Turkey

7.1 INTRODUCTION

Orthopedic surgery is concerned with conditions involving the musculoskeletal system. Musculoskeletal disorders are the principal cause of disability all over the world and are responsible from more than half of all chronic conditions in elderly people in developed countries (Weinstein, 2000; Leveille, 2004). In engineering, anything used for manufacturing is called material. The material used in the biological environment to support or treat damaged organs or body parts is called biomaterial. Biomaterials can be grouped under four headings: metallic, ceramics, polymeric, and composite materials. The biomaterials most commonly used in orthopedic surgery are metallic implants such as steel or titanium alloys because they provide satisfactory mechanical performance. Screws, plates, and nails used for the treatment of fractures, hip and knee prostheses used in joint diseases, artificial bones, and spine implants are typical examples of biomaterials used in orthopedic surgery. Advances in orthopedic surgery have led to increasing dependence on a variety of medical devices, and the number of patients in need of orthopedic implants is growing rapidly. In 2005, nearly 800,000 primary total hip and knee replacements were performed in the United States (Praemer et al., 1992; Iorio et al., 2008). One million hips are replaced by prostheses worldwide every year (Holzwarth and Cotogno, 2012). The rate of orthopedic implant use is increasing, and this trend is expected to continue in the next decades because of an aging population and improvements in medical care (Iorio et al., 2008; Holzwarth and Cotogno, 2012).

7.2 ORTHOPEDIC IMPLANTS AND INFECTIONS

Orthopedic implants, especially hip and knee joint replacements, have a limited lifetime because of implant failure and the need for

Nanotechnology in Diagnosis, Treatment and Prophylaxis of Infectious Diseases.
DOI: http://dx.doi.org/10.1016/B978-0-12-801317-5.00007-4

revision surgeries within certain time periods in patients who might require frequent, complicated, and expensive surgeries. For this reason, lengthening the life span of implants for several decades would prevent considerable patient suffering and save health care costs (Sculco, 1995). Long-term survival and favorable outcome of orthopedic implant use are mainly determined by bone–implant osseointegration and absence of infection near the implants. Bone is a biocomposite material, and its main components are mixture of calcium and phosphate minerals in the form of hydroxyapatite, proteins with mostly type I collagen fibrils, and water. The dimensions of the mineral and organic components of bone tissue are on the nanometer scale (Rho et al., 1998). Ceramics including calcium phosphates, hydroxyapatite, and calcium sulfate dihydrate are classified as reabsorbable bioceramics. Reabsorbable bioceramics degrade over time and are replaced by endogenous tissues, resulting in normal, functional bone when used in orthopedic applications. Usually, orthopedists use mixtures of calcium phosphates formed by hydroxyapatite and tricalcium phosphates. Because of their chemical and structural similarities to the inorganic part of bone, hydroxyapatite has been used as a coating material on orthopedic implants to obtain better bioactivity of metallic surfaces, which combines the strength of the metals with the bioactivity of the ceramics. A hydroxyapatite coating, in addition to providing a mechanism to augment osseointegration, has the function of blocking the interface from wear particles and macrophages associated with periprosthetic osteolysis (Rahbek et al., 2001; Geesink, 2002). Plasma spraying, magnetron sputtering, sol-gel, electrostatic spraying, and biomimetic coating techniques have been used to apply hydroxyapatite coatings on implant surfaces (Ducheyne et al., 1986; De Groot et al., 1987; Klein et al., 1991; Leeuwenburgh et al., 2005). Currently, hydroxyapatite coatings have been used not only for their osteoconductive effects but also as a method for providing bioactive molecules (Saran et al., 2011). These methods of delivery of bioactive molecules expand the function of hydroxyapatite as a coating to increase new bone formation on orthopedic implants. The appropriate timing, dose, and their release kinetics from the hydroxyapatite have to be carefully determined for optimal outcome regarding the biological features added to hydroxyapatite (Goodman et al., 2013).

The use of prophylactic antibiotics and other procedures such as the use of laminar airflow and body exhaust systems in the operating room have decreased the frequency of implant-related infections, but they have not eliminated the risk (Nasser, 1992). It has been reported that the incidence of implant-related surgery is increasing. Despite progress in surgical techniques, up to 5% of implant-related surgeries may become complicated by infection (Goodman et al., 2013). Infection is also known to hinder healing in bone tissue (Petty et al., 1985). *Staphylococcus aureus* and *Staphylococcus epidermidis* are responsible for more than 50% of infections. However, frequencies of methicillin-resistant microbial infections have been increasing. Infection of the implant is one of the most frightening clinical complications for both the patient and the surgeon (Garvin and Hanssen, 1995). Implant-related infections are mostly treated by surgical debridement and systemic administration of antibiotics. Deep infection of implants cannot be managed by systemic administration of antibiotics. Many of these infected implants need removal. In addition to economic aspects, many difficulties may be associated with implant-related infection that may include the need for multiple operations, a long period of disability for the patient, and, occasionally, suboptimal outcome and a higher incidence of mortality. As with most diseases, prevention is the better approach. It is very important to develop novel methods for preventing implant-related infection (Van de Belt et al., 2001; Arciola et al., 2012).

Bacterial adhesion and colonization to the implant surface is the first step in the establishment of an implant-related infection (Gristina and Costerton, 1984). Bacteria colonize metallic implants and form biofilms that slow the penetration of antibiotics to the underlying infection. Biofilm-producing bacteria are known to be much more resistant to antibiotics than planktonic bacteria. It has been reported that the minimum inhibitory concentrations (MIC) for bacteria in biofilm is many times higher than for planktonic bacteria (Davies, 2003; Kirby et al., 2012). Systemic perioperative antibiotics and local antimicrobials are used commonly for implant-related infections. The use of systemic antimicrobials has limited success in reducing infectious complications. Systemic toxicity and poor penetration into infected tissues reduce the success of oral or parenteral antibiotic treatment. Local antibiotic therapy has become a recognized and standard supplement to systemic antibiotics for both prophylaxis and existing infections. Antimicrobials can be locally administered in various forms, including skin antisepsis, use of antimicrobial carriers, and soaking of surgical implants in antimicrobial solutions (Ostermann et al., 1993; Isiklar et al., 1996). Antimicrobial irrigation of the surgical wound and debridement can also be performed. Animal studies show that irrigation and debridement, especially in early cases with antibiotics, may prevent infection, but clinical studies show high failure rates with this approach if the site is already infected (Shah et al., 2013).

7.3 LOCAL DELIVERY OF ANTIMICROBIALS

Local delivery of antimicrobial agents is a smart choice for the prevention of infection. Adequate dose of antibiotics can be delivered directly to the site of infection and antibiotic concentration in the infected area is more likely to be in the therapeutic range, preventing development of antibiotic-resistant strains. With this method, systemic side effects are reduced because antibiotics should not reach high serum concentrations. Many materials have been investigated as potential local delivery vehicles (Darouiche, 1999, 2004; Ewald et al., 2006; Darouiche et al., 2007). Polymers such as collagen and chitosan are simply and rapidly loaded by dipping them in a solution of the chosen antibiotic. These polymers release antibiotics in a quick "burst," and discharge is normally complete within 4 days (Wachol-Drewek et al., 1996; Noel et al., 2008). This limited time period for release of kinetics is not suitable for surgically implanted devices. To provide antibiotics locally, antibiotic-loaded polymethylmethacrylate (PMMA) beads have been used for decades (Van de Belt et al., 2001; Hendriks et al., 2004; Kent et al., 2006). They are produced at the time of implantation by combining the preferred antibiotic with PMMA and shaping into beads. They might treat acute infections in long bones (Wenke et al., 2006) but, because beads are not biodegradable, antimicrobials loaded into the inside of the bead may not be released completely. Actually, 75% of total antibiotics incorporated may not be released from PMMA beads and additional surgery is necessary to remove the beads once antibiotic release has finished. New studies have also indicated that bacterial biofilms can stay on antimicrobial-releasing bone cement beads when implanted into infected areas (Anagnostakos et al., 2008). Biodegradable polymers such as polylactic acid and polylactic-glycolic acid have been used to locally transport antibiotics. Antibiotic-loaded biodegradable polymers are a promising medium for local delivery because of biodegradable nature and biocompatibility of biodegradable polymers, but the release kinetics of antibiotics from biodegradable polymers is not ideal.

7.4 ANTIMICROBIAL IMPLANT COATINGS

The characteristics of biomaterials can be generally divided into two categories: bulk and surface. The surface properties of a material describe the interactions with its environment that occur at the interface. A potential solution to prevent early bacterial adhesion may be to modify the implant surface by coating the surface with an antimicrobial. Infection-resistant coatings are a relatively recent addition to the science of implant and device development. Coatings can be classified as passive or active, depending on whether there are antibacterial agents distributed locally. Passive coatings do not release bactericidal agents to the surrounding tissues; these coatings prevent bacterial adhesion and kill bacteria on contact. In comparison, active coatings discharge loaded bactericidal agents to treat infection. Active antimicrobial agents were utilized first as infection-resistant materials. Passive coating technologies such as adhesion-resistant coatings are being developed (Goodman et al., 2013). The design of anti-infective bioactive coatings can rely on different strategies: bacteria repelling and antiadhesive surfaces; biomaterials delivering antimicrobials; bioactive molecules interfering with the production of bacterial biofilm; intrinsically bioactive materials with antibacterial properties; and nanostructured materials (Campoccia et al., 2013). Nanotechnology is an emerging technology involving the tailoring of materials at atomic levels to obtain unique properties for the desired applications (Roco et al., 2010). Most of the natural processes also take place in the nanometer scale. Materials with high performance and unique properties can be produced that could not be created with traditional manufacturing methods. Nanotechnology provides opportunities for promising sophisticated approaches in the area of medical implant technology. It has impacted the field of biomaterials in several areas, including the manipulation of surface characteristics. Calcium phosphate bioceramics was used to improve the biological properties of metallic implants. The logical next step was coating the surface with antimicrobials. This would inhibit bacteria from colonizing the implant surface. Because of the aforementioned issues, biodegradable materials have been investigated as possible antibiotic carriers. Antibiotic-loaded calcium phosphate coating produced by a dipping method causes a burst release of the antibiotics, and total antibiotics may be released within the first 60 min (Radin et al., 1997; Gautier et al., 2001). The development of microbial resistance to ordinary antibiotics is a major clinical challenge (Spellberg et al., 2008). To eliminate these undesirable issues, research into alternative antimicrobials such as silver compounds and newer areas, such as nanoparticle formulations, have gained attraction (Bai et al., 2010). Hydroxyapatite nanoparticles can be produced through wet synthesis, which proceeds at relatively low temperatures. Hydroxyapatite nanoparticles can be used as drug carriers and show improved performances compared with conventional materials because of their large surface-to-volume ratios.

7.5 IMPLANT COATING WITH NANO-SILVER

To prevent implant-related infections, implants can be coated with different antimicrobials that comprise antibiotics, such as vancomycin, tobramycin, and gentamicin, and silver and silver-containing compounds (Bosetti et al., 2002; Alt et al., 2006). Silver has been described as one of the earliest materials to be intentionally used in surgery for its bactericidal properties (Klasen, 2000). It has remained a relevant antimicrobial and is now used mainly for burn treatments, implant coatings, and dental work (Saint et al., 1998; Darouiche, 1999; Klasen, 2000; Bosetti et al., 2002; Huh and Kwon, 2011). Silver nanoparticles are gaining popularity as an antimicrobial with important potential because they have been shown to be effective again bacteria, viruses, and eukaryotes (Furno et al., 2004; Gong et al., 2007). Silver nanoparticles have been used as a local antimicrobial in coatings of medical implants and wound dressings (Rai et al., 2009). Many authors focused primarily on silver because it has a broad antibacterial spectrum, strong antibacterial activity, and low toxicity. A number of methods can be used to add silver onto the surfaces of medical implants. Although direct coating methods of applying metallic silver to implant surfaces have been described, silver is biologically more active when it is in its ionic form ($Ag+$) (Gosheger et al., 2004; Hardes et al., 2007). The metallic silver did not show enough antimicrobial activity and also deteriorated quickly (Atiyeh et al., 2007; Rai et al., 2009). Metallic silver in a moist environment will react and results in the formation and release of silver ions. These ions are highly potent as antimicrobial agents, but they are difficult to use in combination with medical devices (Rai et al., 2009; Meakins, 2009). The coating of silver comprising biodegradable material could discharge markedly more silver ions than metallic silver (Noda et al., 2009). Silver ions display bactericidal activity at very low concentrations; bactericidal activity has been reported at silver concentrations as low as 35 ppb (Noda et al., 2009). The mechanism of bactericidal activity of silver relies on a series of actions; inactivation of critical enzymes of the respiratory chain by metal binding to thiol groups and induction of hydroxyl radicals appear to play a major role (Klasen, 2000; Feng et al., 2000; Jung et al., 2008). Currently, there is substantial research regarding silver-coated medical devices (Alt et al., 2004; Bai et al., 2010; Johnston et al., 2010). This is mainly driven by the fact that silver is active against both Gram-positive and Gram-negative bacteria. The incidence of antibiotic-resistant bacteria is a serious problem, but silver has advantages over antibiotics; resistance has not been yet been convincingly demonstrated for clinically relevant pathogens.

The potential of silver nanoparticles as an antimicrobial agent has been confirmed by many studies (Furno et al., 2004; Alt et al., 2006) Furno et al. (2004) showed that silicon discs impregnated with silver nanoparticles inhibited bacterial adhesion and growth. Silver nanoparticles are used as a repository of silver ions, which are released from the coating. Silver nanoparticles exhibited potential as an antimicrobial agent, but maintaining this effect for longer periods is dependent on the way the silver particles are anchored to the surface layer. Hydroxyapatite nanoparticles can be used as a carrier of silver particles. There are several methods for introducing silver into calcium phosphate–based ceramic coatings, such as plasma spraying (Chen et al., 2008; Zheng et al., 2009), ion beam-assisted deposition (Bai et al., 2010), magnetron sputtering (Chen et al., 2006), micro-arc oxidation (Song et al., 2009), and sol-gel technology (Chen et al., 2007). Chen et al. (2007) examined antibacterial and osteogenic properties of silver-containing hydroxyapatite coatings produced using a sol-gel process and determined that fixing hydroxyapatite with silver inhibited the adhesion of bacterial on its surface. Chen et al. (2008) fabricated plasma-sprayed silver-containing hydroxyapatite coatings and showed that the coating exhibited a distinct antibacterial effect against *Escherichia coli*, *Pseudomonas aeruginosa*, and *S. aureus*. They observed no major cytotoxicity and hemolysis for the silver-containing coatings.

The use of silver in thin nanocoatings in the form of nanoparticles has been progressively developing (Grunlan et al., 2005; Knetsch and Koole, 2011; Liu et al., 2013; Xie et al., 2014). Peksen et al. (2007) have described the development of a new silver ion–doped calcium phosphate–based ceramic nano-powder. Antibacterial powder was synthesized by using the wet chemical method. Calcium oxide and orthophosporic acid were used as a source

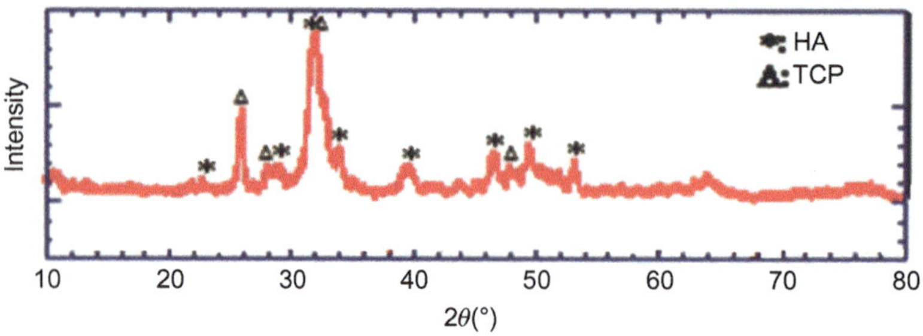

FIGURE 7.1 X-ray diffractometer patterns of powder.

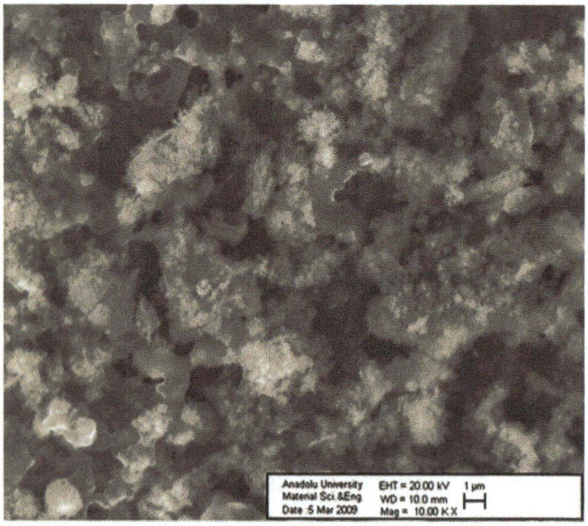

FIGURE 7.2 Morphology of electrosprayed hydroxyapatite-coated implants.

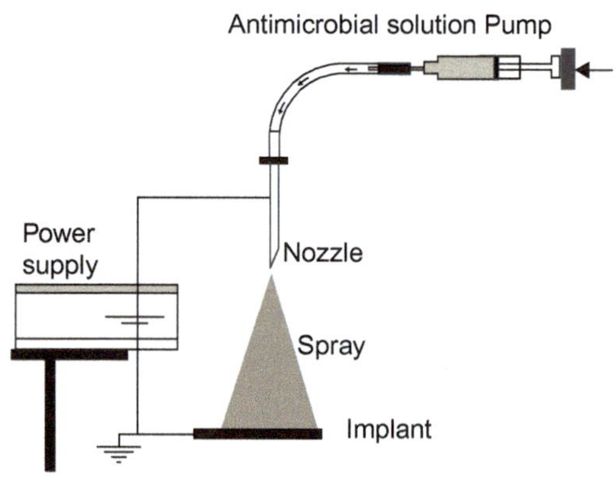

FIGURE 7.3 Schematic view of the ESD setup.

of Ca and PO ions. Silver ions were used as an antibacterial agent. Crystal structures of the synthesized powders were characterized with X-ray diffractometer (Figure 7.1). The particle sizes of powders were determined by laser diffraction method. Powder morphology was studied by scanning electron microscopy (Figure 7.2). Antibacterial effects of this ceramic nano-powder against *E. coli*, *P. aeruginosa*, *S. aureus*, and *Candida albicans* were reported. The powder did not disturb angiogenesis, which plays a key role in bone–implant osseointegration, and at low concentrations it was not cytotoxic. At present, silver is one of the most widely used anti-infective substances (Rai et al., 2009). Incorporating silver into hydroxyapatite coatings is an effective method to provide the coatings with antibacterial properties. However, uniform distribution of nano-sized silver in the coatings is one of the critical factors determining the successful application of hydroxyapatite—silver coatings. Silver particles in the coatings are easy to agglomerate, which in turn affects the applications of the coatings. Kose et al. (2013) used an electrospray method to add silver to hydroxyapatite. The electrospray method was used for the coating process because this technique is a new method for the preparation of uniform ceramic thin films in a controllable way (Figure 7.3). The main advantages of electrospray deposition over other common deposition processes are

FIGURE 7.4 EDX electron image and element (Ag) map taken from antimicrobial-coated pins.

suitability for less soluble materials, lack of solvent evaporation effects, possibility for realizing multilayered devices, and efficient usage of the materials (Balachandrana et al., 2001; Taniguchi and Schoonman, 2002; Jaworek, 2007; Jaworek and Sobczyk, 2008; Thian et al., 2011). Kose et al. (2013) used the same silver ion–doped calcium phosphate–based ceramic nano-powder for implant coating to provide not only biocompatibility but also antibacterial activity to the orthopedic implants. The use of nanoparticulate silver may represent a promising new possibility by embedding nanoparticles of silver in bioceramic coatings; the active surface area is maximized while keeping the total amount of silver low. Therefore, it is possible to tailor the antibacterial activity of the silver ions while remaining below the toxicity threshold for other cell types. Figure 7.4 shows the EDX electron image and element (Ag) map taken from antimicrobial-coated pins. There are some advantages associated with immobilizing silver ions directly into implant coating matrices versus antimicrobials that are eluted from a coating matrix. Anti-infective surfaces involving the use of nonleachable substances, such as hydroxyapatite—silver, can kill bacteria on contact because the bactericidal substances are not released and are active after direct interaction with the bacterial cells. A potential benefit of immobilization is longer-lasting activity. The electrospray method was used for coating the implants. A radiofrequency energy source was used to sinter the coated samples. After sintering, the surface morphology of the pins was observed using scanning electron microscopy. They reported that homogeneous, dense coating layers were obtained. To evaluate silver ion release from antimicrobial powder to the aqueous system, powder was kept in simulated body fluid for 21 days. There was a controlled release of silver ion from powder to an aqueous system. The starting value was 5% weight, and this value was decreased to 4.2% weight after 21 days. However, no free silver ions were detected to the aqueous system with coated samples using an atomic absorption method, which has a detection threshold as low as 0.02 ppm (mg/L). Silver ions doped into the calcium phosphate–based ceramic nano-powder are accepted as non-leachable coating.

To determine if the silver-doped nano-powder–coated implants prevent implant-related infection, 27 titanium pins were implanted in rabbit knees. The pins were divided into three groups of nine. The first group received uncoated implants into their knee joints, the second group received hydroxyapatite-coated implants, and the third group received silver-doped hydroxyapatite-coated implants. To create the experimental infection, methicillin-resistant *S. aureus* was used. After 6 weeks of implantation, bone and implants were radiographically, microbiologically, and histopathologically analyzed. Osteolysis around the implant was investigated on AP and lateral knee radiographs. Each implant was sonicated and the number of organisms was calculated. Bone was swabbed to obtain a specimen for culture. The effects of silver coating on bone tissue surrounding the implants were noted histopathologically. The femoral bone fragments were evaluated to determine the numbers of bacterial colonies. Radiology, microbiology, and histology findings were quantified to define the infection. Plain radiographic findings showed no osteolysis, periosteal bone formation, or focal lysis adjacent to the implant in any groups. Histologic evaluation of sections of the bone surrounding the silver-coated group had healthy cortical osteons and minimal or no periosteal reaction with no cellular inflammation or foreign body granuloma. Silver particulate was not observed on surrounding bone tissue. Cellular inflammation was severe in uncoated samples. Microbiological tests showed silver-coated implants had a lower rate of colonization, and they were associated with lower rates of osteomyelitis.

However, when silver coating is applied to the implants, the safety of silver as well as antibacterial activity should be considered. Although ordinary blood silver concentrations are considered to be less than 10 ppb, toxic side effects of silver were described for a blood concentration of 300 ppb in the form of argyrosis, leucopenia, and liver and kidney damage (Vik et al., 1985). Because silver toxicity is a dose-dependent process, it is necessary to find the optimum silver concentration region for antibacterial activity within the nontoxic region. Nanoparticles chemical composition, size, shape, concentration, rate of dissolution/degradation, and surface properties are all critical variables capable of influencing the toxicological profile. Although silver in small percentages can have an antibacterial effect, larger amounts can be toxic; therefore, optimization of the silver concentration in the coating is critical to guarantee an optimum antibacterial effect without cytotoxicity. The literature and our research support that the presence of 2–5% weight silver in HA coating can have significant antibacterial effects without cytotoxicity. However, even within the optimum silver concentration region, release of silver ions for a long period may lead to the accumulation of silver in the organs. Concerns exist regarding the still incomplete knowledge of the toxicology of nanomaterials and silver (Chopra, 2007; Rai et al., 2009; Johnston et al., 2010; Knetsch and Koole, 2011; Sullivan et al., 2014).

7.6 CONCLUSION AND FUTURE PERSPECTIVES

Infection is a disease state that knows no limits and can affect any patient who undergoes surgical treatment. Despite the many scientific development and technological advances, such as the advent of antibiotics and the use of sterile techniques, infection continues to be a problem for orthopedic surgeons, and it continues to cause suffering to patients. Advances in orthopedic surgery have caused increasing dependence on a variety of medical devices and the number of patients in need of orthopedic implants growing rapidly. Although the use of prophylactic antibiotics and other procedures have helped reduce the incidence of implant-related infections, they have not eliminated the risk. The systemic treatment of infected implants with antibiotics is often poor because access of antibiotics to the infection site is inadequate. The need for new bioactive materials to overcome the growing problems of drug resistance in infections is of increasing importance. Anti-infective biomaterials have become a primary strategy to prevent implant-associated infections. The research focusing on nanotechnology and new materials is promising for the field of medicine. The use of silver in thin nanocoatings in the form of nanoparticles has been developed. There are many studies showing that silver-hydroxyapatite-coated implants had a lower rate of colonization and they were associated with lower rates of infections. Nevertheless, there is incomplete knowledge of the toxicology of nanomaterials and silver.

Acknowledgments

The authors thank Fethiye Gümüştekin for help with the manuscript preparation.

References

Alt, V., Bechert, T., Steinrucke, P., Wagener, M., Seidel, P., Dingeldein, E., et al., 2004. An *in vitro* assessment of the antibacterial properties and cytotoxicity of nanoparticulate silver bone cement. Biomaterials 25, 4383–4391.

Alt, V., Bitschnau, A., Osterling, J., Sewing, A., Meyer, C., Kraus, R., et al., 2006. The effects of combined gentamicin-hydroxyapatite coating for cementless joint prostheses on the reduction of infection rates in a rabbit infection prophylaxis model. Biomaterials 27, 4627–4634.

Anagnostakos, K., Hitzler, P., Pape, D., Kohn, D., Kelm, J., 2008. Persistence of bacterial growth on antibiotic-loaded beads: is it actually a problem? Acta Orthop. 79, 302–307.

Arciola, C.R., Campoccia, D., Speziale, P., Montanaro, L., Costerton, J.W., 2012. Biofilm formation in *Staphylococcus* implant infections. A review of molecular mechanisms and implications for biofilm-resistant materials. Biomaterials 33 (26), 5967–5982.

Atiyeh, B.S., Costagliola, M., Hayek, S.N., Dibo, S.A., 2007. Effect of silver on burn wound infection and healing: review of the literature. Burns 33, 139–148.

Bai, X., More, K., Rouleau, C.M., Rabiei, A., 2010. Functionally graded hydroxyapatite coatings doped with antibacterial components. Acta Biomater. 6, 2264–2273.

Balachandrana, W., Miaoa, P., Xiaob, P., 2001. Electrospray of fine droplets of ceramic suspensions for thin-film preparation. J. Electrostat. 50, 249–263.

Bosetti, M., Masse, A., Tobin, E., Cannas, M., 2002. Silver coated materials for external fixation devices: *In vitro* biocompatibility and genotoxicity. Biomaterials 23, 887–892.

Campoccia, D., Montanaro, L., Arciola, C.R., 2013. A review of the biomaterials technologies for infection-resistant surfaces. Biomaterials 34, 8533–8554.

Chen, W., Liu, Y., Courtney, H.S., Bettenga, M., Agrawal, C.M., Bumgardner, J.D., et al., 2006. *In vitro* antibacterial and biological properties of magnetron co-sputtered silver-containing hydroxyapatite coating. Biomaterials 27, 5512–5517.

Chen, W., Oh, S., Ong, A.P., Oh, N., Liu, Y., Courtney, H.S., et al., 2007. Antibacterial and osteogenic properties hydroxyapatite coatings produced using of silver-containing a sol gel process. J. Biomed. Mater. Res. 82A, 899–906.

Chen, Y., Zheng, X., Xie, Y., Ding, C., Ruan, H., Fan, C., 2008. Anti-bacterial and cytotoxic properties of plasma sprayed silver-containing HA coatings. J. Mater. Sci. Mater. Med. 19, 3603–3609.

Chopra, I., 2007. The increasing use of silver-based products as antimicrobial agents: a useful development or a cause for concern?. J. Antimicrob. Chemother. 59 (4), 587–590.

Darouiche, R.O., 1999. Anti-infective efficacy of silver-coated medical prostheses. Clin. Infect. Dis. 29, 1371–1377.

Darouiche, R.O., 2004. Treatment of infections associated with surgical implants. N. Engl. J. Med. 350, 1422–1429.

Darouiche, R.O., Mansouri, M.D., Zakarevicz, D., Alsharif, A., Landon, G.C., 2007. *In vivo* efficacy of antimicrobial-coated devices. J. Bone Joint Surg. Am. 89, 792–797.

Davies, D., 2003. Understanding biofilm resistance to antibacterial agents. Nat. Rev. Drug Discov. 2, 114–122.

De Groot, K., Geesink, R., Klein, C.P.A.T., Serekian, P., 1987. Plasma sprayed coatings of hydroxyapatite. J. Biomed. Mater. Res. 21, 1375–1381.

Ducheyne, P., Van Raemdonck, W., Heughebaert, J.C., Heughebaert, M., 1986. Structural analysis of hydroxyapatite coatings on titanium. Biomaterials 7, 97–103.

Ewald, A., Gluckermann, S.K., Thull, R., Gbureck, U., 2006. Antimicrobial titanium/silver PVD coating on titanium. Biomed. Eng. Online. 5, 22.

Feng, Q.L., Wu, J., Chen, G.Q., Cui, F.Z., Kim, T.N., Kim, O.J., 2000. A mechanistic study of the antibacterial effect of silver ions on *Escherichia coli* and *Staphylococcus aureus*. J. Biomed. Mater. Res. 52, 662–668.

Furno, F., Morley, K.S., Wong, B., Sharp, B.L., Arnold, P.L., Howdle, S.M., et al., 2004. Silver nanoparticles and polymeric medical devices: a new approach to prevention of infection? J. Antimicrob. Chemother. 54, 1019–1024.

Garvin, K.L., Hanssen, A.D., 1995. Current concepts review: infection after total hip arthroplasty. J. Bone Joint Surg. Am. 77, 1576–1588.

Gautier, H., Daculsi, G., Merle, C., 2001. Association of vancomycin and calcium phosphate by dynamic compaction: *in vitro* characterization and microbiological activity. Biomaterials 22, 2481–2487.

Geesink, R.G., 2002. Osteoconductive coatings for total joint arthroplasty. Clin. Orthop. Relat. Res. 395, 53–65.

Gong, P., Li, H., He, X., Wang, K., Hu, J., Tan, W., et al., 2007. Preparation and antibacterial activity of Fe_3O_4–Ag nanoparticles. Nanotechnology 18, 285604.

Goodman, S.B., Yao, Z., Keeney, M., Yang, F., 2013. The future of biologic coatings for orthopaedic implants. Biomaterials 34 (13), 3174–3183.

Gosheger, G., Hardes, J., Ahrens, H., Streitburger, A., Buerger, H., Erren, M., et al., 2004. Silver-coated megaendoprostheses in a rabbit model—an analysis of the infection rate and toxicological side effects. Biomaterials 25 (24), 5547–5556.

Gristina, G., Costerton, J.W., 1984. Bacterial adherence and the glycocalyx and their role in musculoskeletal infection. Orthop. Clin. North Am. 15, 517–535.

Grunlan, J.C., Choi, J.K., Lin, A., 2005. Antimicrobial behavior of polyelectrolyte multilayer films containing cetrimide and silver. Biomacromolecules 6, 1149–1153.

Hardes, J., Ahrens, H., Gebert, C., Streitbuerger, A., Buerger, H., Erren, M., et al., 2007. Lack of toxicological side-effects in silver coated megaprostheses in humans. Biomaterials 28, 2869–2875.

Hendriks, J.G.E., van Horn, J.R., van der Mei, H.C., Busscher, H.J., 2004. Backgrounds of antibiotic-loaded bone cement and prosthesis-related infection. Biomaterials 25, 545–556.

Holzwarth, U., Cotogno, G., 2012. Total Hip Arthroplasty. State of the Art, Challenges and Prospects. European Commission Joint Research Centre, Luxembourg, July.

Huh, A.J., Kwon, Y.J., 2011. "Nanoantibiotics": a new paradigm for treating infectious diseases using nanomaterials in the antibiotic resistant era. J. Control. Release 156, 128–145.

Iorio, R., Robb, W.J., Healy, W.L., Berry, D.J., Hozack, W.J., Kyle, R.F., et al., 2008. Orthopaedic surgeon workforce and volume assessment for total hip and knee replacement in the United States: preparing for an epidemic. J. Bone Joint Surg. Am. 90, 1598–1605.

Isiklar, Z.U., Darouiche, R.O., Landon, G.C., Beck, T., 1996. Efficacy of antibiotics alone for orthopaedic device related infections. Clin. Orthop. Relat. Res. 332, 184–189.

Jaworek, A., 2007. Electrospray droplet sources for thin film deposition. J. Mater. Sci. 42, 266–297.

Jaworek, A., Sobczyk, A.T., 2008. Electrospraying route to nanotechnology: an overview. J. Electrostat. 66, 197–219.

Jung, W.K., Koo, H.C., Kim, K.W., Shin, S., Kim, S.H., Park, Y.H., 2008. Antibacterial activity and mechanism of action of the silver ion in *Staphylococcus aureus* and *Escherichia coli*. Appl. Environ. Microbiol. 74, 2171–2178.

Johnston, H.J., Hutchison, G., Christensen, F.M., Peters, S., Hankin, S., Stone, V., 2010. A review of the *in vivo* and *in vitro* toxicity of silver and gold particulates: Particle attributes and biological mechanisms responsible for the observed toxicity. Crit. Rev. Toxicol. 40, 328–346.

Kent, M.E., Rapp, R.P., Smith, K.M., 2006. Antibiotic beads and osteomyelitis: here today, what's coming tomorrow? Orthopedics 29, 599–603.

Kirby, A.E., Garner, K., Levin, B.R., 2012. The relative contributions of physical structure and cell density to the antibiotic susceptibility of bacteria in biofilms. Antimicrob. Agents Chemother. 56, 2967–2975.

Klasen, H.J., 2000. Historical review of the use of silver in the treatment of burns. I. Early uses. Burns 26, 117–130.

Klein, C.P.A.T., Patka, P., van der Lubbe, H.B.M., Wolke, J. G.C., de Groot, K., 1991. Plasma-sprayed coatings of tetracalcium phosphate, hydroxy-apatite, and -TCP on titanium alloy: an interface study. J. Biomed. Mater. Res. 25, 53–65.

Knetsch, M.L.W., Koole, L.H., 2011. New strategies in the development of antimicrobial coatings: the example of increasing usage of silver and silver nanoparticles. Polymers 3, 340–366.

Kose, N., Otuzbir, A., Pekşen, C., Kiremitçi, A., Doğan, A., 2013. A silver ion-doped calcium phosphate-based ceramic nanopowder-coated prosthesis increased infection resistance. Clin. Orthop. Relat. Res. 471 (8), 2532–2539.

Leeuwenburgh, S., Wolke, J., Schoonman, J., Jansen, J.A., 2005. Influence of deposition parameters on chemical properties of calcium phosphate coatings prepared by using electrostatic spray deposition. J. Biomed. Mater. Res. A. 74, 275–284.

Leveille, S.G., 2004. Musculoskeletal aging. Curr. Opin. Rheumatol. 16 (2), 114–118.

Liu, X., Mou, Y., Wu, S., Man, H.C., 2013. Synthesis of silver-incorporated hydroxyapatite nanocomposites for antimicrobial implant coatings. Appl. Surf. Sci. 273, 748–757.

Meakins, J.L., 2009. Silver and new technology: dressings and devices. Surg. Infect. 10, 293–296.

Nasser, S., 1992. Prevention and treatment of sepsis in total hip replacement surgery. Orthop. Clin. North Am. 23, 265–277.

Noda, I., Miyaji, F., Ando, Y., Miyamoto, H., Shimazaki, T., Yonekura, Y., et al., 2009. Development of novel thermal sprayed antibacterial coating and evaluation of release properties of silver ions. J. Biomed. Mater. Res. B Appl. Biomater. 89 (2), 456–465.

Noel, S.P., Courtney, H., Bumgardner, J.D., Haggard, W.O., 2008. Chitosan films: a potential local drug delivery system for antibiotics. Clin. Orthop. Relat. Res. 466, 1377–1382.

Ostermann, P.A.W., Henry, S.L., Seligson, D., 1993. The role of local antibiotic therapy in the management of compound fractures. Clin. Orthop. Relat. Res. 295, 102–111.

Pekşen, C., Koparal, A.S., Doğan, A., 2007. Antibacterial Activity of Ag + ion doped calcium phosphate based ceramic power and assessment of its cytotoxicity, 10th International Conference and Exhibition of European Ceramic Society Proceedings Book, pp. 17–21.

Petty, W., Spanier, S., Shuster, J.J., Silverthorne, C., 1985. The influence of skeletal implants on incidence of infection: experiment in a canine model. J. Bone Joint Surg. Am. 67, 1236–1244.

Praemer, A., Furner, S., Rice, D.P., 1992. Musculoskeletal Conditions in the United States. American Academy of Orthopaedic Surgeons, Park Ridge, IL, 27–41.

Radin, S., Campbell, J.T., Ducheyne, P., Cuckler, J.M., 1997. Calcium phosphate ceramic coatings as carriers of vancomycin. Biomaterials 18, 777–782.

Rahbek, O., Overgaard, S., Lind, M., Bendix, K., Bunger, C., Soballe, K., 2001. Sealing effect of hydroxyapatite coating on peri-implant migration of particles. An experimental study in dogs. J. Bone Joint Surg. Br. 83, 441–447.

Rai, M., Yadav, A., Gade, A., 2009. Silver nanoparticles as a new generation of antimicrobials. Biotechnol. Adv. 27, 76–83.

Rho, J.Y., Kuhn-Spearing, L., Zioupos, P., 1998. Mechanical properties and the hierarchical structure of bone. Med. Eng. Phys. 20 (2), 92–102.

Roco, M.C., Mirkin, C.A., Hersam, M.C., 2010. Nanotechnology Research Directions for Societal Needs in 2020 Retrospective and Outlook, WTEC report, September.

Saint, S., Elmore, J.G., Sullivan, S.D., Emerson, S.S., Koepsell, T.D., 1998. The efficacy of silver alloy-coated urinary catheters in preventing urinary tract infection: A meta-analysis. Am. J. Med. 105, 236–241.

Saran, N., Zhang, R., Turcotte, R.E., 2011. Osteogenic protein-1 delivered by hydroxyapatite-coated implants improves bone ingrowth in extracortical bone bridging. Clin. Orthop. Relat. Res. 469, 1470–1478.

Sculco, T.P., 1995. The economic impact of infected joint arthroplasty. Orthopedics 18, 871–873.

Shah, S.R., Kasper, F.K., Mikos, A.G., 2013. Perspectives on the prevention and treatment of infection for orthopedic tissue engineering applications. Chin. Sci. Bull. 58 (35), 4342–4348.

Song, W.H., Ryu, H.S., Hong, S.H., 2009. Antibacterial properties of Ag (or Pt)-containing calcium phosphate coating formed by micro-arc oxidation. J. Biomed. Mater. Res. 88A, 246–254.

Spellberg, B., Guidos, R., Gilbert, D., Bradley, J., Boucher, H.W., Scheld, W.M., et al., 2008. The epidemic of antibiotic-resistant infections: a call to action for the medical community from the Infectious Diseases Society of America. Clin. Infect. Dis. 46155–46164.

Sullivan, M.P., McHale, K.J., Parvizi, J., Mehta, S., 2014. Nanotechnology: current concepts in orthopaedic surgery and future directions. Bone Joint J. 96-B (5), 569–573.

Taniguchi, I., Schoonman, J., 2002. Electrostatic spray deposition of perovskite-type oxide thin films with porous microstructure. J. Mater. Synth. Process 10, 267–275.

Thian, E.S., Li, X., Huang, J., Edirisinghe, M.J., Bonfield, W., Best, S.M., 2011. Electrospray deposition of nanohydroxyapatite coatings: a strategy to mimic bone apatite mineral. Thin Solid Films 519, 2328–2331.

Van de Belt, H., Neut, D., Schenk, W., van Horn, J.R., van der Mei, H.C., Busscher, H.J., 2001. Infection of orthopedic implants and the use of antibiotic-loaded bone cements. A review. Acta Orthop. Scand. 72, 557–571.

Vik, H., Andersen, K.J., Julshamn, K., Todnem, K., 1985. Neuropathy caused by silver absorption from arthroplasty cement. Lancet 325, 872.

Wachol-Drewek, Z.Z., Pfeiffer, M.M., Schol, E.E., 1996. Comparative investigation of drug delivery of collagen implants saturated in antibiotic solutions and a sponge containing gentamicin. Biomaterials 17, 1733–1738.

Wenke, J.C., Owens, B.D., Svoboda, S.J., Brooks, D.E., et al., 2006. Effectiveness of commercially-available antibiotic-impregnated implants. J. Bone Joint Surg. Br. 88, 1102–1104.

Weinstein, S.J., 2000. The bone and joint decade. Bone Joint Surg. Am. 82, 1–3.

Xie, C.M., Lu, X., Wang, K.F., Meng, F.Z., Jiang, O., Zhang, H.P., et al., 2014. Silver nanoparticles and growth factors incorporated hydroxyapatite coatings on metallic implant surfaces for enhancement of osteoinductivity and antibacterial properties. ACS Appl. Mater. Interfaces 6 (11), 8580–8589.

Zheng, X., Chen, Y., Xie, Y., Ji, H., Huang, L., Ding, C., 2009. Antibacterial property and biocompatibility of plasma sprayed hydroxyapatite/silver composite coatings. J. Therm. Spray Technol. 18, 463.

CHAPTER

8

The Potential of Metal Nanoparticles for Inhibition of Bacterial Biofilms

Krystyna I. Wolska, Anna M. Grudniak, Konrad Kamiński and Katarzyna Markowska

Department of Bacterial Genetics, Institute of Microbiology, Faculty of Biology, University of Warsaw, Warsaw, Poland

8.1 INTRODUCTION

It is well-established that in the wide variety of natural habitats, the majority of microbes does not exist as free-living organisms but rather forms a structured biofilm ecosystem in which the microbes are attached to abiotic or biotic surfaces. In a biofilm, microorganisms function as a cooperative consortium that allows survival in hostile environments (Davey and O'Toole, 2000). Mixed-species biofilms are the dominant form in nature, including human hosts; for example, they are found in the oral cavity and in the lungs of cystic fibrosis (CF) patients (Elias and Banin, 2012). Biofilms, including those formed by pathogens, are present in every niche of the human body, such as skin, nose, lung, and intestine, as well as on medical devices, for example, artificial heart valves, catheters, and artificial limbs (Karatan and Watnick, 2009). The formation of the biofilms creates several beneficial phenotypes mainly because of metabolic cooperation, but the competitive relationship is also observed (Rendueles and Ghigo, 2012). Interspecies interactions involve communication, typically via quorum sensing (QS) (Antunes and Ferreira, 2009; Kotler, 2010). It was also well-documented that the bacterial biofilms exhibit an increased resistance to chemical disinfection, human immune responses, and antimicrobial therapy (Hoiby et al., 2010), which create a great medical problem.

In this review, we describe the diseases caused by several bacterial species–forming biofilms, the main mechanisms of biofilm resistance to conventional antibiotics, and novel antibacterial agents and strategies. The main focus is placed on the antibiofilm activity of metal nanoparticles (NPs).

Nanotechnology in Diagnosis, Treatment and Prophylaxis of Infectious Diseases.
DOI: http://dx.doi.org/10.1016/B978-0-12-801317-5.00008-6

8.2 DISEASES CAUSED BY BACTERIAL BIOFILMS

It is estimated that the majority of medical infections are caused by bacterial biofilms (Costerton et al., 1999). Here, we describe biofilms formed by some pathogenic organisms, indicating their impact on the virulence of these species. The exact processes by which biofilm-associated organisms cause diseases in humans is not entirely understood. The most probable strategies include detachment of the cell aggregates from indwelling medical devices, production of endotoxins, resistance to the immune system of the host, and resistance to plasmid exchange (Donlan and Costerton, 2002).

8.2.1 *Pseudomonas aeruginosa* Biofilms

P. aeruginosa, a Gram-negative bacterium, is an opportunistic human pathogen that frequently causes infections in hospitalized patients (Hancock and Speert, 2001). This bacterium is capable of growth in numerous environmental niches because of its large genome and flexible metabolism. *P. aeruginosa* is responsible for lung diseases of patients with CF (Hutchison and Govan, 1999), the infection that leads to lung tissue damage as a result of the combined action of bacteria and increased inflammation, and also for nosocomial tract pneumonia and sepsis of burn wounds (Overhage et al., 2008). The ability to form biofilms is crucial in the infections caused by *P. aeruginosa* and has made this bacterium a model organism with respect to biofilm formation (Hoiby et al., 2001). The three major secreted polysaccharides are the main components of the biofilm matrix: alginate composed of the uronic acid, mammuronic acid, and guluronic acid; Ps1 rich in rhamnose, mannose, and glucose monomers; and Pel (Parsek and Tolker-Nielsen, 2008). Extracellular DNA (eDNA) was also shown to be present in high concentrations in the biofilm matrix (Allesen-Holm et al., 2006). Recently, the loss of social behavior of *P. aeruginosa* colonizing the lung was confirmed. This feature is important for biofilm development and has crucial implications for the treatment of CF infections (Jiricny et al., 2014). The interactions in the respiratory tract infections between the clinical isolates of *P. aeruginosa* and *Staphylococcus aureus*, another prevalent pathogen, were described, and it was proven that these interactions are executed through a small interspecies molecule (Fugere et al., 2014).

P. aeruginosa biofilms are also formed on implanted and indwelling devices, such as catheters and ventilator tubes. These biofilms can be monomicrobial or polymicrobial, usually combined with *S. aureus*, and even aggressive therapy fails to eradicate the infection of the latter (Vandecandelaere et al., 2012). Several animal models of biofilm infections have been developed as a potential guide for clinical treatment. Watters et al. (2013) developed a murine model of chronic diabetic wounds. In turn, Gurjala et al. (2011) elaborated a rabbit model useful for studying nonhealing, subcutaneous wounds.

The reduced susceptibility of *P. aeruginosa* biofilm to antimicrobial agents depends on the restricted penetration of antimicrobials into the biofilm structure (Nichols et al., 1989) and the protection from the immune system, including the humoral host response and innate immune system (Jensen et al., 2010). The biofilm tolerance to antimicrobial agents (Xu et al., 2000) and the formation of persister cells (Lewis, 2010) also play a substantial role in the *P. aeruginosa* resistance to the known antimicrobials. This severe therapeutic problem is described elsewhere. Despite the high resistance to antibiotics for *P. aeruginosa*, they are still used in therapy. For many types of catheter infections, antibiotic lock therapy (ALT) utilizing very high concentrations of antibiotics is clinically recommended. It was also found that sharp debridement followed by antibiotic treatment is very promising for healing the wounds (Dowd et al., 2011). In the treatment of CF, antibiotics combination therapies, for example, fosfomycin and tobramycin, are recommended (Hoffman and Ramsey, 2013). It was also shown that mannitol enhances antibiotic susceptibility of the persister bacteria in *P. aeruginosa* biofilms (Barraud et al., 2013). However, it was generally assumed that the lungs of CF patients, even those receiving improved antibiotic therapy, are persistently colonized by *P. aeruginosa* (Moreau-Marquis et al., 2008).

8.2.2 *Staphylococcus aureus* Biofilms

S. aureus is a ubiquitous Gram-positive bacteria species; 20–25% of the human population is persistently colonized by this bacterium and its niche in the human body is anterior nares (Kluymans et al., 1997). The multilayered *S. aureus* biofilm is embedded with glycocalyx or slime, composed mainly of teichioic acid and bacterial proteins, including clumping factor, fibronectin binding protein A, coagulase, and host proteins. Cell wall-anchored proteins belonging to four distinct classes, as shown by structural and functional analyses, which also contain adhesive matrix molecules, participate in the biofilm formation (Foster et al., 2014). Polysaccharide intracellular antigen (PIA) (Mack et al., 1996) and eDNA (Whitchurch et al., 2002) also play an important role in the development of biofilms. The regulatory factors involved in *S. aureus* biofilm formation, maintenance, and detachment form a complicated molecular network. Biofilm is regulated by the global regulators SarA/Agr, SarA induce cell attachment to the surface, and early biofilm formation; in turn, Agr is involved in biofilm detachment (Beenken et al, 2010). The biofilm is also positively regulated by the alternative sigma factor, σ^S (Nicholas et al., 1999).

S. aureus biofilms present a 25% mortality rate for hospitalized patients in the United States (Archer et al., 2011). The diseases caused by the *S. aureus* biofilm comprise osteomyelitis, a severe bone infection that is especially dangerous for older individuals. Children, when subjected to antibiotic therapy, often resorb the dead bone, which results in the elimination of the bacterial attachment surface (Lew and Waldvogel, 2004). Other severe diseases caused by *S. aureus* biofilms comprise peridontitis (inflammation and infection of the ligaments and bones that support teeth) (Heitz-Mayfield and Lang, 2000), chronic wound infections (for example, venous leg ulcers) (Gjodsbol et al., 2006), and endocarditis (infection of heart valves with a high rate of morbidity and mortality) (Roder et al., 1999). Native valve endocarditis (NVE) results from the interaction between the vascular endothelium, generally of the mitral, aortic, tricuspid, and pulmonic valves of the heart, and bacteria circulating in the bloodstream. It was estimated that the coagulase-positive staphylococci are responsible for approximately 20% of NVE cases (Donlan and Costerton, 2002). Other diseases caused by staphylococci include chronic rhinosinusitis (Ferguson and Stolz, 2005), ocular infections, including keratitis, endophtalmitis, and conjunctivitis (Leid et al., 2002), and otitis media, which is a disease of the middle ear that involves the inflammation of the mucoperiosteal lining (Giebink et al., 1982). Moreover, *S. aureus* and *Staphylococcus epidermidis* cause infections of indwelling medical devices, such as orthopedic implants, artificial heart valves, and cosmetic surgical implants (Costerton et al., 2005).

It should be mentioned that staphylococcal biofilm-associated infections are very resistant to the immune system of the host. This is attributable to several factors, such as the production of staphylococcal complement inhibitors, leading to a substantial decrease in phagocytosis of bacteria by neutrophils (Kobayashi and DeLeo, 2009). IGg1 antibodies may also be less effective because of the heterogeneous antigen character of the biofilm (Archer et al., 2011). It was proposed that the staphylococcal biofilm changes the host immune response from a proinflammatory bactericidal phenotype toward an anti-inflammatory, pro-fibrotic response favoring bacterial persistence.

8.2.3 Biofilms Formed by *E. coli* and Other Pathogens of Gastrointesinal Tract

In this chapter, only three biofilm-formed gastrointestinal pathogens are described that are causes of well-defined diseases.

E. coli is a predominant, facultative anaerobic bacterium of the gastrointestinal tract. Several factors including fimbriae, adhesins, and polysaccharides, such as poly-β-1, 6-*N*-acetyl-glucosamine, and lipopolysaccharides, are crucial for biofilm formation by *E. coli*. Small molecules, such as cyclic-di-GMP, acetyl phosphate, ppGpp, *N*-acetyl-glucosamine, *N*-acetyl-glucosamine-6-P, and QS molecules, play an important role in the switch from planktonic to sessile life (Beloin et al., 2008). The zoonotic Shiga toxin-producing *E. coli* (STEC) O157:H7 is a food-borne and water-borne pathogen that causes diarrhea, hemorrhagic colitis, and hemolytic-uremic syndrome in humans (Chekabab et al., 2013). It was shown that the patients fed by percutaneous endoscopic gastroscopy (PEG), a common form of enteral nutrition in which the feeding tube passes through the abdomen to the stomach, are especially susceptible to *E. coli* diarrhea (Macfarlane and Dillon, 2007). The genome sequence of O157:H7 strains are approximately 75% homologous to *E. coli* chromosome; the remaining 25% are O157:H7-specific sequences, most of which have been horizontally transferred and carry pathogenicity determinants, the most important of which is locus of the enterocyte effacement (LEE) island (McDaniel et al., 1995). Another pathogenicity island codes for additional virulence genes (Croxen and Finlay, 2010). In addition to causing infection in humans through the consumption of contaminated food, *E. coli* is able to survive for a long time in water, which is an environmental threat to humans (Chekabab et al., 2013).

Among other diseases caused by *E. coli*, the most dangerous are urinary tract infections (UTIs) and chronic bacterial prostatitis. *E. coli* causes 80–90% of community-acquired UTIs and 30–50% of nosocomically acquired UTIs; a substantial amount of these can be considered recurrent (RUTI). The species that is the cause of persistent or relapsed infections belongs to the phylogenetic group B2. Until now, no distinct virulence profile could predict recurrence of UTI. Recently, it was observed that *E. coli* can invade and replicate within the murine bladder-forming biofilm-like structures that may represent stable reservoirs for RUTIs (Elimaes, 2011). *E. coli* is only one of several bacterial species isolated in cases of chronic bacterial prostatitis. The prostatic gland may be infected by bacteria that have ascended from the urethra or by the reflux of infected urine into the prostatic ducts. The infecting bacteria form biofilms firmly adhered to the ductal and acinar mucosal layers and can be eradicated only on delivery of high concentrations of antibiotics directly into the prostatic ducts (Nickel et al., 1994).

The members of another genus of the Enterobacteriaceae family, *Salmonella* spp., are the primary enteric pathogens that reach the intestinal epithelium and trigger gastrointestinal diseases. In some patients *Salmonella* is internalized within phagocytes and subsequently disseminates. The two major clinical syndromes caused by *Salmonella* infections are enteric or typhoid fever and colitis/diarrheal disease. The most important virulence genes are located within five *Salmonella* pathogenicity islands (SPIs) (Fábrega and Vila, 2013). Despite that different serovars of *Salmonella* can cause food-borne illness outbreaks, the most common causal agents are *Salmonella enterica* serovar Typhimurium and *S. enterica* serovar Enteritidis (Kozak et al., 2013).

The ability to form biofilm highly contributes to the virulence of *Salmonella*, mainly by promoting its survival within the host. Efficient biofilm formation depends on exopolysaccharide cellulose and curli fimbriae (Austin et al., 1998). *Salmonella*, especially *Salmonella typhi*, can enter the carrier state, which leads to persistent colonization. Bacteria shed into the gallbladder and form a biofilm on the surface of gallstones, which protects *Salmonella* against high concentrations of the bile (Prouty et al., 2002). The ability to form biofilm is strain-dependent, and this dependence is well-documented in the case of *S. enterica* (Lianou and Koutsoumanis, 2013).

Yet another member of Enterobacteriaceae, genus *Shigella*, is an important bacterial cause of infectious diarrhea, mainly in endemic regions. *Shigella* infection is facilitated by a low-infective dose and fecal–oral route of transmission, thereby allowing spread of the infection through contaminated food or by contact. A large 220-kb plasmid that harbors the genes required for invasion, dissemination inside the host cells, and induction of the host inflammatory response is crucial for *Shigella* pathogenicity (Schroeder and Hilbi, 2008). Moreover, pathogenicity of *Shigella* is enhanced by the presence of the antivirulence genes, inhibiting, for example, oxidative stress survival and transepithelial migration (Bilven and Maurelli, 2012) and also integron-mediated resistance to antibiotics (Ke et al., 2011). The number of literature positions on *Shigella* biofilm and its involvement in pathogenesis is scarce, even though the members of this genus are distinguished biofilm formers. For example, it was shown that *Shigella flexneri* homologue of *shf* gene is required for biofilm formation by enteroaggregative *E. coli* 042 containing the virulence plasmid pAA2 (Fujiyama et al., 2008).

8.3 BIOFILM RESISTANCE TO CONVENTIONAL ANTIBIOTICS AND NEW ALTERNATIVE STRATEGIES TO COMBAT BACTERIAL BIOFILMS

Biofilms are known to be adaptively resistant multicellular structures that are difficult to eradicate with the currently available antimicrobials. It was proven that bacterial cells forming a biofilm are able to survive 10-times to 1,000-times higher doses of antibiotics than planktonic cells of the same bacterial species (Mah and O'Toole, 2001). Difficulties in the treatment of biofilms with conventional antibiotics are caused by diverse factors. The certain cause of biofilm resistance is the presence of extracellular matrix that protects cells from different agents and may cause failure of the antimicrobial to penetrate the biofilm (de la Fuente-Núñez et al., 2013). Heterogeneity of the population within a biofilm is caused by decreasing oxygen and nutrient gradients that exist between the surface and deeper layers of the biofilm. This leads to the differentiation of the cells when they metabolize (Mah and O'Toole, 2001). General stress responses cause changes in the expression of many genes and enable bacteria to resist harmful conditions. The biofilm resistance may also be caused by persister cells that can withstand the presence of high concentrations of antimicrobials (de la Fuente-Núñez et al., 2013). Some reports describe specific mechanisms, for example, efflux pumps from *P. aeruginosa* that influence the resistance to aminoglycosides and fluoroquinolones only in bacterial biofilms (Zhang and Mah, 2008). Because of the need for more effective biofilm dissolution treatments, a number of alternative strategies have been devised. Here, we represent some of the most recent advances in the strategies designed to hamper biofilm formation by killing bacterial cells or targeting different stages of biofilm development (Figure 8.1).

Plant-derived compounds may be considered new antimicrobials because of their proven and substantial antibacterial effects. Recent studies describe biological activity of the secondary plant metabolites such as polyphenols, which effectively reduced biofilm formation by *Streptococcus mutans* (Percival et al., 2006). Another interesting group of compounds are terpenoids; in this group, two pentacyclic triterpenoids (oleanolic

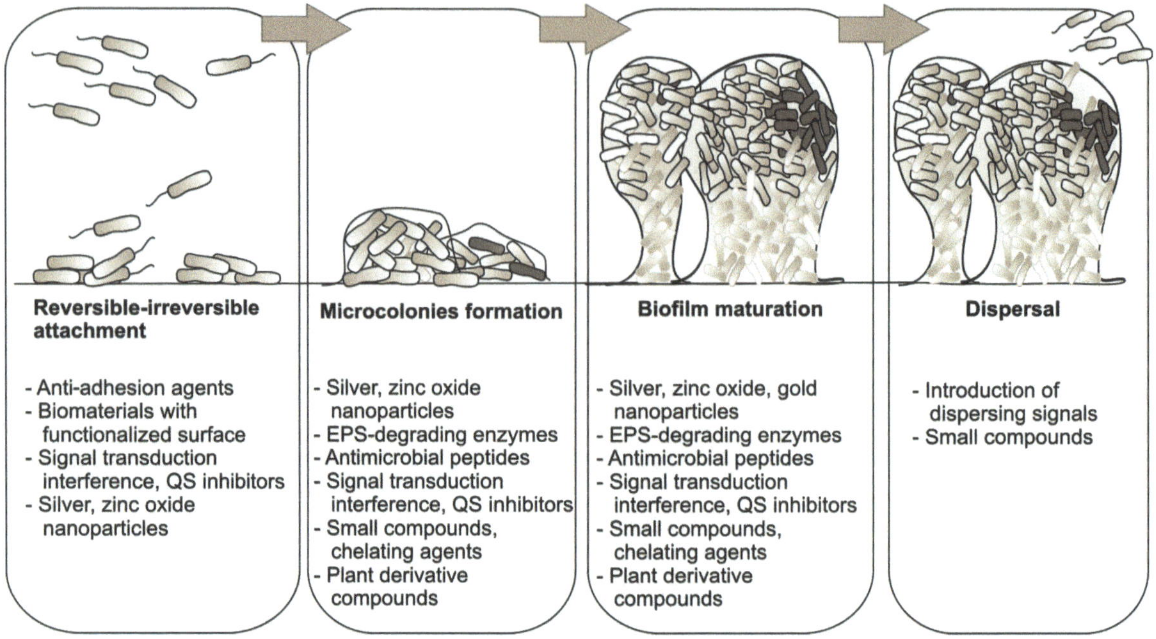

FIGURE 8.1 Exemplary strategies used for inhibiting or disrupting bacterial biofilm formation at specific stages of its development.

acid and ursolic acid) and their derivatives are of special interest. It was shown that ursolic acid inhibits biofilm formation in several bacterial species and other terpenoids have potent antibiofilm activity against staphylococci (for a review see Kurek et al., 2011; Wolska et al., 2012).

High-throughput screens have recently revealed small compounds with antibiofilm properties. Small molecules may modify cell envelopes, directly disassemble the extracellular matrix, or trigger and initiate dispersal of biofilms (for a review see Oppenheimer-Shaanan et al., 2013). Novel bacterial biofilm formation inhibitors need further examination; however, they can be used as therapeutic agents that may be considered alternatives to antibiotics.

Because of the important role of cell-to-cell communication in the process of biofilm formation, molecules able to inhibit this system are currently under investigation. It is suggested that QS inhibitors are effective in controlling biofilm-related bacterial infections while having no negative effect on human cells (Defoirdt et al., 2013). Interestingly, many plant-derived molecules can act as QS inhibitors. Examples may be the compounds found in green tea extract, which modulate QS-related antivirulent activities of *P. aeruginosa* (Mihalik et al., 2008), or ajoene, a molecule from garlic that inhibits genes controlled by QS of the same bacterial species (Jakobsen et al., 2012).

Another approach to target biofilms is the use of synthetic cationic peptide variants derived from natural antimicrobial peptides. Modifications of synthetic peptides by an addition of cationic residues or modification of the proportion of hydrophobic residues enable the modulation of the antibiofilm and antiplanktonic activity of these molecules. Such peptides that can effectively prevent biofilm formation by Gram-positive and Gram-negative bacteria have been described (de la Fuente-Núñez et al., 2012).

TABLE 8.1 Examples of Anti-Biofilm Compounds and Strategies

Compound/strategy	Exemplary literature
Plant metabolites, for example, phenolics, terpenoids	Percival et al. (2006); Ren et al. (2005)
QS inhibitors	Defoirdt et al. (2013); Mihalik et al. (2008); Jakobsen et al. (2012)
Antimicrobial peptides	de la Fuente-Núñez (2012)
Artificial biomaterials	Desrousseaux et al. (2013)
AgNPs	Mohanty et al. (2012); Ramalingam et al. (2013); Roe et al. (2008)
AuNPs	Zhou et al. (2012); Raftery et al. (2013)
Other metal NPs	Applerot et al. (2012)
Photodynamic therapy	de Melo et al. (2013)
Combined therapy	Ammons et al. (2011); Garo et al. (2007)

Antimicrobial photodynamic therapy (aPDT) involves the use of a combination of a nontoxic dye and low-intensity visible light that, in the presence of oxygen, produces cytotoxic reactive oxygen species. It has been demonstrated that many biofilms are susceptible to aPDT, particularly in dental diseases (for a review see de Melo et al., 2013).

Recent reports showed that the newly designed biomaterials may inhibit biofilm formation by impeding bacterial adhesion. Therefore, the latest approaches concern the designing of new devices with surfaces that are capable of limiting bacterial adhesion and/or viability (for a review see Desrousseaux et al., 2013). The majority of the new nanotechnological approaches to combat biofilm formation are based on the use of NPs to functionalize the surface of biomaterials by coating (Roe et al., 2008; Applerot et al., 2012), impregnation (Shi et al., 2006), or by embedding nanomaterials (Beyth et al., 2008). The antibacterial compound and strategies described are listed in Table 8.1.

8.4 ANTIBIOFILM ACTIVITY OF METAL NPs

Metal NPs are composed of clusters of atoms and have a size ranging from 1 to 100 nm. These compounds are different from colloids that contain NPs as well as some particles in the size range of 100 to 2,500 nm (Lemire et al., 2013). Some metals, like silver or zinc, are known for their natural antibacterial activity. It was established that in nano sizes, biological properties of metals are stronger, making NPs very interesting from the medical point of view.

The antibacterial mechanism of all of the varieties of NPs is not fully understood. It has already been stated that NP activity arises from the increased surface-to-volume ratio. The surface area of a dose of NPs increases as the particle size decreases, allowing for a greater material interaction with the surrounding environment (Seil and Webster, 2012). Small NPs appear to be the most capable of penetrating into bacterial cells. Besides the particle size, their shape, zeta potential (the electrical potential at the interface between the particle surface and the bulk solution), and chemistry are also among the most relevant variables affecting antibacterial activity. A strongly positive zeta potential of an NP promotes NPs interactions with the cell membranes, membrane disruption, bacteria flocculation, and a reduction in viability. Therefore, it was suggested that zeta potential, along with particle size and chemistry, are highly relevant parameters responsible for antimicrobial effects (Seil and Webster, 2012).

8.4.1 Silver NPs

Because of novel antibacterial properties of silver NPs (AgNPs), more and more recent studies are concentrating on their antibiofilm characteristics. However, the number of these reports is still incomparable with the work showing AgNP activity toward planktonic cells. Moreover, the mechanisms of AgNP antibacterial activity remain not fully understood, certainly because of the pleiotropic effect, and further studies are needed to be undertaken to reveal this in detail (for a review see Markowska et al., 2013). The literature concerning the mechanisms of AgNP antibiofilm activity is so far scarce. Kalishwaralal et al. (2010) reported potential antibiofilm activity of AgNPs that was tested *in vitro* on 24-h biofilms formed by *P. aeruginosa* and *S. epidermidis*. Treating these bacteria with AgNPs (spherical in shape, with a mean diameter of 50 nm) resulted in more than 95% inhibition of biofilm formation. Interestingly, the observed inhibition was invariable of the species tested. This result is inconsistent with the numerous previous studies conducted on planktonic bacterial cells that showed higher effectiveness of AgNPs toward Gram-negative species (Seil and Webster, 2012).

The antibiofilm activity of AgNPs on various microbial strains was reported. For example, antibiofilm activity of AgNPs (average size of 35.5 nm) was shown against primary biofilm-forming bacteria *P. aeruginosa* and *S. aureus* (Ramalingam et al., 2013). Another recent work showed that AgNPs (average diameter of 12.6 ± 5.7 nm) are also effective against the *Mycobacterium* sp. biofilms. Nanosilver at a concentration of 100 μM almost completely inhibited survival of tested bacteria (Islam et al., 2013). In yet another study, AgNPs (spherical in shape with diameters ranging from 15 to 34 nm) were also successfully applied in the inhibition of marine biofilms (Inbakandan et al., 2013).

However, not all of the studies show high antibiofilm activity of AgNPs. As stated, biofilms remain resistant to various antimicrobials. Choi et al. (2010) suggested that the biofilm resistance to nanosilver could be at least partially caused by NP aggregation and slowed silver ion/particle diffusion. They showed that the nanosilver (average particle size of 15–21 nm) was aggregated in the presence of biofilm-forming *E. coli* cells, resulting in a 40-times increase of the average particle size (Choi et al., 2010).

Many reports show the activity of AgNPs that was tested on biofilms generated under static conditions, but there is also a study involving a bioreactor. It was shown that AgNPs effectively prevented the formation of biofilms and killed bacteria in established biofilms. Authors recorded a 4-log reduction in the number of colony-forming units of *P. aeruginosa* under turbulent fluid conditions in the CDC reactor on exposure to 100 mg/mL of AgNPs (average particle size of 25.2 ± 4 nm) (Martinez-Gutierrez et al., 2013). Other studies were conducted with the use of catheters, because biofilms covering biomaterial surfaces remain a serious medical problem. Roe et al. (2008) used nanosilver-coated (average diameter of 10 nm) catheters that exhibited almost complete inhibition of biofilm formation for *E. coli*, *S. aureus*, and *Candida albicans*, and more than 50% inhibition for *Enterococcus* sp., coagulase-negative staphylococci, and *P. aeruginosa* after 72 h of incubation (Roe et al., 2008). Paladini et al. (2013) used temporary catheters coated with AgNPs for hemodialysis. These studies showed the efficacy of the obtained silver coating in inhibiting adhesion and biofilm formation by *S. aureus* and *E. coli*, even in flow conditions, for the whole working life of the catheters. Noteworthy, Besinis et al. (2014) showed that the application of a silver nano-coating directly on dentine can successfully prevent the biofilm formation on dentine surfaces and can inhibit bacterial growth in the surrounding media.

Some researchers propose using stabilized or modified AgNPs. Such amendments may impede NP aggregation. An example may be starch-stabilized AgNPs (the average particle size of

20 nm) that disrupt *P. aeruginosa* and *S. aureus* biofilm formation at very low concentrations of 1–2 mM and may exhibit higher antibacterial activity compared with human cationic antimicrobial peptide LL-37 (Mohanty et al., 2012). Park et al. (2013) showed that citrate-capped AgNPs (average particle size of 7.9 nm) efficiently inactivated *P. aeruginosa* biofilms. Interestingly, Hartmann et al. (2013) used a new calorimetric method that allows for noninvasive and real-time investigation of the effects of NPs on bead-grown biofilms by the detection of bacterial metabolic changes. Their investigation showed antibacterial effects of AgNPs on *Pseudomonas putida* biofilms grown on agarose beads. Another novel antibiofilm strategy is associated with the usage of a monolayer of AgNPs anchored to an amino-silanized glass surface that showed antibiofilm activity against *S. epidermidis* RP62A. However, authors suggested that the observed effect was caused by prolonged release and a high local concentration of Ag^+ without any detachment of AgNPs (Taglietti et al., 2014). This approach has potential in biomaterial designing and producing, for example, new antibacterial implants.

8.4.2 Zinc NPs

Strong antibiofilm properties of zinc NPs were reported. However, all conducted studies with bacterial biofilms were concerned with zinc oxide (ZnO) particles. Applerot et al. (2012) evaluated the ability of glass slides coated with zinc oxide NPs to restrict the biofilm formation by *E. coli* and *S. aureus*. The glass slides were challenged with bacterial cultures for 10 consecutive days using a continuous flow chamber. The ZnO-coated surfaces dramatically restricted bacterial colonization (to ~20 and ~0 cfu/cm^2 for *E. coli* and *S. aureus*, respectively, on the last day of the study). It was also shown that other Gram-positive bacteria may be more susceptible to the reduction in viability by ZnO NPs than Gram-negative bacteria (Seil and Webster, 2012). Other studies of ZnO NPs also involved the use of a bacterial model of dental biofilm. Antibacterial effectiveness of ZnO NPs was assessed against *Streptococcus sobrinus* ATCC 27352 grown as biofilms on composites. The results revealed a statistically significant suppression of the biofilm growth. Although 20% of the bacterial population survived and could form a biofilm layer again, 10% of ZnO NP-containing composites maintained at least some inhibitory activity even after the third generation of the biofilm growth (Aydin Sevinç and Hanley, 2010). Another study also showed that zinc oxide NPs retain the antibacterial properties even after aging for 90 days. Here, NPs were effective in bacterial biofilm reduction and disruption of biofilm structure (Shrestha et al., 2010).

It was also demonstrated that ZnO NPs reduced the growth of *Enterobacter* sp. by 50%, whereas 80% reduction was observed in halophilic *Marinobacter* sp. In case of halophiles, it was suggested that such susceptibility to NPs may be attributed to the higher content of negatively charged cardiolipins on their cell surface (Sinha et al., 2011).

8.4.3 Gold NPs

There are reports concerning the strong antibacterial activity of gold NPs (AuNPs). The research of Zhou et al. (2012) demonstrated that AuNPs (spherical, 20 – 30 nm in diameter) showed excellent antibacterial potential against Gram-negative bacteria *E. coli* and Gram-positive bacteria *Bacillus* Calmette–Guerin. This study proposed that the antibacterial activity could result from the uptake of a single AuNP by bacteria and their rearrangement inside the cytoplasm (Zhou et al., 2012). However, the literature concerning antibiofilm properties of AuNPs is scarce. One example may be the study performed by Sathyanarayanan et al. (2013), who evaluated the effect of AuNPs (less than 10 nm in size) on biofilm-forming pathogens such as *S. aureus* and *P. aeruginosa*. In this work, the biofilm growth was significantly reduced only at higher concentrations of AuNPs (0.05, 0.10, and 0.15 mg/mL) compared with that in the absence of NPs. Another work showed that melanin-derived NPs synthesized by cold-adapted yeast also displayed antibiofilm activity (Nair et al., 2013).

Noteworthy is a study by Raftery et al. (2013), who demonstrated significant changes in *Legionella pneumophila* biofilm morphology after exposure to very low AuNPs concentration of 0.7 μg/L. In further studies it was stated that these morphological changes alter host–bacteria interactions (using *Acanthamoeba polyphaga* as a model host species). The biofilms exposed to AuNPs significantly altered the grazing ability of *A. polyphaga*, which was not observed in the biofilms exposed to 24-nm polystyrene beads.

Interestingly, Khan et al. (2012) showed that AuNPs enhance methylene blue-induced photodynamic therapy that can be a novel therapeutic approach to inhibit *C. albicans* biofilm. There are results showing a significant reduction of *Candida* biofilm in the presence of the conjugate. Authors suggest that this approach may be used against nosocomially acquired refractory *C. albicans* biofilms.

There is some work that compares the results of antibiofilm activity of AuNPs and AgNPs in which AuNPs are much less effective in microbial biofilms reduction or do not show any remarkable antibiofilm activity (Vijayan et al., 2014).

Therefore, despite the reports confirming the antibiofilm properties of AuNPs, more perspective regarding the usage of AuNPs in molecular imaging or as carrier molecules for other therapeutics is needed. Because of better antibiofilm properties of the first two types of NPs, nanogold is less promising in this field of research.

8.5 CONCLUSIONS

The potency of metal NPs to inhibit biofilms of many bacterial species, including pathogens, is well-proven by much research. AgNPs are most active in preventing biofilm formation, killing bacterial cells in biofilm, and also in the eradication of already formed biofilms. AuNPs are good carriers of other antibiofilm compounds, including antibiotics. The wide use of NPs as antimicrobials, including antibiofilm therapy, is so far restricted because of their side effects on eukaryotic cells, limited knowledge about their cellular target(s), and the molecular mechanisms of their activity. Considering the great scientific effort put forth to study these problems, they should be resolved in the near future.

References

Allesen-Holm, M., Barken, K.B., Yang, L., Klausen, M., Webb, J.S., Kjelleberg, S., et al., 2006. A characterization of DNA release in *Pseudomonas aeruginosa* cultures and biofilms. Mol. Microbiol. 59, 1114–1128.

Ammons, M.C., Ward, L.S., James, G.A., 2011. Anti-biofilm efficacy of a lactoferrin/xylitol wound hydrogel used in combination with silver wound dressings. Int. Wound J. 8, 268–273.

Antunes, L.C.M., Ferreira, R.B.R., 2009. Intracellular communication in bacteria. Crit. Rev. Microbiol. 35, 69–80.

Applerot, G., Lellouche, J., Perkas, N., Nitzan, Y., Gedanken, A., Banin, E., 2012. ZnO nanoparticle-coated surfaces inhibit bacterial biofilm formation and increase antibiotic susceptibility. RSC Advances 2, 2314.

Archer, N.K., Mazaitis, M.J., Costerton, W., Leid, J.G., Powers, M.E., Shirtiff, M.E., 2011. *Staphylococcus aureus* biofilms. Properties, regulation and roles in human disease. Virulence 25, 445–459.

Austin, J.W., Sanders, G., Kay, W.W., Collinson, S.K., 1998. Thin aggregative fimbriae enhance *Salmonella enteritidis* biofilm formation. FEMS Microbiol. Lett. 162, 295–301.

Aydin Sevinç, B., Hanley, L., 2010. Antibacterial activity of dental composites containing zinc oxide nanoparticles. J. Biomed. Mater. Res. 94B, 22–31.

Barraud, N., Buson, A., Jarolimek, W., Rice, S.A., 2013. Mannitol enhances antibiotic sensitivity of persister bacteria in *Pseudomonas aeruginosa* biofilms. PLoS One 8, e84220.

Beenken, K.E., Mrak, L.N., Griffin, L.M., Zielinska, A.K., Shaw, L.N., Rice, K., et al., 2010. Epistatic relationship between *sarA* and *agr* in *Staphylococcus aureus* biofilm formation. PLos One 5, 10790.

Beloin, C., Roux, A., Ghigo, J.-M., 2008. *Escherichia coli* biofilms. Curr. Top. Microbiol. Immunol. 322, 249–289.

Besinis, A., De Peralta, T., Handy, R.D., 2014. Inhibition of biofilm formation and antibacterial properties of a silver nano-coating on human dentine. Nanotoxicology 8, 745–754.

Beyth, N., Houri-Haddad, Y., Baraness-Hadar, L., Yudovin-Farber, I., Domb, A.J., Weiss, E.I., 2008. Surface antimicrobial activity and biocompatibility of incorporated polyethylenimine nanoparticles. Biomaterials 29, 4157–4163.

Bilven, K.A., Maurelli, A.T., 2012. Antivirulence genes: insights into pathogen evolution through gene loss. Infect. Immun. 80, 4061–4070.

Chekabab, S.M., Paquin-Veillette, J., Dozois, C.M., Harel, J., 2013. The ecological habitat and transmission of *Escherichia coli* O157:H7. FEMS Microbiol. Lett. 341, 1–12.

Choi, O., Yu, C.-P., Esteban Fernández, G., Hu, Z., 2010. Interactions of nanosilver with *Escherichia coli* cells in planktonic and biofilm cultures. Water Res. 44, 6095–6103.

Costerton, J.W., Steward, P.S., Greenberg, E.P., 1999. Bacterial biofilms: a common cause of persistent infections. Science 284, 1318–1322.

Costerton, J.W., Montanaro, I., Arciola, C.R., 2005. Biofilm in implant infections: its production and regulation. Int. J. Artif. Organs 28, 1062–1068.

Croxen, M.A., Finlay, B.B., 2010. Molecular mechanisms of *Escherichia coli* pathogenicity. Nat. Rev. Microbiol. 8, 26–38.

Davey, M.E., O'Toole, G.A., 2000. Microbial biofilms: from ecology to molecular genetics. Microbiol. Mol. Biol. Rev. 64, 847–867.

Defoirdt, T., Brackman, G., Coenye, T., 2013. Quorum sensing inhibitors: how strong is the evidence? Trends Microbiol. 21, 619–624.

de la Fuente-Núñez, C., Korolik, V., Bains, M., Nguyen, U., Breidenstein, E.B.M., Horsman, S., et al., 2012. Inhibition of bacterial biofilm formation and swarming motility by a small synthetic cationic peptide. Antimicrob. Agents Chemother. 56, 2696–2704.

de la Fuente-Núñez, C., Reffuveille, F., Fernández, L., Hancock, R.E., 2013. Bacterial biofilm development as a multicellular adaptation: antibiotic resistance and new therapeutic strategies. Curr. Opin. Microbiol. 16, 580–589.

de Melo, W.C., Avci, P., de Oliveira, M.N., Gupta, A., Vecchio, D., Sadasivam, M., et al., 2013. Photodynamic inactivation of biofilm: taking a lightly colored approach to stubborn infection. Expert Rev. Anti. Infect. Ther. 11, 669–693.

Desrousseaux, C., Sautou, V., Descamps, S., Traoré, O., 2013. Modification of the surfaces of medical devices to prevent microbial adhesion and biofilm formation. J. Hosp. Infect. 85, 87–93.

Donlan, R.M., Costerton, J.W., 2002. Biofilms: survival mechanisms of clinically relevant microorganisms. Clin. Microbiol. Rev. 15, 167–193.

Dowd, S.E., Wolcott, R.D., Kennedy, J., Jones, C., Cox, S.B., 2011. Molecular diagnosis and personalized medicine in wound care: assessment of outcomes. J. Wound Care 20, 232–239.

Elias, S., Banin, E., 2012. Multi-species biofilms: living with friendly neighbors. FEMS Microbiol. Rev. 36, 990–1004.

Elimaes, K., 2011. Bacterial characteristics of importance for recurrent urinary tract infections caused by *Escherichia coli*. Dan. Med. Bull. 58, B4187.

Fábrega, A., Vila, J., 2013. *Salmonella enterica* serovar typhimurium skills to succeed in the host: virulence and regulation. Clin. Microbiol. Rev. 26, 308–341.

Ferguson, B.J., Stolz, D.B., 2005. Demonstration of biofilm in human bacterial chronic rhinosinusitis. Am. J. Rhinol. 19, 452–457.

Foster, T.J., Geoghegan, J.A., Ganesh, V.K., Höök, M., 2014. Adhesion, invasion and evasion: the many functions of the surface proteins of *Staphylococcus aureus*. Nat. Rev. Microbiol. 12, 49–62.

Fugere, A., Séguin, D.L., Mitchell, G., Déziel, E., Dekimpe, V., Cantin, A.M., et al., 2014. Interspecies small molecule interactions between clinical isolates of *Pseudomonas aeruginosa* and *Staphylococcus aureus* from adult cystic fibrosis patients. PLoS One 9, e86705.

Fujiyama, R., Nishi, J., Imuta, N., Tokuda, K., Manago, K., Kawano, Y., 2008. The *shf* gene of *Shigella flexneri* homologue on the virulent plasmid pAA2 of enteroaggregative *Escherichia coli* 042 is required for biofilm formation. Curr. Microbiol. 56, 474–480.

Garo, E., Eldridge, G.R., Goering, M.G., DeLancey Pulcini, E., Hamilton, M.A., Costerton, J.W., et al., 2007. Asiatic acid and corosolic acid enhance the susceptibility of *Pseudomonas aeruginosa* biofilms to tobramycin. Antimicrob. Agents Chemother. 51, 1813–1817.

Giebink, G., Junk, S.S.K., Weber, M.I., Le, C.T., 1982. The bacteriology and cytology of chronic otitis media with effusion. Pediatr. Infecet. Dis. 1, 98–103.

Gjodsbol, K., Christensen, J.J., Karlsmark, T., Jorgensen, B., Klein, B.M., Krogfeld, K.A., 2006. Multiple bacterial species reside in chronic wounds: a longitudinal study. Int. Wound J. 3, 225–231.

Gurjala, A.N., Geringer, M.R., Seth, A.K., Hong, S.J., Smeltzer, M.S., Galiano, R.D., et al., 2011. Development of novel, highly quantitative *in vivo* model for the study of biofilm-impaired cutaneous wound healing. Wound Rep. Regen. 19, 400–410.

Hancock, R.E., Speert, D.P., 2001. Antibiotic resistance in *Pseudomonas aeruginosa*: mechanism and impact of treatment. Drug Resist. Updat. 3, 247–255.

Hartmann, T., Mühling, M., Wolf, A., Mariana, F., Maskow, T., Mertens, F., et al., 2013. A chip-calorimetric approach to the analysis of Ag nanoparticle caused inhibition and inactivation of beads-grown bacterial biofilms. J. Microbiol. Methods 95, 129–137.

Heitz-Mayfield, I.J., Lang, N.P., 2000. Comparative biology of chronic and aggressive peridondontitis. Peridontology 53, 167–181.

Hoffman, L.R., Ramsey, B.W., 2013. Cystic fibrosis therapeutics: the road ahead. Chest. 143, 207–213.

Hoiby, N., Krough Johansen, H., Moser, C., Song, Z., Ciofu, O., Kharazmi, A., 2001. *Pseudomonas aeruginosa* and the *in vitro* and *in vivo* biofilm mode of growth. Microbes Infect. 3, 23–35.

Hoiby, N., Bjarnsholt, T., Givskov, M., Molin, S., Ciofu, O., 2010. Antibiotic resistance of bacterial biofilms. Int. J. Antimicrob. Agents 35, 322–332.

Hutchison, M.L., Govan, J.R., 1999. Pathogenicity of microbes associated with cystic fibrosis. Microbes Infect. 1, 1005–1014.

Inbakandan, D., Kumar, C., Abraham, L.S., Kirubagaran, R., Venkatesan, R., Khan, S.A., 2013. Silver nanoparticles with anti microfouling effect: a study against marine biofilm forming bacteria. Colloids Surf. B. 111C, 636–643.

Islam, M.S., Larimer, C., Ojha, A., Nettleship, I., 2013. Antimycobacterial efficacy of silver nanoparticles as deposited on porous membrane filters. Mater. Sci. Eng. C. Mater. Biol. Appl. 33, 4575–4581.

Jakobsen, T.H., van Gennip, M., Phipps, R.K., Shanmugham, M.S., Christensen, L.D., Alhede, M., et al., 2012. Ajoene, a sulfur-rich molecule from garlic, inhibits genes controlled by quorum sensing. Antimicrob. Agents Chemother. 56, 2314–2325.

Jensen, P.O., Givskov, M., Bjarnsholt, T., Moser, C., 2010. The immune system vs. *Pseudomonas aeruginosa* biofilms. FEMS Immunol. Med. Microbiol. 59, 292–305.

Jiricny, N., Molin, S., Foster, K., Diggle, S.P., Scanian, P.D., Ghoul, M., et al., 2014. Loss of social behaviours in populations of *Pseudomonas aeruginosa* infecting lungs patients with cystic fibrosis. PLoS One 9, e83124.

Kalishwaralal, K., BarathManiKanth, S., Pandian, S.R.K., Deepak, V., Gurunathan, S., 2010. Silver nanoparticles impede the biofilm formation by *Pseudomonas aeruginosa* and *Staphylococcus epidermidis*. Colloids Surf. B. 79, 340–344.

Karatan, E., Watnick, P., 2009. Signals, regulatory networks, and materials that build and break bacterial biofilms. Microbiol. Mol. Biol. Rev. 73, 310–347.

Ke, X., Gu, B., Pan, S., Tong, M., 2011. Epidemiology and molecular mechanism of intergron-mediated antibiotic resistance in *Shigella*. Arch. Microbiol. 193, 767–774.

Khan, S., Alam, F., Azam, A., Khan, A.U., 2012. Gold nanoparticles enhance methylene blue-induced photodynamic therapy: a novel therapeutic approach to inhibit *Candida albicans* biofilm. Int. J. Nanomed. 7, 3245–3257.

Kluymans, J., van Belkum, A., Verbrugh, H., 1997. Nasal carriage of *Staphylococcus aureus*: epidemiology underlying mechanisms and associated risks. Clin. Microbiol. Rev. 10, 505–520.

Kobayashi, S.D., DeLeo, F.R., 2009. Role of neutrophils in innate immunity: a systems biology-level approach. Wiley Interdiscip. Rev. Syst. Biol. Med. 1, 309–333.

Kotler, R., 2010. Biofilms in lab and nature: a molecular geneticist's voyage to microbial ecology. Int. Microbiol. 13, 1–7.

Kozak, G.K., Macdonald, D., Landry, I., Farber, I.M., 2013. Foodborne outbreaks in Canada linked to produce: 2001 through 2009. J. Food Prot. 76, 173–183.

Kurek, A., Grudniak, A.M., Kraczkiewicz-Dowjat, A., Wolska, K.I., 2011. New antibacterial therapeutics and strategies. Pol. J. Microbiol. 60, 3–12.

Lemire, J.A., Harrison, J.J., Turner, R.J., 2013. Antimicrobial activity of metals: mechanisms, molecular targets and applications. Nat. Rev. Microbiol. 11, 371–384.

Lew, D.P., Waldvogel, F.A., 2004. Osteomyelitis. Lancet 364, 369–379.

Lewis, K., 2010. Persister cells. Annu. Rev. Microbiol. 64, 357–372.

Leid, J.G., Costerton, J.W., Shirtliff, M.E., Gilmore, M.S., Engelbert, M., 2002. Immunology of staphylococcal biofilm infection in the eye: new tools to study biofilm endophthalmitis. DNA Cell Biol. 21, 405–413.

Lianou, A., Koutsoumanis, K.P., 2013. Strain variability of the behavior of foodborne bacterial pathogens: a review. Int. J. Food Microbiol. 167, 310–321.

Mack, D., Fischer, W., Krokotsch, A., Leaopold, K., Hartman, R., Egge, H., 1996. The intracellular adhesin involved in biofilm accumulation of *Staphyloccocus epidermidis* is a liner beta-1,6-linked glucosaminoglycan: purification and structural analysis. J. Bacteriol. 178, 175–185.

Macfarlane, S., Dillon, J.F., 2007. Microbial biofilms in the human gastrointestinal tract. J. Appl. Microbiol. 192, 1187–1196.

Mah, T.-F.C., O'Toole, G.A., 2001. Mechanisms of biofilm resistance to antimicrobial agents. Trends Microbiol. 9, 34–39.

Markowska, K., Grudniak, A.M., Wolska, K.I., 2013. Silver nanoparticles as an alternative strategy against bacterial biofilms. Acta Biochim. Pol. 60, 523–530.

Martinez-Gutierrez, F., Boegli, L., Agostinho, A., Sánchez, E.M., Bach, H., Ruiz, F., et al., 2013. Anti-biofilm activity of silver nanoparticles against different microorganisms. Biofouling 29, 651–660.

McDaniel, T.K., Jarvis, K.G., Donnenberg, M.S., Kaper, J.B., 1995. A genetic locus of enterocyte effacement conserved among diverse enterobacterial pathogens. Proc. Natl. Acad. Sci. USA 92, 1664–1668.

Mihalik, K., Chung, D.W., Crixell, S.H., McLean, R.J.C., Vattem, D.A., 2008. Quorum sensing modulators of *Pseudomonas aeruginosa* characterized in *Camellia sinensis*. Asian J. Trad. Med. 3, 12–23.

Mohanty, S., Mishra, S., Jena, P., Jacob, B., Sarkar, B., Sonawane, A., 2012. An investigation on the antibacterial, cytotoxic, and antibiofilm efficacy of starch-stabilized silver nanoparticles. Nanomedicine 8, 916–924.

Moreau-Marquis, S., Stanton, B.A., O'Toole, G.A., 2008. *Pseudomonas aeruginosa* biofilm formation in the cystic fibrosis airway. A short review. Pulm. Pharmacol. Ther. 21, 595–599.

Nair, V., Sambre, D., Joshi, S., Bankar, A., Ravi Kumar, A., Zinjarde, S., 2013. Yeast-derived melanin mediated synthesis of gold nanoparticles. J. Bionanosci. 7, 159–168.

Nicholas, R.O., Li, T., McDevitt, D., Marra, A., Sucoloski, S., Demarsh, P.L., et al., 1999. Isolation and characterization of *sigB* deletion mutant of *Staphyloccocus aureus*. Infect. Immun. 67, 3667–3669.

Nichols, W.W., Evans, M.J., Slack, M.P., Walmsley, H.L., 1989. The penetration of antibiotics into aggregates of mucoid and non-mucoid *Pseudomonas aeruginosa*. J. Gen. Microbiol. 135, 1291–1303.

Nickel, J.C., Costerton, J.W., McLean, R.J.C., Olson, M., 1994. Bacterial biofilms: influence on the pathogenesis, diagnosis and treatment of urinary tract infections. J. Antimicrob. Chemother. 33 (Suppl. A), S31–S41.

Oppenheimer-Shaanan, Y., Steinberg, N., Kolodkin-Gal, I., 2013. Small molecules are natural triggers for the disassembly of biofilms. Trends Microbiol. 21, 594–601.

Overhage, J., Campisano, A., Bains, M., Torfs, E.C.W., Rehm, B.H.A., Hancock, R.E.W., 2008. Human host defensive peptide LL-37 prevents bacterial biofilm formation. Infect. Immun. 76, 4176–4182.

Paladini, F., Pollini, M., Deponti, D., Di Giancamillo, A., Peretti, G., Sannino, A., 2013. Effect of silver nanocoatings on catheters for haemodialysis in terms of cell viability, proliferation, morphology and antibacterial activity. J. Mater. Sci. Mater. Med. 24, 1105–1112.

Park, H.-J., Park, S., Roh, J., Kim, S., Choi, K., Yi, J., et al., 2013. Biofilm-inactivating activity of silver nanoparticles: a comparison with silver ions. J. Ind. Eng. Chem. 19, 614–619.

Parsek, M.R., Tolker-Nielsen, T., 2008. Pattern formation in *Pseudomonas aeruginosa* biofilms. Curr. Opin. Microbiol. 11, 560–566.

Percival, R.S., Devine, D.A., Duggal, M.S., Chartron, S., Marsh, P.D., 2006. The effect of cocoa polyphenols on the growth, metabolism, and biofilm formation by *Streptococcus mutans* and *Streptococcus sanguinis*. Eur. J. Oral Sci. 114, 343–348.

Prouty, A.M., Schwesinger, W.H., Gunn, J.S., 2002. Biofilm formation and the interaction with the surface of gallstones by *Salmonella* spp. Infect. Immun. 70, 2610–2619.

Raftery, T.D., Lindler, H., McNealy, T.L., 2013. Altered host cell–bacteria interaction due to nanoparticle interaction with a bacterial biofilm. Microb. Ecol. 65, 496–503.

Ramalingam, V., Rajaram, R., Premkumar, C., Santhanam, P., Dhinesh, P., Vinothkumar, S., et al., 2013. Biosynthesis of silver nanoparticles from deep sea bacterium *Pseudomonas aeruginosa* JQ989348 for antimicrobial, antibiofilm, and cytotoxic activity. J. Basic Microbiol. 53, 1–9.

Ren, D., Zuo, R., Gonzalez Barrios, A.F., Bedzyk, L.A., Eldridge, G.R., Pasmore, M.E., et al., 2005. Differential gene expression for investigation of *Escherichia coli* biofilm inhibition by plant extract ursolic acid. Appl. Environ. Microbiol. 71, 4022–4034.

Rendueles, O., Ghigo, J.M., 2012. Multi-species biofilms: how to avoid unfriendly neighbors. FEMS Microbiol. Rev. 36, 972–989.

Roder, B.I., Wandall, D.A., Frimod-Moller, N., Espersen, F., Skinhoj, P., Rosdahl, V.T., 1999. Clinical features of *Staphylococcus aureus* endocarditis: a 10-year experiments in Denmark. Arch. Intern. Med. 159, 462–469.

Roe, D., Karandikar, B., Bonn-Savage, N., Gibbins, B., Roullet, J.B., 2008. Antimicrobial surface functionalization of plastic catheters by silver nanoparticles. J. Antimicrob. Chemother. 61, 869–876.

Sathyanarayanan, M.B., Balachandranath, R., Genji Srinivasulu, Y., Kannaiyan, S.K., Subbiahdoss, G., 2013. The effect of gold and iron-oxide nanoparticles on biofilm-forming pathogens. ISRN Microbiol. 2013, e272086.

Schroeder, G.N., Hilbi, H., 2008. Molecular pathogenesis of *Shigella* spp.: controlling host cell signaling, invasion, and death by type III secretion. Clin. Microbiol. Rev. 21, 134–156.

Seil, J.T., Webster, T.J., 2012. Antimicrobial applications of nanotechnology: methods and literature. Int. J. Nanomed. 7, 2767–2781.

Shi, Z., Neoh, K.G., Kang, E.T., Wang, W., 2006. Antibacterial and mechanical properties of bone cement impregnated with chitosan nanoparticles. Biomaterials 27, 2440–2449.

Shrestha, A., Shi, Z., Neoh, K.G., Kishen, A., 2010. Nanoparticulates for antibiofilm treatment and effect of aging on its antibacterial activity. J. Endod. 36, 1030–1035.

Sinha, R., Karan, R., Sinha, A., Khare, S.K., 2011. Interaction and nanotoxic effect of ZnO and Ag nanoparticles on mesophilic and halophilic bacterial cells. Bioresour. Technol. 102, 1516–1520.

Taglietti, A., Arciola, C.R., D'Agostino, A., Dacarro, G., Montanaro, L., Campoccia, D., et al., 2014. Antibiofilm activity of a monolayer of silver nanoparticles anchored to an amino-silanized glass surface. Biomaterials 35, 1779–1788.

Vandecandelaere, I., Matthijs, N., Van Nieuwerburgh, F., Deforce, D., Vosters, P., De Bus, L., et al., 2012. Assessment of microbial diversity in biofilms recovered from endotracheal tubes using culture dependent and independent approaches. PLoS One 7, e38401.

Vijayan, S.R., Santhiyagu, P., Singamuthu, M., Kumari Ahila, N., Jayaraman, R., Ethiraj, K., 2014. Synthesis and characterization of silver and gold nanoparticles using aqueous extract of seaweed, *Turbinaria conoides*, and their antimicrofouling activity. Scientific World J. 2014, 938272.

Watters, C., DeLeon, K., Trivedi, U., Griswold, J.A., Lyte, M., Hampel, K.J., et al., 2013. *Pseudomonas aeruginosa* biofilms perturb wound resolution and antibiotic tolerance in diabetic mice. Med. Microbiol. Immunol. 202, 131–141.

Whitchurch, C.B., Tolker-Nielsen, T., Raggas, P.C., Mattick, J.S., 2002. Extracellular DNA required for bacterial biofilm formation. Science 295, 1487.

Wolska, K.I., Grześ, K., Kurek, A., 2012. Synergy between novel antimicrobials and conventional antibiotics or bacteriocins. Pol. J. Microbiol. 61, 95–104.

Xu, K.D., McFeters, G.A., Stewart, P.S., 2000. Biofilm resistance to antimicrobial agents. Microbiology. 146, 547–549.

Zhang, L., Mah, T.-F., 2008. Involvement of a novel efflux system in biofilm-specific resistance to antibiotics. J. Bacteriol. 190, 4447–4452.

Zhou, Y., Kong, Y., Kundu, S., Cirillo, J.D., Liang, H., 2012. Antibacterial activities of gold and silver nanoparticles against *Escherichia coli* and *Bacillus* Calmette–Guérin. J. Nanobiotechnol. 10, article 19.

CHAPTER

9

Tackling the Problem of Tuberculosis by Nanotechnology: Disease Diagnosis and Drug Delivery

Mahendra Rai[1], Avinash P. Ingle[1], Sunita Bansod[1] and Kateryna Kon[2]

[1]Nanobiotechnology Laboratory, Department of Biotechnology, Sant Gadge Baba Amravati University, Amravati, Maharashtra, India [2]Department of Microbiology, Virology and Immunology, Kharkiv National Medical University, Kharkiv, Ukraine

9.1 INTRODUCTION

9.1.1 Disease Severity

Tuberculosis (TB) remains a major public health problem all over the world. India contributes 26% of the global TB burden. Since ancient times, TB has been a leading cause of morbidity and mortality. According to the report of World Health Organization (WHO), approximately 8.6 million people have been infected with TB and 1.3 million died from the disease in 2012 (WHO TB Report, 2013). The earliest references to TB can be found in *Sanskrit* (an Indian language). TB is not new at all; in fact, there is also reference to this disease in Ayurveda. It was also described in Chinese and Arabic literature. The word "tuberculosis" was derived from the Latin word *tubercula* (meaning "small lump") (Rubin, 1995; Sharma and Mohan, 2013), and the causative agent (tubercle bacillus) for the disease was discovered by Robert Koch in 1882 (Frank and Tabrah, 2011). After thousands of years, *Mycobacterium tuberculosis* (MTB) was established as the cause of TB in humans. No other disease in history matches TB in terms of morbidity and mortality in humans. Historically, even though several other diseases like smallpox and plague have killed millions of people, they were relatively short-lived. TB has been ever-present and is becoming difficult to treat because of the emergence of multidrug-resistant (MDR) strains.

Nanotechnology in Diagnosis, Treatment and Prophylaxis of Infectious Diseases.
DOI: http://dx.doi.org/10.1016/B978-0-12-801317-5.00009-8

In many developing countries, most notably in Africa where the human immunodeficiency virus (HIV) epidemic is particularly severe, individuals infected with HIV are initiated on antiretroviral therapy (ART) only when their $CD4^+$ T-cell count is less than $200/mm^3$. At this stage, an HIV-infected individual is likely to be coinfected with MTB because of a new MTB infection or latent MTB reactivated attributable to suppression of the immune system. Treatment of HIV-associated TB represents a serious problem (Dean et al., 2002; Kwara et al., 2005; McIlleron et al., 2007; Jiang et al., 2014). However, recent reports suggest new hope for the discovery of new drugs (Thacher et al., 2014).

9.1.2 Distribution of TB

It is reported that the average annual risk of infection (ARI) for TB, as computed from the estimated prevalence, is 1.5%. In India, the ARI showed regional variations; it was higher in the northern (1.9%) and western (1.8%) zones compared with the eastern (1.3%) and southern (1%) zones (Chadha et al., 2005; Sharma and Mohan, 2013). However, in house-based tuberculin surveys conducted among children aged 1 to 9 years in statistically selected clusters during 2000 to 2003 and 2009 to 2010, it was observed that ARI rates decreased by, respectively, 6% and 11.7% per year in the north and west zones; however, no change was evident in the south and east zones. In India, ARI decreased by 4.5% per year between 1998 and 2007 (Chadha et al., 2013).

HIV infection is a potent risk factor for TB. HIV increases the risk of reactivating latent MTB infection and also increases the risk of rapid TB progression soon after infection or reinfection with MTB (Duffin and Tullis, 2002; Garira et al., 2005). In the persons infected with MTB, the lifetime risk of developing TB ranges from 10% and 20% (Kirschner and Webb, 1996, 1997). In the persons coinfected with MTB and HIV, however, the annual risk can exceed 10% (Magombedze et al., 2008). The TB burden in different countries has increased rapidly over the past decade with the generalized HIV epidemic, especially in the severely affected countries of eastern and southern Africa (Raviglione et al., 1997; Wilkinson and Davies, 1997; Corbett et al., 2003). TB is one of the most common causes of morbidity and mortality in HIV-positive adults living in less developed countries (Witten and Perelson, 2004; Garira et al., 2005; Magombedze et al., 2008; De Boer et al., 2010); however, it is a preventable and treatable disease. India is among the 22 countries with a high TB rate and has accounted for an estimated one-quarter (26%) of all TB cases worldwide. Observations from reliable accredited mycobacteriology laboratories in India suggest that the prevalence of MDR-TB is quite low in new TB cases (<3%) compared with previously infected cases (15–30%) (http://www.newtbdrugs.org/pipeline.php, accessed on June 10, 2014).

9.1.3 Aim of the Chapter

Public health experts declared that "virtual elimination of the disease as a public health problem" was in sight (Keshavjee, 2012). In the United States, federal funding for TB provides limited funding for research on TB, thus affecting drug discovery, development of diagnostics, and vaccine research. The first decade of the 21st century has been ravaged by extensively drug-resistant TB (XDR-TB). Recently, concern has been expressed regarding the occurrence of extremely drug-resistant TB (XXDR-TB), super XDR-TB, and totally drug-resistant TB (TDR-TB) in some parts of the world because of current TDR bacteria. The role of nanotechnology for predicting a lasting cure and discovery of newer anti-TB drugs, and development of newer drug delivery and vaccines continue to help to achieve the goal of eliminating TB altogether by 2050.

9.2 THE PRESENT SCENARIO OF ANTIBIOTICS USED AGAINST TB

TB treatment has been available for the past 50 years. On average, active TB infection occurs in approximately 10 to 15 people every year. If active TB is not treated, it can be transmitted to others. TB is treated with antibiotics; however, antibiotic treatment therapy is lengthy and it takes 6 to 12 months to destroy the MTB bacteria. The treatment duration and the drug type needed are determined according to age, overall health, results of susceptibility tests, and whether the TB infection is active. During the past 10 years, the researchers have made significant progress regarding treatment for MTB. Regimens have been optimized and directly observed therapy short-course (DOTS) initiatives have been implemented (Jalhan et al., 2013). Currently, TB chemotherapy comprises a cocktail of first-line drugs, Isoniazid (INH), Rifampin (RIF), Pyrazinamide (PZA), and Ethambutol (EMB), administered for 6 months. If treatment fails because of drug resistance, then second-line drugs are the alternative. Drugs such as para-aminosalicylate (PAS), kanamycin, fluoroquinolones, capreomycin, ethionamide, and cycloserine can be used, but they have serious side effects (Dheda et al., 2008; Keshavjee and Farmer, 2010a; Udwadia et al., 2012; Mani et al., 2014).

Because of resistance to antibiotics, the incidence of MDR-TB has increased (Han et al., 2005). Isoniazid drug initially kills approximately 95% of organisms during the first 2 days of treatment; some other effective drugs for TB include rifampicin (RIF) and pyrazinamide (Tahaoğlu et al., 2001; Mitchison, 2003; Vasquez-Campos et al., 2004; Kim et al., 2008; Madan et al., 2013; Mani et al., 2014).

9.2.1 The Problem of Drug Resistance in TB Strains

Discovery of streptomycin, para-amino salicylic acid (PAS), and the availability of isoniazid ushered in the modern era of effective treatment of TB in the mid 1940s. With the emergence of short-course in the late 1970s in India, there was optimism in the developed world that TB may cease to be a public health problem. Recognizing the impact of TB globally, WHO declared TB to be a "global emergency" in April 1993. The late 1990s also witnessed the resurgence of drug-resistant TB (DR-TB), with MDR-TB emerging as a major threat (Hwang et al., 2009; Kim et al., 2010; Shubladze et al., 2013). Development of XDR-TB occurred during the first decade of the 21st century, and the report of the occurrence of DR-TB in India has raised concern and consternation (World Health Organization, 2011). TB occurs in the rich and poor alike, with equal disdain.

In recent survey studies, the choice of drugs has been driven by the actual or presumed (in view of past failed treatment) resistance characteristics of the strains of MTB (Tahaoğlu et al., 2001; Vasquez-Campos et al., 2004; Leimane et al., 2005; Keshavjee and Farmer, 2010b). In order of preference, the following drugs can be chosen:

1. First-line drugs: Isoniazid, Rifampin, Pyrazinamide, and Ethambutol.
2. First-line drugs followed by injectable drugs: Streptomycin, Kanamycin, Amikacin, Capreomycin, or Viomycin/tuberactinomycin B, and the related tuberactinomycins A, N, and O.
3. Antibacterial fluoroquinolones, such as Ciprofloxacin, Ofloxacin, Levofloxacin, or the more recent Sparfloxacin, Gatifloxacin, Moxifloxacin, and Sitafloxacin should be included in the regimen. This class of antibiotics has been proven as an indispensable treatment for MDR-TB. Moreover, some of these drugs may lead to shorter anti-TB regimens, although their use in immunotherapy also leads to the occurrence of fluoroquinone-resistant strains of MTB. An actual preference of

fluoroquinolones, especially during the latest generations, for the specific treatment of MDR-TB is still a matter of preclinical and clinical research.

4. Second-line drugs: Ethionamide, Cycloserine, and PAS.
5. Other drugs are also considered. Their use is the subject of debate and only time and proper observations will provide the necessary data. Clofazimine is among these compounds, and it is also used against *Mycobacterium leprae*. The combination of amoxicillin and the penicillinase inhibitor clavulanic acid has an antimycobacterial effect *in vitro*. The same is true for clarithromycin, although its clinical efficacy remains to be established.

9.2.2 New Drugs for MDR-TB

The US FDA approved bedaquiline as a novel diarylquinoline drug for MDR-TB (Chahine et al., 2014). However, anti-TB drugs are generally given in combination, but extensive studies are required to overcome the problem of drug resistance and to develop most efficient drugs. From the literature available, it was reported that no one drug is 100% effective for MDR-TB. However, in some cases where the use of TB drugs is regular, the success rate was reported to be up to 70–80%. It can also be increased in some patients, up to 80–90%, with surgical resection and standard drugs. For the patients with XDR-TB, five drugs are generally used, including Linezolid and Clofazimine. Hence, there is the possibility for the development of new drug molecules with novel mechanisms of action that can provide a maximum success rate (Shim and Jo, 2013).

9.2.3 Side Effects of Chemotherapy

Antimicrobial resistance is one of the most serious health threats. Infections from TB strains resistant to antibiotics are increasing at an alarming rate. Unfortunately, some pathogens have even become resistant to multiple types of antibiotics. The loss of efficacy of antibiotics and the decrease in their ability to fight infectious diseases and manage complications common in vulnerable patients are matters of great concern.

Treatment for HIV has some side effects, but these side effects have been found to increase because of overlapping toxicity profiles and development of drug-resistant strains of both MTB and HIV when drugs for HIV are combined with anti-TB drugs. Drug interaction may lead to diminished therapeutic results, depending on the choice of the drugs. For example, some anti-TB drugs reduce the concentration of certain antiretroviral drugs by as much as 90% (McIlleron et al., 2007). After ART, the recovery of the immune system may result in immune reconstitution inflammatory syndrome (IRIS), which is especially problematic for an individual with TB. Because of these problems, the timing of ART relative to TB treatment for coinfection is an important question that needs to be addressed. Therefore, Abdool-Karim et al. (2011) rightly stated that "the optimal timing for the initiation of ART in relation to TB therapy remains controversial."

ART for TB can be performed in three phases, which may also be called as arms. One is the sequential arm: ART is performed after TB treatment with standard drugs. However, the other two arms are integrated: ART is performed before TB treatment or during TB treatment (Ramkissoon et al., 2012). TB may be caused by drug-susceptible or drug-resistant strains. In such settings many of the social determinants of TB, including extreme poverty, severe malnutrition, and overcrowded living conditions, become the exception rather than the norm. Public health experts declared that "virtual elimination of the disease as a public health problem" was in sight.

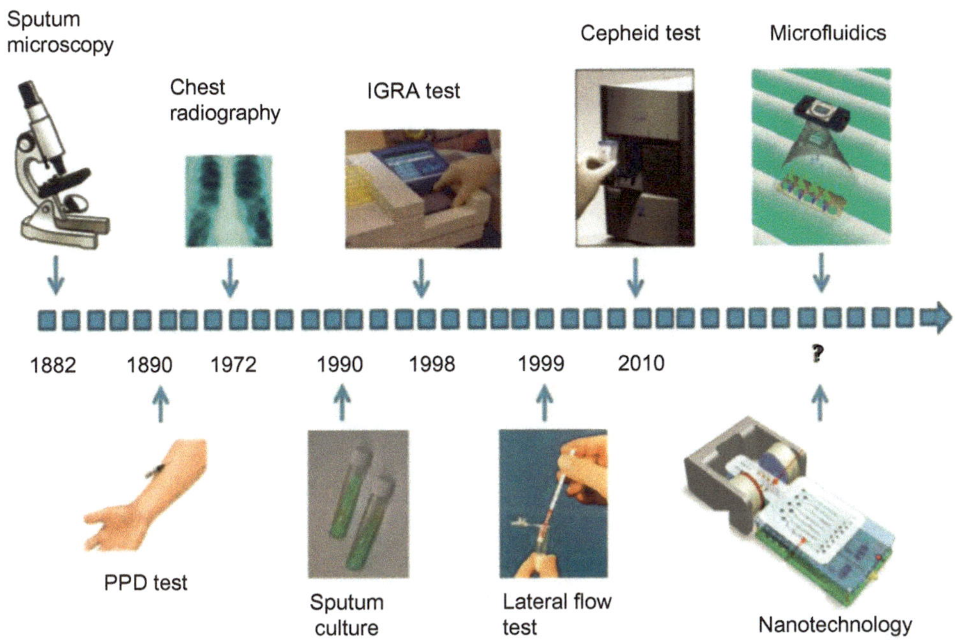

FIGURE 9.1 Different methods generally used for diagnosis of TB. *(Reprinted with permission of publisher, from Wang et al., 2013).*

9.3 NANOTECHNOLOGY AS A NOVEL APPROACH IN DRUG DISCOVERY

The drug resistance in TB presents major problems for the effective control of TB. The TB drugs currently in use were developed 40 years ago, and there is a great need for a new generation of TB drugs. Nanotechnology is a multidisciplinary field that has recently emerged, and it is extremely necessary for TB treatment.

Before the discovery of different nanomaterials (or the science "nanotechnology"), many traditional methods like sputum microscopy, chest radiography, IGRA test, and PPD test were routinely used for the diagnosis of TB. Now, many nanotechnological methods have been developed for the same use. Hence, nanotechnology is important for diagnosis, treatment, and prevention of TB. Figure 9.1 shows the schematic representation of different diagnostic methods used in past and now.

Nanotechnology-based drug delivery will help to deliver even those drug molecules that are poorly soluble in water, and the intracellular specificity with regard to various tissues and cells is an added advantage (Farokhzad and Langer, 2009; Mamo et al., 2010). Nanotechnology-based systemic delivery of anti-TB drugs have many advantages such as controlled release of drugs, which helps to keep the drugs working for longer period of time. It also helps to enhance and modulate the distribution of different drugs such as hydrophobic and hydrophilic into and within different tissues because of their small size (Mamo et al., 2010).

Nanoparticles were found to be promising for carrying synthetic anti-TB drugs worldwide. Also, natural drugs can be carried and released in infected macrophages for anti-TB chemotherapy. The possible mechanism for

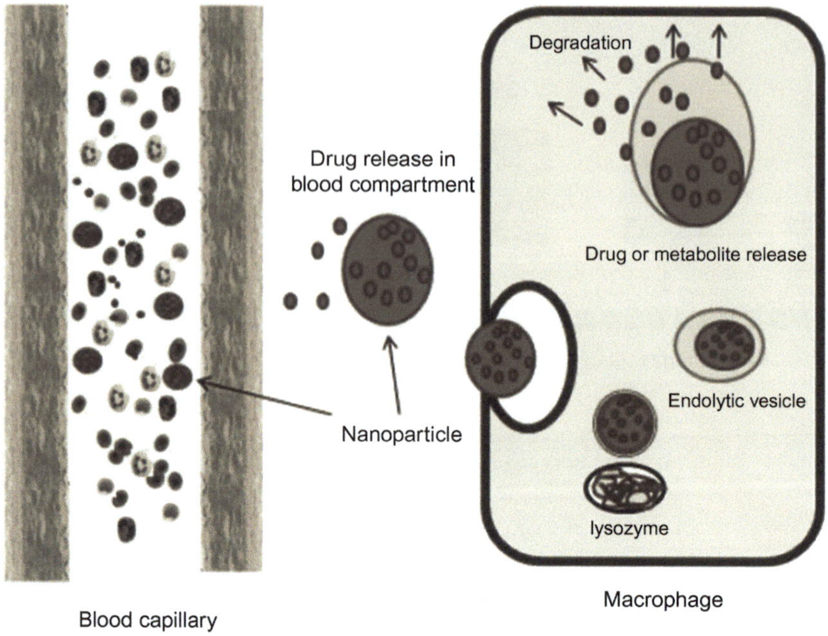

FIGURE 9.2 Possible mechanism for nanoparticles to carry both natural and synthetic drugs and the release of encapsulated drug in infected macrophage for anti-TB chemotherapeutic agents. *(Reprinted with permission of publisher, from Cheepsattayakorn and Cheepsattayakorn, 2013).*

nanoparticle release of encapsulated drugs to the infected macrophages is shown in Figure 9.2. Hence, these nano-based drug delivery systems hold the most promise for their use in clinical treatment and prevention of TB.

9.3.1 Nanotechnology-Based Drug Delivery for TB Treatment

9.3.1.1 Polymeric Nano-Carrier: Dendrimers

Dendrimers are synthetic nanomaterials that are approximately 100 nm in diameter and comprise layers of polymers surrounding a central core. Dendrimers are regularly hyperbranched and three-dimensional macromolecules with low molecular weight and polydispersity; they also have highly adjustable functionality. Drug encapsulation is performed by virtue of the dendrimeric core and complexation and conjugation on the surface (Duncan and Izzo, 2005). Functionality of dendrimers archetypically comprise three different topological components of chemical significance: a poly-functional core, interior layers, and a multivalent surface. The poly-functional focal core can encapsulate various chemical species and exhibits unparalleled properties because of the special nano-environment surrounded by extensive dendritic branching. Because of this structure, they are attractive candidates for the encapsulation and delivery of anti-TB agents for diverse administration routes. RIF-loaded mannosylated fifth-generation polypropyleneimine (PPI) dendrimeric nano-carriers have been developed (Bosman et al., 1999; Kumar et al., 2007). Surface modification with sugar molecules (e.g., mannose) recognizable by lectin receptors located on the surface of phagocytic cells improved the selective uptake of the drug-loaded nano-carriers by cells of the immune system.

The binding efficacy of RIF with core is approximately 37% and occurs through the hydrophobic interactions and hydrogen bonding. The solubility of RIF within unmodified dendrimers was 52 mg/mL, whereas the superficial mannose molecules sterically hindered the complexation and encapsulation of the drug, and the solubilization of RIF was substantially less efficient at approximately 5 mg/mL (two-fold when compared with the aqueous solubility of RIF). Surface modification with sugar molecules (e.g., mannose) recognizable by lectin receptors located on the surface of phagocytic cells improved the selective uptake of the drug-loaded nano-carriers by cells of the immune system. Increased hemolysis levels shown by amine-terminated dendrimers preclude their clinical application. Mannosylation significantly reduced the hemolytic toxicity of the nano-carrier materials from 15.6% to 2.8%. The use of RIF-containing dendrimers as a carrier was found to be very beneficial and it was reported that the intrinsic hemolytic effect can be reduced from 9.8% to 6.5%. The phagocytic uptake of RIF and RIF-loaded dendrimers was investigated with alveolar macrophages harvested from rat lungs. A clear increase in the intracellular concentration of the antibiotic was apparent. Using a similar approach, more recent work investigated the suitability of RIF-containing fourth-generation and fifth-generation polyethylene glycosylated (PEGylated) PPI dendrimers to sustain the delivery of RIF (Kumar et al., 2007). PEG-grafted dendrimers showed a minimal hemolytic activity (1–3%) as opposed to the NH2-terminated ones (14–20%).

9.3.1.2 *Cyclodextrins*

Cyclodextrins (CD) are a group of structurally related natural products formed during bacterial digestion of cellulose. CD are cyclic oligosaccharides that consist of (α-1,4)-linked α-D-glucopyranose units and contain a somewhat hydrophobic central cavity and a hydrophilic outer surface, and thus are able to host other hydrophobic molecules (Pitha et al., 1986). Several researchers reported on the complexation of RIF by means of different CD molecules, although results regarding the efficiency of this approach are ambiguous. Ferreira et al. (2004) prepared inclusion complexes of RIF with hydroxypropyl-β-cyclodextrin (HPβCD). In this context, poorly water-soluble nitroimidazole P-824, a new anti-TB drug, has shown activity against drug-sensitive and MDR bacilli. *In vivo* experiments in a short-course murine infection model were conducted, a complex with HP-γ-CD was developed, and a CD/lecithin formulation was prepared (Lenaerts et al., 2005). A reduction in the bacterial load in the lungs was observed with 50 and 100 mg/kg doses. CD has been also investigated as a carrier for local delivery to the lung (Evrard et al., 2004).

9.3.1.3 *Polymeric Micelles*

Polymeric micelles are used for nanotechnology-based drug delivery systems for pulmonary administration and pulmonary targeting. Polymeric micelles have promising applications for drug delivery, cancer targeting, and tumor imaging (Croy and Kwon, 2006). Regarding the use of polymeric micelles as a carrier for drug delivery systems for respiratory and nonrespiratory diseases (pulmonary administration of drugs) (Smola et al., 2008) as well as for systemic targeted treatment of lung diseases (Goel et al., 2013), a novel approach in the field of polymeric drug delivery systems was introduced by the formation of polymeric micelles and subsequently by functionalized polymeric micelles (Jhaveri and Torchilin, 2014). Polymeric micelles are expected to find a wide application in the fields of drug delivery and diagnosis because the of the possibility of coupling to bioactive substances. Nanospheric particles as drug delivery systems are gaining increasing interest in the biomedical field. Nanospheres and microspheres are also found to be efficient drug delivery systems for intravenous administration because of their comparatively long bloodstream circulation (Hire and Derle, 2014; Wang et al., 2014). It was expected that when the drugs are released after internalization of micelle, by cleavage, they act as pro-drugs and enhance solubility of hydrophobic drugs, reduce drug toxicity, increase bioavailability, specificity, and systematic release of drugs, and prolong the drug action (Moghimi et al., 1993; Croy and Kwon, 2006; Dadwal, 2014). Micelle-forming polymer derivatives of the initial-phase Anti-TB drugs—pyrazinamide, thioridazine, isoniazid, and rifampin—showed potential activity against drug-resistant bacteria (Silva et al., 2007; Amaral and Viveiros, 2012). This potentiates the drug activity when conjugated to the polymer and, hence, is promising with regard to the possible reductions in the dose. The aforementioned derivatives of different drugs were also found to be effective against several virulent strains of mycobacteria such as MTB and *Mycobacterium avium* when the activity was determined by critical micelle concentration (CMC) (Emanuele and Attwood, 2005; Silva et al., 2006; Chen et al., 2007). Similarly, some other drug derivatives like copolymer PEG–PASP, containing pyrazinamide, isoniazid, and rifampin, were successfully synthesized using various processes that can be used as effective drugs for TB (Francis et al., 2004; Hans et al., 2005).

9.3.1.4 Nanosuspensions

Submicron colloidal dispersions of pure drugs stabilized with surfactants are nanosuspensions. Reduction of the average size of solid drug particles to the nanoscale, generally by top milling or grinding, is a useful methodology to improve the solubility of drugs. Dimethyl sulfoxide methanol, ethanol, and ether solvents are used to solubilize the TB drugs for nanonization (Hari et al., 2010). Dimethyl sulfoxide was found to be the most excellent solvent. Spherical particles of various sizes (mean diameters between 400 nm and 3 mm) are needed and sizes are tuned by changing the conditions of the process. This may support TB treatment, especially the local delivery of anti-TB drugs to the lungs. Nanocrystalline suspensions of poorly soluble drugs such as riminophenazines and clofazimine are easy to prepare and to lyophilize for extended storage and represent a promising new drug formulation for intravenous therapy of mycobacterial infections (Peters et al., 2000).

9.3.1.5 Nanoemulsions

Nanoemulsions are thermodynamically stable oil-in-water (o/w) dispersions displaying drop sizes between 10 and 100 nm (Constantinides et al., 2008). Advantages of nanoemulsion are that they are generated spontaneously, can be produced in a large scale without the need of high homogenization energy, and can be sterilized by filtration. The enhanced uptake of nanoemulsions (lipid emulsion) by cells of the phagocytic system reported and they have a potential role as a novel antimicrobial agent (Seki et al., 2004).

9.3.1.6 Niosomes

Niosomes are biocompatible, nonimmunogenic, and biodegradable in nature and exhibit flexibility in their structured characterization (Junyaprasert et al., 2008). Stable liposome-like vesicles produced by different methods like hydration of cholesterol, charge-inducing components like charged phospholipids, and nonionic surfactants have advantages such as higher stability, entrapment of more substances, and no need for handling or storing in special conditions. Niosomes have the ability to hold hydrophilic drugs within the core and to hold lipophilic drugs by entrapment in hydrophobic domains.

The prepared microsized (8–15 mm) RMP-loaded niosomes contain Span 85 as the surfactant. *In vivo* studies have revealed that by adjusting the size of the carrier, up to 65% of the drug can be localized in the lungs (Junyaprasert et al., 2008). Niosomes have been used for improving the stability of entrapped drug RMP.

9.3.1.7 Polymeric and Nonpolymeric Nanoparticles

Polymeric nanoparticles (PNPs) have been extensively used for the drug solubilization, stabilization, and targeting (Delie and Blanco-Prieto, 2005). High stability, given the high loading capacity for drugs and feasibility of administration by different routes, has made PNPs one of the most popular approaches for drug encapsulation (Brannon-Peppas, 1995). Depending on application, two types of system can be developed, namely, nanocapsules and nanospheres. The drug solubilized in aqueous or oily solvents is surrounded by a polymeric membrane. There are several biomaterials available for the production of PNP. PNPs are removed from the body by opsonization and phagocytosis (Owens and Peppas, 2005).

To prevent recognition by the host immune system and to prolong circulation time in the blood, the modification of the surface with highly hydrophilic chains (e.g., PEG) has been performed (Hari et al., 2010). This approach was one of the most extensively investigated with respect to anti-TB drug delivery systems (Kataoka et al., 2001). Du Toit et al. (2008) developed INH-loaded polymer-based nanosystems by means of a salting-out approach (nanoprecipitation). Hari et al. (2010) reported that encapsulation of different TB drugs within poly(*n*butylcyanoacrylate) (PBCA) and poly (isobutylcyanoacrylate) (PIBCA) nanoparticles increase in the concentration of such drugs when tested for the accumulation in human blood monocytes *in vitro*.

9.3.1.8 Liposomes

Liposomes are nano-sized to microsized vesicles comprising a phospholipid bilayer that surrounds an aqueous core (Cheepsattayakorn and Cheepsattayakorn, 2013). In liposomes, the core encapsulates the water-soluble drugs and the hydrophobic domain is responsible for entrapping insoluble agents. After administration, liposomes are usually recognized by phagocytic cells and are expelled from the blood rapidly. To prevent better efficacy of liposomes, they are usually PEGylated. In more recent investigations, pyrazinamide and rifabutin-containing liposomes were also produced. Reports of INH and rifampin encapsulated in lung-specific stealth liposomes against MTB infection revealed that liposome-encapsulated drugs at and below therapeutic concentrations were more effective than free drugs against TB (Pandey et al., 2004).

9.4 NANO-BASED DNA VACCINES FOR TB

Bacillus Calmette–Guerin (BCG) is the only vaccine that has been discovered for TB. Unfortunately, its efficacy varies from 0% to 80% (Fine, 1995; Lin et al., 2007). Clinical trials of BCG have shown total lack of protection in regions of the world where the disease is common, and thus BCG vaccination is considered ineffective. In addition, BCG is a live vaccine and can cause disseminated disease in immunocompromised individuals. Therefore, these disadvantages indicate the urgent need to develop more effective vaccines against TB (Dhanasooraj et al., 2013). BCG is used in many countries with a high prevalence of TB to prevent childhood TB meningitis and other diseases like miliary disease. However, BCG vaccine is not generally recommended in countries with a low risk of infection with MTB, like the United States (www.vaccines.gov/

diseases/tb/ accessed on June 15, 2014). BCG also showed variable effectiveness against adult pulmonary TB and its potential interference with tuberculin skin test reactivity. According to TB experts, the BCG vaccine can be used for very selective individuals who meet specific criteria (www.vaccines.gov/diseases/tb/ accessed on June 15, 2014). Development of a novel and effective vaccine against MTB for preventing TB infection is a challenge (Feng et al., 2013). Therefore, the concept of nano-based vaccines like chitosan-based DNA vaccine came into existence. A few studies have been performed concerning the use of nanomaterials (mostly chitosan nanoparticles) for the development of nano-based DNA vaccines, which are reviewed in this chapter.

Bivas-Benita et al. (2004) contributed extensively to the use of nanomaterials in different forms for the control of TB infections. The authors formulated the conjugate of DNA plasmid from MTB encoding for different restricted T-cell epitopes with chitosan nanoparticles. Later, they investigated the effects of these conjugates on pulmonary delivery and reported that chitosan–DNA conjugates induce the maturation of dendritic cells, which cannot be achieved by chitosan nanoparticles alone. This indicates that release of DNA from nanoparticles stimulates the dendritic cells and increases levels of interferon-gamma secretion compared with delivery of plasmid in intramuscular immunization routes. Hence, these findings proved that use of DNA vaccines (DNA encapsulated in chitosan nanoparticles) against TB would be more efficient than intramuscular immunization because it increases immunogenicity. In another study, they reported that pulmonary delivery of DNA encoding MTB antigen Rv1733c conjugated with poly (D,L-lactide-co-glycolide) (PLGA)-polyethyleneimine (PEI) nanoparticles (NP) (PLGA-PEI) augmented the T-cell responses in a DNA prime/protein that boosts the vaccination regimen in mice (Bivas-Benita et al., 2009). From both studies it can be concluded that conjugates of DNA encoding for different antigens from MTB with various nanoparticles would be helpful to boost the immune response against TB.

Heuking et al. (2013) investigated the role of TLR-1/TLR-2 agonist–functionalized pDNA nanoparticles on human bronchial epithelium. They reported that chitosan-based DNA delivery permits the uptake into monocyte-derived dendritic cells, which are the most important cells of human immune systems. For example, in the human lung it induces antigen-specific immunity. From these findings they proposed that such a DNA delivery approach was attractive for potential DNA vaccination against intracellular pathogens in the lung (e.g., MTB or influenza virus). Another attempt has been made regarding vaccine delivery system for TB using nano-sized hepatitis B virus core protein particles. According to Dhanasooraj et al. (2013), nano-sized hepatitis B virus core protein particles (HBc-VLP) were suitable and can be easily taken up by antigen-presenting cells.

It was well-known that the antigen culture filtrate protein 10 (CFP-10) is an important vaccine candidate against TB. However, it was reported that without any adjuvant, these antigens showed very low immune response and, hence, has low protective efficacy. However, when these proteins (CFP-10) were used in combination with HBc nanoparticles, it provided higher protection compared with the native antigen alone.

Feng et al. (2013) developed a novel nanoparticle-based recombinant DNA vaccine. It was a complex of Esat-6 three T-cell epitopes (Esat-6/3e) and fms-like tyrosine kinase 3 ligand (FL) genes (Esat-6/3e-FL) enveloped with chitosan (CS) nanoparticles (nano-chitosan). This complex is termed nano-Esat-6/3e-FL). Further, they demonstrated the immunologic and protective efficacy of these nano-chitosan-based DNA vaccines (nano-Esat-6/3e-FL) in C57BL/6 mice after intramuscular prime vaccination with the plasmid DNA and nasal boost with the Esat-6/3e peptides. The findings showed that the immunized mice had significantly enhanced T-cell responses and protection against MTB. These findings indicate that the nano-chitosan can significantly elevate the immunologic and protective effects of the DNA vaccine and would be useful vaccine against TB.

All these reports collectively proposed the efficient use of nano-based DNA vaccines for the control of TB infections in mice. Further extensive studies are required for the development of novel and 100% efficient nano-based vaccines against TB infections in humans.

9.5 ROLE OF NANOBIOSENSORS IN DIAGNOSTICS OF TB

The tubercle bacterium is sluggish in growth, taking 1–2 months for *in vitro* growth (Tortoli et al., 1997; Davies et al., 1999). Therefore, it is difficult to find the presence of infection at its early stage. Ziehl-Neelsen staining is the conventional method for its identification. This staining is needed for preliminary identification of the causative organism, but it lacks the sensitivity (Moore and Curry, 1998; Mahaisavariya et al., 2005). The conventional methods of cultivation of mycobacteria are time-consuming and need several weeks. Polymerase chain reaction (PCR) is a sensitive method for early detection of mycobacterium, but the amplification process requires ample processing time, chemicals, and reagents, which contribute to the high cost of the assay. Additionally, it is labor-intensive and expensive (Tombelli et al., 2000; Minnuni et al., 2005). Therefore, there is an urgent requirement for the development of a rapid, low-cost, and convenient diagnostic method for detection of TB. In this respect, biosensors appear to be a good option. There is increasing demand for biosensor technology for fast and precise detection of TB with high affinity and specificity. Biosensor technology has the potential to provide a qualitative and quantitative analysis and is free from radioactive or fluorescent tags (Tombelli et al., 2000, Zhou et al., 2001; Yao et al., 2008).

In recent years, in view of the benefits of various remarkable studies performed in nanotechnology, important efforts have been made to combine it with highly sensitive and accurate biosensor technology to develop nanobiosensors. Nanobiosensors systems are efficiently used in diagnosis of diseases, environmental monitoring, food quality control, and defense as a smart approach (Zhou et al., 2011; Rai et al., 2012; Singh et al., 2014). For instance, by using gold-coated nanobiosensors, Duman et al. (2009) demonstrated detection of target molecules (synthetic and PCR products) very effectively at nanomolar levels. A single-stranded oligodeoxynucleotide carrying a thiol group at the end and complementary of the target characteristic sequence of the MTB complex was used as the probe immobilized on the gold-coated surface of the surface plasmon resonance slides. It is interesting to note that the sensor platform is reusable and has long shelf life. A quartz crystal microbalance (QCM) biosensor in combination with AuNPs has been developed for the detection of MTB (Kaewphinit et al., 2012). According to the study, AuNPs improved the sensitivity of immobilized gold electrode of quartz crystal using the specific thiol-modified oligonucleotide probe. The QCM has been shown to detect up to 5 pg of MTB genomic DNA without showing any cross-hybridization with other mycobacteria.

During the past few years, many techniques have exploited the materials at the nano-scale level for designing biosensors with high specificity and efficacy. Among all nanomaterials, metal oxides are of particular interest because of their unique physical, chemical, and catalytic properties (Shi et al., 2014). Das et al. (2010) have made an important attempt to diversify the application of such metal oxides in the generation of nanobiosensors for detection of mycobacteria. They deposited nanoscale zinc oxide on the indium-tin-oxide (ITO)-coated glass plate. The presence of nanostructured ZnO films allowed an increase in the electro-active surface area for DNA molecule loading and for detecting genomic target DNA up to 100 pM, which enables the direct detection of pathogens in clinical samples at point of care. The main characteristics of the technique are: (i) the covalent immobilization of the sensor without using any cross-linker that might limit its sensitivity; (ii) detection limit of 0.065 ng/μL; (iii) detection process requires only 60 s; (iv) can be reused up to 10 times; and (v) stable up to 4 months at 4°C. Therefore, it is an efficient nanobiosensor for rapid and accurate diagnosis of mycobacteria (Das et al., 2010). Zirconium oxide (ZrO_2) is an important oxide of metal with greater stability and inertness. Moreover, it has affinity toward groups containing oxygen. Das et al. (2011) developed zirconium oxide and carbon nanotube ($NanoZrO_2$-CNT) nanocomposite-based nucleic acid nanobiosensors deposited on ITO. The group utilized this electrode ($NanoZrO_2$-CNT/ITO) for immobilization of single-stranded probe DNA (ssDNA) specific for MTB to reveal its application to biosensor for nuclei acid detection.

Thiruppathiraja et al. (2011) fabricated and evaluated the DNA electrochemical biosensor for genomic DNA of *Mycobacterium* sp. using a signal amplifier as dual-labeled gold nanoparticles (AuNPs). The method involves the sandwich detection strategy comprising two types of DNA probes: the probes of enzyme ALP and the detector probe conjugated on AuNPs. Both of these probes were specific for *Mycobacterium* sp. genomic DNA. The study claimed that under optimized conditions, the detection limit of the method was 1.25 ng/mL genomic DNA. The said nanobiosensors were also promising and evaluation of the clinical sputum samples showed the higher sensitivity and specificity. Another study (Torres-Chavolla and Alocilja, 2011) with different approaches also fabricated the DNA-based biosensor encompassing AuNPs and amine-terminated magnetic particles (MPs) to detect the mycobacteria. The study made use of thermophilic helicase-dependent isothermal amplification (tHDA) and dextrin-coated AuNPs as electrochemical reporters. The AuNPs and MPs were functionalized independently with different DNA probes that specifically hybridize with a fragment within a gene of mycobacteria. Later, that group separated the MP-target–AuNPs complex magnetically from the solution and detected AuNPs electrochemically. Torres-Chavolla and Alocilja (2011) claimed the sensitivity of this method was 0.01 ng/μL of isothermally amplified target of 105 bp. Such a sensor thus can be used to regularly analyze the clinical samples suspected to have mycobacteria.

In addition to metal oxide nanoparticles, porous silicon is also getting more attention regarding biosensor applications, mostly in label-free applications. Wu et al. (2012) made use of a nanoscale porous silicon micro-cavity biosensor for fast sero-diagnosis of MTB. Through a series of experiments, the study testified the feasibility of this biosensor for the detection of interaction between 16 kDa antigen and 16 kDa antibody. The detection of MDR-TB is of utmost importance. In a recent study, Li et al. (2014) reported development of a DNA sensor for the specific detection of the rpoB gene of MDR-TB by using ruthenium (II) complex–functionalized grapheme oxide (Ru-GO) as a suspension-sensing interface and ferrocene-labeled single-stranded DNA (FC-ssDNA) as an electrochemiluminescence (ECL) intensity controller. The assay relies on the principle that when mutant ssDNA target hybridizes with FC-ssDNA, it is released from the Ru-Go surface, leading to recovery of ECL. The assay is reported to have a detection range from 0.1 to 100 nM and 0.04 nM sensitivity.

More studies are needed that mainly focus on the fabrication of various nanobiosensors for diagnosis of such a dreadful disease. Because this technique is highly sensitive, it requires little sample preparation and is fast, specific, cheap, and easy to use; it has great potential for the clinical diagnosis of TB.

9.6 CONCLUSION AND FUTURE PERSPECTIVES

It is evident that TB is still a major public health problem because of the emergence of MDR strains of *Mycobacterium*. The emergence of MDR strains of TB in HIV patients in developing countries like Africa has made the problem more complicated and, thus, it is a matter of great concern. In fact, the disease should be eliminated as a major public health problem. WHO emphasizes that there is a "global emergency" to eradicate drug-resistant strains of TB. Taking these facts into consideration, there is a pressing need to develop newer anti-TB drugs, drug delivery systems, and development of vaccines to combat the grave problem of MDR mycobacterium by 2050. In this context, nanotechnology may play a vital role in fighting the MDR strains of TB. Development of nanobiosensors for early diagnosis and use of nano-based drugs in combination with the existing antibiotics and delivery system may provide new ways to combat the MDR problem.

References

Abdool-Karim, S.S., Naidoo, K., Grobler, A., Padayatchi, N., Baxter, C., 2011. Integration of antiretroviral therapy with tuberculosis treatment. N. Engl. J. Med. 365 (16), 1492–1501.

Amaral, L., Viveiros, M., 2012. Why thioridazine in combination with antibiotics cures extensively drug-resistant *Mycobacterium tuberculosis* infections. Int. J. Antimicrob. Agents 39, 376–380.

Bivas-Benita, M., van Meijgaarden, K.E., Franken, K.L., Junginger, H.E., Borchard, G., Ottenhoff, T.H., et al., 2004. Pulmonary delivery of chitosan-DNA nanoparticles enhances the immunogenicity of a DNA vaccine encoding HLA-A*0201-restricted T-cell epitopes of *Mycobacterium tuberculosis*. Vaccine 22 (13–14), 1609–1615.

Bivas-Benita, M., Lin, M.Y., Bal, S.M., van Meijgaarden, K. E., Franken, K.L., Friggen, A.H., et al., 2009. Pulmonary delivery of DNA encoding *Mycobacterium tuberculosis* latency antigen Rv1733c associated to PLGA-PEI nanoparticles enhances T cell responses in a DNA prime/protein boost vaccination regimen in mice. Vaccine 27 (30), 4010–4017.

Bosman, A.W., Janssen, H.M., Meijer, E.W., 1999. About dendrimers: structure, physical properties, and applications. Chem. Rev. 99, 1665–1688.

Brannon-Peppas, L., 1995. Recent advances on the use of biodegradable microparticles and nanoparticles in controlled drug delivery. Int. J. Pharm. 116, 1–9.

Chadha, V.K., Kumar, P., Jagannatha, P.S., Vaidyanathan, P.S., Unnikrishnan, K.P., 2005. Average annual risk of tuberculous infection in India. Int. J. Tuberc. Lung. Dis. 9, 116–118.

Chadha, V.K., Sarin, R., Narang, P., John, K.R., Chopra, K. K., Jitendra, R., et al., 2013. Trends in the annual risk of tuberculous infection in India. Int. J. Tuberc. Lung. Dis. 17, 312–319.

Chahine, E.B., Karaoui, L.R., Mansour, H., 2014. Bedaquiline: a novel diarylquinoline for multidrug-resistant tuberculosis. Ann. Pharmacother. 48 (1), 107–115.

Cheepsattayakorn, A., Cheepsattayakorn, R., 2013. Roles of nanotechnology in diagnosis and treatment of tuberculosis. J. Nanotechnol. Diagn. Treat. 1, 19–25.

Chen, L., Xie, Z., Hu, J., Chen, X., Jing, X., 2007. Enantiomeric PLA–PEG block copolymers and their stereocomplex micelles used as rifampin delivery. J. Nanopart. Res. 9, 777–785.

Constantinides, P.P., Chaubal, M.V., Shorr, R., 2008. Advances in lipidnanodispersions for parenteral drug delivery and targeting. Adv. Drug Deliv. Rev. 60, 757–767.

Corbett, E.L., Watt, C.J., Walker, N., Maher, D., Williams, B.G., Raviglione, M.C., et al., 2003. The growing burden of tuberculosis: global trends and interactions with the HIV epidemic. Arch. Intern. Med. 163, 1009–1021.

Croy, S.R., Kwon, G.S., 2006. Polymeric micelles for drug delivery. Curr. Pharm. Des. 12, 4669–4684.

Dadwal, M., 2014. Polymeric nanoparticles as promising novel carriers for drug delivery: an overview. J. Adv. Pharm. Edu. Res. 4 (1), 20–30.

Das, M., Sumana, G., Nagarajan, R., Malhotra, B.D., 2010. Zirconia based nucleic acid sensor for *Mycobacterium tuberculosis* detection. Appl. Phys. Lett. 96 (13), 133703 (3 pp.).

Das, M., Dhand, C., Sumana, G., Srivastava, A.K., Vijayan, N., Nagarajan, R., et al., 2011. Zirconia grafted carbon nanotubes based biosensor for *M. tuberculosis* detection. Appl. Phys. Lett. 99, 143702.

Davies, A.P., Newport, L.E., Billington, O.J., Gillespie, S.H., 1999. Length of time to laboratory diagnosis of *Mycobacterium tuberculosis* infection: comparison of in-house methods with reference laboratory results. J. Infect. 39, 205–208.

De Boer, R.J., Ribeiro, R.M., Perelson, A.S., 2010. Current estimates for HIV-1 production imply rapid viral clearance in lymphoid tissues. PLoS Comput. Biol. 6 (9), e1000906.

Dean, G.L., Edwards, S.G., Ives, N.J., Matthews, G., Fox, E. F., 2002. Treatment of tuberculosis in HIV-infected persons in the era of highly active antiretroviral therapy. AIDS. 16, 75–83.

Delie, F., Blanco-Prieto, M.J., 2005. Polymeric particulates to improve oral bioavailability of peptide drugs. Molecules 10, 65–80.

Dhanasooraj, D., Kumar, R.A., Mundayoor, S., 2013. Vaccine delivery system for tuberculosis based on nano-sized hepatitis B virus core protein particles. Int. J. Nanomed. 8, 835–843.

Dheda, K., Shean, K., Badri, M., 2008. Extensively drug-resistant tuberculosis. N. Engl. J. Med. 359, 2390.

Du Toit, L.C., Pillay, V., Choonara, Y.E., Iyuke, S.E., 2008. Formulation and evaluation of a salted-out isoniazid-loaded nanosystem. AAPS Pharm. Sci. Technol. 9, 174–181.

Duffin, R.P., Tullis, R.H., 2002. Mathematical models of the complete course of HIV infection and AIDS. J. Theor. Med. 4 (4), 215–221.

Duman, M., Çağlayan, Demirel, G., Pişkin, E., 2009. Detection of *Mycobacterium tuberculosis* complex using surface plasmon resonance based sensors carrying self-assembled nano-overlayers of probe oligonucleotide. Sens. Lett. 7 (4), 535–542.

Duncan, R., Izzo, L., 2005. Dendrimers biocompatibility and toxicity. Adv. Drug Deliv. Rev. 57, 2215–2237.

Emanuele, A.D., Attwood, D., 2005. Dendrimer–drug interactions. Adv. Drug Deliv. Rev. 57, 2147–2162.

Evrard, B., Bertholet, P., Gueders, M., Flament, M.P., Piel, G., Delattre, L., 2004. Cyclodextrins as a potential carrier in drug nebulization. J. Control. Release 96, 403–410.

Farokhzad, O.C., Langer, R., 2009. Impact of nanotechnology on drug delivery. ACS Nano. 3 (1), 16–20.

Feng, G., Jiang, Q., Xia, M., Lu, Y., Qiu, W., Zhao, D., et al., 2013. Enhanced immune response and protective effects of nano-chitosan-based DNA vaccine encoding T cell epitopes of esat-6 and FL against *Mycobacterium tuberculosis* infection. PLos One. 8 (4), e61135. Available from: <http://dx.doi.org/10.1371/journal.pone.0061135>.

Ferreira, D.A., Ferreira, A.G., Vizzotto, L., Federman, A.N., Gomes, A., 2004. Analysis of the molecular association of rifampicin with hydroxypropyl-β- cyclodextrin. Braz. J. Pharm. Sci. 1, 43–51.

Fine, P.E., 1995. Variation in protection by BCG: implications of and for heterologous immunity. Lancet 346, 1339–1345.

Francis, M.F., Cristea, M., Winnik, F.M., 2004. Polymeric micelles for oral drug delivery: why and how. Pure Appl. Chem. 76, 1321–1335.

Frank, L., Tabrah, M.D., 2011. Koch's postulates, carnivorous cows and tuberculosis today. Hawaii Med. J. 70, 144–148.

Garira, W., Musekwa, S.D., Shiri, T., 2005. Optimal control of combined therapy in a single strain HIV-1 model. Electron. J. Differ. Equ. 52, 1–22.

Goel, A., Baboota, S., Sahni, J.K., Ali, J., 2013. Exploring targeted pulmonary delivery for treatment of lung cancer. Int. J. Pharm. Investig. 3 (1), 8–14.

Han, L.L., Sloutsky, A., Canales, R., 2005. Acquisition of drug resistance in multidrug-resistant *Mycobacterium tuberculosis* during directly observed empiric retreatment with standardized regimens. Int. J. Tuberc. Lung Dis. 9, 818–821.

Hans, M.L., 2005. Synthesis, characterization, and application of biodegradable polymeric prodrug micelles for longterm drug delivery. Thesis. Faculty of Drexel University.

Hari, B.N.V., Chitra, K.P., Bhimavarapu, R., Karunakaran, P., Muthukrishnan, N., Rani, B.S., 2010. Novel technologies: a weapon against tuberculosis. Indian J. Pharmacol. 42 (6), 338–344.

Heuking, S., Rothen-Rutishauser, B., Raemy, D.O., Gehr, P., Borchard, G., 2013. Fate of TLR-1/TLR-2 agonist functionalized pDNA nanoparticles upon deposition at the human bronchial epithelium *in vitro*. J. Nanobiotechnol. 11, 29 (10 pp.).

Hire, N.N., Derle, D.V., 2014. Microsphere as drug carrier: a review. Int. J. Adv. Res. 2 (3), 901–913.

Hwang, S.S., Kim, H.R., Kim, H.J., Kim, M.J., Lee, S.M., Yoo, C.G., et al., 2009. Impact of resistance to first-line and injectable drugs on treatment outcomes in MDR-TB. Eur. Respir. J. 33, 581–585.

Jalhan, S., Jindal, A., Aggarwal, S., Gupta, A., Hemraj, 2013. Review on current trends and advancement in drugs trends and drug targets for tuberculosis therapy. Int. J. Pharm. Bio. Sci. 4 (1), 320–333.

Jhaveri, A.M., Torchilin, V.P., 2014. Multifunctional polymeric micelles for delivery of drugs and siRNA. Front. Pharmacol. 5 (77), 1–26.

Jiang, H.Y., Zhang, M.N., Chen, H.J., Yang, Y., Deng, M., Ruan, B., 2014. Nevirapine versus efavirenz for patients co-infected with HIV and tuberculosis: a systematic review and meta-analysis. Int. J. Infect. Dis. 25, 130–135.

Junyaprasert, V.B., Teeranachaideekul, V., Supaperm, T., 2008. Effect of charged and non-ionic membrane additives on physicochemical properties and stability of niosomes. APS Pharm. Sci. Technol. 9, 851–859.

Kaewphinit, T., Santiwatanakul, S., Chansiri, K., 2012. Gold nanoparticle amplification combined with quartz crystal microbalance DNA based biosensor for detection of *Mycobacterium tuberculosis*. Sens. Trans. 146 (11), 156–163.

Kataoka, K., Harada, A., Nagasaki, Y., 2001. Block copolymer micelles for drug delivery: design, characterization and biological significance. Adv. Drug Deliv. Rev. 47, 113–131.

Keshavjee, S., 2012. Tuberculosis, drug resistance, and the history of modern medicine. N. Engl. J. Med. 367, 931–936.

Keshavjee, S., Farmer, P.E., 2010a. Time to put boots on the ground: making universal access to MDR-TB treatment a reality. Int. J. Tuberc. Lung Dis. 14, 1222–12225.

Keshavjee, S., Farmer, P.E., 2010b. Picking up the pace-scale-up of MDR tuberculosis treatment programs. N. Engl. J. Med. 363, 1781–1784.

Kim, D.H., Kim, H.J., Park, S.K., Kong, S.J., Kim, Y.S., Kim, T.H., et al., 2008. Treatment outcomes and long-term survival in patients with extensively drug-resistant tuberculosis. Am. J. Respir. Crit. Care Med. 178, 1075–1082.

Kim, D.H., Kim, H.J., Park, S.K., Kong, S.J., Kim, Y.S., Kim, T.H., et al., 2010. Treatment outcomes and survival based on drug resistance patterns in multidrug-resistant tuberculosis. Am. J. Respir. Crit. Care Med. 182, 113–119.

Kirschner, D., Webb, G.F., 1996. A model for treatment strategy in the chemotherapy of AIDS. Bull. Math. Biol. 58 (2), 367–390.

Kirschner, D.E., Webb, G.F., 1997. A mathematical model of combined drug therapy of HIV infection. J. Theor. Med. 1, 25–34.

Kumar, P.V., Agashe, H., Dutta, T., Jain, N.K., 2007. PEGylated dendritic architecture for development of a prolonged drug delivery system for an antitubercular drug. Curr. Drug Deliv. 4, 11–19.

Kwara, A., Flanigan, T.P., Carter, E.J., 2005. Highly active antiretroviral therapy (HAART) in adults with tuberculosis: current status. Int. J. Tuberc. Lung Dis. 9 (3), 248–257.

Leimane, V., Riekstina, V., Holtz, T.H., 2005. Clinical outcome of individualized treatment of multidrug-resistant tuberculosis in Latvia: a retrospective cohort study. Lancet 365, 318–326.

Lenaerts, A.J., Gruppo, V., Marietta, K.S., Johnson, C.M., Driscoll, D.K., Tompkins, N.M., 2005. Preclinical testing of the nitroimidazopyran PA-824 for activity against *Mycobacterium tuberculosis* in a series of *in vitro* and *in vivo* models. Antimicrob. Agents Chemother. 49, 2294–2301.

Li, F., Yu, Y., Li, Q., Zhou, M., Cui, H., 2014. A homogeneous signal-on strategy for the detection of rpoB genes of *Mycobacterium tuberculosis* based on electrochemiluminescent graphene oxide and ferrocene quenching. Anal. Chem. 86 (3), 1608–1613.

Lin, M.Y., Geluk, A., Smith, S.G., Stewart, A.L., Friggen, A. H., Franken, K.L.M.C., et al., 2007. Lack of immune responses to *Mycobacterium tuberculosis* DosR regulon proteins following *Mycobacterium bovis* BCG vaccination. Infect. Immun. 75 (7), 3523–3530.

Madan, K., Singh, N., Das, A., Behera, D., 2013. Pleural tuberculosis following lung cancer chemotherapy: a report of two cases proven pathologically by pleural biopsy. BMJ Case Rep. Available from: <http://dx.doi.org/10.1136/bcr-2012-008196>.

Magombedze, G., Garira, W., Mwenje, E., 2008. Modeling the immunopathogenesis of HIV-1 infection and the effect of multidrug therapy: the role of fusion inhibitors in HAART. Math. Biosci. Eng. 5, 485–504.

Mahaisavariya, P., Chaiprasert, A., Manonukul, J., Khemngern, S., Tingtoy, N., 2005. Detection and identification of *Mycobacterium* species by polymerase chain reaction (PCR) from paraffin embedded tissue compare to AFB staining in pathological sections. J. Med. Assoc. Thai. 88, 108–113.

Mamo, T., Moseman, E.A., Kolishetti, N., Salvador-Morales, C., Shi, J., Kuritzkes, D.R., et al., 2010. Emerging nanotechnology approaches for HIV/AIDS treatment and prevention. Nanomedicine (Lond). 5 (2), 269–285.

Mani, V., Wang, S., Inci, F., De Liberoa, G., Singhal A., Demirci, U., 2014. Emerging technologies for monitoring drug-resistant tuberculosis at the point-of-care. Adv. Drug Deliv. Rev. 78, 105–117.

McIlleron, H., Meintjes, G., Burman, W.J., Maartens, G., 2007. Complications of antiretroviral therapy in patients with tuberculosis: drug interactions, toxicity, and immune reconstitution inflammatory syndrome. J. Infect. Dis. 196, 63–75.

Minnuni, M., Tombelli, S., Fonti, J., Spiriti, M.M., Mascini, M., Bogani, P., et al., 2005. Detection of fragmented genomic DNA by PCR-free piezoelectric sensing using a denaturation approach. J. Am. Chem. Soc. 127, 7966–7967.

Mitchison, D.A., 2003. Role of individual drugs in the chemotherapy of tuberculosis. Int. J. Tuberc. Lung Dis. 4 (9), 796–806.

Moghimi, S.M., Muir, I.S., Illum, L., Davis, S.S., Kolb-Bachofen, V., 1993. Coating particles with a block co-polymer (poloxamine-908) suppresses opsonisation but permits the activity of dysopsonins in the serum. Biochim. Biophys. Acta 1179, 157–165.

Moore, D.F., Curry, J.I., 1998. Detection and identification of *Mycobacterium tuberculosis* directly from sputum sediments by ligase chain reaction. J. Clin. Microbiol. 36, 1028–1031.

Owens, D.E., Peppas, N.A., 2005. Opsonization, biodistribution, and pharmacokinetics of polymeric nanoparticles. Int. J. Pharm. 307, 93–102.

Pandey, R., Sharma, S., Khuller, G.K., 2004. Reports with INH and rifampin encapsulated in lung-specific stealth liposomes against MTB infection. Indian J. Exp. Biol. 42, 562–566.

Peters, K., Leitzke, S., Diederichs, J.E., Borner, K., Hahn, H., Muller, R.H., 2000. Preparation of a clofazimine nanosuspension for intravenous use and evaluation of its therapeutic efficacy in murine *Mycobacterium avium* infection. J. Antimicrob. Chemother. 45, 77–83.

Pitha, J., Milecki, J., Fales, H., Pannell, L., Uekama, K., 1986. Hydroxypropyl-β-cyclodextrin: preparation and characterization; effects on solubility of drugs. Int. J. Pharma. 29 (1), 73–82.

Rai, M., Gade, A., Gaikwad, S., Marcato, P.D., Duran, N., 2012. Biomedical applications of nanobiosensors: the state-of-the-art. J. Braz. Chem. Soc. 23 (1), 14–24.

Ramkissoon, S., Mwambi, H.G., Matthews, A.P., 2012. Modelling HIV and MTB co-infection including combined treatment strategies. PLoS One 7 (11), e49492.

Raviglione, M.C., Harries, A.D., Msiska, R., Wilkinson, D., Nunn, P., 1997. Tuberculosis and HIV: current status in Africa. AIDS 11 (Suppl. B), S113–S123.

Rubin, S.A., 1995. Tuberculosis: captain of all these men of death. Radiol. Clin. North. Am. 33 (4), 619–639.

Seki, J., Sonoke, S., Saheki, A., Fukui, H., Sasaki, H., Mayumi, T., 2004. A nanometer lipid emulsion, lipid nano-sphere (LNS), as a parenteral drug carrier for passive drug targeting. Int. J. Pharm. 273, 75–83.

Sharma, S.K., Mohan, A., 2013. Tuberculosis: from an incurable scourge to a curable disease—journey over a millennium. Indian. J. Med. Res. 137 (3), 455–493.

Shi, X., Gu, W., Li, B., Chen, N., Zhao, K., Xian, Y., 2014. Enzymatic biosensors based on the use of metal oxide nanoparticles. Microchim. Acta 181 (1–2), 1–22.

Shim, T.S., Jo, K.W., 2013. Medical treatment of pulmonary multidrug-resistant tuberculosis. Infect. Chemother. 45 (4), 367–374.

Shubladze, N., Tadumadze, N., Bablishvili, N., 2013. Molecular patterns of multidrug resistance of *Mycobacterium tuberculosis* in Georgia. Int. J. Mycobact. 2 (2), 73–78.

Silva, M., Ricelli, N.L., Valentim, C.S., Ferreira, A.G., Sato, D., Leite, C.Q.F., et al., 2006. Potential tuberculostatic agent: micelle-forming pyrazinamide prodrug. Arch. Pharm. 39, 283–290.

Silva, M., Ferreira, E.I., Leite, C.Q.F., Sato, D.N., 2007. Preparation of polymeric micelles for use as carriers of tuberculostatic drugs. Trop. J. Pharma. Res. 6 (4), 815–824.

Singh, R., Mukherjee, M.D., Sumana, G., Gupta, R.K., Soode, S., Malhotra, B.D., 2014. Biosensors for pathogen detection: a smart approach towards clinical diagnosis. Sens. Actuators B Chem. 197, 385–404.

Smola, M., Vandamme, T., Sokolowski, A., 2008. Nanocarriers as pulmonary drug delivery systems to treat and to diagnose respiratory and non-respiratory diseases. Int. J. Nanomed. 3 (1), 1–19.

Tahaoğlu, K., Törün, T., Sevim, T., 2001. The treatment of multidrug-resistant tuberculosis in Turkey. N. Engl. J. Med. 345, 170–174.

Thacher, E.G., Cavassini M., Audran, R., Thierry, A.C., Bollaerts, A., Cohen, J., et al., 2014. Safety and immunogenicity of the M72/AS01 candidate tuberculosis vaccine in HIV-infected adults on combination antiretroviral therapy: a phase I/II, randomized trial. AIDS 28 (12), 1769–1781.

Thiruppathiraja, C., Kamatchiammal, S., Adaikkappan, P., Santhosh, D.J., Alagar, M., 2011. Specific detection of *Mycobacterium* sp. genomic DNA using dual labeled gold nanoparticle based electrochemical biosensor. Anal. Biochem. 417 (1), 73–79.

Tombelli, S., Mascini, M., Sacco, C., Turner, A.P.F., 2000. A DNA piezoelectric biosensor assay coupled with a polymerase chain reaction for bacterial toxicity determination in environmental samples. Anal. Chem. Acta 418, 1–9.

Torres-Chavolla, E., Alocilja, E.C., 2011. Nanoparticle based DNA biosensor for tuberculosis detection using thermophilic helicase-dependent isothermal amplification. Biosens. Bioelectron. 26 (11), 4614–4618.

Tortoli, E., Lavinia, F., Simonetti, M.T., 1997. Evaluation of a commercial ligase chain reaction kit (Abbott LCx) for direct detection of *Mycobacterium tuberculosis* in pulmonary and extrapulmonary specimens. J. Clin. Microbiol. 35, 2424–2426.

Udwadia, Z.F., Amale, R.A., Ajbani, K.K., Rodrigues, C., 2012. Totally drug-resistant tuberculosis in India. Clin. Infect. Dis. 54, 579–581.

Vasquez-Campos, L., Asencios-Solis, L., Leo-Hurtado, E., 2004. Drug resistance trends among previously treated tuberculosis patients in a national registry in Peru, 1994–2001. Int. J. Tuberc. Lung Dis. 8, 465–472.

Wang, S., Inci, F., De Libero, G., Singhal, A., Demirci, U., 2013. Point-of-care assays for Tuberculosis: role of nanotechnology/microfluidics. Biotechnol. Adv. 31 (4), 438–449.

Wang, H., Xu, Y., Zhou, X., 2014. Docetaxel-loaded chitosan microspheres as a lung targeted drug delivery system: *in vitro* and *in vivo* evaluation. Int. J. Mol. Sci. 15 (3), 3519–3532.

Wilkinson, D., Davies, G.R., 1997. The increasing burden of tuberculosis in rural South Africa: impact of the HIV epidemic. S. Afr. Med. J. 87, 447–450.

Witten, G.Q., Perelson, A.S., 2004. Modelling the cellular-level interaction between the immune system and HIV. S. Afr. J. Sci. 100, 447–451.

World Health Organization, 2011. Guidelines for the Programmatic Management of Drug-Resistant Tuberculosis—2011 Update. WHO, Geneva, Switzerland.

World Health Organization. TB report 2013, http://www.who.int/tb/publications/global_report/2013/pdf/report_without_annexes.pdf.

Wu, B., Rong, G., Zhao, J., Zhang, S., Zhu, Y., He, B., 2012. A nanoscale porous silicon microcavity biosensor for novel label-free tuberculosis antigen–antibody detection. Nano. 7 (6), 1250049, http://dx.doi.org/10.1142/S179329201250049X. www.vaccines.gov/diseases/tb/.

Yao, C., Zhu, T., Tang, J., Wu, R., Chen, Q., Chen, M., et al., 2008. Hybridization assay of hepatitis B virus by QCM peptide nucleic acid biosensor. Biosens. Bioelectron. 23, 879–885.

Zhou, L., He, X., He, D., Wang, K., Qin, D., 2011. Biosensing technologies for *Mycobacterium tuberculosis* detection: status and new developments. Clin. Dev. Immunol. 193963. Available from: <http://dx.doi.org/10.1155/2011/193963>.

Zhou, X.C., Huang, L.Q., Li, S.F., 2001. Microgravimetric DNA sensor based on quartz crystal microbalance: comparison of oligonucleotide immobilization methods and the application in genetic diagnosis. Biosens. Bioelectron. 16, 85–95.

CHAPTER

10

Influence of Physicochemical Properties of Nanomaterials on Their Antibacterial Applications

Hemant Kumar Daima[1] *and Vipul Bansal*[2]

[1]Department of Biotechnology, Siddaganga Institute of Technology, Tumkur, Karnataka, India

[2]Ian Potter NanoBioSensing Facility and NanoBiotechnology Research Laboratory (NBRL), School of Applied Sciences, RMIT University, Melbourne, Australia

10.1 INTRODUCTION

Nanomaterials are engineered constructions of minuscule size wherein at least one dimension of the material is less than or equivalent to 100 nm. These engineered nanomaterials have significant importance because they are now regularly used as fillers, catalysts, semiconductors, cosmetics, microelectronics, and biomedical devices. Materials in the nanoscale domain can be engineered to show certain specific physical and chemical properties of their controllable interactions with the surrounding environment. Therefore, properties of engineered nanomaterials differ considerably from the bulk materials of the similar composition, sanctioning nanomaterials to accomplish remarkable feats of conductivity, reactivity, and optical sensitivity. However, coexistence of nanomaterials with the biological or environmental systems may create undesirable harmful interactions that may lead to toxic effects on living bodies (Nel et al., 2006). Therefore, it is imperative to realize that these interactions of engineered nanomaterials with various biological entities such as deoxyribonucleic acid (DNA), proteins, membranes, organelles, cells, tissues, and organs generate a series of communications between nanomaterial and biological boundaries. These communications are driven by colloidal forces and depend on the dynamic biophysicochemical properties of the nano–bio interface (Nel et al., 2009).

In recent times, the importance of nano–bio interfacial interactions have been highlighted to promote innovative design strategies for the fabrication of engineered nanomaterials with unique physicochemical properties, which eventually determine nanomaterial cellular uptake, transport, and their ultimate fate in the biological systems (Zhu et al., 2013). To utilize

Nanotechnology in Diagnosis, Treatment and Prophylaxis of Infectious Diseases.
DOI: http://dx.doi.org/10.1016/B978-0-12-801317-5.00010-4

the full spectrum of nanobiotechnology and nanomedicine, an in-depth understanding of nanomaterial physicochemical properties is imperative, because it will guide the researchers to control nano–bio interfacial interactions or communication of nanomaterials with the biological world. Furthermore, designing nanomaterials with distinctive properties will be useful for overcoming their potential hazard and using them for safe therapeutic applications. As discussed, physicochemical properties of nanomaterials may influence their bioavailability, transport, fate, cellular uptake, and catalysis of injurious biological responses (Zhu et al., 2013). Therefore, the understanding of the interactions between physicochemical properties of nanomaterials and biological entity will also allow the development of predictive relationships at the nano–bio interface, which is important from the perspective of safe use of nanomaterials (Hobson, 2009).

From this discussion, it is crystal clear that to succeed in the relatively new area of nanomedicine or nanobiotechnology, the effects of nanomaterials must be foreseeable and precise, and nanomaterials must deliver the preferred therapeutic outcomes with minimal toxicity. In this context, several nanomaterials have been engineered with noble inorganic metal cores, their alloys, oxides, and semiconductors for their vast potential toward biomedicine applications from diagnostics to treatment of diseases (Niemeyer and Mirkin, 2004; Fortina et al., 2005; Mirkin and Niemeyer, 2007; Dykman and Khlebtsov, 2012; Rai et al., 2014). Tailored nanomaterials have such a huge potential influence in biomedical sciences because of their dimensional similarities with the basic biological components. Biological components and biomolecules like enzymes, DNA, membrane, and proteins are functional configurations at nanometer and sub-micrometer levels; therefore, atomic or molecular level tailoring of physiochemical properties of materials have immense potential to attain the desired level of sensitivity with improved bio-functionality, bio-efficiency, and bio-specificity (Suh et al., 2009; Farokhzad and Langer, 2009; Albanese et al., 2012).

On the contrary, nanotoxicology is in its embryonic stage of development, but it is a vital part of nanomedicine and nanobiotechnology. Nanotoxicology discusses interactions of engineered nanomaterials with biological systems and the environment while particular emphasis is placed on the correlation between the physicochemical properties of nanomaterials with induction of toxic or adversarial biological responses (Fischer and Chan, 2007). Moreover, nanotoxicology aims to discover favorable physicochemical characteristics of various nanomaterials, which may render them more responsive toward inner biological environment for therapeutic benefits (Holl, 2009; Mahmoudi et al., 2011a). From this perspective, the principal footstep toward understanding the broader implications of the impact of designer-made nanomaterials for any biomedical application is to characterize and recognize how novel nanomaterials interact with simple biological organisms, such as microbial cells.

In the milieu of nanobiotechnology, nanomedicine, and nanotoxicology, several nanomaterials have been developed to demonstrate strong antimicrobial activities through diverse mechanisms (Allahverdiyev et al., 2011). Predominantly, engineered nanomaterials illustrate antimicrobial action by inducing photocatalytic production of reactive oxygen species (ROS), interruption of energy transduction, inhibition of enzyme activity or DNA synthesis, and by damaging/disrupting/penetrating the cell envelop (Sondi and Salopek-Sondi, 2004; Li et al., 2008; Rai et al., 2009; Daima et al., 2011, 2013, 2014; Hajipour et al., 2012; Ingle et al., 2014). This chapter demonstrates how various interrelated physicochemical properties of engineered nanomaterials influence their antibacterial activities.

10.2 PHYSICOCHEMICAL PROPERTIES OF NANOMATERIALS AND THEIR INFLUENCE ON ANTIBACTERIAL PERFORMANCE

The toxic potentials of nanomaterials are rather unclear or still not well-understood, but it is evident from past research that physicochemical characteristics of nanomaterials play a fundamental role in governing their interactions with bacteria, leading to antibacterial actions due to various physiological consequences. Presently, there is ambiguity about how different physicochemical parameters of nanomaterials influence their antibacterial potential but, in general, it is suggested that smaller particles with higher specific surface area (surface-to-volume ratio) must have greater antibacterial activity. The issue becomes particularly complex because published research studies cannot necessarily always be compared because different physicochemical experimental conditions such as bacterial concentration/strains and external factors including light intensity may have a significant influence on antibacterial activity of nanomaterials, which are mostly not taken into account for evaluation. Although numerous characteristics of nanomaterials may affect their complex interactions with bacterial cells to exhibit antibacterial activities, in this section the most important physicochemical characteristics of the engineered nanomaterials influencing their antibacterial activities are discussed in detail. Depicted in Figure 10.1 is a schematic representation summary of various physicochemical properties of engineered nanomaterials, including their shape, size, surface charge, composition, hydrophilic or hydrophobic nature, oxidative dissolution, presence of functional groups, or ligands or biomolecules in the surface corona or surface chemistry of nanomaterials, that have the potential to influence their antibacterial activities.

10.2.1 Influence of Nanomaterial Size, Surface Area, Composition, and Aggregation on Their Antibacterial Performance

The first and foremost important characteristic of nanomaterials is their size, which falls

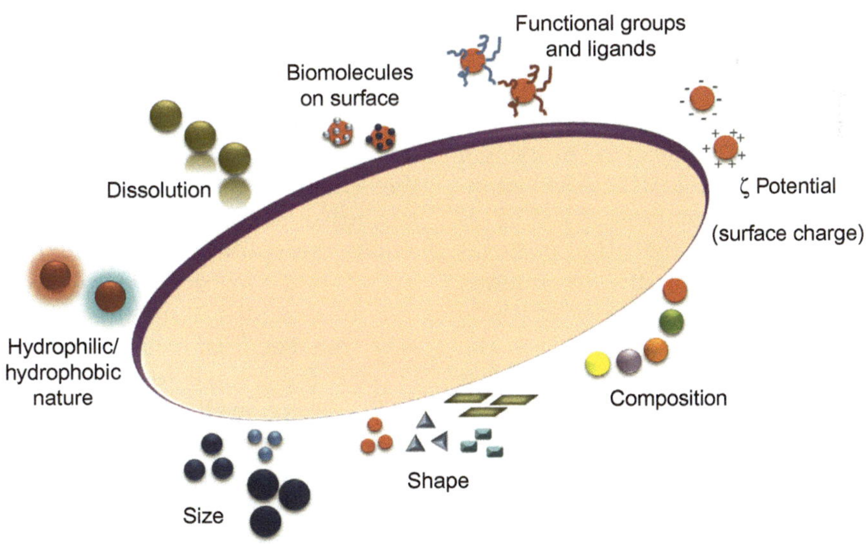

FIGURE 10.1 Schematic representation of various physicochemical properties of nanomaterials that may influence their interaction with bacterial cells.

between the atomic or molecular zone and the bulk of the corresponding material of the same composition. In this transition zone, physicochemical properties of materials will be modified significantly and their uptake/interaction with biological entity opens new opportunities because with reduced nanosize, the number of particles per unit of mass increases considerably. This consideration may produce destructive or hazardous effects in living cells due to the nanoscale size domain. Moreover, the nanomaterials are expected to cross all the biological barriers, gaining entry to the body; consequently, size may also govern their kinetics, absorption, distribution, metabolism, and excretion that may otherwise not be likely with the bulk material of similar composition (Whitesides, 2003; Nel et al., 2006). Theoretical aspects suggest that smaller particles with greater specific surface area should be more toxic in nature. The nanoscale size may possibly promote interactions with the surface of bacterial cells (followed by the ions, active biomolecules, or any other functional groups, if any), which has the potential to disrupt normal cellular functioning, leading to bacterial cell death.

There are numerous studies that have clearly established that size does matter with respect to toxicological consequences. In fact, investigations of the effect of silver nanoparticles in the 1–100 nm size range have been evaluated on Gram-negative bacterium using a high-angle annular dark field (HAADF) scanning transmission electron microscopy (STEM) technique to demonstrate that the particles with a diameter of 1–10 nm directly interact with the bacterial cells and, because of these interactions, hostile effects are created on the bacterial cells. This investigation confirms that the antibacterial properties of nanoparticles are size-dependent (Morones et al., 2005). Likewise, in a recent study in 2014, Adams and colleagues showed size-dependent antibacterial properties of palladium nanoparticles wherein three different well-constrained sub-10 nm sizes, 2.0 ± 0.1, 2.5 ± 0.2, and 3.1 ± 0.2 nm, were investigated. All these palladium nanoparticles of different sizes were incubated with two different bacterial strains, namely, *Escherichia coli* (Gram-negative) and *Staphylococcus aureus* (Gram-positive). Captivatingly, in the case of *S. aureus* mid-sized (2.5 ± 0.2 nm) palladium nanoparticles were found to be the most noxious, whereas smaller (2.0 ± 0.1 nm) and larger (3.1 ± 0.2 nm) nanoparticles were comparatively less toxic. However, in *E. coli* tiny palladium nanoparticles (2.0 ± 0.1 nm) exhibited the highest antibacterial effects, followed by mid-size and larger nanoparticles. These observations were interesting and confirmatory that even the fine scale of 1-nm differences in size can amend their antibacterial activity, and the effects will depend on the strains of the tested bacterial species (Adams et al., 2014).

Highly purified single-walled carbon nanotubes (SWNTs) and multiwalled carbon nanotubes (MWNTs) have also been exploited to demonstrate their comparative antibacterial activities and to reveal a significant role of size in terms of diameter. It has been discovered that the diameter or size of carbon nanotubes (CNTs) is the fundamental factor leading to their antibacterial effects. Moreover, it has been projected and established that the main CNT toxicity mechanism relates to cell membrane damage by direct contact with CNTs. Further, it has been claimed that in the case of SWNT, the enhanced antibacterial activity may be attributed to: smaller SWNT diameter, which assists the splitting and partial penetration of SWNT into the bacterial cell wall; larger surface area of SWNT for interaction and contact with the bacterial cell surface that seriously impacts the cellular membrane integrity, metabolic activity, and morphology; and/or unique chemical and electronic characteristics conveying higher chemical reactivity. Also, as depicted in Figure 10.2, SEM observations show severe cellular destruction of *E. coli* cells after their incubation with MWNT and SWNT

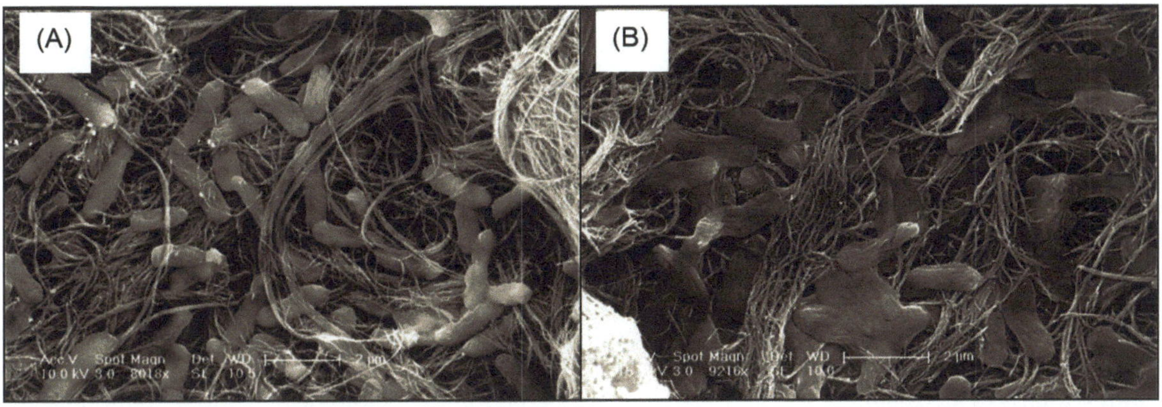

FIGURE 10.2 SEM micrographs of *E. coli* cells treated with (A) MWNTs and (B) SWNTs for 60 min. The scale bars correspond to 2 μm. *Adapted from Kang et al. (2008).*

for 60 min. In the same study, the release of nucleic acids into the solution has further supported the severe destruction of bacteria through the expression of genes related to cell damage. Molecular-level studies of DNA microarrays and gene expression profiling confirm that in the presence of both MWNTs and SWNTs, *E. coli* cells express high levels of stress-related gene products related to cell membrane damage and oxidative stress; however, their degree of expression is considerably higher in the presence of SWNTs (Kang et al., 2008).

In another study, a series of metallic silver (Ag) and gold (Au) nanoparticles was designed by *in situ* reduction and stabilization engaging hyper-branched poly(amidoamine) with terminal dimethylamine groups [HPAMAM-N$(CH_3)_2$]. In this study, by changing the molar ratio of N/Ag (or N/Au) during synthesis of metal nanoparticles, the size and their dispersion stability control could be attained. Both the silver and the gold nanoparticles with smaller particle sizes exhibited higher antimicrobial efficiency because of more effective surface contact with the bacterial cells, as reported in other studies. Moreover, the cationic terminal dimethylamine groups also contribute to a certain extent to the antimicrobial activity through strong ionic interactions with the negatively charged bacterial membranes (Zhang et al., 2008), suggesting that along with the size of nanomaterials, surface functionalization or surface chemistry also plays a role in their antibacterial activities.

In an experimental setup and physiological conditions, many nanomaterials have the tendency to agglomerate because of their intrinsic high reactivity. Therefore, when nanomaterials are exposed to bacterial cells in physiological reaction conditions, it is highly likely that the nanomaterials may create aggregates rather than being individual units. Consequently, the observed antibacterial activities will be of an agglomerated form of nanomaterial and may not provide perfect size-dependent activity. This phenomenon has often been overlooked or not considered for many antibacterial studies, which may lead to misleading interpretations. To address this, a comparative antibacterial study of three photosensitive nanomaterials has been reported (Adams et al., 2006). During this study, all the experiments were performed in water suspensions using two different bacterial strains. By using titanium dioxide (TiO_2), silicon dioxide (SiO_2), and zinc oxide (ZnO) nanoparticles in this study, the authors claimed that although the advertised nanomaterials size did not correspond to the true particle size due to potential aggregation of nanomaterials, apparently aggregation produced similar sized particles that had similar antibacterial activity at a given concentration. However, it is believed that the antibacterial activity may be significantly higher than the observed if the aggregation of nanomaterials did not take place. From this, it may be concluded that the particle size may typically have higher influence than the other physicochemical properties of materials; however, it would not be appropriate to state this as a generalized claim (Adams et al., 2006).

To enhance antibacterial potential by controlling aggregation of nanomaterials, hybrid nanocomposites of Ag-SiO_2 and Cu-SiO_2 have been prepared, wherein SiO_2 nanoparticles synthesized by the "Stober process" were used as seeds to immobilize silver or copper on their surface. In Ag-SiO_2 and Cu-SiO_2 composites, silver or copper nanoparticles were found homogeneously distributed on the surface of SiO_2 nanoparticles without any sign of aggregation and exhibited excellent antibacterial abilities due to lack of aggregation (Kim et al., 2006, 2007). Similarly, sol–gel Ag-SiO_2 thin film was prepared using high concentrations of silver nanoparticles in which metallic silver nanoparticles were immobilized throughout in a well-ordered fashion at the depth of 6 nm on the surface. The synthesized nanocomposite thin film of Ag–SiO_2 displayed high bactericidal activity against both *E. coli* and *S. aureus*, and $\sim 10^5$ cfu/mL bacterial cells were completely destroyed 5 and 7 h after contact with the antibacterial film, respectively. During this process, considerable reduction in the silver ion (Ag^+) release rate was observed because of silver nanoparticles embedded within the silica thin film. Therefore, it has been realized that the discharge process of Ag^+ ions was controlled by water diffusion in the surface pores of the silica film. In HNO_3 solution, the controlled Ag^+ release was measured to be 11 nM/mL after 10 days, indicating that the synthesized Ag–SiO_2 sol–gel composite thin film can be used as efficient and durable material for antibacterial and biomedical applications by controlling nanoparticle aggregation and Ag^+ ion release (Akhavan and Ghaderi, 2009).

10.2.2 Influence of Nanomaterial Shape on Their Antibacterial Performance

In additional to the nanomaterial size, surface area, composition, and its aggregation behavior in biological fluids, nanomaterial shape or morphology is also related to adverse effects on bacterial systems. However, a few investigations focusing on the toxicological relationship associated with this parameter alone are available in the literature. Different physicochemical characteristics including electromagnetic, optical, and catalytic properties of metallic nanomaterials are strongly influenced by their shape and size. Therefore, a variety of synthesis routes have been developed to gain better control over the shape and size of nanoparticles for various applications (Mulvaney, 1996; Narayanan and El-Sayed, 2004; Burda et al., 2005; Bansal et al., 2010). Consequently, it is believed that along with size of the nanoparticles, shape of the nanomaterial also has substantial potential to influence nano–bio interactions. In addition to leading uptake within the cells, size and morphology of a nanomaterial are important factors that are associated with the surface area for a particular mass dosage. In general, to the overall surface area, contribution of the shape of nanomaterials will be important; for instance, an octagonal structure will have a different surface area than spheres of the same size. The higher catalytic activity of the nanomaterial with greater surface area is well-documented to enhance its reactivity because surface atoms have a tendency to hold unsatisfied high-energy bonds. Therefore, these unsatisfied high-energy surface atoms will easily react with other molecules to achieve stabilization: the higher the surface area, the better the nanomaterials reaction potential. Therefore, inside the cellular environment (after successful entrance), these nanomaterials will have a much greater chance compared with their counterpart micron-sized particles to interact with biomolecules of cells, causing direct cellular damage and promoting oxidative stress (Mahmoudi et al., 2011a, b).

Various research reports have demonstrated that the shape of a nanomaterial can greatly influence their rate of uptake by the cells. Spherical-shaped nanoparticles illustrate higher uptake than nanorods, whereas internalization of these nanorods is also highly dependent on their dimensions. Nanorods with a high aspect ratio will be internalized considerably faster than low-aspect ratio rods (Mahmoudi et al., 2011a). In 2007, antibacterial properties of silver nanoparticles of different shapes against *E. coli* were investigated. It was reported that truncated triangular nanoplates of silver accomplished the highest antibacterial action in comparison with spherical and rod-shaped silver nanoparticles. Furthermore, substantial changes in the bacterial cell membrane were reported on treatment with triangular silver nanoplates with a (111) lattice plane, resulting in cell death. Moreover, this study suggested that nanoscale size and the presence of (111) lattice plane combine to promote antimicrobial property, and nanoparticles undertook shape-dependent interaction with *E. coli* (Pal et al., 2007).

10.2.3 Influence of Surface Chemistry of Nanomaterials and Their Exterior Corona on Antibacterial Performance

In contrast to other physicochemical parameters, the influence of nanoparticles surface chemistry and presence of surface corona on their antimicrobial activity is rather unclear or have not been evaluated systematically. Stereotypically, antibacterial activity of nanomaterials is assigned to the nanomaterial composition, surface charge, size, and/or shape; however, influence of complex surface chemistry and peripheral corona stabilizing the nanomaterial is frequently ignored, if not completely overlooked. However, this parameter is gaining considerable attention because of its outstanding potential for controlling pathogenic bacteria without harming normal cells. Because of the existence of different surface chemistries or coronas nanomaterials of similar size, shape, and composition and those prepared using diverse synthesis routes or in different laboratories show dissimilar biological performance. In literature, it is assumed that exterior surface coatings on nanomaterials surface may support governing the surface charge of nanomaterials and consequently allow appropriate electrostatic interactions between nanomaterials and the biological surface (Wiesner et al., 2006; Sapsford et al., 2013). In recent times, through molecular dynamic simulation, the importance of amphiphilic amino acids in adsorbing proteins on nanopatterned surfaces has been established. Furthermore, it has been discovered that presence of such biomolecules on the exterior surface corona of nanomaterials may influence their biological profile because of altered surface chemistry (Hung et al., 2013). Recently, we experimentally demonstrated the essential role of surface chemistry and organic surface coronas surrounding metal nanoparticles in controlling their antibacterial and anticancer properties. Furthermore, it has been anticipated that the tailored modification of chemical properties of nanomaterials surface may generate new and remarkable opportunities to develop efficient antimicrobial agents (Daima et al., 2013, 2014).

Our group developed a new synthetic scheme and confirmed improvement in the antibacterial profile of silver nanoparticles by their surface modification through creating a controlled surface corona by tailoring the surface chemistry of biologically active polyoxometalates (POMs). The stable surface corona of two different POMs, such as phosphotungstic acid (PTA) and phosphomolybdic acid (PMA), has been accomplished by using zwitterionic amino acid tyrosine as a pH-switchable reducing and capping agent around silver nanoparticles. Further, antibacterial investigations revealed that through conjugation with silver nanoparticles, the surface corona of POMs enhances the degree of physical damage to *E. coli* and *Staphylococcus albus* cells because of synergistic antibacterial action of silver nanoparticles and POMs, whereas these nanomaterials indicated biocompatibility toward mammalian cells. Depicted in Figure 10.3 is the antimicrobial activity of silver nanoparticles before and after their functionalization with PTA and PMA molecules. Surface-functionalized silver nanoparticles exhibited enhanced antibacterial activity due to the presence of bioactive PTA and PMA molecules in the outer corona, wherein PTA-functionalized nanomaterials displayed higher activity over PMA-functionalized nanomaterials. At a fixed silver (Ag) dose of 1 μM, silver nanoparticles caused 36% bacterial cell death, which increased to 66% and 85% after PMA and PTA functionalization, respectively. Furthermore, treatment of bacterial cells for 15 min with various nanomaterials revealed distinctive morphological changes indicative of significant damage to the cellular integrity. The higher level of physical damage caused to the bacterial cells by POM-functionalized nanomaterials in comparison with those treated only with silver nanoparticles was also evident from SEM micrographs. The capability of surface corona with a controllable surface chemistry to cause selective toxicity against bacterial cells without causing hostile effects to mammalian cells offers opportunities to utilize them for topical wound healing applications, wherein the control of bacterial infections to allow the growth of new epithelial cells is considered essential (Daima et al., 2014).

Furthermore, our group has also proposed sequential surface functionalization strategies wherein tyrosine-capped gold nanoparticles

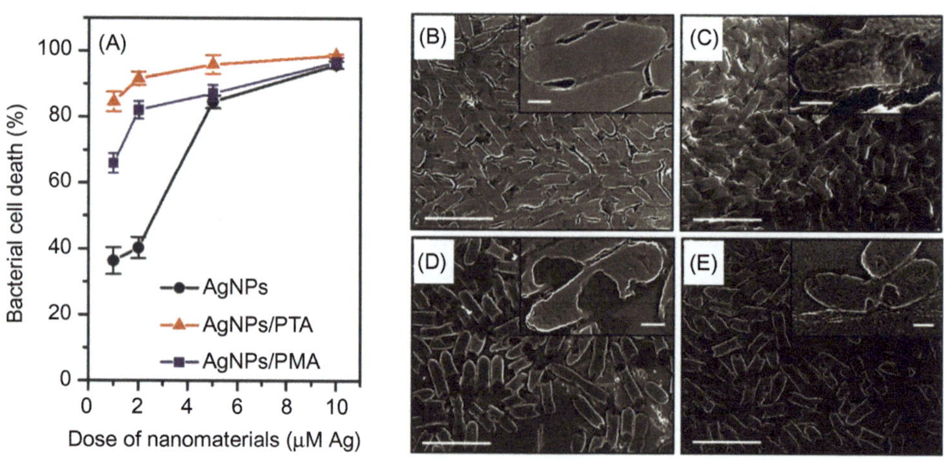

FIGURE 10.3 (A) Antibacterial activity of pristine silver nanoparticles and those functionalized with PTA and PMA against *E. coli*. SEM micrographs of (B) untreated *E. coli* cells and after their treatments with (C) silver nanoparticles, (D) silver/PTA nanoparticles, and (E) silver/PMA nanoparticles. Insets show the higher magnification of the respective SEM images. Scale bars in the main figures and insets correspond to 5 μm and 500 nm, respectively. *Adapted from Daima et al. (2014).*

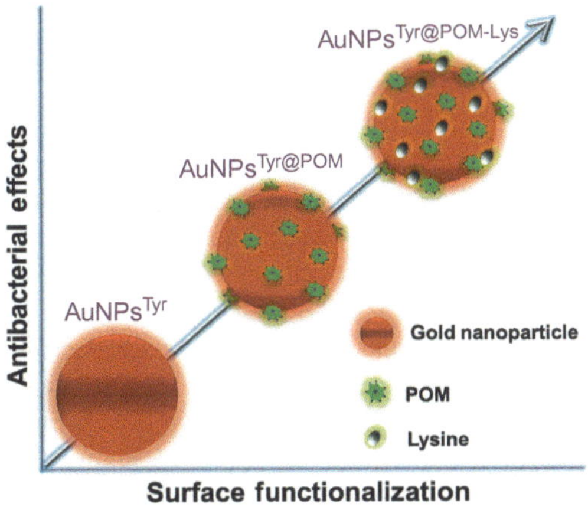

FIGURE 10.4 Schematic representation of increase in antibacterial activity of tyrosine (Tyr)-reduced gold nanoparticles after their stepwise surface functionalization with POM and lysine (Lys). *Adapted from Daima et al. (2013).*

were used as a core material and POMs (PTA and PMA) along with cationic amino acid lysine were decorated on their surface to explore controlled chemical functionality–driven antibacterial action. The antibacterial studies on *E. coli* validated the importance and antibacterial potential of stepwise surface-controlled nanomaterials, confirming that the nanomaterial toxicity and biological applicability are strongly governed by their surface corona, as shown in Figure 10.4. Nevertheless, it was even more interesting that these investigations discovered that highly biocompatible gold nanoparticles can be regulated to be a strong antibacterial agent by fine-tuning their surface corona or chemistry in a controllable manner (Daima et al., 2013).

Along with surface corona and surface chemistry, surface charge of nanomaterial is an important characteristic that has the potential to influence their antibacterial performance. The surface charge of nanomaterials is critical for providing insight into their biological behavior under different experimental conditions. Based on the solution pH and ionic strengths, the surface charge of nanomaterials may also dictate the potential aggregation of nanomaterials in aqueous environments. Surface charge of a nanomaterial also plays an important role in governing the initial electrostatic interaction and the cellular uptake by bacteria. For instance, cationic surface–charged nanomaterials have been reported to have an influence on toxicity of living organisms by posing threats to membrane barrier integrity (Goodman et al., 2004). It has also been discovered that many cationic antimicrobial peptides and nanomaterials do not have a specific target in bacteria and usually they interact with the bacterial cell wall through an electrostatic interaction, which causes physical damage to bacterial cells by forming pores (Shai, 2002; Brogden, 2005). This physical action prevents microbes from developing resistance against nanomaterials; in fact, it is proven that cationic antimicrobial peptides and nanomaterials can overcome bacterial resistance through this physical damage mechanism (Hancock and Lehrer, 1998; Brogden, 2005; Li et al., 2008; Nederberg et al., 2011). Furthermore, peripheral surface coatings can impart a selective charge to the nanomaterials that can stabilize them against aggregation while controlling their outer surface corona. Some reports are available to explore toxic effects of positively charged nanomaterial, but these effects were not observed when the same nanomaterial was coated with negatively charged functional groups (Wiesner et al., 2006; Ozay et al., 2010). Additionally, DNA is negatively charged; thus, cationic nanomaterials are more likely to interact with the genetic material causing genotoxicity. If the surface charge on antibacterial nanomaterials is similar to the surface charge of the bacterium cell (often negative), this may induce repulsion and prevent nanomaterial–bacteria contact. In contrast, providing a positive surface charge to the formulated nanomaterials using different surface coatings may considerably improve the antibacterial performance of such formulations (Hamouda and

Baker, 2000). Therefore, as discussed in previous sections, antibacterial activity of nanomaterials involves direct contact between bacterial surface and nanomaterials, suggesting that the surface charge of nanomaterials could play an influential role in controlling nanomaterial toxicity against bacterial strains.

10.2.4 Influence of Nanomaterial Dissolution into Metal Ions on Their Antibacterial Performance

In 2000, Feng et al. performed a mechanistic study to demonstrate antibacterial effects of silver ions (Ag^+) using $AgNO_3$ on two microorganisms, namely *E. coli* and *S. aureus*, by using combined electron microscopy and X-ray microanalysis. The Ag^+ treatment exhibited similar morphological changes to both bacterial strains, wherein the cytoplasmic membrane was found to be separated from the cell wall and an electron-light region seemed to be appearing in the center of bacterial cells. This central electron-light region was found to contain a condensed form of DNA molecules, whereas many other small electron-dense regions deposited inside the cells or surrounding the cell wall were also observed. X-ray microanalysis detected elements of silver (Ag) and sulfur (S) in the electron-dense granules and cytoplasm, suggesting possible antibacterial mechanism of ionic silver through protein inactivation and loss of DNA replication by Ag^+ treatment (Feng et al., 2000). Since 2000, after this report, scientists have been constantly debating about the role of Ag^+ ions produced by oxidative dissolution of silver nanoparticles in aqueous environments to understand the possible mechanisms by which they exert toxicity against bacteria and other organisms. Recently, this mystery regarding whether the silver nanoparticles exert direct "particle-specific" activity beyond the known antimicrobial activity of released Ag^+ ions has been attempted to be solved, and Ag^+ ions were found to be the definitive molecular toxicant. It has been demonstrated that under strictly anaerobic conditions, by preventing oxidation of silver nanoparticles (Ag^0) and therefore preventing Ag^+ ion release, antibacterial activity can be hindered. Furthermore, it has been established that the antibacterial activity of different coatings of silver nanoparticles such as PEG-/PVP-coated and different sizes of each coating precisely follow the dose–response pattern of *E. coli* exposed to Ag^+, which was added in the form of $AgNO_3$. This work suggested that morphological properties of nanomaterials, which have been identified to affect antibacterial activity, are indirect effectors that primarily influence Ag^+ discharge. Therefore, this study further suggested that the antibacterial action of nanomaterials could be controlled by modifying Ag^+ release. This discharge of Ag^+ control is possibly through manipulation of oxygen availability, particle size, shape, and/or type of coating (Xiu et al., 2012).

Recently, our group discovered silver-tetracyanoquinodimethane (AgTCNQ), an organic semiconducting charge transfer complex, as a new antibacterial material (Davoudi et al., 2014). In this study, extremely high-aspect ratio (>3,000) AgTCNQ nanowires of sub-millimeter lengths and ~100 nm diameter were directly grown on the surface of a fabric to develop antibacterial textiles, as shown in Figure 10.5. It was discovered that these fabrics functionalized with AgTCNQ nanowires show outstanding antibacterial performance against both Gram-positive and Gram-negative bacteria, which was even better than their Ag nanoparticle counterparts. Interestingly, the outcomes of this study reflected on a fundamentally important aspect that the antibacterial performance of Ag-based nanomaterials may not necessarily be solely due to the amount of Ag^+ ions released from these materials; instead, the nanomaterial itself may play a direct role in causing antibacterial action. This study therefore

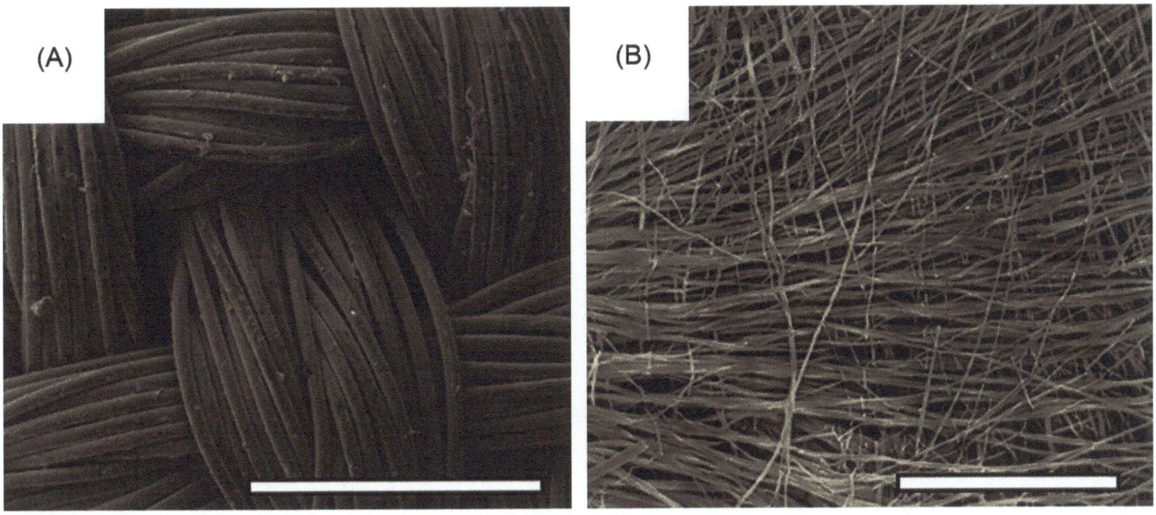

FIGURE 10.5 SEM micrographs of a silver fabric (A) before and (B) after growth of AgTCNQ nanowires with antibacterial properties. Scale bars in (A) correspond to 500 μm and that in (B) correspond to 100 μm. *Adapted from Davoudi et al. (2014).*

challenges the existing paradigm by questioning whether the observed antibacterial effects of nanomaterials are due to metal ion leaching through metal oxidation or whether nanomaterial may also contribute directly to the antibacterial activity (Davoudi et al., 2014).

10.2.5 Influence of ROS Generation Ability of Photosensitive Nanomaterials on Their Antibacterial Performance

Along with other physicochemical characteristics of photosensitive nanomaterials, external factors such as presence of light may significantly influence the exhibition of antibacterial activity because of their ROS production capability. In 2006, Alvaraz and colleagues conducted a comparative study against *Bacillus subtilis* and *E. coli* using three photosensitive nanomaterials for their potential antibacterial activity in water suspensions. Three photosensitive nanomaterials, namely, TiO_2 (titanium dioxide), SiO_2 (silicon dioxide), and ZnO (zinc oxide), have been shown to possess toxicity with varying degrees, and antibacterial activity of these materials was found to be particle concentration–dependent. Although all of the nanomaterials showed antibacterial activity, ZnO nanoparticles demonstrated the highest level of activity, whereas SiO_2 revealed minimum activity relative to other studied materials. Moreover, it was interesting to note that *B. subtilis* species (Gram-positive bacteria) was found more susceptible to these nanomaterials (Adams et al., 2006). Along with intrinsic physicochemical properties of these photosensitive nanomaterials, external factors may also have a noteworthy impact on their antibacterial performance, which can be concluded from the same study. This study claims that the presence of light had remarkable influence under most of the examined experimental conditions, which is probably associated with its role in stimulating production of ROS. Further, it is interesting to note that the bacterial growth inhibition was also observed in the dark, demonstrating involvement of other undetermined mechanisms in addition to photocatalytic ROS production (Adams et al., 2006).

In the presence of light, the antibacterial activity of photosensitive TiO_2 nanomaterials against both the Gram-positive and Gram-negative bacterial species have been found to be considerably higher than in the dark. The degree of inhibition for Gram-positive *B. subtilis* was found to be 2.5-fold greater in the presence of light than in the dark compared with 1.8-fold inhibition of Gram-negative *E. coli* (Adams et al., 2006). The higher inhibition of both the bacterial strains in the presence of light suggested that the antibacterial activity of TiO_2 was related to photocatalytic ROS creation. When Maness et al. (1999) irradiated near-UV light on TiO_2, they observed significant bactericidal activity against *E. coli* K-12 cells. In this research, evidence of antibacterial mechanisms by lipid peroxidation reactions of UV-irradiated TiO_2 was revealed against *E. coli* by estimating production of malondialdehyde (MDA) as an index to assess cell membrane damage by lipid peroxidation after 30 min of illumination. During these investigations, significant reductions of the cell respiratory movements were identified and measured by both oxygen uptake and reduction of 2,3,5-triphenyltetrazolium chloride from succinate as the electron donor. The lipid peroxidation and responsive damage to membrane-dependent respiratory activity and cell viability in *E. coli* depend on both the incident light and TiO_2. It has further been established that photocatalysis by TiO_2 primarily promotes peroxidation of the polyunsaturated phospholipid constituent of the lipid membrane, leading to lysis of bacterial cell membrane. Consequently, vital cellular functions that rely on intact cell membrane, such as respiratory activity, were lost, leading to cell death (Maness et al., 1999). Nevertheless, cell death with TiO_2 in the dark occurred, although it was less pronounced, indicating additional undetermined mechanisms involved in antibacterial action that remain to be established (Adams et al., 2006). The possible mechanisms that may be explored include oxidative stress to bacteria via ROS formation, organic radicals generated in the absence of light, and the direct role of nanomaterials in the disruption of membrane integrity.

The antibacterial activity and mode of action of quantum dots (QDs) such as CdTe, CdSe, and CdSe/ZnS have also been investigated. Often, to render QDs water-soluble for biomedical applications, passivation using several coatings of inorganic and/or organic layers is used. Nevertheless, these surface coverings significantly increase the size of the particle, making uptake by bacterial cells difficult. It is proposed that QDs or conjugated QDs <5 nm can enter the cells, possibly by means of oxidative damage to the cell membrane (Kloepfer et al., 2005; Lu et al., 2008). For example, in the case of antimicrobial activity of CdTe QDs, the proposed mechanism of action involves the QD–bacteria association followed by ROS-mediated cell death. Various spectrophotometry and microscopy analyses revealed that after binding on the bacterial surface, QDs can adversely affect the cellular function by damaging the antioxidative systems, downregulating antioxidative genes, and decreasing the antioxidative enzyme activities. Moreover, with thiobarbituric reacting substances and protein carbonyl assays, the oxidative damage caused to the cellular proteins and lipids, respectively, has also been observed. Further, the survival of bacteria and the level of oxidative destruction to various biomolecules were found to be reliant on QD concentration, because it was noted that QDs could efficiently inhibit the growth of the bacteria in a concentration-dependent manner (Lu et al., 2008).

Furthermore, a recent study evaluated the photogeneration of ROS of various uncoated metallic and single-element nanomaterials and their linkage to the antibacterial activity. This study used the same weight of different materials to facilitate a quantitative comparison of ROS generation ability of various nanomaterials under photochemical excitation conditions performed under UV irradiation. Uncoated silver, gold, nickel, and silicon nanoparticles in aqueous suspension were analyzed under 365 nm of UV light to determine their prospective ROS photogeneration mechanisms and to disclose their associated antibacterial effects. The experimental outcomes indicated that silver nanoparticles generate superoxide and hydroxyl radicals, whereas gold-, nickel-, and silicon-based nanoparticles generate only singlet oxygen species. The redox potentials and electronic structure of silicon nanoparticles were found to mediate ROS generation, whereas ROS generation on silver, gold, and nickel nanoparticles was predominantly found due to their surface plasmon resonance properties. Enhanced antibacterial activity of these nanoparticles was therefore associated with their potential to generate ROS and release of metal ions due to oxidative stress. Under UV irradiation, the order of antibacterial activity toward *E. coli* was found to be silver nanoparticles (strongest) > silicon nanoparticles > nickel nanoparticles > gold nanoparticles (Zhang et al., 2013).

10.2.6 Influence of Other Often Neglected Physicochemical Parameters of Nanomaterials on Their Antibacterial Performance

Purity is one of the most important characteristics of any nanomaterial, but it has often been neglected for its biomedical role, which needs to be considered for its role in therapeutic and antibacterial activities. Presence of residual contaminating foreign metals, unreduced metal ions, chemicals, or other agents (from the precursor material) may actually be responsible for antibacterial responses rather than the actual nanomaterials itself, and the quantity of contaminating materials are fully reliant on the synthesis procedure used. Various research groups use several nanomaterial processing methods, postproduction, to remove most of these precursor metal catalysts and chemical agents; however, purified nanomaterials may still contain certain amounts of residual substances. Therefore, the effects of such chemical impurities, residual metals, and presence of counter ions on their antibacterial activity cannot be overlooked for their potential damaging effects. Hydrophobicity/hydrophilicity, electron transfer capability, smoothness/roughness, surface defects, crystallinity, oxidizability of nanomaterials in aqueous physiological conditions, and counter ion effects are other important physicochemical parameters of diverse nanomaterials that need to be considered while assigning antibacterial potential to physicochemical characteristics of any nanomaterial. Figure 10.6 provides a schematic representation of various possible mechanisms by which different nanomaterials may exhibit antibacterial action.

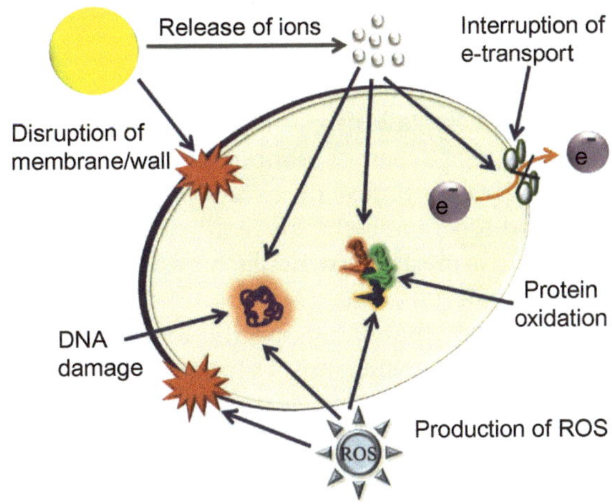

FIGURE 10.6 Schematic representation of potential mechanisms through which different nanomaterials may exhibit antibacterial activities.

10.3 CONCLUSIONS

This chapter has provided an overview of different physicochemical parameters of nanomaterials, each of which may play an important role in contributing to the overall antibacterial performance of nanomaterials. It is shown that by tailoring some of these physicochemical properties of nanomaterials, the antibacterial performance of nanomaterials can be controllably enhanced. The control of physicochemical properties, particularly surface corona of nanomaterials, has the potential to design nanomaterials that show specific toxicity of bacteria without causing any cytotoxicity to mammalian cells. In context of metal nanoparticles, recent research has started providing evidence that the antibacterial effects of metal nanoparticles may not necessarily be due to oxidation of metal into metal ions, but instead metal nanostructures may contribute directly to the antibacterial activity. Furthermore, it is discussed that great attention must be given while assigning the mode of antibacterial action to a particular physicochemical property of a nanomaterial, because a combination of these properties typically work in synergy to dictate the final mode of antibacterial action of nanomaterials. Future research in this area should focus on the mechanistic understanding of the complex relationship between different physicochemical parameters controlling the antibacterial activity of nanomaterials.

References

Adams, C.P., Walker, K.A., Obare, S.O., Docherty, K.M., 2014. Size-dependent antimicrobial effects of novel palladium nanoparticles. PLoS One 9, 1–12.

Adams, L.K., Lyon, D.Y., Alvarez, P.J.J., 2006. Comparative eco-toxicity of nanoscale TiO_2, SiO_2, and ZnO water suspensions. Water Res. 40, 3527–3532.

Akhavan, O., Ghaderi, E., 2009. Bactericidal effects of Ag nanoparticles immobilized on surface of SiO_2 thin film with high concentration. Curr. Appl. Phys. 9, 1381–1385.

Albanese, A., Tang, P.S., Chan, W.C.W., 2012. The effect of nanoparticle size, shape, and surface chemistry on biological systems. Annu. Rev. Biomed. Eng. 14, 1–16.

Allahverdiyev, A.M., Kon, K.V., Abamor, E.S., Bagirova, M., Rafailovich, M., 2011. Coping with antibiotic resistance: combining nanoparticles with antibiotics and other antimicrobial agents. Expert Rev. Anti Infect. Ther. 9, 1035–1052.

Bansal, V., Li, V., O'mullane, A.P., Bhargava, S.K., 2010. Shape dependent electrocatalytic behaviour of silver nanoparticles. CrystEngComm. 12, 4280–4286.

Brogden, K.A., 2005. Antimicrobial peptides: pore formers or metabolic inhibitors in bacteria? Nat. Rev. Microbiol. 3, 238–250.

Burda, C., Chen, X., Narayanan, R., El-sayed, M.A., 2005. Chemistry and properties of nanocrystals of different shapes. Chem. Rev. 105, 1025–1102.

Daima, H.K., Selvakannan, P.R., Homan, Z., Bhargava, S. K., Bansal, V. 2011. Tyrosine mediated gold, silver and their alloy nanoparticles synthesis: antibacterial activity toward Gram positive and Gram negative bacterial strains. 2011 International Conference on Nanoscience, Technology and Societal Implications, NSTSI11.

Daima, H.K., Selvakannan, P.R., Shukla, R., Bhargava, S.K., Bansal, V., 2013. Fine-tuning the antimicrobial profile of biocompatible gold nanoparticles by sequential surface functionalization using polyoxometalates and lysine. PLoS One 8, 1–14.

Daima, H.K., Selvakannan, P.R., Kandjani, A.E., Shukla, R., Bhargava, S.K., Bansal, V., 2014. Synergistic influence of polyoxometalate surface corona towards enhancing the antibacterial performance of tyrosine-capped Ag nanoparticles. Nanoscale 6, 758–765.

Davoudi, Z.M., Kandjani, A.E., Bhatt, A.I., Kyratzis, I.L., O'mullane, A.P., Bansal, V., 2014. Hybrid antibacterial fabrics with extremely high aspect ratio Ag/AgTCNQ nanowires. Adv. Funct. Mater. 24, 1047–1053.

Dykman, L., Khlebtsov, N., 2012. Gold nanoparticles in biomedical applications: recent advances and perspectives. Chem. Soc. Rev. 41, 2256–2282.

Farokhzad, O.C., Langer, R., 2009. Impact of nanotechnology on drug delivery. ACS Nano. 3, 16–20.

Feng, Q., Wu, J., Chen, G., Cui, F., Kim, T., Kim, J., 2000. A mechanistic study of the antibacterial effect of silver ions on *Escherichia coli* and *Staphylococcus aureus*. J. Biomed. Mater. Res. 52, 662–668.

Fischer, H.C., Chan, W.C.W., 2007. Nanotoxicity: the growing need for *in vivo* study. Curr. Opin. Biotechnol. 18, 565–571.

Fortina, P., Kricka, L.J., Surrey, S., Grodzinski, P., 2005. Nanobiotechnology: the promise and reality of new approaches to molecular recognition. Trends Biotechnol. 23, 168–173.

Goodman, C., Mccusker, C., Yilmaz, T., Rotello, V., 2004. Toxicity of gold nanoparticles functionalized with cationic and anionic side chains. Bioconjug. Chem. 15, 897–900.

Hajipour, M.J., Fromm, K.M., Akbar Ashkarran, A., Jimenez De Aberasturi, D., Larramendi, I.R.D., Rojo, T., et al., 2012. Antibacterial properties of nanoparticles. Trends Biotechnol. 30, 499–511.

Hamouda, T., Baker, J.R., 2000. Antimicrobial mechanism of action of surfactant lipid preparations in enteric Gram-negative bacilli. J. Appl. Microbiol. 89, 397–403.

Hancock, R.E.W., Lehrer, R., 1998. Cationic peptides: a new source of antibiotics. Trends Biotechnol. 16, 82–88.

Hobson, D.W., 2009. Commercialization of nanotechnology. Wiley Interdiscip. Rev. Nanomed. Nanobiotechnol. 1, 189–202.

Holl, M.M.B., 2009. Nanotoxicology: a personal perspective. Wiley Interdiscip. Rev. Nanomed. Nanobiotechnol. 1, 353–359.

Hung, A., Mager, M., Hembury, M., Stellacci, F., Stevens, M.M., Yarovsky, I., 2013. Amphiphilic amino acids: a key to adsorbing proteins to nanopatterned surfaces? Chem. Sci. 4, 928–937.

Ingle, A., Duran, N., Rai, M., 2014. Bioactivity, mechanism of action, and cytotoxicity of copper-based nanoparticles: a review. Appl. Microbiol. Biotechnol. 98, 1001–1009.

Kang, S., Herzberg, M., Rodrigues, D.F., Elimelech, M., 2008. Antibacterial effects of carbon nanotubes: size does matter! Langmuir 24, 6409–6413.

Kim, Y.H., Lee, D.K., Cha, H.G., Kim, C.W., Kang, Y.C., Kang, Y.S., 2006. Preparation and characterization of the antibacterial Cu nanoparticle formed on the surface of SiO_2 nanoparticles. J. Phys. Chem. B. 110, 24923–24928.

Kim, Y.H., Lee, D.K., Cha, H.G., Kim, C.W., Kang, Y.S., 2007. Synthesis and characterization of antibacterial Ag-SiO_2 nanocomposite. J. Phys. Chem. C. 111, 3629–3635.

Kloepfer, J.A., Mielke, R.E., Nadeau, J.L., 2005. Uptake of CdSe and CdSe/ZnS quantum dots into bacteria via purine-dependent mechanisms. Appl. Environ. Microbiol. 71, 2548–2557.

Li, Q., Mahendra, S., Lyon, D.Y., Brunet, L., Liga, M.V., Li, D., et al., 2008. Antimicrobial nanomaterials for water disinfection and microbial control: potential applications and implications. Water Res. 42, 4591–4602.

Lu, Z., Li, C.M., Bao, H., Qiao, Y., Toh, Y., Yang, X., 2008. Mechanism of antimicrobial activity of CdTe quantum dots. Langmuir 24, 5445–5452.

Mahmoudi, M., Azadmanesh, K., Shokrgozar, M.A., Journeay, W.S., Laurent, S., 2011a. Effect of nanoparticles on the cell life cycle. Chem. Rev. 111, 3407–3432.

Mahmoudi, M., Lynch, I., Ejtehadi, M.R., Monopoli, M.P., Bombelli, F.B., Laurent, S., 2011b. Protein–nanoparticle interactions: opportunities and challenges. Chemi. Rev. 111, 5610–5637.

Maness, P.C., Smolinski, S., Blake, D.M., Huang, Z., Wolfrum, E.J., Jacoby, W.A., 1999. Bactericidal activity of photocatalytic TiO(2) reaction: toward an understanding of its killing mechanism. Appl. Environ. Microbiol. 65, 4094–4098.

Mirkin, C.A., Niemeyer, C.M. (Eds.), 2007. Nanobiotechnology II: More Concepts and Applications. Wiley-VCH Verlag GmbH & Co. KGaA.

Morones, J.R., Elechiguerra, J.L., Camacho, A., Holt, K., Kouri, J.B., Ramirez, J.T., et al., 2005. The bactericidal effect of silver nanoparticles. Nanotechnology. 16, 2346–2353.

Mulvaney, P., 1996. Surface plasmon spectroscopy of nanosized metal particles. Langmuir. 12, 788–800.

Narayanan, R., El-sayed, M.A., 2004. Shape-dependent catalytic activity of platinum nanoparticles in colloidal solution. Nano Lett. 4, 1343–1348.

Nederberg, F., Zhang, Y., Tan, J.P.K., Xu, K., Wang, H., Yang, C., et al., 2011. Biodegradable nanostructures with selective lysis of microbial membranes. Nat. Chem. 3, 409–414.

Nel, A., Xia, T., Madler, L., Li, N., 2006. Toxic potential of materials at the nanolevel. Science. 311, 622–627.

Nel, A.E., Madler, L., Velegol, D., Xia, T., Hoek, E.M.V., Somasundaran, P., et al., 2009. Understanding biophysicochemical interactions at the nano–bio interface. Nat. Mater. 8, 543–557.

Nanobiotechnology: concept. In: Niemeyer, C.M., Mirkin, C.A. (Eds.), Application and Perspectives. Wiley-VCH Verlag GmbH & Co. KGaA.

Ozay, O., Akcali, A., Otkun, M.T., Silan, C., Aktas, N., Sahiner, N., 2010. P(4-VP) based nanoparticles and composites with dual action as antimicrobial materials. Colloids Surf. B Biointerfaces. 79, 460–466.

Pal, S., Tak, Y.K., Song, J.M., 2007. Does the antibacterial activity of silver nanoparticles depend on the shape of the nanoparticle? A study of the Gram-negative bacterium *Escherichia coli*. Appl. Environ. Microbiol. 73, 1712–1720.

Rai, M., Yadav, A., Gade, A., 2009. Silver nanoparticles as a new generation of antimicrobials. Biotechnol. Adv. 27, 76–83.

Rai, M., Kon, K., Ingle, A., Duran, N., Galdiero, S., Galdiero, M., 2014. Broad-spectrum bioactivities of silver nanoparticles: the emerging trends and future prospects. Appl. Microbiol. Biotechnol. 98, 1951–1961.

Sapsford, K.E., Algar, W.R., Berti, L., Gemmill, K.B., Casey, B.J., Oh, E., et al., 2013. Functionalizing nanoparticles with biological molecules: developing chemistries that facilitate nanotechnology. Chem. Rev. 113, 1904–2074.

Shai, Y., 2002. Mode of action of membrane active antimicrobial peptides. Biopolymers. 66, 236–248.

Sondi, I., Salopek-sondi, B., 2004. Silver nanoparticles as antimicrobial agent: a case study on *E. coli* as a model for Gram-negative bacteria. J. Colloid Interface Sci. 275, 177–182.

Suh, W.H., Suslick, K.S., Stucky, G.D., Suh, Y.-H., 2009. Nanotechnology, nanotoxicology, and neuroscience. Prog. Neurobiol. 87, 133–170.

Whitesides, G.M., 2003. The 'right' size in nanobiotechnology. Nat. Biotechnol. 21, 1161–1165.

Wiesner, M.R., Lowry, G.V., Alvarez, P., Dionysiou, D., Biswas, P., 2006. Assessing the risks of manufactured nanomaterials. Environ. Sci. Technol. 40, 4336–4345.

Xiu, Z.-M., Zhang, Q.-B., Puppala, H.L., Colvin, V.L., Alvarez, P.J.J., 2012. Negligible particle-specific antibacterial activity of silver nanoparticles. Nano Lett. 12, 4271–4275.

Zhang, W., Li, Y., Niu, J., Chen, Y., 2013. Photogeneration of reactive oxygen species on uncoated silver, gold, nickel, and silicon nanoparticles and their antibacterial effects. Langmuir 29, 4647–4651.

Zhang, Y., Peng, H., Huang, W., Zhou, Y., Yan, D., 2008. Facile preparation and characterization of highly antimicrobial colloid Ag or Au nanoparticles. J. Colloid Interface Sci. 325, 371–376.

Zhu, M., Nie, G., Meng, H., Xia, T., Nel, A., Zhao, Y., 2013. Physicochemical properties determine nanomaterial cellular uptake, transport, and fate. Acc. Chem. Res. 46, 622–631.

CHAPTER

11

Nanocarriers Against Bacterial Biofilms: Current Status and Future Perspectives

Noha Nafee

Department of Pharmaceutics and Biopharmacy, Philipps University, Marburg, Germany; Department of Pharmaceutics, Faculty of Pharmacy, Alexandria University, Alexandria, Egypt

11.1 INTRODUCTION

Health care–associated infection (HAI) encompasses almost all clinically evident infections that do not originate from a patient's original admitting diagnosis. According to the Center of Disease Control and Prevention (CDC) in the United States, a prevalence survey regarding HAIs estimated that on a given day, approximately 1 in 25 hospital patients has at least one HAI; there were approximately 722,000 HAIs in US acute care hospitals in 2011. Approximately 75,000 hospital patients with HAIs died during their hospitalizations. More than half of all HAIs occurred outside of the intensive care unit (Magill et al., 2014). These infections result in additional costs in excess of $4.5 billion to the health care sector, patients, and those who care for them.

A major complication involves development of bacterial biofilms in which bacteria encase themselves in sessile communities of hydrated polymeric matrices of their own synthesis. The inherent resistance of biofilms to antimicrobials is at the root of many persistent bacterial infections. Therefore, in the past decade, biofilm development and control became areas of intense research. More insight has been focused on the components involved in biofilm development as possible targets for biofilm control.

Broad-spectrum antibiotics such as tobramycin, cephamycins, and gentimicin remain the first-line treatment for bacterial biofilms. However, the widespread imprudent use of antibiotics has provoked an exponential increase in the incidence of antibiotic resistance in several bacterial groups in recent years. New antibiotics such as peptide antibiotics and/or new ways of using old antibiotics are necessary to overcome bacterial resistance.

Other strategies for effective antibiofilm therapy involve the use of nanomedicine and the development of novel anti-infectives with different modes of action. Research efforts regarding the use of nanoparticles (NPs) in the prevention of "medical" biofilm formation and the eradication of existing biofilms have been

Nanotechnology in Diagnosis, Treatment and Prophylaxis of Infectious Diseases.
DOI: http://dx.doi.org/10.1016/B978-0-12-801317-5.00011-6

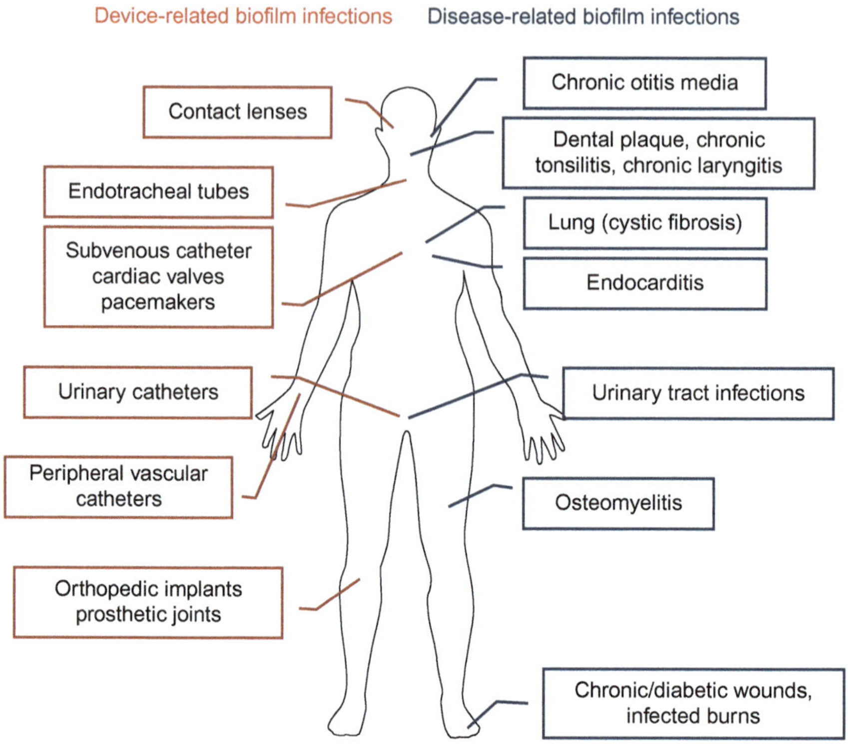

FIGURE 11.1 Device-related and disease-related biofilm infections.

growing steadily. Nanomaterials often show unique and superior physical, chemical, and biological properties compared with their macroscale counterparts. Several studies have shown the "nanofunctionalization" of surfaces through coating, impregnation, or embedding to inhibit bacterial adhesion and biofilm formation. Meanwhile, novel anti-infectives including antimicrobial peptides, quorum-sensing (QS) inhibitors, and some plant extracts were recently discovered and represent a new era for antibiotic-free therapeutics.

In this chapter, the application of nanomedicine as a viable approach in antibiofilm therapy is addressed. The concept of biofilm formation and challenges in biofilm eradication are introduced. Insights regarding the different types of nanocarriers for the delivery of antimicrobials as well as metallic/inorganic NPs exerting antimicrobial and/or photodynamic therapy (PDT) are discussed in detail. In parallel, recent advances toward the discovery of novel anti-infectives and innovative strategies for biofilm targeting are explored.

11.2 BIOFILMS—HEALTH AND ECONOMIC BURDENS

Different areas in the human body can be vulnerable to biofilm infections as a result of either a preexisting disease or a hospital-acquired infection (Figure 11.1). Approximately 60% of all hospital-associated infections, more than 1 million cases per year, are attributable to biofilms that have formed on indwelling medical devices. The central problem with microbial biofilm infections is their propensity to resist clearance by the host immune system and all antimicrobial agents tested to date. In fact, compared with their free-floating planktonic counterparts, microbes within a biofilm are 50- to 500-times more resistant to antimicrobial agents (Otto, 2014). *Staphylococcus aureus* is able to quickly develop antibiotic resistance and methicillin-resistant strains (MRSA) are considered endemic in hospitals. This bacterium features a myriad of virulence factors that allow it to colonize and damage the host and perfectly avoid the host immune response (Brady et al., 2009). Therefore, achieving therapeutic and nonlethal dosing regimens within the human host is impossible.

The end result is a conversion from an acute infection to one that is persistent, chronic, and recurrent, most often requiring device removal to eliminate the infection. Because the use and the types of indwelling medical devices commonly used in modern health care are continuously expanding, especially with an aging population, the incidence of biofilm infections will also continue to increase (Otto, 2014). Prosthetic implant infection can be caused by direct inoculation of bacteria to the implant or by seeding from the blood (hematogenous). Because the implant is quickly coated by host connective tissue on implantation, bacteria such as *S. aureus* and *Staphylococcus epidermidis* are able to readily gain a foothold in the host. Therefore, prosthetic implant infection has a high rate of morbidity and mortality for patients, as well as an extreme economic burden on the US health care system (Brady et al., 2009).

As a consequence, the economic burden due to biofilm infections is substantial. Device-related biofilm infections increase hospital stays and add more than $1 billion/year to US hospitalization costs (Otto, 2014). For example, catheter-related sepsis costs an additional $28,000 per case. Nosocomial urinary tract infections, which are a subset of these catheter-related infections, account for approximately 900,000 admissions annually in the United States.

11.3 BIOFILMS—DEFINITION, COMPOSITION, AND DEVELOPMENT

A common definition of a biofilm involves the formation of a community of microorganisms attached to a surface (Costerton et al., 1999; Lewis, 2001). Biofilm formation enables bacteria to assume a temporary multicellular lifestyle in which "group behavior" facilitates survival in adverse environments and raises resistance level. Within the biofilm, bacteria are cocooned in a self-produced extracellular matrix, which accounts for approximately 90% of the biomass (Flemming and Wingender, 2010). The matrix (Figure 11.2) is composed of extracellular polymeric substances (EPS) that, along with carbohydrate-binding proteins, pili, flagella, other adhesive fibers and extracellular DNA (eDNA), act as a stabilizing scaffold for the three-dimensional biofilm structure (Cegelski et al., 2009; Flemming and Wingender, 2010; Vilain et al., 2009). In the matrix, nutrients are trapped for metabolic utilization by the resident bacteria and water is efficiently retained through H-bond interactions with hydrophilic polysaccharides (Flemming and Wingender, 2010). The robust biomass protects bacteria from desiccation, predation, oxidizing molecules, radiation, and other damaging agents (Flemming and Wingender, 2010).

Biofilms of *Pseudomonas aeruginosa* persistent in the lungs of cystic fibrosis (CF) patients represent a famous example. *P. aeruginosa* grows in small colonies with biofilm-like characteristics in the hypoxic mucopurulent environment of alginate and other EPS within the stationary mucus, where the bacterial cells elaborate a quorum-sensing system to control gene expression specifically for growth as a biofilm (Moreau-Marquis et al., 2008). Alginate production is mostly stimulated in such hypoxic conditions, converting nonmucoid cultures to mucoid (Meers et al., 2008).

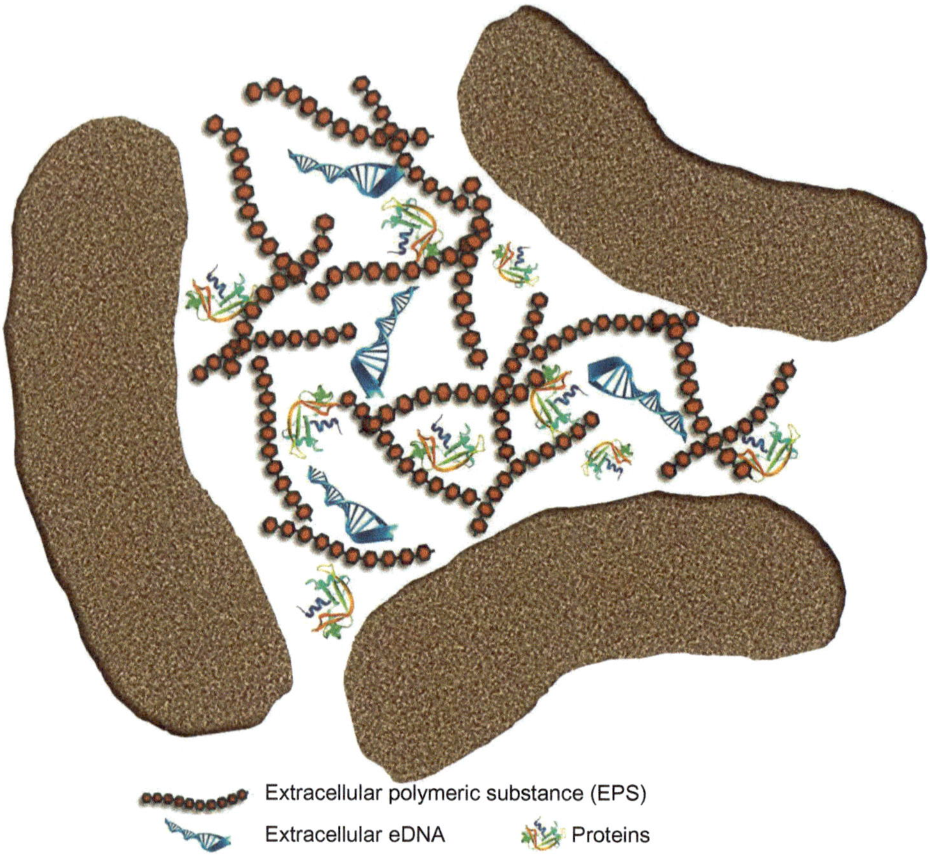

FIGURE 11.2 Composition of biofilm matrix.

Biofilm growth is recognized as a complex developmental process that is multifaceted and dynamic in nature. The transition from planktonic growth to biofilm occurs in response to environmental changes and involves multiple regulatory networks, which translate signals to concerted gene expression changes, thereby mediating the spatial and temporal reorganization of the bacterial cell (Lenz et al., 2008; Monds and O'Toole, 2009). This cellular reprogramming alters the expression of surface molecules, nutrient utilization, and virulence factors, enabling bacterial survival in unfavorable conditions (Klebensberger et al., 2009; Lewis, 2008; Zhang and Mah, 2008). For instance, enzymes secreted by the bacteria modify EPS composition in response to changes in nutrient availability, thereby tailoring biofilm architecture to the specific environment (Ma et al., 2009).

The pattern of development involves initial attachment to a solid surface, followed by the formation of microcolonies on the surface and, finally, differentiation of microcolonies into EPS-encased, mature biofilms (Costerton et al., 1999). Bacteria have a sense of touch that enables detection of a surface and the expression of specific genes. Some of the cells in a biofilm adopt a distinct and protected biofilm phenotype that is a biologically programmed response to growth on a surface. Both the surface of the cell and the surface of the substratum obviously determine the effectiveness of adhesion in biofilm formation. Numerous surface adhesins of pathogens facilitate binding to host cells and/or abiotic surfaces (Stewart, 1996). During this attachment phase, perhaps after microcolony formation, the transcription of specific genes is activated. In particular, studies with *P. aeruginosa algC, algD,* and *algU::lacZ* reporter constructs show that the transcription of these genes, which are required for synthesis of the extracellular alginate, is activated after attachment to a solid surface (Lenz et al., 2008). Research on QS in Gram-negative bacteria has shown that acylhomoserine lactone signals are produced by individual bacterial cells. At a critical cell density, these signals can accumulate and trigger the expression of specific sets of genes (for reviews see Hassett et al., 2002; Häussler, 2010; and Ma et al., 2009).

Bacteria in a sessile biofilm tend to colonize new areas; therefore, there must be some mechanisms for dispersion. Pieces of biofilms can break off in the flow and may colonize new surfaces. Even in case of the nonpathogenic, photosynthetic bacterium *Rhodobacter sphaeroides*, an acylhomoserine lactone quorum-sensing signal is required for dispersal of individual cells from community structures. It has been also suggested that escape of *P. aeruginosa* cells from the biofilm matrix involves the action of an enzyme that digests alginate (Costerton et al., 1999).

11.4 CHALLENGES IN ANTIMICROBIAL TREATMENT OF BIOFILMS

The ability of opportunistic bacteria to grow in biofilms is decisive in the pathogenesis of chronic infectious diseases. Bacteria perfectly understand that their residence in biofilms protects them from innate host immune system (e.g., opsonization and phagocytosis) and effects of antimicrobials. Lewis defined biofilm resistance to antimicrobials as increased resistance of cells to killing, which does not mean that biofilm cells grow better than planktonic cells in presence of antimicrobials (Stewart, 1996). Compared with planktonic cells, bacteria encapsulated within biofilms withstand diverse environmental perturbations, such as increased tolerance and resistance to antimicrobial therapy (Forier et al., 2014). Alone or in combination, many factors related to the bacteria, the surrounding environment, or the antimicrobial itself can contribute to biofilm resistance (Figure 11.3). Among the bacterial biofilm-related factors, restricted penetration of antimicrobials into bacterial biofilm, decreased growth rate, and expression of possible restriction genes are commonly reported (Forier et al., 2014). Restricted penetration can be manifested as limited diffusion of large antimicrobial proteins (e.g., lysozymes) as well as smaller antimicrobial peptides (e.g., defensins and analogues), or as binding of the negatively charged exopolysaccharide with the positively charged aminoglycoside antibiotics (Espuelas et al., 2002; Hunter et al., 2012). Mathematical models predict that a formidable penetration barrier should be established if the antimicrobial agent is deactivated in the outer layers of the biofilm faster than it diffuses (Stewart, 1996). This is true for reactive oxidants such as hypochlorite and hydrogen peroxide (Xu et al., 1996). These antimicrobial oxidants are products of the oxidative burst of phagocytic cells, and poor penetration of reactive oxygen species (ROS) may partially account for the inability of phagocytic cells to destroy biofilm microorganisms.

In other cases, biofilm only slows the diffusion of antimicrobials (e.g., fluoroquinolone), thus enhancing degradation of upcoming antibiotic by β-lactamase (Abeylath and Turos, 2008; Wissing and Muller, 2003). In CF patients, the absence of a chloride channel

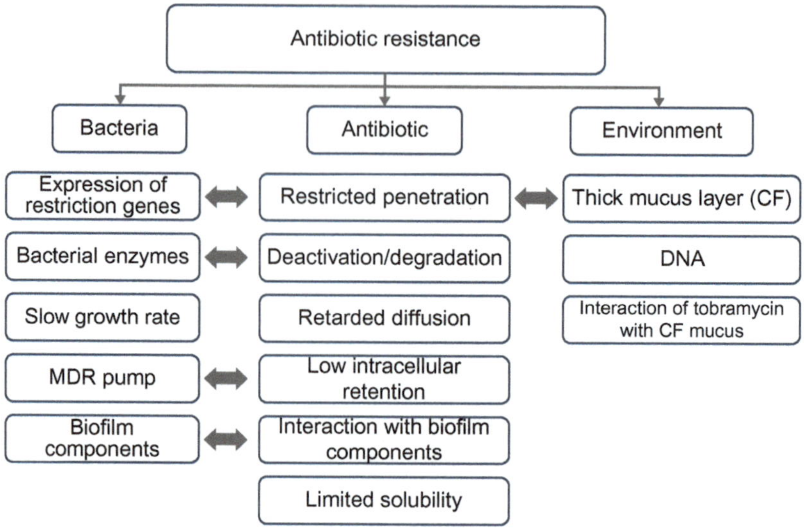

FIGURE 11.3 Factors contributing to antibiotic resistance.

leads to an elevated salt content in the airway surface fluid. The salt inhibits the activity of antimicrobial peptides and proteins involved in the innate immunity of the airways (Costerton et al., 1999). Surprisingly, bacteria did not come up with a general mechanism of antibiotic degradation; nevertheless, any mechanism of antibiotic destruction coupled with diffusion barrier of biofilm provides effective resistance (Cheow et al., 2010b).

Slow growth undoubtedly contributes to biofilm resistance to killing; virtually all antibiotics are more effective against growing, rapidly dividing cells (Pandey and Khuller, 2005). Multiple drug resistance (MDR) pumps are reported to play a role in biofilm resistance; unknown MDR pumps are overexpressed in *P. aeruginosa* biofilms (Souto et al., 2004). However, the presence of metabolically inactive nondividing persister cells within biofilms, which are tolerant to a number of antibiotics despite being genetically identical to the rest of the bacterial population, makes the treatment more challenging; these dormant persisters are believed to be responsible for the reseeding of biofilms on cessation of antibiotic treatment in the clinical setting (Lewis, 2008).

In addition, the environment in which biofilm resides defines the antimicrobial efficacy; in the CF lung, clusters of *P. aeruginosa* are found in the thick layer of mucus overlying airway epithelial cells. The attachment of bacteria to the host mucin could be one of the steps leading to increased antimicrobial resistance associated with biofilms. Also, amino acids play a potential role in the expression of OprF, a porin required for the anaerobic respiration and biofilm formation on abiotic surfaces by *P. aeruginosa* (Cavalli et al., 2002). The restricted motility of *P. aeruginosa* in the CF-like mucus layer apparently resulted in the formation of bacterial cell clusters, in contrast with the normal, hydrated mucus layer, whose larger mesh pore size likely prevented the accumulation of quorum-sensing molecules and the formation of biofilms (Bargoni et al., 2001). DNA—a common component of the CF lung sputum—facilitates the formation of biofilms (Sanna et al., 2007).

Apart from the biofilm component and environment, the nature of antimicrobials also

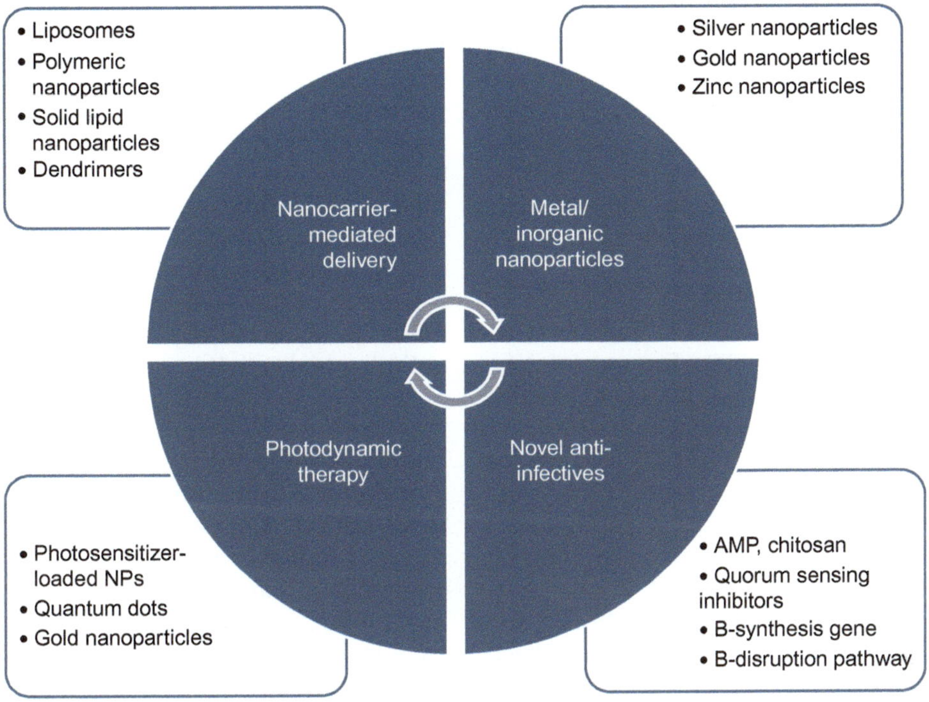

FIGURE 11.4 Current approaches for efficient anti-infective therapy.

contributes to challenging antibiofilm therapy. Many antimicrobials are difficult to administer because of limited solubility, cytotoxicity to healthy tissue, rapid degradation, and clearance from the body (Zhang et al., 2010). For instance, β-lactams and aminoglycosides, have restricted cellular penetration because of high hydrophilicity, whereas fluoroquinolones and macrolides display low intracellular retention, although perfectly diffused in the cells (Gelperina et al., 2005). Further unwanted interactions with biofilm components (e.g., the binding of aminoglycoside antibiotics to the alginate matrix of mucoid *P. aeruginosa* biofilms) or the surrounding environment (e.g., the inhibition of the activity of tobramycin by lung mucus in CF patients) definitely lead to ineffective antimicrobial therapy (Cheow et al., 2011; Costerton et al., 1999; Cui et al., 2012).

11.5 CURRENT APPROACHES FOR EFFICIENT ANTI-INFECTIVE THERAPY

Overcoming biofilm infections necessitates the development of various treatment strategies. Such approaches include the use of nanocarriers for improved delivery and efficacy of antibiotics and other anti-infectives, the antimicrobial potential of metal/inorganic NPs, antimicrobial photodynamic therapy (APDT), as well as the discovery of novel antibiofilm agents (Figure 11.4).

11.5.1 Nanocarrier-Mediated Delivery of Antimicrobials

Nanocarriers give promises not only of efficient drug delivery to the infection site (i.e., intracellular) but also of controlled amount and frequency of dosage, thereby preventing toxicities related to therapy (Abeylath and Turos, 2008; Espuelas et al., 2002). Protection of antimicrobials from the aforementioned unwanted interactions with biofilm offered by nanoencapsulation would be of ultimate impact (Abeylath and Turos, 2008).

11.5.1.1 Liposomes

One of the distinguishing features of *liposomes* is the lipid bilayer structure, which mimics cell membranes and can readily fuse with infectious microbes, typically referred to as "*fusogenic liposomes*." By directly fusing with bacterial membranes, the drug payloads of liposomes can be released either to the cell membranes or to the interior of the bacteria. Fusogenic liposomes consist of lipids containing a phosphoethanolamine (PE) moiety such as 1,2-dioleoyl-sn-glycero-3-phosphoethanolamine (DOPE) that makes the lipid bilayer more fluid (Cheow et al., 2010b; Turos et al., 2007). The fluidity of liposomes can be tuned by the choice of phospholipids; shorter or asymmetric acyl chains of the phospholipids as well as the addition of cholesterol tend to lower the phase transition temperature (Tc) of the liposomes by disturbing the packing of the phospholipid bilayer (Pandey and Khuller, 2005; Wissing and Muller, 2003). An interesting example is the liposomal encapsulation of polymyxin B that was reported to dramatically diminish the drug's side effects and improve its antimicrobial activity against resistant strains of *P. aeruginosa* (Souto et al., 2004). Studies have revealed lipid reorganizations in bacterial membranes on incubation with polymyxin B-loaded liposomes (Sanna et al., 2007). The action mechanism has thus been recognized as rapid and spontaneous membrane fusion driven by noncovalent forces such as van der Waals force and hydrophobic interactions that minimize the system's free energy.

Antibiotic efflux is a well-known mechanism of microbial drug resistance in which proteinaceous transports located in bacterial membranes preferentially pump antimicrobial agents out of the cells (Cavalli et al., 2002). As a consequence of liposomal fusion with cell membranes, a high dosage of drug contents is immediately delivered to the bacteria, which can potentially suppress the antimicrobial resistance of the bacteria by overwhelming the efflux pumps, thereby improving the drug's antimicrobial activity.

The unique structure of liposomes, a lipid membrane surrounding an aqueous cavity, enables them to carry both hydrophobic and hydrophilic compounds without chemical modification. Detailed reviews reported the superior antimicrobial activity of liposomal ampicillin, benzyl penicillin, ciprofloxacin, gentamicin, streptomycin, vancomycin, and tecoplanin formulations against *Salmonella typhimurium*, *S. aureus*, MRSA, *Salmonella dublin*, and *Brucella* sp. (Bargoni et al., 2001; Forier et al., 2014; Gelperina et al., 2005; Zhang et al., 2010). In addition, lower minimal inhibitory concentrations (MIC) and/or lower minimal biofilm inhibitory concentrations (MBIC) were recorded for antibiotics in liposomes compared with the free antibiotic *in vitro* against clinically relevant biofilm-forming organisms. In CF sputum, tobramycin or polymyxin B incorporated in DMPC:Chol or DPPC:Chol liposomes was shown to be protected from binding to polyanionic polymers commonly found in CF mucus. The liposome-encapsulated antibiotics demonstrated a significant increase in antimicrobial activity toward *P. aeruginosa* bacteria in the presence of DNA, F-actin, lipopolysaccharides, and lipoteichoic acid (Huh and Kwon, 2011).

Another approach for increased antimicrobial effect is to prolong the contact times between the antibiotic and the biofilm bacteria. With this approach, positively charged liposomes were able to eradicate biofilms at a lower concentration than the negatively charged ones (Stewart, 1996; Xu et al., 1996).

11.5.1.2 Polymeric NPs

Polymeric NPs possess several unique characteristics for antimicrobial drug delivery. First, unlike liposomes, polymeric NPs are physically stable and can be synthesized with a sharper size distribution. Second, the colloidal properties and drug release profiles can be precisely tuned by selecting different polymer lengths, surfactants, and organic solvents during the synthesis. Third, the surface of polymeric NPs typically contain functional groups that can be chemically modified with either drug moieties or targeting ligands (Zhang et al., 2010).

NP-mediated delivery of a variety of antimicrobial agents have shown great therapeutic efficacy for the treatment of various infectious diseases. For example, the antimicrobial activity of amphotericin B-loaded poly(ε-caprolactone) nanospheres have shown greater therapeutic efficacy against *Leishmania donovani* as compared with the free drug counterparts (Espuelas et al., 2002). Recently, chitosan-alginate NPs loaded with benzoyl peroxide, an antimicrobial against *Propionibacterium acne*, have been developed. Particles induced disruption of the *P. acnes* cell membrane also exhibiting anti-inflammatory properties as they inhibited *P. acnes*-induced inflammatory cytokine production in human monocytes and keratinocytes (Friedman et al., 2013).

Penicillin incorporated in polyacrylate NPs was able to retain its antimicrobial activity against MRSA even in the presence of β-lactamase at high concentrations (Abeylath and Turos, 2008). *N*-thiolated β-lactam antibiotics covalently conjugated onto the polymer network of polyacrylate NPs demonstrated potent antibacterial properties against MRSA with improved bioactivity relative to the free drug (Turos et al., 2007). Formulations of ciprofloxacin and levofloxacin encapsulated in either PLGA or poly(caprolactone) were evaluated in *Escherichia coli* biofilms (Cheow et al., 2010b). Ciprofloxacin-loaded PLGA NPs were identified as the most ideal formulation because of their high drug encapsulation efficiency, high antibacterial efficacy at a low dose against biofilm cells, and biofilm-derived planktonic cells of *E. coli* even for concentrations of ciprofloxacin as low as one-sixteenth of the MBIC. Meanwhile, in case of levofloxacin PLGA particles, resistant bacteria survived the initial antibiotic treatment and reformed the biofilm, even in the presence of levofloxacin above the MBIC.

The surface properties of the NPs dramatically affect their behavior and biodistribution. Inhalable tobramycin-loaded PLGA NPs were coated with different hydrophilic polymers. PVA and chitosan were essential to optimize the size and modulate the surface properties, whereas alginate allowed efficient drug entrapment. *In vivo* biodistribution studies showed that PVA-modified alginate/PLGA NPs reached the deep lung, whereas chitosan-modified NPs were found in great amounts in the upper airways, lining lung epithelial surfaces (Ungaro et al., 2012). In a similar study, the interaction with mucin on the surface of Calu-3 cells reduced the penetration of chitosan-coated and PVA-coated PLGA NPs, whereas the hydrophilic pluronic F68-coated NPs diffused across the mucus barrier, leading to a higher intracellular accumulation (Mura et al., 2011).

11.5.1.3 Solid Lipid NPs

Even though the development history of SLN-based antimicrobial drug delivery systems is relatively shorter than that of other NP systems such as liposomes and polymeric NPs, SLNs have shown great therapeutic potentials (Mehnert and Maeder, 2012). SLNs offer combined advantages of polymeric NPs and liposomes while avoiding some of their drawbacks (Müller et al., 2000; Wissing and Muller, 2003).

Unlike liposomes and polymeric NPs, SLNs are stable, have high drug entrapment, and avoid risk of retaining residual organic solvents (Müller et al., 2000). Improved bioavailability and targeted delivery of antimicrobial drugs using SLNs have been reported (Pandey and Khuller, 2005). SLNs increased transdermal diffusion of encapsulated water-insoluble azole antifungal drugs (e.g., clotrimazole, miconazole, econazole, oxiconazole, and ticonazole) (Sanna et al., 2007; Souto et al., 2004; Zhang et al., 2010).

SLNs contain occlusive excipients that readily form a thin occlusive film on application to the skin to reduce water evaporation and retain skin moisture, thus promoting drug penetration into the skin. For this reason, SLNs encapsulated antimicrobial agents such as retinol and retinyl palmitate have shown better drug penetration rates and slower drug expulsion than the free drug counterparts (Müller et al., 2002).

Tobramycin-laden SLNs provided significantly higher bioavailability in the aqueous humor compared with standard eye drops and may replace the advantages of subconjunctival injections for pseudomonal keratitis and preoperative prophylaxis (Cavalli et al., 2002). In comparison, the absorption rate of orally administered tobramycin by the intestinal cells is poor because of P-glycoproteins (P-gp), an ATP-dependent drug efflux pump, on the brush border of small intestine. In contrast, tobramycin-loaded SLNs can significantly suppress the P-gp efflux pump as they penetrate the intestinal linings via endocytosis rather than passive diffusion (Bargoni et al., 2001). After internalization, SLNs are carried away from the transmembrane drug efflux pumps and release tobramycin payloads inside the cells. This represents an effective oral antimicrobial therapy against *P. aeruginosa* parasitizing in the GIT of CF patients.

SLNs are assumed to be phagocyted by alveolar macrophages in the lungs and subsequently transported to the lymphoid tissues (Bargoni et al., 2001; Huh and Kwon, 2011; Müller et al., 2000). Accordingly, tubercle bacilli were not detected in the lungs and spleens after nebulization of rifampicin, isoniazid, and pyrazinamide-encapsulating SLNs to infected guinea pigs every 7 days. In comparison, daily oral administrations of the free drugs for the same period were required to obtain equivalent therapeutic effects (Gelperina et al., 2005; Müller et al., 2000). Such results represent a cost-effective and patient-friendly approach for tuberculosis treatment using antimicrobial drug–loaded SLNs.

11.5.1.4 Lipid–Polymer Hybrid

Lipid–polymer hybrid (LPH) nanoformulation composed of PLGA core loaded with levofloxacin and coated with phosphatidylcholine (PC) lipid ensured higher antibiofilm efficacy compared with PLGA NPs against *P. aeruginosa* biofilms (Cheow et al., 2011).

11.5.1.5 Dendrimers

Dendrimers possess several unique properties that make them an interesting platform for antimicrobial drug delivery; the highly branched nature of dendrimers provides enormous surface area-to-size ratio enabling better *in vivo* reactivity with bacteria. Besides, the structure of dendrimers allows binding of drugs with hydrophobic or hydrophilic nature (Klebensberger et al., 2009; Lenz et al., 2008). On this basis, many antimicrobial drugs have been successfully loaded into dendrimer NPs and have shown improved solubility and therapeutic efficacy (Muñoz-Bonilla and Fernández-García, 2012). Interestingly, by using antimicrobial drugs as a building block, the synthesized dendrimers themselves can become a potent antimicrobial such as "*dendrimer biocides*" functionalized with quaternary ammonium salts (Zhang et al., 2010). The polycationic structure of dendrimer biocides facilitates the initial electrostatic adsorption to negatively charged bacteria and increases membrane permeability, leading to complete disintegration of the bacterial membrane (Chen and Cooper, 2002). In particular, quaternary ammonium salt prepared with the dimethyldodecyl amine exhibited antimicrobial efficacy against *Staphylococcus* and *E. coli* bacteria (Charles et al., 2012).

11.5.2 Antimicrobial Activity of Metal/Inorganic NPs

Metals have been used for centuries as antimicrobial agents. Silver, copper, gold, titanium, and zinc have attracted particular attention, each having different properties and spectra of activity (Morones et al., 2005). Bacteria are far less likely to acquire resistance to metal NPs than they are to other conventional and narrow-spectrum antibiotics (Lara et al., 2011). With respect to this, the antimicrobial properties of silver NPs have received the most attention. Nanosilver is an effective killing agent against a broad spectrum of Gram-negative and Gram-positive bacteria, including antibiotic-resistant strains as previously reviewed by Morones et al. (2005), Kim et al. (2007), Allaker and Memarzadeh (2014), and Muñoz-Bonilla and Fernández-García (2012). The mechanism of antimicrobial activity of silver is not completely understood but is likely to involve multiple targets, in contrast to the more defined targets of antibiotics. Studies have shown that the positive charge on the Ag^+ ion is critical for antimicrobial activity, which allows the electrostatic attraction between the negative charge of the bacterial cell membrane and positively charged NPs (Jung et al., 2008). Structural changes, effect on membrane-bound respiratory enzymes, and damage to bacterial membranes resulting in cell death were reported (Bragg and Rainnie, 1974). With regard to molecular mechanisms, it has been shown that DNA loses its ability to replicate (Feng et al., 2000), while the expression of ribosomal subunit proteins and other cellular proteins and enzymes necessary for ATP production becomes inactive (Yamanaka et al., 2005). These particular studies suggest that sulfur-containing proteins in the membrane or inside the cells as well as phosphorus-containing elements such as DNA are likely to be the preferential binding sites for silver NPs.

An inverse relationship between the size of silver NPs and antimicrobial activity has been clearly demonstrated, where particles in the size range of 1–10 nm have been shown to have the greatest killing activity against bacteria compared with larger particles (Morones et al., 2005). Sotiriou and Pratsinis (2010) attributed the antimicrobial activity of small (<10 nm) nanosilver particles to Ag^+ ions released from the NP surface. For larger silver particles (>15 nm), the contributions of Ag^+ ions and particles to the antibacterial activity are comparable, with the Ag^+ ion release being proportional to the exposed nanosilver surface area. Shape also had an impact on the activity of NPs, as demonstrated with the shape of silver NPs and antimicrobial activity against *E. coli* (Pal et al., 2007). Truncated triangular silver nanoplates with a(Anderson) lattice plane as the basal plane showed the greatest biocidal activity compared with spherical and rod-shaped NPs. The differences appear to be explained by the proportion of active facets present in NPs of different shapes.

The antimicrobial activity of two silver dressings, a silver-containing Hydrofiber® (SCH) dressing and a nanocrystalline silver-containing dressing (NCS), was evaluated on a variety of antibiotic-sensitive and resistant bacteria (Percival et al., 2007). The SCH dressing was most effective against strains of *P. aeruginosa*, *Candida albicans*, and *S. aureus*; the NCS was most effective against strains of *Klebsiella pneumoniae*, *Enterococcus faecalis*, and *E. coli*. Similar to the antibiotic-susceptible microorganisms, 9 of 10 antibiotic-resistant bacterial strains when grown on agar were more susceptible to the SCH dressing compared with the NCS.

Unfortunately, the microorganisms tested were universally less susceptible to the silver dressings when in their biofilm state. Nevertheless, in the majority of cases, the SCH dressing demonstrated greater biofilm-inhibiting activity than the NCS (Percival et al., 2007).

Alternatively, the use of silver salt NPs instead of elemental silver or complex silver compounds to prevent biofilm formation has been investigated. Surfaces comprising silver bromide (AgBr) polymer–nanocomposites were shown to resist biofilm formation. Through controlling the size of the embedded AgBr, it was possible to modify the release of biocidal Ag^+ ions (Sambhy et al., 2006).

Metallic and other NPs are now being combined with polymers and other base materials and are being coated onto surfaces to provide a variety of potential antimicrobial and antiadhesive applications within the oral cavity (Allaker and Memarzadeh, 2014). The efficacy of a silver–silica nanocomposite material as an antimicrobial was compared with the efficacy of conventional silver nitrate and silver zeolite (Egger et al., 2009). The nanocomposite consists of silver NPs embedded in a matrix of amorphous silicon dioxide (SiO_2). On contact with moisture, the pure silver particles act as a source that releases silver ions, which represent the active antimicrobial. Authors reported some key advantages of this nanocomposite, including the dispersion of the discrete silver particles throughout the silica (which prevents agglomeration of the silver particles), the small diameter of the silver particles (which results in a large surface area and release of a large amount of Ag^+), and the small size of the silver–silica composite (approximately 1 μm), allowing the material to be uniformly dispersed and readily incorporated into a variety of substrates. The material exhibited very good antimicrobial activity against a wide range of microorganisms. However, Gram-positive bacteria appeared to be more tolerant to silver than Gram-negative cells (Egger et al., 2009).

Gold NPs may become useful in the development of antibacterial strategies because of their nontoxicity, versatility in surface modification, polyvalent effects, and photothermal effects (Stewart, 1996; Hassett et al., 2002; Lewis, 2008; Klebensberger et al., 2009; Ma et al., 2009; Häussler, 2010). Unlike most antibiotics and nanomaterials, the bactericidal action of gold NPs does not include ROS-related mechanism. Gold NPs exert their antibacterial activities mainly by collapsing membrane potential, inhibiting ATPase activities to decrease the ATP level (by inducing the down-regulation of oxidative phosphorylation pathway), and also by inhibiting the subunit of ribosome from binding tRNA (Cui et al., 2012). In comparison, the proposed mechanisms of antibacterial activity with respect to nano zinc oxide include generation of ROS and damage to the cell membrane with subsequent interaction of the NP with the intracellular contents.

11.5.3 Nanocarriers for Improved APDT

PDT is an interesting approach for oncological applications as well as various diseases (Allison et al., 2004; Babilas et al., 2010; Rai et al., 2010). APDT utilizes the ability of a photosensitizer (PS) in combination with visible light and molecular oxygen to kill many pathogens, including Gram-positive (e.g., *S. aureus*) and Gram-negative bacteria (e.g., *P. aeruginosa*) (Cassidy et al., 2009). Multidrug-resistant strains are just as susceptible to this treatment as their naïve counterparts. As a result of photosentization reactions, free radicals react with molecular oxygen–generating ROS such as $^1O_2$ that, in turn, can readily react with a variety of biological molecules such as unsaturated lipids, cholesterol, and nucleic acid bases in DNA/RNA (Cheng and Burda, 2011).

A PS encapsulated in NPs gives promises of improved intracellular delivery, higher selectivity, and controlled release rate (Cassidy et al., 2009; Chatterjee and Yong, 2008; Ideta et al., 2005). In some cases, APDT can be achieved without PS release from the NPs (Cheng and Burda, 2011). In addition, light-harvesting NPs (e.g., two-photon absorption NPs and upconverting NPs) expand the applicable excitation wavelengths for PDT (Chatterjee and Yong, 2008; Dayal and Burda, 2008). Some NPs have dual functions for PDT; for example, quantum dot (QD)-based NPs with tunable absorption and size can transfer energy to activate the PS or directly to oxygen and generate singlet oxygen (Juzenas et al., 2008; Samia et al., 2006). An interesting dendrimeric approach to deliver PDT drugs was to use PS as the dendrimer core, such as dendrimer porphyrin, with better PDT and lower dark toxicity (Zhang et al., 2003). Further enhancement of this entity was observed after encapsulation in poly (ethylene glycol)-poly (L-lysine) block copolymer (Ideta et al., 2005).

The use of PDT in the treatment of antibiotic-resistant biofilms, antibiotic-resistant wound infections, and azole-resistant oral candidiasis using methylene blue–based PDT has been reviewed by Biel and coworkers (Biel, 2010; Biel et al., 2013). In a study investigating the potential use of APDT in the treatment of *P. aeruginosa* CF pulmonary infection, delivery of red light (635 nm) and two photosensitizers (toluidine blue O [TBO] and meso-tetra (*N*-methyl-4-pyridyl) porphine tetra tosylate [TMP]) across artificial CF mucus was successfully achieved. TMP diffused more efficiently across artificial CF mucus than TBO. TMP required significantly higher concentrations (2.5 mg/mL) than TBO to achieve high killing rates (>99%) for *P. aeruginosa* isolates growing planktonically. Higher concentrations (5.0 mg/mL) of both photosensitizers were required to achieve high rates of killing (>99%) of *P. aeruginosa* isolates growing in biofilms (Donnelly et al., 2007).

The antimicrobial photoactivity of hypericin-laden NPs (45 nm) was assessed *in vitro* against biofilm and planktonic MRSA cells clinical isolates and *in vivo* on infected wounds in rats. Hypericin NPs demonstrated superior inhibition of biofilm over planktonic cells. *In vivo* wound healing studies in rats revealed faster healing, better epithelialization, keratinization, and development of collagen fibers when hypericin NPs were applied. Determination of growth factors and inflammatory mediators in the wound area confirmed superior healing potential of nanoencapsulated hypericin (Nafee et al., 2013).

The effectiveness of 5-aminolevulinic acid–mediated photodynamic therapy (ALA-PDT) on methicillin-resistant *S. aureus* biofilms and methicillin-resistant *S. epidermidis* biofilms has been studied. The drastic reduction in cell survival within biofilms and the disruption of biofilms were confirmed microscopically (Li et al., 2013).

Exploitation of nanoparticulate metals and metal oxides producing ROS when exposed to ultraviolet (UV) light, such as titanium dioxide (TiO_2) and zinc oxide (ZnO), are finding increased use in antimicrobial applications, with silver metal NPs (5–40 nm) having been reported to inactivate most microorganisms, including human immunodeficiency virus type 1 (HIV-1) (Allaker and Memarzadeh, 2014). Incubation of *P. aeruginosa* and *S. epidermidis* with $AgNO_3$-containing NPs inhibited the amount of biofilm formed by 98%. Silver NPs reduced exopolysaccharide content, indicating that biofilm formation was inhibited, although bacterial viability was unaffected (Kalishwaralal et al., 2010).

Chitosan-based silver NPs reduced *S. aureus* and *P. aeruginosa* biofilms by 22% and >65%, respectively. Scanning electron microscopy confirmed the destruction of the *P. aeruginosa* cell membrane, but no cytotoxic effects toward macrophages were observed (Jena et al., 2012). Targeting of liposomes with WGA was successfully combined with PDT (Babilas et al., 2010). It was found that WGA-modified DPPC: DOTAP:DSPE-PEG2000 liposomes were able to deliver more sensitizer to the bacterial cells compared with nontargeted liposomes. This resulted in the complete eradication of MRSA and an increased antimicrobial effect on *P. aeruginosa* bacteria in suspension.

11.5.4 Nano-Based Delivery of Novel Anti-Infectives

Although an individual is helped by antibiotic treatment, the future public is hurt because the treatment naturally selects for the evolution of more prevalent and increased resistance in the environment (Chatterjee et al., 2008). Limiting antibiotic use can control the evolution of resistance; such a policy might translate to improved health outcomes through the discovery of novel anti-infectives with different modes of action.

The *QS* of the *Pseudomonas quinolone signal (pqs)* is a potential target in *P. aeruginosa*. *PqsR* is a key DNA-binding receptor of this *pqs QS* system that is specific to *P. aeruginosa* and a critical regulator that fine-tunes a large set of genes that encode for virulence factors, such as pyocyanin, elastase B, and hydrogen cyanide. Thus, *PqsR* is considered as an attractive target for the development of novel *QS inhibitors* (Anderson, 2010; Lu et al., 2014). Moreover, the anti-QS activity of the *Lagerstroemia speciosa* extract affected tolerance to tobramycin and reduced the expression of virulence factors such as LasA protease, LasB elastase, and pyoverdin (Singh et al., 2012).

The exopolysaccharide synthesis genes seem like a promising potential choice as targets because these components are probably required for the maintenance of biofilm formation and not just for the initial steps of biofilm formation. However, redundancy of polysaccharides and the differences between the biosynthesis genes in various species are a serious limitation for possible drug development by use of this pathway as a target. The antibiofilm activity of some plant extracts (e.g., *Rhodiola crenulata* [arctic root], *Epimedium brevicornum* [rowdy lamb herb], and *Polygonum cuspidatum* [Japanese knotweed]) have been shown (Coenye et al., 2012).

Another and possibly more productive approach would be to develop specific drugs that interact directly with the components of the biofilm self-destruction pathway. Disruption of the biofilm structure could be achieved via the degradation of individual biofilm compounds by various *enzymes* such as deoxyribonuclease I, lysostaphin (a natural staphylococcal endopeptidase that can penetrate bacterial biofilms), α-amylases, lyase, and lactonase (Taraszkiewicz et al., 2013). In an experiment that could serve as a model for this approach, expression of alginate lyase from a controllable promoter increased sloughing of cells from a colony of mucoid *P. aeruginosa* cells that overproduced alginate (Boyd and Chakrabarty, 1994). Genes controlling biofilm self-destruction might appear to be of more use than genes involved in biofilm formation.

Antimicrobial peptides, also called host defense peptides, are an important fraction of the innate immune response and are produced by all living organisms. These peptides, such as defensins, cathelicidins (LL-37), and magainins, typically show broad and effective antimicrobial activity against bacteria, viruses, and fungi. These evolutionarily conserved peptides have both hydrophobic and hydrophilic sides that enable the molecule to be soluble in aqueous environments, and also to pass through the lipid-rich membranes (Muñoz-Bonilla and Fernández-García, 2012). Considerable research has been devoted to the preparation of peptides maintaining the natural peptide skeleton as well as non-naturally occurring structures. Béven et al. (2003) obtained peptides composed of leucyl and lysyl residues (LK peptides) with different compositions and sequences. The antibacterial activity of the amphiphilic α-helical peptides varied in size, with the optimal length being 15 residues. Its activity is similar to that of melittin.

Chitosan also exhibits antibiofilm properties (Chávez de Paz et al., 2011). Chitosan NPs were analyzed against 24-h formed biofilms of *Streptococcus mutans*. The antimicrobial effect of chitosan was tested against the three biofilm layers that could be identified within the mature biofilm structure: the upper (20 μm), middle (15 μm), and lower (2 μm) biofilm layers. High-molecular-weight chitosan displayed biofilm reductions of 21.4% (upper layer), 7.5% (middle layer), and 1.2% (low layer). Low-molecular-weight chitosan reduced 24-h formed biofilms by 93.6–96.7% in each biofilm layer (Chávez de Paz et al., 2011).

11.6 BIOFILM TARGETING

Passive targeting of biofilms can be achieved by hydrogen bonding, as in case of phosphatidylinositol liposomes with teichoic acids in the glycocalix of bacteria (Jones and Kaszuba, 1994). More efficient binding takes place using liposomes decorated with positively charged entities such as stearylamine, dimethyldioctadecyl ammonium bromide, or 3β-(*N*(*N*1*N*1-dimethylaminoethane) carbamoyl) cholesterol (Kaszuba et al., 1997). Cationic liposomes effectively targets several oral and topical infections; however, cytotoxicity of cationic molecules like stearylamine should be considered (Kaszuba et al., 1997; Yoshihara and Nakae, 1986).

Various approaches to actively target biofilms have been reported. Immunoliposomes carrying *Streptococcus oralis* antibody were strongly and specifically adsorbed to *S. oralis* biofilms compared with other commensal bacteria (Robinson et al., 2000). Lectins such as concanavalin-A (con-A) selectively bind to α-mannopyranosyl and α-glucopyranosyl residues that can be found in the extracellular polysaccharide matrix of *S. mutans* in the oral cavity (Strathmann et al., 2002). Con-A-functionalized liposomes showed higher growth inhibition compared with untargeted metronidazole-loaded liposomes; the latter were more effective than the free drug that was protected from β-lactamases. Interestingly, less effective targeting was recognized in case of bacteria lacking Con-A-binding sites such as *S. epidermidis* and *Proteus vulgaris* (Kaszuba et al., 1995). Wheat germ agglutinin was reported to similarly bind *N*-acetylglucosamine and *N*-acetylneuramic acid residues in the extracellular matrix of many biofilms (Strathmann et al., 2002).

Lectin-decorated polymeric NPs represent a promising approach for targeted antimicrobial therapy (Zhang et al., 2010). Recently, folate-conjugated PAMAM dendrimers targeting *Chlamydia* were significantly localized in the inflamed tissues where folate receptors are overexpressed (Benchaala et al., 2014).

11.7 EXPERIMENTAL EVALUATION OF NANOCARRIER–BIOFILM INTERACTION

11.7.1 Microscopical Investigation

Binding of nanocarriers to bacterial membrane can be visualized by various microscopical techniques. For instance, adsorption of targeted liposomes to *S. epidermidis* and *S. oralis* was detected by *TEM*, whereas internalization of immunogold labeled antibiotics encapsulated in Fluidosomes™ within bacteria was observed by *SEM* (Kaszuba et al., 1997).

11.7.2 Confocal Microscopy

Confocal microscopy has the advantage of observing adsorption of fluorescent liposomes to living, fully hydrated biofilms; however, TEM allows higher resolution at the nanocarrier–biofilm interface (Ahmed et al., 2002).

11.7.3 Flow Cytometry

The increase in fluorescence of a bacterial population after incubation with fluorescent nanocarriers, for example, Fluidosomes™, can be detected by flow cytometry and can be taken as a measure for higher membrane binding and/or internalization.

11.7.4 Fluorescence Correlation Spectroscopy

This method allows monitoring of fluorescence fluctuations due to diffusion of fluorescent particles in and out of the focused laser beam of a confocal microscope. Determination of the diffusion coefficient could be taken as a measure for particle diffusion in *Pseudomonas fluorescens* biofilms (Peulen and Wilkinson, 2011).

11.7.5 Fluorescence Resonance Energy Transfer

With this technique, nanocarriers were labeled with two membrane-inserting fluorescent probes. The change in fluorescence resonance energy transfer (FRET) after incubation with bacteria is indicative of nanocarrier fusion with cell membrane (Forier et al., 2014).

11.7.6 Single Particle Tracking

Another method for the determination of the diffusion coefficient of NPs inside biofilms can be achieved by recording a time-lapse video of the movement of the NPs inside the biofilm, from which the motion trajectories of individual particles could be calculated using image processing. The diffusion coefficient and mode of motion (free diffusion, anomalous diffusion, or directed motion) can be derived on a particle-by-particle basis (Saxton, 1997).

11.7.7 Multiple Particle Tracking

The dynamics of particles in CF sputum were quantified using multiple particle tracking. Briefly, 20-s movies were analyzed with Metamorph software (Universal Imaging, Glendale, WI) to extract x and y positional data over time. Time-averaged mean square displacement (MSD) and effective diffusivity (D_{eff}) for each particle were calculated as a function of time scale (τ). Bulk transport properties were calculated by geometric ensemble-averaging of individual transport rates (Suk et al., 2009).

11.7.8 Optical Tweezers

Optical tweezer instruments use the forces of laser radiation pressure to trap small particles. These trapped particles can then be manipulated and forces of approximately 1–100 piconewtons (pN) on the objects in the trap can be measured (Kirch et al., 2012). This concept was applied to investigate the mobility of NPs in mucus and similar hydrogels as model systems in an attempt to elucidate the link between microscopic diffusion behavior and macroscopic penetration of such gels (Kirch et al., 2012).

11.8 PHARMACEUTICAL APPLICATION OF NANOANTIMICROBIALS

11.8.1 Topical

Treatment of skin infections that develop in traumatic and surgical wounds or burns is very challenging. Chitosan and silver NPs were combined as antimicrobial burn dressings. Synergistic killing of Gram-positive (MRSA) and Gram-negative (*P. aeruginosa, Proteus mirabilis, Acinetobacter baumannii*) bacteria was observed after 30 min (Huang et al., 2011).

The incorporation of the antibacterial agent triclosan into polymeric micelles of poloxamine proved to be active *in vitro* against MRSA and vancomycin-resistant *E. faecalis* (VREF) biofilms compared with the free triclosan (Chiappetta et al., 2008).

Penicillin-conjugated polyacrylate NPs have shown *in vitro* antibacterial activity against MRSA with no cytotoxicity toward human dermal cells and *in vivo* when applied to a dermal abrasion murine model (Greenhalgh and Turos, 2009).

The use of nitric oxide (NO)-releasing silica NPs to eradicate biofilm growth in wound infections has been described. Rapid diffusion of NO into the biofilm matrix probably provides improved efficacy against biofilm-embedded bacteria. *In vitro* grown biofilms of *P. aeruginosa, E. coli, S. aureus, S. epidermidis,* and *C. albicans* were exposed to NO-releasing silica NPs. More than 99% of cells from each type of biofilm were killed as a result of NO release. Compared with small-molecule NO donors, the physicochemical properties, for example, hydrophobicity, charge, and size, of NPs can be altered to increase antibiofilm efficacy (Hetrick et al., 2009).

11.8.2 Oral

The potential of NPs to control the formation of biofilms within the oral cavity, as a function of their biocidal, antiadhesive, and delivery capabilities, is receiving close attention. Dental caries and periodontal disease involve the adherence of bacteria and development of biofilms both on the natural and the restored tooth surface (Allaker and Memarzadeh, 2014).

Micelles of Pluronic 123-alendronate and Pluronic P85 containing the hydrophobic antibacterial agent farnesol were proposed for the prevention of dental caries. These micelles were capable of preventing *S. mutans* biofilm formation on hydroxyapatite discs, even after extensive washing. The high affinity of alendronate to hydroxyapatite enhanced the strong binding (Louie et al., 2009). In another study, nisin-encapsulated liposomes inhibited glucans production by cariogenic *Streptococci* on dental enamel over prolonged periods relative to the free counterparts (Yamakami et al., 2012).

11.8.3 Pulmonary

To improve the therapeutic efficacy of inhalable antibiotics, polymeric and lipid-based NPs have been developed. Increased antibacterial activity of inhalable levofloxacin-loaded polymeric NPs over the free antibiotic was demonstrated against *E. coli* biofilm cells (Cheow et al., 2010a). In addition, the potential of spray-dried antibiotic-loaded polycaprolactone NP aggregates was studied for inhaled antibiofilm therapy (Kho et al., 2010). *P. aeruginosa* is one of the main infections in CF lungs, antibiotic-loaded liposomes efficiently reduced the viability of *P. aeruginosa* biofilms *in vitro*, and complete eradication of *P. aeruginosa* has been shown *in vivo* in murine models (Beaulac et al., 1999).

11.8.4 Urinary Infections

The indwelling urinary catheter is the most commonly deployed prosthetic medical device. The risk of infection is related to the length of time the catheter is in place; for the many patients catheterized for periods longer than 4 weeks, it is inevitable that bacterial communities will establish themselves in the bladder. The biofilms produced by urease-positive bacteria, such as *P. mirabilis*, pose particular threats to the health of catheterized patients. The urease generates ammonia and creates alkaline conditions under which calcium and magnesium phosphates crystallize in the urine and the biofilm. A strategy to deliver the biocide triclosan to the catheterized bladder have been developed. Triclosan prevented catheter encrustation by *P. mirabilis* and biofilm formation by several other common pathogens of the catheterized urinary tract. However, a small effect on urease-producing *P. aeruginosa*, *S. marcescens*, or *M. morganii* was shown (Jones et al., 2006).

In a similar approach, the prevention of catheter-related infections could be achieved by embedding liposomal ciprofloxacin into a gelatin–polyethylene glycol (PEG) hydrogel that can be applied to silicone catheter material. Ciprofloxacin released over 7 days allowed complete inhibition of *P. aeruginosa* (DiTizio et al., 1998).

11.9 CLINICAL STUDIES AND MARKETED PRODUCTS

The number of products clearly claiming efficacy against bacterial biofilms is still limited. However, some liposomal formulations are developed for the treatment of biofilm infections and are currently being evaluated in clinical trials (Table 11.1).

11.10 CONCLUSION AND FUTURE PERSPECTIVES

Advances in the field of biofilm therapy are tremendous. Nanocarrier-mediated delivery of antibiotics and/or novel antimicrobials play a distinct role in terms of improved efficacy and reduced side effects. Nevertheless, there is still much to be done. Combinatorial drug therapy is expected to have higher potency; multiple drugs acting by different modes of therapy can achieve synergistic effects and overwhelm microbial defense mechanisms. For instance, a combination of antibiotics with novel compounds interfering with biofilm formation can improve the action of antibiotics and lower bacterial resistance. Meanwhile, the synergy between certain antibiotics and specific phospholipids that

TABLE 11.1 Marketed Liposomal Formulations for the Treatment of Biofilm Infections

Antibiotic	Microorganism	Trade name	Company	Clinical trials
Amikacin	*P. aeruginosa*	Arikase	Insmed	Phase III clinical trials, placed on hold by FDA due to carcinogenicity
Ciprofloxacin	*P. aeruginosa*	Pulmaquin™ Lipoquin™	Aradigm, Hayward, CA, USA	
Tobramycin	*Burkholderia cepacia*	Fluidosomes™	Axentis Pharma, Zurich, Switzerland	
Amikacin	complicated urinary tract infections	MiKasome	NeXstar Pharmaceuticals, Inc., Boulder, CO, USA	stopped at Phase II clinical trials
Amphotericin B	*C. albicans*	AmBisome®	Astellas, Northbrook, IL, USA	
		Abelcet®	Sigma-Tau PharmaSource, Inc., Indianapolis, IN, USA	
		Amphotec®	BenVenue Laboratories, Inc., Bedford, OH, USA	

increase cell membrane permeability of bacteria sounds of interest. Second, decorating NP surface with biofilm and/or bacterial-targeting moieties would lead to superior selective binding and, therefore, more effective therapy. Furthermore, the development of NPs ensuring microenvironment-sensitive drug release can minimize premature drug loss before reaching its target.

References

Abeylath, S.C., Turos, E., 2008. Drug delivery approaches to overcome bacterial resistance to beta-lactam antibiotics. Expert Opin. Drug Deliv. 5, 931–949.

Ahmed, K., Gribbon, P.N., Jones, M.N., 2002. The application of confocal microscopy to the study of liposome adsorption onto bacterial biofilms. J. Liposome Res. 12, 285–300.

Allaker, R.P., Memarzadeh, K., 2014. Nanoparticles and the control of oral infections. Int. J. Antimicrob. Agents 43, 95–104.

Allison, R.R., Downie, G.H., Cuenca, R., Hu, X.-H., Childs, C.J.H., Sibata, C.H., 2004. Photosensitizers in clinical PDT. Photodiag. Photodyn. Ther. 1, 27–42.

Anderson, P., 2010. Emerging therapies in cystic fibrosis. Ther. Adv. Respir. Dis. 4, 177–185.

Babilas, P., Schreml, S., Landthaler, M., Szeimies, R.M., 2010. Photodynamic therapy in dermatology: state-of-the-art. Photodermatol. Photoimmunol. Photomed. 26, 118–132.

Bargoni, A., Cavalli, R., Zara, G.P., Fundaro, A., Caputo, O., Gasco, M.R., 2001. Transmucosal transport of tobramycin incorporated in solid lipid nanoparticles (SLN) after duodenal administration to rats. Part II—tissue distribution. Pharmacol. Res. 43, 497–502.

Beaulac, C., Sachetelli, S., Lagace, J., 1999. *In vitro* bactericidal evaluation of a low phase transition temperature liposomal tobramycin formulation as a dry powder preparation against gram negative and gram positive bacteria. J. Liposome Res. 9, 301–312.

Benchaala, I., Mishra, M.K., Wykes, S.M., Hali, M., Kannan, R.M., Whittum-Hudson, J.A., 2014. Folate-functionalized dendrimers for targeting *Chlamydia*-infected tissues in a mouse model of reactive arthritis. Int. J. Pharm. 466, 258–265.

Béven, L., Castano, S., Dufourcq, J., Wieslander, Å., Wróblewski, H., 2003. The antibiotic activity of cationic linear amphipathic peptides: lessons from the action of leucine/lysine copolymers on bacteria of the class mollicutes. Eur. J. Biochem. 270, 2207–2217.

Biel, M.A., 2010. Photodynamic therapy of bacterial and fungal biofilm infections. Methods Mol. Biol. 635, 175–194.

Biel, M.A., Pedigo, L., Gibbs, A., Loebel, N., 2013. Photodynamic therapy of antibiotic-resistant biofilms in a maxillary sinus model. Int. Forum Allergy Rhinol. 3, 468–473.

Boyd, A., Chakrabarty, A.M., 1994. Role of alginate lyase in cell detachment of *Pseudomonas aeruginosa*. Appl. Env. Microbiol. 60, 2355–2359.

Brady, R., Calhoun, J., Leid, J., Shirtliff, M., 2009. Infections of orthopaedic implants and devices. In: Shirtliff, M., Leid, J. (Eds.), The Role of Biofilms in Device-Related Infections. Springer, Berlin, Heidelberg, pp. 15–55.

Bragg, P., Rainnie, D., 1974. The effect of silver ions on the respiratory chain of *Escherichia coli*. Can. J. Microbiol. 20, 883–889.

Cassidy, C.M., Tunney, M.M., McCarron, P.A., Donnelly, R.F., 2009. Drug delivery strategies for photodynamic antimicrobial chemotherapy: from benchtop to clinical practice. J. Photochem. Photobiol. B. 95, 71–80.

Cavalli, R., Gasco, M.R., Chetoni, P., Burgalassi, S., Saettone, M.F., 2002. Solid lipid nanoparticles (SLN) as ocular delivery system for tobramycin. Int. J. Pharm. 238, 241–245.

Cegelski, L., Pinkner, J.S., Hammer, N.D., Cusumano, C.K., Hung, C.S., Chorell, E., et al., 2009. Small-molecule inhibitors target *Escherichia coli* amyloid biogenesis and biofilm formation. Nat. Chem. Biol. 5, 913–919.

Charles, S., Vasanthan, N., Kwon, D., Sekosan, G., Ghosh, S., 2012. Surface modification of poly(amidoamine) (PAMAM) dendrimer as antimicrobial agents. Tetrahedron Lett. 53, 6670–6675.

Chatterjee, D.K., Yong, Z., 2008. Upconverting nanoparticles as nanotransducers for photodynamic therapy in cancer cells. Nanomedicine 3, 73–82.

Chatterjee, D.K., Fong, L.S., Zhang, Y., 2008. Nanoparticles in photodynamic therapy: an emerging paradigm. Adv. Drug Deliv. Rev. 60, 1627–1637.

Chávez de Paz, L.E., Resin, A., Howard, K.A., Sutherland, D.S., Wejse, P.L., 2011. Antimicrobial effect of chitosan nanoparticles on *Streptococcus mutans* biofilms. Appl. Environ. Microbiol. 77, 3892–3895.

Chen, C.Z., Cooper, S.L., 2002. Interactions between dendrimer biocides and bacterial membranes. Biomaterials 23, 3359–3368.

Cheng, Y., Burda, C., 2011. Editors-in-Chief: David, L.A., Gregory, D.S., Gary, P.W. Eds., 2.01—Nanoparticles for photodynamic therapy. In: Comprehensive Nanoscience and Technology. Academic Press, Amsterdam, pp. 1–28.

Cheow, W., Chang, M., Hadinoto, K., 2010a. Antibacterial efficacy of inhalable levofloxacin-loaded polymeric nanoparticles against *E. coli* biofilm cells: the effect of antibiotic release profile. Pharm. Res. 27, 1597–1609.

Cheow, W.S., Chang, M.W., Hadinoto, K., 2010b. Antibacterial efficacy of inhalable antibiotic-encapsulated biodegradable polymeric nanoparticles against *E. coli* biofilm cells. J. Biomed. Nanotechnol. 6, 391–403.

Cheow, W.S., Chang, M.W., Hadinoto, K., 2011. The roles of lipid in anti-biofilm efficacy of lipid–polymer hybrid nanoparticles encapsulating antibiotics. Col. Surf. A: Physicochem. Eng. Asp. 389, 158–165.

Chiappetta, D.A., Degrossi, J., Teves, S., D'Aquino, M., Bregni, C., Sosnik, A., 2008. Triclosan-loaded poloxamine micelles for enhanced topical antibacterial activity against biofilm. Eur. J. Pharm. Biopharm. 69, 535–545.

Coenye, T., Brackman, G., Rigole, P., De Witte, E., Honraet, K., Rossel, B., et al., 2012. Eradication of *Propionibacterium acnes* biofilms by plant extracts and putative identification of icariin, resveratrol and salidroside as active compounds. Phytomedicine 19, 409–412.

Costerton, J.W., Stewart, P.S., Greenberg, E.P., 1999. Bacterial biofilms: a common cause of persistent infections. Science 284, 1318–1322.

Cui, Y., Zhao, Y., Tian, Y., Zhang, W., Lü, X., Jiang, X., 2012. The molecular mechanism of action of bactericidal gold nanoparticles on *Escherichia coli*. Biomaterials 33, 2327–2333.

Dayal, S., Burda, C., 2008. One- and two-photon induced QD-based energy transfer and the influence of multiple QD excitations. Photochem. Photobiol. Sci. 7, 605–613.

DiTizio, V., Ferguson, G.W., Mittelman, M.W., Khoury, A. E., Bruce, A.W., DiCosmo, F., 1998. A liposomal hydrogel for the prevention of bacterial adhesion to catheters. Biomaterials 19, 1877–1884.

Donnelly, R.F., McCarron, P.A., Cassidy, C.M., Elborn, J.S., Tunney, M.M., 2007. Delivery of photosensitisers and light through mucus: Investigations into the potential use of photodynamic therapy for treatment of *Pseudomonas aeruginosa* cystic fibrosis pulmonary infection. J. Control. Release 117, 217–226.

Egger, S., Lehmann, R.P., Height, M.J., Loessner, M.J., Schuppler, M., 2009. Antimicrobial properties of a novel silver–silica nanocomposite material. Appl. Environ. Microbiol. 75, 2973–2976.

Espuelas, M.S., Legrand, P., Loiseau, P.M., Bories, C., Barratt, G., Irache, J.M., 2002. *In vitro* antileishmanial activity of amphotericin B loaded in poly(epsilon-caprolactone) nanospheres. J. Drug Target 10, 593–599.

Feng, Q.L., Wu, J., Chen, G.Q., Cui, F.Z., Kim, T.N., Kim, J. O., 2000. A mechanistic study of the antibacterial effect of silver ions on *Escherichia coli* and *Staphylococcus aureus*. J. Biomed. Mater. Res. 52, 662–668.

Flemming, H.C., Wingender, J., 2010. The biofilm matrix. Nat. Rev. Microbiol. 8, 623–633.

Forier, K., Raemdonck, K., De Smedt, S.C., Demeester, J., Coenye, T., Braeckmans, K., 2014. Lipid and polymer nanoparticles for drug delivery to bacterial biofilms. J. Control. Release 190, 607–623

Friedman, A.J., Phan, J., Schairer, D.O., Champer, J., Qin, M., Pirouz, A., et al., 2013. Antimicrobial and anti-inflammatory activity of chitosan-alginate nanoparticles: a targeted therapy for cutaneous pathogens. J. Invest. Dermatol. 133, 1231–1239.

Gelperina, S., Kisich, K., Iseman, M.D., Heifets, L., 2005. The potential advantages of nanoparticle drug delivery systems in chemotherapy of tuberculosis. Am. J. Respir. Crit. Care Med. 172, 1487–1490.

Greenhalgh, K., Turos, E., 2009. *In vivo* studies of polyacrylate nanoparticle emulsions for topical and systemic applications. Nanomedicine 5, 46–54.

Hassett, D.J., Cuppoletti, J., Trapnell, B., Lymar, S.V., Rowe, J.J., Sun Yoon, S., et al., 2002. Anaerobic metabolism and quorum sensing by *Pseudomonas aeruginosa* biofilms in chronically infected cystic fibrosis airways: rethinking antibiotic treatment strategies and drug targets. Adv. Drug Deliv. Rev. 54, 1425–1443.

Häussler, S., 2010. Multicellular signalling and growth of *Pseudomonas aeruginosa*. Int. J. Med. Microbiol. 300, 544–548.

Hetrick, E.M., Shin, J.H., Paul, H.S., Schoenfisch, M.H., 2009. Anti-biofilm efficacy of nitric oxide-releasing silica nanoparticles. Biomaterials 30, 2782–2789.

Huang, L., Dai, T., Xuan, Y., Tegos, G.P., Hamblin, M.R., 2011. Synergistic combination of chitosan acetate with nanoparticle silver as a topical antimicrobial: efficacy against bacterial burn infections. Antimicrob. Agents Chemother. 55, 3432–3438.

Huh, A.J., Kwon, Y.J., 2011. "Nanoantibiotics": a new paradigm for treating infectious diseases using nanomaterials in the antibiotics resistant era. J. Control. Release 156, 128–145.

Hunter, R.C., Klepac-Ceraj, V., Lorenzi, M.M., Grotzinger, H., Martin, T.R., Newman, D.K., 2012. Phenazine content in the cystic fibrosis respiratory tract negatively correlates with lung function and microbial complexity. Am. J. Respir. Cell Mol. Biol. 47, 738–745.

Ideta, R., Tasaka, F., Jang, W.-D., Nishiyama, N., Zhang, G.-D., Harada, A., et al., 2005. Nanotechnology-based photodynamic therapy for neovascular disease using a supramolecular nanocarrier loaded with a dendritic photosensitizer. Nano Lett. 5, 2426–2431.

Jena, P., Mohanty, S., Mallick, R., Jacob, B., Sonawane, A., 2012. Toxicity and antibacterial assessment of chitosan-coated silver nanoparticles on human pathogens and macrophage cells. Int. J. Nanomed. 7, 1805–1818.

Jones, G.L., Muller, C.T., O'Reilly, M., Stickler, D.J., 2006. Effect of triclosan on the development of bacterial biofilms by urinary tract pathogens on urinary catheters. J. Antimicrob. Chemother. 57, 266–272.

Jones, M.N., Kaszuba, M., 1994. Polyhydroxy-mediated interactions between liposomes and bacterial biofilms. Biochim. Biophys. Acta 1193, 48–54.

Jung, W.K., Koo, H.C., Kim, K.W., Shin, S., Kim, S.H., Park, Y.H., 2008. Antibacterial activity and mechanism of action of the silver ion in *Staphylococcus aureus* and *Escherichia coli*. Appl. Environ. Microbiol. 74, 2171–2178.

Juzenas, P., Chen, W., Sun, Y.-P., Coelho, M.A.N., Generalov, R., Generalova, N., et al., 2008. Quantum dots and nanoparticles for photodynamic and radiation therapies of cancer. Adv. Drug Deliv. Rev. 60, 1600–1614.

Kalishwaralal, K., BarathManiKanth, S., Pandian, S.R.K., Deepak, V., Gurunathan, S., 2010. Silver nanoparticles impede the biofilm formation by *Pseudomonas aeruginosa* and *Staphylococcus epidermidis*. Colloids Surf. B Biointerfaces 79, 340–344.

Kaszuba, M., Lyle, I.G., Jones, M.N., 1995. The targeting of lectin-bearing liposomes to skin-associated bacteria. Colloids Surf. B Biointerfaces 4, 151–158.

Kaszuba, M., Robinson, A.M., Song, Y.H., Creeth, J.E., Jones, M.N., 1997. The visualisation of the targeting of phospholipid liposomes to bacteria. Colloids Surf. B Biointerfaces 8, 321–332.

Kho, K., Cheow, W.S., Lie, R.H., Hadinoto, K., 2010. Aqueous re-dispersibility of spray-dried antibiotic-loaded polycaprolactone nanoparticle aggregates for inhaled anti-biofilm therapy. Powder Technol. 203, 432–439.

Kim, J.S., Kuk, E., Yu, K.N., Kim, J.-H., Park, S.J., Lee, H.J., et al., 2007. Antimicrobial effects of silver nanoparticles. Nanomedicine 3, 95–101.

Kirch, J., Schneider, A., Abou, B., Hopf, A., Schaefer, U.F., Schneider, M., et al., 2012. Optical tweezers reveal relationship between microstructure and nanoparticle penetration of pulmonary mucus. Proc. Natl. Acad. Sci. U.S.A. 109, 18355–18360.

Klebensberger, J., Birkenmaier, A., Geffers, R., Kjelleberg, S., Philipp, B., 2009. SiaA and SiaD are essential for inducing autoaggregation as a specific response to detergent stress in *Pseudomonas aeruginosa*. Environ. Microbiol. 11, 3073–3086.

Lara, H.H., Garza-Trevino, E.N., Ixtepan-Turrent, L., Singh, D.K., 2011. Silver nanoparticles are broad-spectrum bactericidal and virucidal compounds. J. Nanobiotechnol. 9, 30.

Lenz, A.P., Williamson, K.S., Pitts, B., Stewart, P.S., Franklin, M.J., 2008. Localized gene expression in *Pseudomonas aeruginosa* biofilms. Appl. Environ. Microbiol. 74, 4463–4471.

Lewis, K., 2001. Riddle of biofilm resistance. Antimicrob. Agents Chemother. 45, 999–1007.

Lewis, K., 2008. Multidrug tolerance of biofilms and persister cells. Curr. Top. Microbiol. Immunol. 322, 107–131.

Li, X., Guo, H., Tian, Q., Zheng, G., Hu, Y., Fu, Y., et al., 2013. Effects of 5-aminolevulinic acid-mediated photodynamic therapy on antibiotic-resistant staphylococcal biofilm: an *in vitro* study. J. Surg. Res. 184, 1013–1021.

Louie, T.J., Emery, J., Krulicki, W., Byrne, B., Mah, M., 2009. OPT-80 eliminates *Clostridium difficile* and is sparing of bacteroides species during treatment of *C. difficile* infection. Antimicrob. Agents Chemother. 53, 261–263.

Lu, C., Maurer, C.K., Kirsch, B., Steinbach, A., Hartmann, R.W., 2014. Overcoming the unexpected functional inversion of a PqsR antagonist in *Pseudomonas aeruginosa*: an *in vivo* potent antivirulence agent targeting pqs quorum sensing. Angew. Chem. 126, 1127–1130.

Ma, L., Conover, M., Lu, H., Parsek, M.R., Bayles, K., Wozniak, D.J., 2009. Assembly and development of the *Pseudomonas aeruginosa* biofilm matrix. PLoS Pathog. 5, e1000354.

Magill, S.S., Edwards, J.R., Bamberg, W., Beldavs, Z.G., Dumyati, G., Kainer, M.A., et al., 2014. Multistate point-prevalence survey of health care–associated infections. N. Engl. J. Med. 370, 1198–1208.

Meers, P., Neville, M., Malinin, V., Scotto, A.W., Sardaryan, G., Kurumunda, R., et al., 2008. Biofilm penetration, triggered release and *in vivo* activity of inhaled liposomal amikacin in chronic *Pseudomonas aeruginosa* lung infections. J. Antimicrob. Chemother. 61, 859–868.

Mehnert, W., Maeder, K., 2012. Solid lipid nanoparticles: production, characterization and applications. Adv. Drug Deliv. Rev. 64, 83–101.

Monds, R.D., O'Toole, G.A., 2009. The developmental model of microbial biofilms: ten years of a paradigm up for review. Trends Microbiol. 17, 73–87.

Moreau-Marquis, S., Stanton, B.A., O'Toole, G.A., 2008. *Pseudomonas aeruginosa* biofilm formation in the cystic fibrosis airway. Pulm. Pharmacol. Ther. 21, 595–599.

Morones, J.R., Elechiguerra, J.L., Camacho, A., Holt, K., Kouri, J.B., Ramirez, J.T., et al., 2005. The bactericidal effect of silver nanoparticles. Nanotechnology 16, 2346–2353.

Muñoz-Bonilla, A., Fernández-García, M., 2012. Polymeric materials with antimicrobial activity. Prog. Polym. Sci. 37, 281–339.

Müller, R.H., Mader, K., Gohla, S., 2000. Solid lipid nanoparticles (SLN) for controlled drug delivery—a review of the state of the art. Eur. J. Pharm. Biopharm. 50, 161–177.

Müller, R.H., Radtke, M., Wissing, S.A., 2002. Solid lipid nanoparticles (SLN) and nanostructured lipid carriers (NLC) in cosmetic and dermatological preparations. Adv. Drug Deliv. Rev. 54, S131–S155.

Mura, S., Hillaireau, H., Nicolas, J., Kerdine-Romer, S., Le Droumaguet, B., Delomenie, C., et al., 2011. Biodegradable nanoparticles meet the bronchial airway barrier: how surface properties affect their interaction with mucus and epithelial cells. Biomacromolecules 12, 4136–4143.

Nafee, N., Youssef, A., El-Gowelli, H., Asem, H., Kandil, S., 2013. Antibiotic-free nanotherapeutics: hypericin nanoparticles thereof for improved *in vitro* and *in vivo* antimicrobial photodynamic therapy and wound healing. Int. J. Pharm. 454, 249–258.

Otto, M., 2014. Biofilms in disease. In: Rumbaugh, K.P., Ahmad, I. (Eds.), Antibiofilm Agents. Springer, Berlin, Heidelberg, pp. 3–13.

Pal, S., Tak, Y.K., Song, J.M., 2007. Does the antibacterial activity of silver nanoparticles depend on the shape of the nanoparticle? A study of the gram-negative bacterium *Escherichia coli*. Appl. Environ. Microbiol. 73, 1712–1720.

Pandey, R., Khuller, G.K., 2005. Solid lipid particle-based inhalable sustained drug delivery system against experimental tuberculosis. Tuberculosis 85, 227–234.

Percival, S.L., Bowler, P.G., Dolman, J., 2007. Antimicrobial activity of silver-containing dressings on wound microorganisms using an *in vitro* biofilm model. Int. Wound J. 4, 186–191.

Peulen, T.-O., Wilkinson, K.J., 2011. Diffusion of nanoparticles in a biofilm. Environ. Sci. Technol. 45, 3367–3373.

Rai, P., Mallidi, S., Zheng, X., Rahmanzadeh, R., Mir, Y., Elrington, S., et al., 2010. Development and applications of photo-triggered theranostic agents. Adv. Drug Deliv. Rev. 62, 1094–1124.

Robinson, A.M., Creeth, J.E., Jones, M.N., 2000. The use of immunoliposomes for specific delivery of antimicrobial agents to oral bacteria immobilized on polystyrene. J. Biomater. Sci. 11, 1381–1393.

Sambhy, V., MacBride, M.M., Peterson, B.R., Sen, A., 2006. Silver bromide nanoparticle/polymer composites: dual action tunable antimicrobial materials. J. Am. Chem. Soc. 128, 9798–9808.

Samia, A.C.S., Dayal, S., Burda, C., 2006. Quantum dot-based energy transfer: perspectives and potential for applications in photodynamic therapy. Photochem. Photobiol. 82, 617–625.

Sanna, V., Gavini, E., Cossu, M., Rassu, G., Giunchedi, P., 2007. Solid lipid nanoparticles (SLN) as carriers for the topical delivery of econazole nitrate: *in-vitro* characterization, *ex-vivo* and *in-vivo* studies. J. Pharm. Pharmacol. 59, 1057–1064.

Saxton, M.J., 1997. Single-particle tracking: the distribution of diffusion coefficients. Biophys. J. 72, 1744–1753.

Singh, B.N., Singh, H.B., Singh, A., Singh, B.R., Mishra, A., Nautiyal, C.S., 2012. *Lagerstroemia speciosa* fruit extract modulates quorum sensing-controlled virulence factor production and biofilm formation in *Pseudomonas aeruginosa*. Microbiology 158, 529–538.

Sotiriou, G.A., Pratsinis, S.E., 2010. Antibacterial activity of nanosilver ions and particles. Environ. Sci. Technol. 44, 5649–5654.

Souto, E.B., Wissing, S.A., Barbosa, C.M., Muller, R.H., 2004. Development of a controlled release formulation based on SLN and NLC for topical clotrimazole delivery. Int. J. Pharm. 278, 71–77.

Stewart, P.S., 1996. Theoretical aspects of antibiotic diffusion into microbial biofilms. Antimicrob. Agents Chemother. 40, 2517–2522.

Strathmann, M., Wingender, J., Flemming, H.-C., 2002. Application of fluorescently labelled lectins for the visualization and biochemical characterization of polysaccharides in biofilms of *Pseudomonas aeruginosa*. J. Microbiol. Methods 50, 237–248.

Suk, J.S., Lai, S.K., Wang, Y.-Y., Ensign, L.M., Zeitlin, P.L., Boyle, M.P., et al., 2009. The penetration of fresh undiluted sputum expectorated by cystic fibrosis patients by non-adhesive polymer nanoparticles. Biomaterials 30, 2591–2597.

Taraszkiewicz, A., Fila, G., Grinholc, M., Nakonieczna, J., 2013. Innovative strategies to overcome biofilm resistance. BioMed. Res. Int. 2013, 13.

Turos, E., Shim, J.-Y., Wang, Y., Greenhalgh, K., Reddy, G. S.K., Dickey, S., et al., 2007. Antibiotic-conjugated polyacrylate nanoparticles: New opportunities for development of anti-MRSA agents. Bioorg. Med. Chem. Lett. 17, 53–56.

Ungaro, F., d'Angelo, I., Coletta, C., d'Emmanuele di Villa Bianca, R., Sorrentino, R., Perfetto, B., et al., 2012. Dry powders based on PLGA nanoparticles for pulmonary delivery of antibiotics: modulation of encapsulation efficiency, release rate and lung deposition pattern by hydrophilic polymers. J. Control. Release 157, 149–159.

Vilain, S., Pretorius, J.M., Theron, J., Brözel, V.S., 2009. DNA as an adhesin: *Bacillus cereus* requires extracellular DNA to form biofilms. Appl. Environ. Microbiol. 75, 2861–2868.

Wissing, S.A., Muller, R.H., 2003. Cosmetic applications for solid lipid nanoparticles (SLN). Int. J. Pharm. 254, 65–68.

Xu, X., Stewart, P.S., Chen, X., 1996. Transport limitation of chlorine disinfection of *Pseudomonas aeruginosa* entrapped in alginate beads. Biotechnol. Bioeng. 49, 93–100.

Yamakami, K., Tsumori, H., Sakurai, Y., Shimizu, Y., Nagatoshi, K., Sonomoto, K., 2012. Sustainable inhibition efficacy of liposome-encapsulated nisin on insoluble glucan-biofilm synthesis by *Streptococcus mutans*. Pharm. Biol. 51, 267–270.

Yamanaka, M., Hara, K., Kudo, J., 2005. Bactericidal actions of a silver ion solution on *Escherichia coli*, studied by energy-filtering transmission electron microscopy and proteomic analysis. Appl. Environ. Microbiol. 71, 7589–7593.

Yoshihara, E., Nakae, T., 1986. Cytolytic activity of liposomes containing stearylamine. Biochim. Biophys. Acta. 854, 93–101.

Zhang, G.-D., Harada, A., Nishiyama, N., Jiang, D.-L., Koyama, H., Aida, T., et al., 2003. Polyion complex micelles entrapping cationic dendrimer porphyrin: effective photosensitizer for photodynamic therapy of cancer. J. Control. Release 93, 141–150.

Zhang, L., Mah, T.-F., 2008. Involvement of a novel efflux system in biofilm-specific resistance to antibiotics. J. Bacteriol. 190, 4447–4452.

Zhang, L., Pornpattananangkul, D., Hu, C.-M.J., Huang, C.-M., 2010. Development of nanoparticles for antimicrobial drug delivery. Curr. Med. Chem. 17 (6), 585–594.

CHAPTER

12

Nanomaterials for Antibacterial Textiles

Nabil A. Ibrahim

Textile Research Division, National Research Centre, Giza, Egypt

12.1 INTRODUCTION

Hydrophilic natural and regenerated textile fibers are more susceptible to the growth of pathogenic and odor-generating bacteria, such as Gram-negative and Gram-positive bacteria, and biodeterioration than hydrophobic synthetic ones. Hence, antibacterial finishing of textile materials becomes extremely important to cope with the great demands and pressing needs for functionalized textile products that can offer improved and durable protection for both the textile user and textile substrate itself. Antibacterial finishing agents fall into two categories based on their mode of action on harmful bacteria called bacteriostats and bacteriocides. However, an antibacterial finishing agent can act in two distinct ways, namely by controlled-release mechanisms (i.e., leaching types) and by direct contact (i.e., bacteria have to contact the immobilized antibacterial agent). Many different compounds have been used to impart antibacterial activity to textile materials, including metals and metal salts, quaternary ammonium compounds, triclosan, chistosan, *N*-halamine and peroxyacid, some synthetic and natural dyes, immobilized enzymes, and inorganic nano-structured materials. These antibacterial agents differ in their chemical structure, antibacterial activity, mode of interaction, application methods, and biological activity, taking in consideration the economic, environment, and human health concerns and challenges. Nanomaterials may provide effective solutions to technological and environmental challenges in the areas of imparting new functionalities to textile materials to satisfy the urgent needs for ecologically smart processes and products and to consumer's demands for hygienic clothes and active wear.

This chapter discusses classification of textile fibers, textile wet processes and their environmental concerns, the currently available antibacterial agents, potential application of nanomaterials in functional finishes of textiles, especially antibacterial finish, the green routes for preparation of key nanomaterials, finishing treatments, and mode of action against pathogenic microorganisms such as *Escherichia coli* (G −ve) and *Staphylococcus aureus* (G +ve). Finally, environmental issues related to nanomaterials and standard test methods are highlighted and future trends are discussed.

Nanotechnology in Diagnosis, Treatment and Prophylaxis of Infectious Diseases.
DOI: http://dx.doi.org/10.1016/B978-0-12-801317-5.00012-8

12.2 TEXTILE FIBERS

The main categories of textile fibers are natural and human-made. Natural fibers may be plant fibers such as cotton, flax, or jute, or protein fibers like wool or silk. Human-made fibers are broadly classified into regenerated fibers (e.g., viscose, lyocell) and synthetic fibers (e.g., polyester, polyamide, polyacrylic, polyurethane). Chemical structures of the most common textile fibers are demonstrated in Table 12.1 (Cook, 2001; Bide, 2009). However, blending of textile fibers, especially the natural and synthetic ones, offers the following

TABLE 12.1 Type and Chemical Structure of the Most Common Textile Fibers

Type of textile	Main chemical structure	Functional groups
1. NATURAL FIBERS		
1.1 Cellulose e.g., cotton, flax, jute	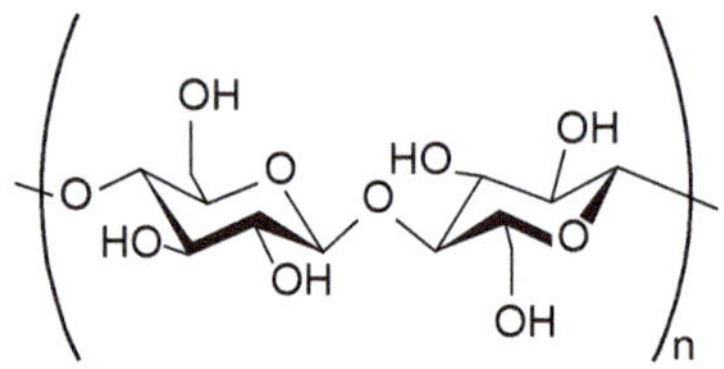	–OH
1.2 Proteinic fibers e.g., wool	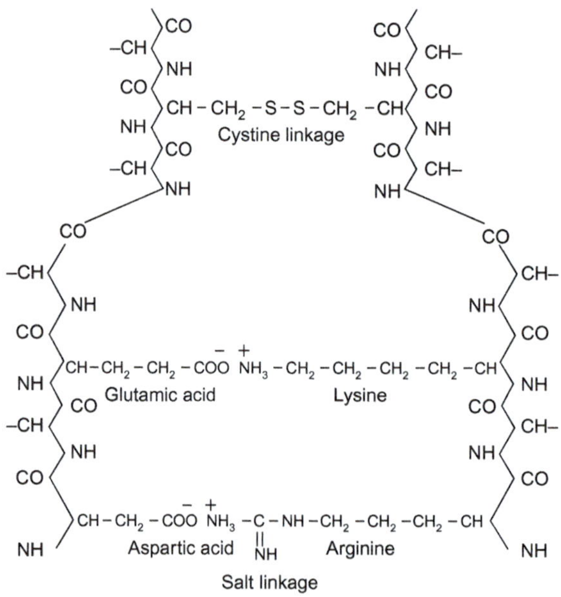	$-NH_2$ –OH –COOH –SH
2. MANMADE FIBERS		
2.1. Synthetic fibers		
Polyester Polyethylene terephthalate (PET)	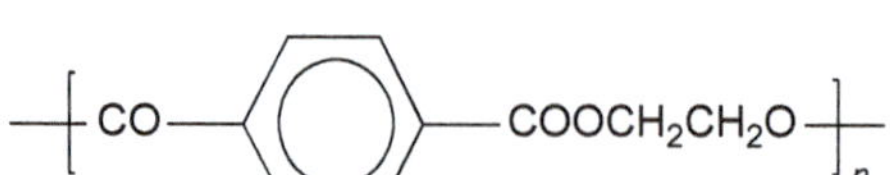	–COOH –OH (end groups)

(*Continued*)

TABLE 12.1 (Continued)

Type of textile	Main chemical structure	Functional groups
Polyamide (nylon)	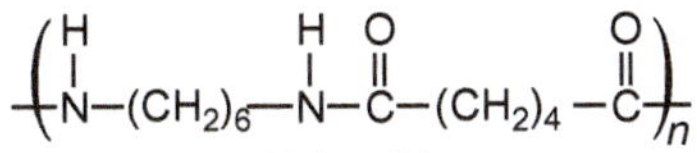 Nylon 66 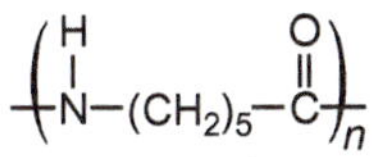 Nylon 6	$-NH_2$ (end groups) $-CO\,NH$
Polyacrylonitrile	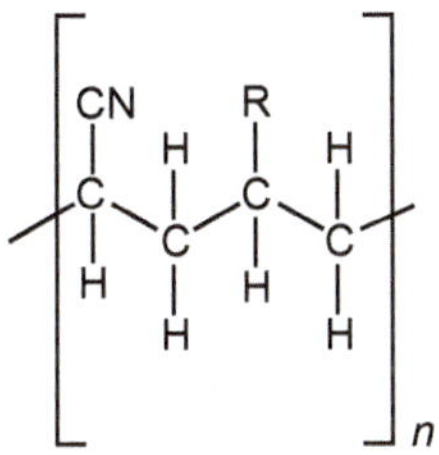 (homopolymer R = CN, copolymer R = COOH)	Anionic sites, e.g., $-COOH$ or sulfonic
L–Poly (lactic acid)	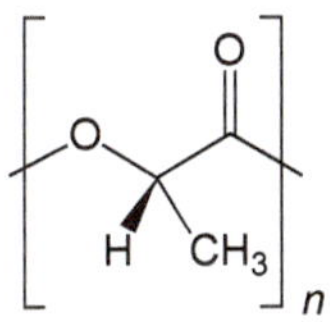	None
Polyurethane	$-O-(CH_2)_n-O-\overset{O}{\overset{\Vert}{C}}-NH-(CH_2)_n-NH-\overset{O}{\overset{\Vert}{C}}-$	$-NH\underset{\underset{O}{\Vert}}{C}O$
2.2. *Regenerated fibers*		
Unmodified e.g., viscose, lyocell	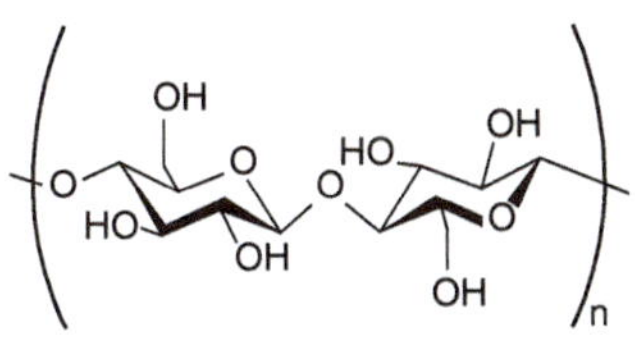	$-OH$
Modified cellulose e.g., acetate, triacetate	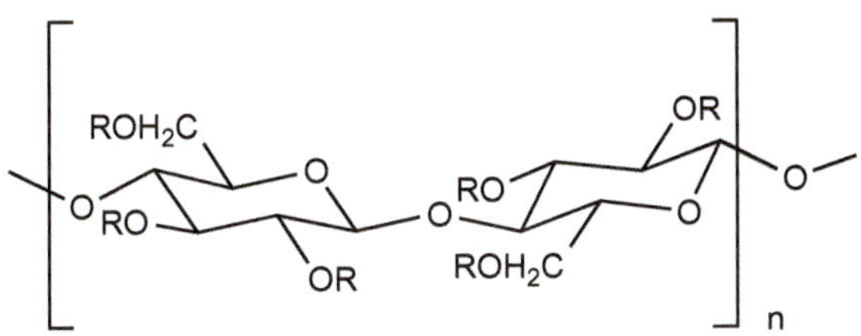 Cellulose triacetate (all R = $COCH_3$) and Cellulose acetate (2/3 R = $COCH_3$, 1/3 R = H)	$-OH$ $-OCOCH_3$

TABLE 12.2 Most Common Textile Fibers, Active Sites, Proper Class of Dyestuff, and Mode of Interaction

Textile fibers	Active sites	Class of dyestuff	Mode of interaction
Cellulosic fibers Wool	–OH groups –OH, NH_2	Reactive	Covalent bonds
Cellulosic fibers	–OH groups	Direct	Hydrogen bonds
Protein fibers, Polyamide Polyurethane	–OH, NH_2 NH_2 end groups –OCO–NH–	Acid	Ionic bonds
Acrylic Cationic-dyeable polyester	–CN SO_3H or COOH	Basic	Ionic bonds
Polyester, cellulose acetates	Terminal –OH and COOH –OH and $OCOCH_3$	Disperse	Hydrophobic, hydrogen bonds and van der Waals forces
Cellulosic and protein fibers	–OH and NH_2	Natural	Complexation with mordant
Cellulosic fibers	–OH	Vat (soluble leuco form)	*In-situ* insolubilization by oxidation
Cellulosic fibers	–OH	Sulfur (soluble leuco form)	*In-situ* insolubilization by oxidation
Cotton, polyester and their blends Synthetic fibers	–OH, –COOH	Pigment	By using suitable binding agent By incorporation during melt spinning

advantages (Shore, 1998; El-Moghazy, 2004; Ibrahim, 2011): (i) avoiding a shortage in natural fibers and their drastically increasing costs; (ii) extending the shelf-life time of textile products; (iii) improving performance properties; and (iv) satisfying the growing needs and demands of textile users.

12.3 PREPARATORY PROCESSES

Eco-friendly, efficient, and uniform preparatory processes are essential to (Choudhury, 2006a,b): (i) eliminate various impurities such as sizing agents, noncellulosic impurities from cotton, vegetable matter, and grease from wool, as well as oils and spin-finishes from synthetic fibers; (ii) enhance the wettability and absorbency especially of natural fibers; (iii) destruct coloring matters to obtain a high degree of whiteness, especially natural fibers; and (iv) insure uniform subsequent coloration and/or finishing processes without seriously affecting the inherent physico-mechanical properties of the treated substrates. Fiber type, yarn and fabric construction, nature, and content of impurities, as well as the targeted performance and functional properties, will dictate the preparatory stages and sequences, treatment formulations and conditions, used auxiliaries, subsequent wet processes, and the production line machineries.

12.4 COLORATION PROCESSES

When applied correctly, the preparatory processes result in marked improvements in dye receptivity, uniform absorption, high color yield and fixation, high fastness ratings, and very good reproducibility. Dyeing processes include the following main steps: dye transportation from the dye bath to the fiber surface; dye adsorption at the fiber surface; and dye diffusion into the interior followed by dye fixation. However, textile dyestuffs may be classified according to their chemical structure and method of application. Table 12.2

demonstrates some of the most common dye classes, textile fibers, and mode of their interactions (Choudhury, 2006a,b, 2011; Ibrahim, 2011; Lewis, 2011; Koh, 2011).

12.5 ENVIRONMENTAL CONCERNS

It is well-known that pretreatment and textile coloration processes create several environmental concerns and negative impacts. Rationalization and optimization of these wet processes conserve a significant amount of water/energy/chemicals and, in turn, reduce the pollution load of the dye house effluent to be treated. Recommended pollution prevention measures in pretreatment and coloration processes of textile materials include the following options (Ibrahim, 2011; Lewis, 2011; Kumar and Choudhury, 2013): (i) better use of resources and good management of wet processes; (ii) optimization and/or modification of the conventional wet processes; (iii) minimizing the use of additives and chemicals and maximizing fixation wash off; (iv) replacement of hazardous chemicals with eco-friendly alternatives; (v) practicing on-site recovery, reuse, and recovery; (vi) replacement of nonbiodegradable textile auxiliaries by biodegradable ones; (vii) selecting environmentally safe dyestuffs for correct first-time dyeing; (viii) application of new emerging technologies (e.g., biotechnology, plasma technology, super critical CO_2 technology); and (ix) development of human resources and raising environmental awareness.

12.6 ANTIBACTERIAL FUNCTION FINISH

Figure 12.1 shows the key steps in textile wet processing of the most widely natural (i. e., cotton, wool) and synthetic (e.g., polyester fabrics for apparel manufacturing) (Sawhney et al., 2008; Lam et al., 2012). Chemical finishing is an important component and an integrated stage of textile wet processes for developing the final fabric properties for different end uses and potential applications. Efficient and eco-friendly pretreatment and coloration processes are essential for optimum finishing results and functional finishing of textiles (Ibrahim, 2011). Recently, eco-friendly functional finishing of textiles has gained great interest to (Holme, 2007; Sawhney et al., 2008; Kumar and Choudhury, 2013; Chen et al., 2012; Lam et al., 2012; Köhler and Som, 2013): (i) upgrade existing textile properties; (ii) develop innovative textile products with high value added (i.e., better competitive edge); (iii) cope with the increasing environmental concerns of textile users; (iv) minimize the ecological impact to the environment and on human health concerns; and (v) decrease production and environmental costs to remain competitive.

Functionalization of textile materials is accompanied by a significant improvement in their functional properties like antibacterial function, UV protection, flame-retardant capability, water repellency, water/oil repellency, self-cleaning status, and wrinkle recovery (Hebeish and Ibrahim, 2007; Holme, 2007; Sawhney et al., 2008; Gowri et al., 2010; Chen et al., 2012; Lam et al., 2012; Kumar and Choudhury, 2013; Köhler and Som, 2013). This chapter focuses mainly on antibacterial functional finish of textiles using nanomaterials by taking into consideration the increasing demand for hygienic, comfortable, and active-wear textile products (Gao and Cranston, 2008) and the significant increase of the antibacterial textiles market share.

12.6.1 Main Objectives

The growth of microorganisms (e.g., bacteria, fungus, yeast, mold) on textile materials is accompanied by detrimental effects not only to

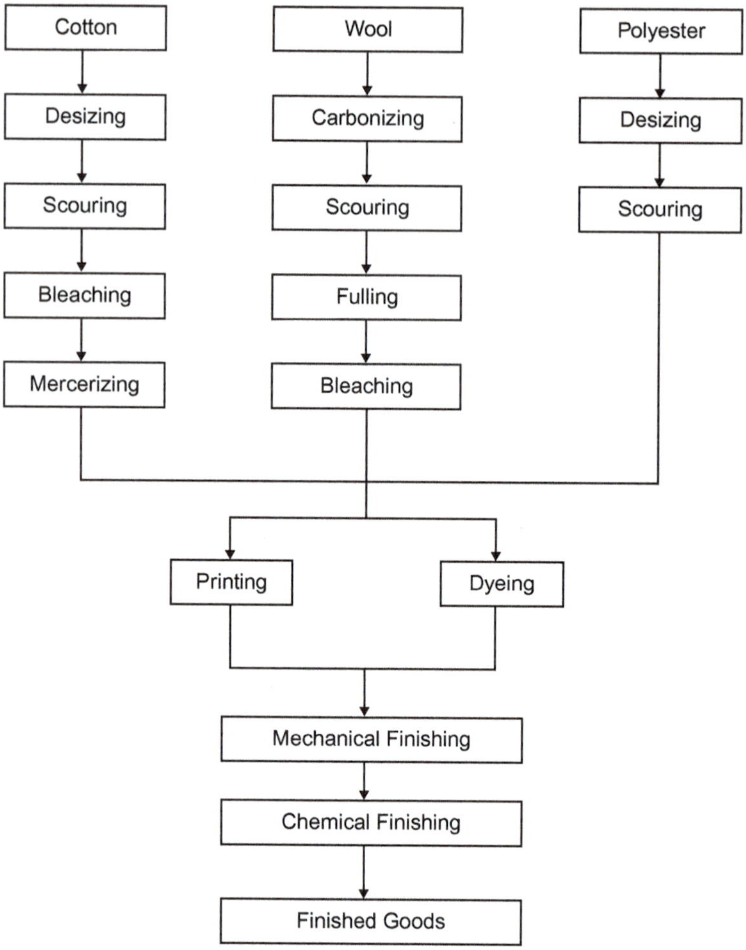

FIGURE 12.1 Flow chart of textile wet processing.

the textile-user (e.g., contamination risks and generation of unpleasant odor) but also to the fabric itself (e.g., quality loss, stains, strength reduction) (Gao and Cranston, 2008; Dastjerdi and Montazer, 2010). Natural fibers (e.g., cotton and wool) can act as proper media to support the growth of microorganisms. However, the extent of contamination and subsequent attack of microorganisms on most synthetic fibers (e.g., polyester) are less than that on natural ones, probably because of their hydrophobic nature. Therefore, the most common objectives of antibacterial functionalization of textiles are (Höfer, 2006): (i) to avoid contamination risk and the generation of unpleasant odors; (ii) to prevent the incidence of harmful bacteria (e.g., *E. coli*, *S. aureus*); and (iii) to minimize or avoid loss in quality and mechanical properties of textiles.

12.6.2 Main Requirements

Antibacterial functional finishes of the textiles must fulfill the following requirements (Purwar and Joshi, 2004; Williams et al., 2005; Höfer, 2006; Gao and Cranston, 2008; Dastjerdi and Montazer, 2010; Lam et al., 2012): (i) must kill or inhibit the growth of pathogenic and/or odor-causing undesirable bacteria; (ii) highly efficient against a broad spectrum of harmful bacteria; (iii) no negative impacts on human heath, friendly microorganisms, nonpathogenic bacteria, and the environment; (iv) no adverse effect on inherent properties and comfort of the fabric; (v) compatibility with other common textile finishing agents and/or wet processes (e.g., dyeing, printing); (vi) cost-effectiveness and ease of application; and (vii) environmentally safe and no toxic effects for manufacturers and textile users.

12.6.3 Application Methods or Techniques

The most important methods or techniques for conferring antibacterial activity to textiles may be broadly classified (Vigo, 1994; Mao and Murphy, 2001; Wallace, 2001; Lee et al., 2003; Zhang et al., 2003; Lim and Hudson, 2004; Mahltig et al., 2004; Purwar and Joshi, 2004; Mahltig et al., 2005; Williams et al., 2005; Höfer, 2006; Gao and Cranston, 2008; Dastjerdi and Montazer, 2010; Ibrahim et al., 2010a) as follows: (i) by incorporation of proper bio-active agents into the polymer melts before extrusion; (ii) by surface application (e.g., chemical grafting); (iii) by exhaustion and/or pad-dry-cure for natural substrates (e.g., resin treatment); (iv) by coating technology; (v) by chemical or physical modification of substrate and/or the active agent for chemical bonding; (vi) by spraying or foaming technique; or (v) by using sol-gel or encapsulation in sol-gel particles. Selection of the most appropriate physicochemical (e.g., coating, plasma, or laser surface treatment, microencapsulation) and/or chemical (e.g., grafting, cross-linking, covalent bonding, chelation) methods or techniques is governed by the chemical nature of the active agent, the fiber type, the fabric structure, the available equipment, and the demanded performance properties.

12.6.4 Chemistry and Mode of Action of Current Antibacterial Agents (Schindler and Hauser, 2004; Holme, 2007; Gao and Cranston, 2008; Lam et al., 2012)

Antibacterial agents can act as follows: (a) by a controlled-release mechanism (leaching type) from the textile materials into their surroundings, thereby attacking harmful bacteria on the fiber surface or in the surrounding environment, and most of these agents are not chemically bonded and show poor durability, or (b) by direct contact mechanism (nonleaching type) (i.e., bacteria have to contact the functionalized fabric surface), and most of these agents are chemically immobilized onto the fabric surface (i.e., good durability, no health problems, no leaching of hazardous materials into surroundings). The actual mechanisms by which antibacterial agents kill or inhibit the growth of harmful bacteria include cell wall damage, disruption of cytoplasmic membranes of bacterial cells via physical and/or ionic phenomena, subsequent release of the cytoplasmic constituents, inhibition of enzyme action, and protein or nucleic acid synthesis (Yao et al., 2003; Schindler and Hauser, 2004; Holme, 2007; Gao and Cranston, 2008; Lam et al., 2012). The most common antibacterial agents, appropriate textile fibers, and their mode of fixation and action are summarized in Table 12.3.

12.7 ANTIBACTERIAL TEXTILES USING NANOMATERIALS

In recent years, more attention has been given to the potential applications of innovative technologies, especially nanotechnology, the wave of future, as well as smart nanomaterials (size range of 1–100 nm) for the following: (i) enhancing the performance and functional properties of the current textile products; (ii) developing smart and intelligent textiles with novel functions; (iii) satisfying the

TABLE 12.3 Some of the Most Common Used Antibacterial Agents, Appropriate Textile Fibers, and Modes of Their Fixation and Action

Antibacterial agent	Chemical structure	Fiber	Mode of fixation	Mode of action	Reference
Metals	e.g., Ag in ultra-fine metallic particles	Synthetic, e.g., polyester, nylon	By incorporation into the polymer melt before spinning	Deactivation of cellular proteins resulting in cell damage or death	Schindler and Hauser (2004), Gao and Cranston (2008), Ibrahim et al. (2010b), Lam et al. (2012), Ibrahim et al. (2012a, 2013a,b)
Metallic ions	e.g., Ag^{+}, Cu^{2+}	Natural, e.g., cotton, wool	By binding sites and chelating abilities		
Quaternary ammonium compounds (QACs)	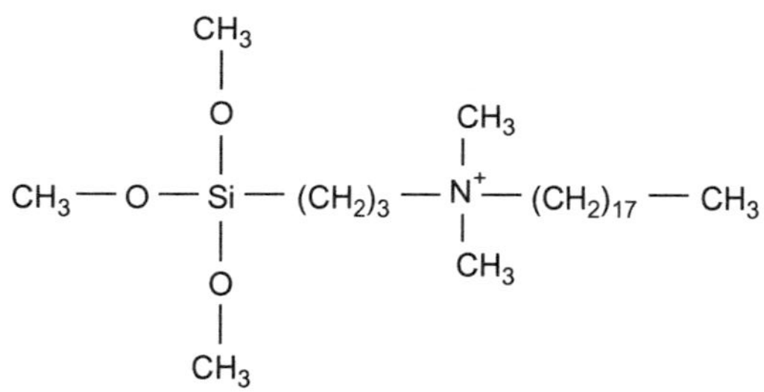	Synthetic, e.g., polyester, nylon	By ionic interaction	Membrane damage and leakage of intra cellular constituents	Kim and Sun (2001), Schindler and Hauser (2004), Zhao and Sun (2006), Son et al. (2006), Gao and Cranston (2008), Lam et al. (2012)
e.g., AEM 5700	3-(Trihydroxysilyl) propy (dimethyl-octdecyl ammonium chloride	Natural, e.g., wool	By covalent bond as in case of modified wool via its thiol group		
PHMB	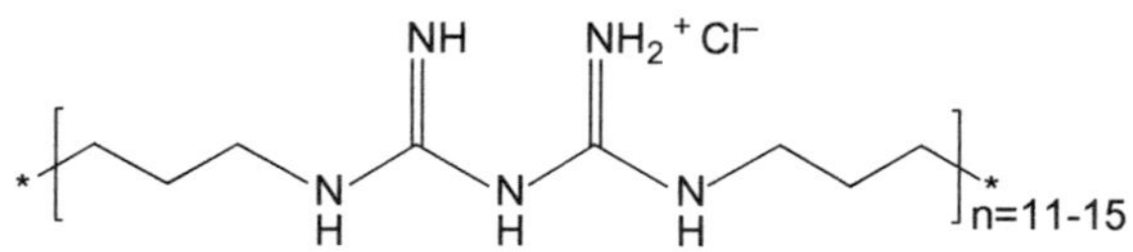	Synthetic	By crosslinking	Via electrostatic attraction with the negatively charged bacterial cell surface and increasing the permeability of the cell walls	Schindler and Hauser (2004), Kawabata and Tylor (2004), Gao and Cranston (2008), Lam et al. (2012)
	Polyhexamethylene biguanide	Natural, e.g., cellulosic	Through ionic and hydrogen bonds		
Triclosan	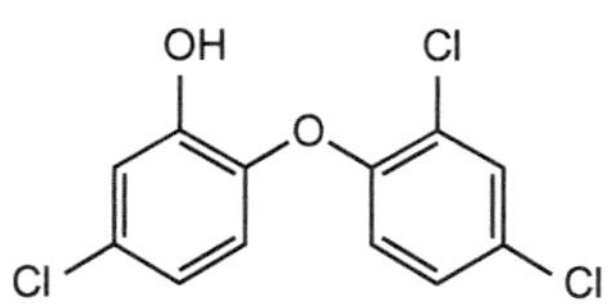	Synthetic, e.g., polyester, nylon	Via utilization of dispersing, binding and/or cross-linking agents	By blocking lipid-bio synthesis and by acting as a barrier to microorganism	Kaylon and Olgun (2001), Mao and Murphy (2001), Lam et al. (2012), Ibrahim et al. (2013a,c)
		Natural, e.g., cotton, wool and blends			

(*Continued*)

TABLE 12.3 (Continued)

Antibacterial agent	Chemical structure	Fiber	Mode of fixation	Mode of action	Reference
Chitosan	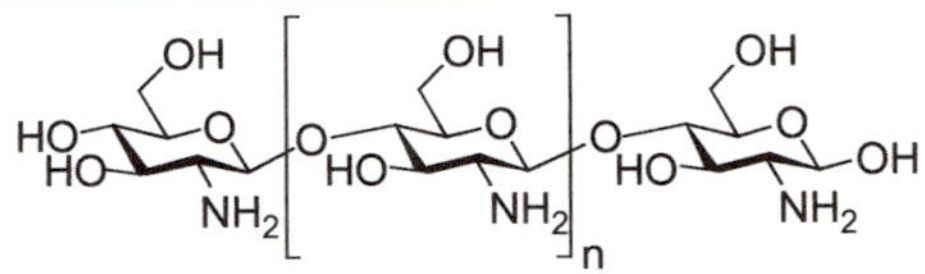	Synthetic, e.g., polyester Natural, e.g., cotton, wool and blends	By micro encapsulation, by reactive bonding and/or by cross-linking or by plasma treatment	By inhibiting the normal metabolism of bacteria and causing death of their cells due to its polycationic nature	Lim and Hudson (2003), Gao and Cranston (2008), Rinaudo (2006), Ibrahim et al. (2013d,e)
N-halamine	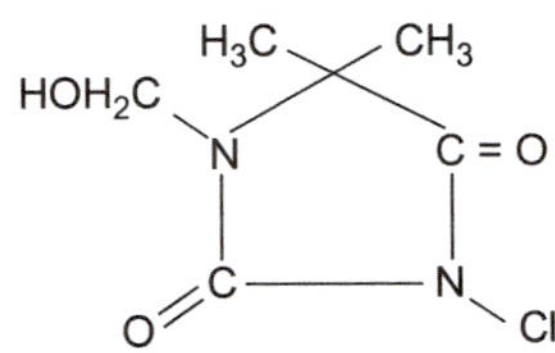 Dimethylol-5,5-dimethylhydantoin	Synthetic, e.g., polyester, nylon Natural, e.g., cotton, wool	Via covalent attachment or grafting onto the textile	Killing of bacteria due to the oxidation properties of the halamine bond (N–Cl)	Sun and Sun (2001), Schindler and Hauser (2004), Qian and Sun (2004), Gouda and Ibrahim (2008), Ibrahim et al. (2008)
Dyestuffs, e.g., basic	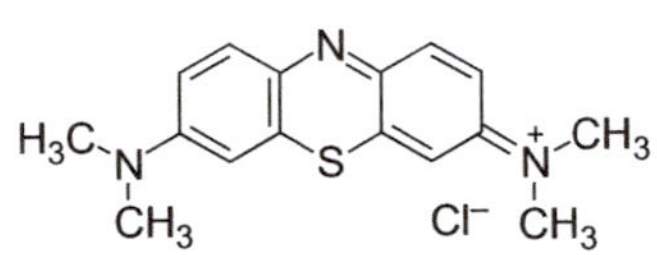 C.I. Basic Blue 9	Natural, e.g., cotton, wool	Ionic links (between positively charged amino groups and negatively charged surface)	Interruption of all essential functions of the cell membrane	Ma et al. (2003), Ibrahim et al. (2010c), Lam et al. (2012), Shahid et al. (2013)
Natural	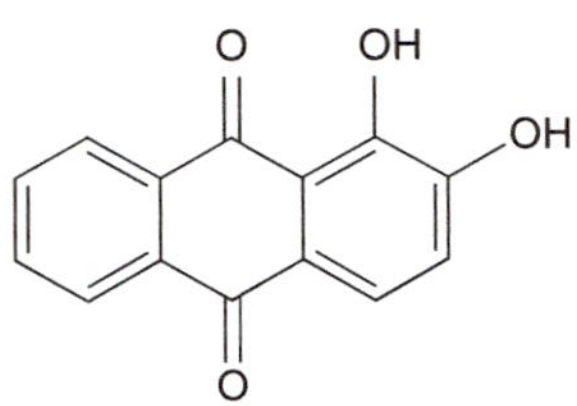 Alizarin 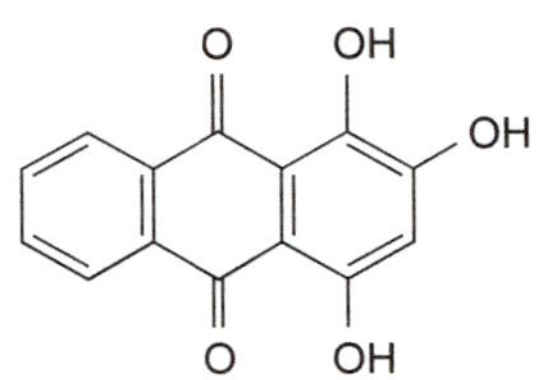 Pupurin Madder	Natural, e.g., cotton, wool	Complexation with mordant	Presence of tannins and protein-binding ability of tannins	Ibrahim et al. (2009, 2013f)

(Continued)

TABLE 12.3 (Continued)

Antibacterial agent	Chemical structure	Fiber	Mode of fixation	Mode of action	Reference
Natural herbal products, e.g., neem oil and neem derivatives	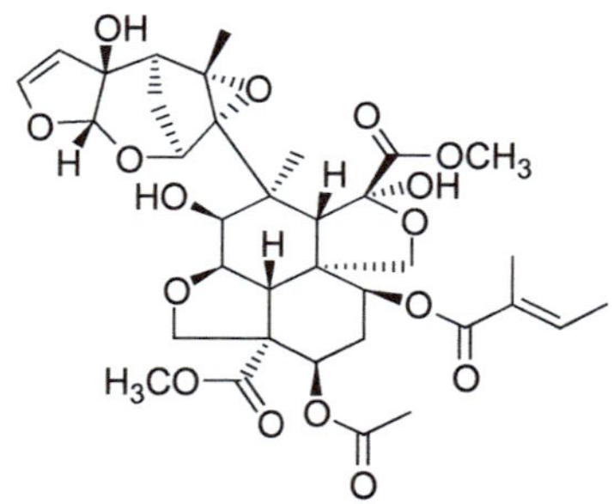 Azadirachtin	Cotton based textiles	By physical bonding By cross-linking or By encapsulation	Biological activity especially azadirachtin	Josh et al. (2007), Ibrahim et al. (2011), Lam et al. (2012)
Immobilized enzymes	Alkaline pectinase, amylase, laccase	Cotton	Immobilization onto the ester crosslinked-postactivated cotton, and coordination on the preaminated Cu-chelated cotton fabrics	The antibacterial activity depends on the type of immobilized enzyme as well as nature and structure of microorganism	Ibrahim et al. (2007), Gulrajani and Gupta (2011)

growing needs of textiles users for hygienic clothing and active wear; and (iv) allowing for great opportunities and options to develop innovative textile processes and products with high value added (Mahlting et al., 2005; Hebeish and Ibrahim, 2007; Sawhney et al., 2008; Black, 2009; Dastjerdi and Montazer, 2010; Gowri et al, 2010; Evans et al., 2012; Harifi and Montazer, 2012). More recently, nanomaterials have used for imparting limitless functional performance properties like antibacterial function, UV protection, flame retardancy, self-cleaning status, electrical conductivity, super-hydrophilicity or hydrophobicity, water/oil repellency, and antifelting, most likely because of their unique and novel properties as well as functions (Gulrajani, 2006; Gao and Cranston, 2008; Sawhney et al., 2008; Black, 2009; Goesmann and Feldmann, 2010; Simoncic and Tomsic, 2010; Joshi and Bhattcharyya, 2011; Evans et al., 2012; Rizzello et al., 2013; Windler et al., 2013).

In this chapter, emphasis is given to the use of nanomaterials as a new generation of antibacterial agents to: (i) impart higher a standard of antibacterial functionality and durability to textiles and (ii) avoid or minimize the drawbacks/risks/possible environmental, ecological, and/or economical concerns associated with the application of the current antibacterial agents (i.e., as an eco-friendly, feasible, and promising alternatives).

12.7.1 Greener Nanomaterial Production (Mahlting et al., 2005; Dastjerdi and Montazer, 2010; Goesmann and Feldmann, 2010; Simoncic and Tomsic, 2010)

The current practices used to synthesize nanomaterials are (i) grinding (e.g., wet, dry, reactive); (ii) gas phase techniques (e.g., laser ablation or chemical vapor deposition); or (iii) liquid phase techniques (e.g., sol-gel, hydrothermal, micro emulsion).

Using homogenous liquid-phase methods results in permitting the formation of high-quality nanoparticles of varying compositions, different ranges of particle sizes and controlled shapes, surface functionality, and material properties. To avoid or overcome the negative impacts of using conventional chemical routes to produce nanomaterials such as the utilization of hazardous and aggressive reducing, capping, and/or stabilizing agents, organic solvent along with higher energy and materials consumption, it is of a great importance to search for and develop easy, clean, nontoxic, and eco-friendly sustainable processes for the preparation and implementation of the demanded nanoparticles relevant to a given application (Bhattacharya and Gupta, 2005; Dahl et al., 2007; Hu B. et al., 2008; Sharma et al., 2009; Simoncic and Tomsic, 2010; Abdel-Aziz et al., 2014). Greener synthesis of nanoparticles based on noble metals and inorganic oxides can be accomplished in the following ways (Bhattacharya and Gupta, 2005; Dahl et al., 2007; Hu B. et al., 2008; Emam et al., 2013; Abdel-Aziz et al., 2014): (i) chemical reduction; (ii) electro-chemical methods; (iii) bio-based approaches; (iv) green energy source alternatives; and/ or (v) utilization of environmentally benign solvents. It is expected that application of these routes, whenever possible and taking into consideration the principle of green chemistry, will result in: (i) full replacement of hazardous chemicals and solvents; (ii) rationalization of materials and energy requirements; (iii) possible higher yields with minimal waste and maintenance of product quality, especially in large scale; and (iv) minimization or prevention of the environmental negative impacts and health awareness (Dahl et al., 2007; Hu B. et al., 2008; Ramirez et al., 2009; Sharma et al., 2009; Abdel-Aziz et al., 2014). The main greener nano-synthesis approaches along with some specific examples are demonstrated in this chapter.

12.7.1.1 *Eco-Friendly Chemical Reduction*

Various greener chemical and biological routes using naturally occurring materials (e.g., biodegradable polymers, sugars, plant extract, microorganism) have been described for preparation of noble metal and metal oxide nanoparticles.

12.7.1.1.1 REDUCING SUGARS

Stable Ag-NP's [Ag^0] (with average particle size of approximately 5.3) were synthesized by using the reducing sugars (β-D-glucose as a reductant and soluble starch as a protecting agent) and water (as a benign solvent) with gentle heating (Raveendran et al., 2003). A general method for the synthesis of different metal nanoparticles (i.e., Au^0, Ag^0, Pt^0, and Pd^0) using $HAuCl_4$, $AgNO_3$, H_2PtCl_6, and $PdCl_2$ metal salts along with glucose, fructose, and sucrose as reducing agents was presented by Panigrahi et al. (2004). Among the nominated sugars, fructose has been found to be the proper one for attaining highly stable and smaller nanoparticles (i.e., Au-NPs [~1 nm], Pt-NPs [~3 nm], Ag-NPs [~10 nm], and Pd-NPs [~20 nm]). Singh et al. recently studied the effect of pH on the size and size distribution of the prepared Ag-NPs using glucose as a reductant, $AgNO_3$ (a precursor starch) as a stabilizing agent, NaOH as a reduction accelerator, and water as a solvent. Two tentative mechanisms were also postulated (Wang et al., 2005; Singh et al., 2009):

a. $$Ag^+ + \underset{\text{Starch}}{St.OH} \rightarrow Ag(St.OH)^+ \quad (12.1)$$

$$2Ag(St.OH)^+ + \underset{\text{Glucose}}{HO.H_2C(CHOH)_4 - CHO} + 2OH^- \rightarrow 2Ag(St.OH)\downarrow + \underset{\text{Gluconic acid}}{HO.H_2C(CHOH)_4 - COOH} + H_2O \quad (12.2)$$

b. $$2Ag^+ + 2OH^- \rightarrow Ag_2O + H_2O \quad (12.3)$$

$$Ag_2O + HO.H_2C(CHOH)_4 - CHO + 2\,St.OH \rightarrow 2Ag(St.OH)\downarrow + HO.H_2C(CHOH)_4 - COOH \quad (12.4)$$

In 2009, size- and shape-controlled Ag-NPs and Au-NPs were successfully prepared by Chairam et al. (2009) using β-D-glucose as an eco-friendly reducing agent, partially hydrolyzed mung bean starch vermicelli as a stabilizing template and green nanoreactor, and $AgNO_3$ and $HAuCl_4 \cdot 3H_2O$ as precursor metal salts with microwave heating. More recently, the green synthesis of core-shell Ag/starch NPs (~20 nm) were synthesized by reducing Ag^+ from $AgNO_3$ by glucose and in the presence of biodegradable starch as a capping stabilizing agent (Gao et al., 2011). In 2011, Ag-NPs (20–45 nm) were successfully prepared by the sol-gel route using $AgNO_3$ as a precursor and glucose as an eco-friendly reducing agent in aqueous medium according to the following reaction (Lkhavajav et al., 2011):

$$2AgNO_3 + C_6H_{12}O_6 + H_2O \xrightarrow[\text{stirring}]{\text{mixing}} \underset{\text{Colloidal Ag-NPs}}{2Ag^0} + \underset{\text{Gluconic acid}}{C_6H_{12}O_7} + 2HNO_3 \quad (12.5)$$

12.7.1.1.2 BIOPOLYMERS

Highly stable Au-NPs and Ag-NPs were prepared by a simple green method using positively charged chitosan and negatively charged heparin polysaccharides as reducing/stabilizing agents by heating aqueous solutions of $HAuCl_4$ and $AgNO_3$, respectively, at 55°C for 2 h and at 70°C for 8 h while stirring in a water bath. After heating, the solutions were turned to red and yellow, indicating the formation of Au-NPs and Ag-NPs, respectively (Huang and Yang, 2004). Both the morphology and size distribution of the nominated nanoparticles were governed by the concentration of the used polysaccharides and the metal salts. Ag-NPs (10–34 nm) were synthesized by developing a novel one-pot green synthetic route using soluble starch as both the reducing and stabilizing agents, along with $AgNO_3$, in an autoclave at 15 psi at 121°C for 5 min. The obtained nanoparticles were stable in aqueous solution for 3 months at room temperature (Vigneshwaran et al., 2006a). Additionally, an eco-friendly and simple method of obtaining Ag-NPs (<15 nm) with a narrow particle size distribution has been reported by Darroudi et al. (2011a) by reducing Ag^+ ions in aqueous gelatin (as a reducing/stabilizing agent) media. The reported method may extend to other noble metals (e.g., Au, Pd), most probably because of the ability of gelatin active sites (e.g., $-NH_2$) to form gelatin-stabilized nanoparticles such as (AuNPs-gelatin) in addition to its reducing power (Zhang et al., 2009). Darroudi et al. (2010) recently reported the green synthesis of Ag-NPs (10–20 nm) by glucose reduction of $AgNO_3$ in the presence of gelatin ($HOOC\text{-}G\text{-}NH_2$) as a stabilizer/matrix and NaOH as an accelerator during the synthesis. The possible mechanism for Ag-NPs preparation is as follows:

$$Ag^+_{(aq.)} + \underset{\text{Gelatin}}{HOOC\text{-}G\text{-}NH_2} \rightarrow \underset{\text{Gelatinous complex}}{[HOOC\text{-}G\text{-}NH_2]^+ Ag_{(aq.)}} \quad (12.6)$$

$$\begin{aligned} &2[HOOC\text{-}G\text{-}NH_2]^+ Ag_{(aq.)} + 2OH^- \\ &+ \underset{\text{Glucose}}{HOH_2C(CHOH)_4-CHO} \xrightarrow{\Delta} \\ &2[HOOC\text{-}G\text{-}NH_2] + \underset{\text{Ag-NPs}}{Ag^0} \\ &+ \underset{\text{Gluconic acid}}{HOH_2C(CHOH)_4-COOH} + H_2O \end{aligned} \quad (12.7)$$

12.7.1.1.3 NA-CITRATE AND TOLLENS REAGENT

Ag-NPs (~100 nm) were prepared by chemical reduction of $AgNO_3$ aqueous solution using Na-citrate ($C_6H_5O_7Na_3$) as a reducing agent according to the following equation (Sileikaite et al., 2006):

$$4Ag^+ + C_6H_5O_7Na_3 + 2H_2O \xrightarrow[\text{vigorous mixing}]{100°C} 4Ag^0 + C_6H_5O_7Na_3H_3 + 3Na^+ \quad (12.8)$$

Highly nano-dispersed Ag-NPs (10–50 nm) were green-synthesized at ambient temperatures using ascorbic acid, sodium citrate, sodium borohydride, or dimethylamine boran as a reducing agent, gum Arabic as a dispersing agent, and $AgNO_3$ as a precursor. Reduction of Ag^+ by using ascorbic acid ($C_6H_8O_6$) is shown (Sondi et al., 2003; Ramirez et al. 2009) as follows:

$$4Ag^+ + \underset{\text{Ascorbic acid}}{C_6H_8O_6} \rightleftharpoons 2Ag^0 + C_6H_6O_7Na_6 + 2H^+ \quad (12.9)$$

The greener synthesis of Ag-NPs of controlled sizes by reduction of Tollens reagent, $Ag(NH_3)_2^+$ (aq), using glucose, galactose, maltose, or lactose as an environmentally benign reducing agent was reported (Panacek et al. 2006; Sharma et al. 2009). Ammonia concentration, pH, as well as type and chemical structure of the reducing agent affect the average size of the obtained Ag-NPs. Using maltose disaccharide as a reducing agent created the smallest Ag-NPs, with an average size of 25 nm, compared with other reductants. The possible chemical reaction for reducing Ag^+ (aq) to Ag^0 is as follows:

$$Ag(NH_3)^+_{2(aq)} + \underset{\text{Maltose}}{M\text{-}CHO_{(aq)}} \xrightarrow{\text{pH } 11.5} Ag^0 \downarrow + M\text{-}COOH_{(aq)} \quad (12.10)$$

Incorporation of polyvinylypyrrolidone (PVP-360) or sodium dodecyl sulfate (SDC-surfactant) had a positive impact on extending the stability of shelf life of the prepared Ag-NPs (Sharma et al., 2009).

12.7.1.2 *Biological Method*

12.7.1.2.1 PLANT EXTRACT

Ag-NPs (~90 nm) and Au-NPs (~150 nm) were prepared by biological reduction of $AgNO_3$ and Au_4Cl_4 precursors using *Mentha piperita* (Lamiaceae) leaf extract in ambient conditions, most probably because of the presence of phyto-chemicals in the extract, thereby reducing Ag^+ and Au^+ into Ag^0 and Au^0, respectively (Ali et al., 2011). A simple green and cost-effective approach for the synthesis of Ag-NPs with an average size of 12 nm using garlic clove extract as a reducing/stabilizing agent in aqueous solution with constant stirring at 50–60°C for 30 min was reported (Ahmed et al., 2011). This was attributed to the presence of antioxidants within garlic extract and their positive role in reducing Ag^+ (aq) to Ag^0. Moreover, a green biological method for the synthesis of Ag-NPs using banana peel extract as a reductant and $AgNO_3$ as precursor in aqueous medium under various conditions was demonstrated (Bankar et al., 2010). The functional groups (e.g., –COOH, NH_2, –OH) associated with the polymeric materials (e.g., lignin, pectin, hemicelluloses) as well as proteinaceous matter components of BP interacted with Ag^+ ions and mediated their reduction to Ag-NPs under proper conditions. Ravindra et al. (2010) developed a new green process for the preparation of Ag-NPs (~20 nm) by using *Eucalyptus citriodora* (Neelagiri) and *Ficus bengalensis* (Marri) fresh leaf extract along with $AgNO_3$ in aqueous solution and without reducing or stabilizing agents at room temperature for 2–5 min. Reduction and stabilization of Ag^+ ions were enhanced by functional groups present in the polysaccharide constituents in leaf extract. Shankar et al. (2004) demonstrated a new process for producing Ag-NPs, Au-NPs, and bimetallic Au/Ag core-shell NPs by the reduction of $AgNO_3$ and/or $HAuCl_4$ in aqueous solutions with neem (*Azadirachta indica*) leaf broth. Reduction of the nominated metal ions to NPs as well as stabilization of these NPs might be facilitated by the presence of reducing sugars, terpeoids, and flavanone constituents in the broth. Using glutathione, an eco-friendly antioxidant, as an efficient reducing and capping agent for the synthesis of Ag-NPs (5–10 nm) and other nanoparticles (e.g., Pd, Pt, Au), using microwave irradiation (50 W/30–60 s) was reported (Baruwati et al., 2009).

Using rice wine as a solvent and reductant along with soda as a base catalyst and protective agent for greener synthesis (at pH 6.5 and 25–55°C) of Au-NPs was reported by Wu and Chen (2007). Xiong et al. (2011) reported a new method for the synthesis of highly stable dispersion of Cu-NPs (<2 nm) using L-ascorbic acid as an environmentally benign reducing and capping agent in aqueous medium as follows:

$$\text{L-ascorbic acid} + Cu^{2+} \rightarrow \text{dehydroascorbic acid} + Cu^0 \qquad (12.11)$$

TiO_2-NPs (25–100 nm) were produced by using aqueous extract (0.3%) prepared from latex of *Jatropha curcas* L (Hudlikar et al., 2012). However, *Aloe barbadensis* Miller leaf extract (Sangeetha et al., 2011) or gelatin by the sol-gel technique (Khorsand et al., 2011) were used for green synthesis of ZnO-NPs (25–60 nm).

12.7.1.2.2 MICROORGANISMS

Microorganisms (such as bacteria, fungi, yeast) have been used as potential environmentally benign nanofactories and as alternatives to the conventional routes to prepare metal nanoparticles either intracellularly or extracellulary (Dahl et al., 2007; Sharma et al., 2009; Narayanan and Sakthivel, 2010; Kharissova et al., 2013). Biosynthesis of metal nanoparticles is governed by localization of the reductive components of the cell. Some notable examples of biosynthesis of metal nanoparticles using microorganisms include (Beveridge and

Murray, 1980; Mukherjee et al., 2001; Ahmad et al., 2005; He et al., 2007; Vigneshwaran et al., 2007; Parikh et al., 2008; Philip, 2009; Pugazhenthiran et al., 2009; Narayanan and Sakthivel, 2010) the following:

a. Intracellular synthesis of Au-NPs by using *Bacillus subtilis* 168 or *Verticillium* sp. (AAT-TS-4) (Beveridge and Murray, 1980; Mukherjee et al., 2001), as well as synthesis of Ag-NPs by using an airborne *Bacillus* sp. or *Aspergillus flavus* (Vigneshwaran et al., 2007; Pugazhenthiran et al., 2009).

b. Extracellular synthesis of Au-NPs by using *Rhodopseudomonas capsulata*, a *prokaryotic* bacterium, or by the fungus *Trichothecium* sp. (Ahmad et al., 2005; He et al., 2007), as well as synthesis of Ag-NPs by using a silver-resistant bacterium *Moganella* sp. or the extract of saprophylic straw mushroom *Volvariella volvacea* fungus (Parikh et al., 2008; Philip, 2009).

However, extracellular production of metal oxides, such as Zn and Cu oxide nanoparticles, using *Streptomyces* sps. in aqueous solutions was reported and the possibility of the reduction of metal ions was ascribed to the reductase enzyme (Usha et al., 2010). Moreover, biosynthesis of TiO_2-NPs (8–35 nm in size) by using *Lactobacillus* sp. and baker's yeast (*Saccharomyces cerevisiae*) at room temperature or by using bacterium *B. subtilis* (size, 66–77 nm) was reported (Jha et al., 2009).

More recently Ag-NPs have been synthesized by using marine bacterial isolate (*Bacillus* sp.), 2–7 nm in size, and used for antibacterial functionalization of cellulosic fabrics (Abdel-Aziz et al., 2014).

12.7.1.3 *Using of Eco-Friendly Alternative Solvents*

Using supercritical fluids (e.g., SC-CO_2) and ionic liquids (ILs) as benign solvents for the green synthesis of inorganic nanoparticles has received increased attention (Antonietti et al., 2004; Dahl et al., 2007). SC-CO_2 has been used for preparation and deposition of Au-NPs into low-defection films (Liu et al., 2006). However, ILs have been used as a reducing and capping material to prepare Au-NPs (Kim et al., 2006), as well as to facilitate and drive the synthesis of ZnO-NPs (Zhou et al., 2005). Careful selection of the IL anions is a must to ensure more environmentally friendly synthesis routes (Dahl et al., 2007). It is expected that utilization of SC fluids and ILs for synthesis and potential applications of nanoparticles will ultimately be useful for large-scale greener production.

12.7.1.4 *Using of Electromechanical Routes*

In an attempt to search for an attractive/eco-friendly alternative to the conventional method for preparing nanoparticles, a number of R&D activities have utilized electrochemical routes for: (i) synthesis of ZnO nanostructure or ZnO thin film (Choi et al., 2002); (ii) synthesis of stable Au complexes (Liu and Chuang, 2003); and (iii) deposition of Au core on titania shell to form a stable structure (Liu and Juang, 2004) on the laboratory scale.

12.7.1.5 *Using Irradiation Methods*

Recently, utilization of microwave, ultrasonic, gamma ray irradiation, UV irradiation, and laser ablation physical techniques were used as alternatives energy sources for facilitating green synthesis of nanoparticles (Dahl et al., 2007; Sharma et al., 2009).

12.7.1.5.1 MICROWAVE IRRADIATION

The microwave technique offers numerous advantages in green preparation of nanoparticles such as lower energy consumption, shorter reaction times, as well as better yield of size-controllable nanoparticles, in addition to rapid and uniform heating. Microwave irradiation has been used in: (i) the preparation of Au-NPs (15–20 nm) as well as Ag nanospheres or nanorods using Na-citrate as a reducing agent (Liu et al., 2003, 2005); (ii) the production of

TiO_2-NPs (Baldassari et al., 2005; Murugan et al., 2006); (iii) the green synthesis of very stabilized Ag-NPs with negligible aggregation using CMS as a reducing/stabilizing agent along with $AgNO_3$ precursor (Chen et al., 2008); and (iv) green synthesis of nearly monodisperse Ag-NPs in an aqueous medium using certain amino acids as a reducing agent and soluble starch as a protecting agent (Hu B. et al., 2008).

12.7.1.5.2 ULTRASOUND IRRADIATION

Ultrasound irradiation has been used in the fabrication and deposition of ZnO-NPs (Perelshtein et al., 2010), Cu-NPs (Perelshtein et al., 2009), or TiO_2-NPs (Perelshtein et al., 2012) in a one-step process via *in situ* generation of the metal oxide NPs and their subsequent immobilization on the cotton fabric surface in a one-step reaction. Sonochemistry has also been used to produce Au-NPs (~20 nm) in an aqueous medium without alcohol or stabilizer (Su et al., 2003). However, Au-NPs (2–15 nm) were prepared by a combination of electrochemical oxidation/reduction and sonochemical reduction in an aqueous environment and without a stabilizer (Liu et al., 2004).

12.7.1.5.3 LASER ABLATION

Laser ablation of Au and Ag metals dipped in water, ethanol, or *n*-hexane generated Au-sols and Ag-sols containing approximately 12–20 nm NPs (Compagnini et al., 2002). Laser irradiation of an aqueous solution of Ag precursor and surfactant has also been used to facilitate production of Ag-NPs with well-defined shape and size distributions and without reducing agents (Abid et al., 2002). Additionally, preparation of Ag-NPs (~9 to 15 nm) in the gelatin matrix using laser ablation was reported (Darroudi et al., 2011b). Factors affecting nanoparticle size include type of liquid (media), energy density, as well as ablation time (Compagnini et al., 2003).

12.7.1.5.4 UV PHOTOACTIVATION

Pal et al. demonstrated the utilization of the UV photoactivation technique for the preparation of Au-NPs in the presence of Na-alginate as a reducing and stabilizing agent along with $HAuCl_4$ (precursor) in aqueous solution (Pal et al., 2005).

12.7.2 Nanocomposites

Twu et al. (2008) developed an eco-friendly potential process for the preparation of Ag/chitosan nanocomposites by using $AgNO_3$ as a precursor and basic chitosan suspension as a reducing/stabilizing agent. Likewise, Hu Z. et al. (2008) developed nanocomposites of chitosan/Ag-oxide as an effective antibacterial agent. However, the synthesis of the methylcellulose–Ag nanocomposite, MC-Ag hybrid nanocomposite, by using MC solution as a stabilizing and reducing agent along with $AgNO_3$ at higher pH was presented in 2012 (Maity et al., 2012). Transmission electron microscope (TEM) analysis demonstrated the presence of Ag-NPs (~22 nm) in MC-Ag nanocomposite films. Vigneshwaran et al. (2006a) also studied the preparation of ZnO-soluble starch nanocomposites (nano-ZnO) to attain functional nanostructures with great potential for functionalization of textiles. The average size of ZnO-NPs embedded in polymer matrices is estimated to be approximately 38 nm.

12.7.3 Antibacterial Functionalization of Textiles

Currently, application of antibacterial nanomaterials to textile fibers and/or onto fabric surfaces has attracted significant attention to: (i) impart antibacterial activity and prevent the hazardous effects of harmful bacteria; (ii) avoid transfer and prevent microbial infection; (iii) prevent formation and generation of offensive odors; and (iv) minimize or hinder the loss of performance properties of textiles such as discoloration, degradation, staining (Höfer, 2006).

A broad range of possible approaches has been suggested and/or used to load, stabilize, and incorporate nanomaterials onto and/within textiles for producing more appealing and achieving permanent antibacterial functionality without adversely affecting the inherent physico-mechanical and comfort properties of the treated substrates. There are two primary options to permit the integration of antibacterial nanomaterials into textiles: (i) impregnation or coating the textile materials with colloidal solution of pre-formed NPs or (ii) *in situ* generation of NPs and their subsequent deposition on the surface of fabrics in a one-step process.

12.7.3.1 *Application Methods*

12.7.3.1.1 BY INCORPORATION

Incorporation of bioactive nanoparticles into the spinning mass of synthetic fibers such as Ag-based nanocomposite fillers Ag/Zn and Ag/TiO_2 is performed to functionalize synthetic fibers (Dastjerdi and Montazer, 2010). Incorporated NPs in the central part of the filaments have no positive impact on the antibacterial performance.

12.7.3.1.2 BY IMMOBILIZATION

Immobilization of functional nanomaterials onto fabric surface is achieved by impregnation or coating with the colloidal solution of nanomaterials. Ibrahim et al. (2013g) have multifunctionalized cotton/spandex woven fabrics by using citric acid (CA) as an eco-friendly cross-linker as well as spacers for fixing Ag-NPs and TiO_2-NPs onto the treated fabric surface, NaH_2PO_2 (SHP) as a catalyst, and the pad-dry-cure technique. Hashemikia and Montazer (2012) have also produced multifunctional cotton/polyester knitted fabric through exhaustion in an ultrasonic bath containing TiO_2-NPs, citric acid, and Na-hyposphite as a catalyst, followed by drying and curing. Montazer et al. (2012) have used Ag-NPs along with 1,2,3,4-butantetracarboxylic acid (BTCA) as a cross-linking agent and SHP as a catalyst by padding technique to impart antibacterial properties to nylon knitted fabric without yellowing. Selvam et al. (2012) have synthesized sulfated β-cyclodextrin cross-linked cotton fabric using EDTA as a cross-linker, with improved antibacterial activities via padding in ZnO, TiO_2, and Ag-NP colloidal solution separately and curing at 120°C for 3 min. A new method to impart durable antibacterial properties to PET fabric via treatment with cross-linkable polysiloxane along with nano-sized colloidal-Ag in one or two separate steps using a pad-dry cure technique has been investigated by Dastjeradi et al. (2009). Enhancement antibacterial activity of cotton fabric by using sericin/TiO_2 nanocomposite in the presence and absence of using CA or BTCA as a cross-linker by pad-dry-technique was reported (Doakhan et al., 2013). Moreover, Vigneshwaran et al. (2006b) have coated fabrics with ZnO-soluble St • OH nanocomposite using the pad-dry-cure technique for attaining an efficient antibacterial functionalization.

12.7.3.1.3 BY PRE-MODIFICATION OF FABRICS SURFACE

Khalil-Abed et al. (2009) have demonstrated the positive effect of pre-cationization of cotton surfaces using 3-chloro-2-hydroxypropyl trimethyl ammonium chloride on enhancing the extent of adsorption of Ag-NPs onto surface of fibers as well as imparting strong antibacterial activity against *E. coli* bacteria. Enzymatic pretreatment of wool/polyester fabric blends with protease and then lipase to hydrolyze the wool and polyester surfaces, respectively, followed by dipping into an ultrasound bath containing TiO_2-NPs along with BTCA and curing at 180°C for 1–5 min to attain remarkable multifunctional properties (i.e., self-cleaning, antibacterial, and UV protection) were also performed by Montazer and Seifollahzadeh (2011). Gorensek et al. (2010) have introduced plasma surface modification using Ar/N_2 (50/50%) plasma to enable greater adhesion of Ag-NPs to raw PET fabric, which in turn enhances the antibacterial activity of the modified substrate. In addition, Abdel-Aziz et al. (2014) have pretreated cellulosic fabrics with O_2 plasma for surface modification followed by treatment with biosynthesized Ag-NPs for attaining remarkable and durable antibacterial activity against *S. aureus* (G +ve) and *E. coli* (G −ve) bacteria.

12.7.3.1.4 BY *IN SITU* SYNTHESIS AND DEPOSITION

Recently, many trials have been focused on the investigation of the *in situ* generation of nanoparticles (e.g., Ag-NPs, Cu-NPs) and their subsequent deposition onto cotton fabrics by using chitosan as a chelating agent and Na-citrate as a reductant, as in the case of Ag-NPs (Thomas et al., 2011), or onto cotton bandages by sonochemical irradiation, as in case of Cu-NPs (Perelshtein et al., 2009).

12.7.3.1.5 BY MULTIFUNCTIONALIZATION IN A ONE-STEP PROCESS

Recently, some studies have focused on imparting multifunctional properties alone or in combination with coloration properties of cellulose-containing fabrics in a one-step process. Some specific examples of these applications in the field of textile wet processing include: (i) multifunctional finishing of cellulosic/polyester blends using Ag-NPs/PVP and ZnO-NPs/HBPAA hybrids along with citric acid as an eco-friendly cross-linking agent (Ibrahim et al., 2013e); (ii) simultaneous functionalization and pigment coloration of cellulosic/wool blends using TiO_2-NPs along a proper polyacrylate binding agent (Ibrahim et al., 2013h); (iii) development of multifunction cellulosic pigment prints using Ag-, ZnO-, ZrO_2-, and TiO_2-NPs individually and in the presence of different binding agents (Ibrahim et al., 2013i); (iv) enhancing antibacterial functionality of pigment printed cotton, linen, and viscose cellulosic fabrics by using an Ag-NPs/PVP hybrid (Ibrahim et al., 2013a); (v) combined antibacterial finishing and pigment printing of cotton/polyester blends using Ag-NPs/HBPAA (Ibrahim et al., 2013j); (vi) reactive dyeing and antibacterial finishing of cellulosic fabrics using Ag-NPs loaded on HBPAA (Ibrahim et al., 2012c) in one step; (vii) imparting multifunctional properties to cellulose-containing fabrics using poly (acrylic acid)/poly(ethylene glycol) adduct along with Ag-NPs or TiO_2-NPs (Ibrahim et al., 2012b); and (viii) upgrading of antibacterial, UV protection, and self-cleaning properties of cotton fabrics by treatment with silver-, silica-, and titania-based sols by the dip-coating method (Onar et al., 2011).

12.7.4 Mode of Action

The antibacterial effect of a biocidal agent against harmful microorganisms such as bacteria is generally a direct consequence of its ability to: (i) interact with the outer cell wall; (ii) interact with cytoplasmic membrane; and/or (iii) interact with cytoplasmic constituents (e.g., nucleic acid, various enzymes, ribosomes). All these interactions lead to the damage of specific target sites within the cell or cause overall damage of the bacterial cell or even death of the microorganism (Maillard, 2002).

12.7.4.1 *Ag-NPs*

Ag-NPs have been known to be effective biocides against a wide range of microorganisms such as: (i) bacteria (e.g., *E. coli, S. aureus* [G + ve]); (ii) fungi (e.g., *Candida albicans, Aspergillus niger*); (iii) viruses (e.g., hepatitis B, HIV-1); and (iv) yeast (Marambio-Jones and Hoek, 2010). The most common mechanisms of the antibacterial activity of Ag-NPs proposed to date are: (i) cell wall damage and increase of cell membrane permeability; (ii) accumulation in the cell membrane, passing through it and disturbing its proper functions; (iii) gradual release of free Ag^+ ions in the bacterial cells followed by generation of reactive oxygen species (ROS) in the presence of dissolved oxygen (Dastjeradi et al., 2009; Ibrahim et al., 2012b; Radetic, 2013):

$$O_{2(aq)} + 4H_3O^+ + 4Ag^0 \rightarrow Ag^+_{(aq)} + 6H_2O \quad (12.12)$$

$$H_2O + \frac{1}{2}O_2 \xrightarrow{Ag^+} H_2O + [O] \quad (12.13)$$

and (iv) interaction of uptaken Ag^+ ions with sulfur-containing proteins in the cell and phosphorous-containing compounds like DNA, thereby disrupting ATP production and DNA replication, and finally leading to cell death (Maillard, 2002; Morones et al., 2005; Marambio-Jones and Hoek, 2010; Rai et al., 2009; Radetic, 2013). However, factors affecting the biocidal activity of AgNPs include particle size, stability, particle shape, surface chemistry, and capping agents (Marambio-Jones and Hoek, 2010).

12.7.4.2 Metal Oxide-NPs

The antibacterial activity of metal oxide NPs (e.g., ZnO, TiO_2) is related to their photocatalytic nature and their ability to photogenerate electrons and holes under UV/vis irradiation and subsequent generation of extremely ROS (e.g., $\cdot OH$, $\cdot O_2^-$, H_2O_2, single oxygen) as follows:

$$\text{MO-NPs} + h\upsilon \xrightarrow[\text{irradiation}]{\text{UV/Vis}} h^+ + e^- \quad (12.14)$$

$$H_2O + h^+ \rightarrow {}^\bullet OH + H^+ \quad (12.15)$$

$$O_2 + e^- \rightarrow {}^\bullet O_2^- + H^+ \rightarrow HO_2^\bullet \quad (12.16)$$

$$2HO_2^\bullet \rightarrow O_2 + H_2O_2 \quad (12.17)$$

thereby inhibiting the normal metabolism of bacteria, oxidizing organic components in the bacteria cell, and finally leading to the death of bacteria (Erem et al., 2011; Montazer and Pakdel, 2011; Selvam et al., 2012; Ibrahim et al., 2013c). However, the abrasion effect of MO-NPs on contact with the membrane cell wall results in cell damage, alteration of cell wall permeability, and release of Mn^+ ions, which cause the death of bacteria (Erem et al., 2011; Ibrahim et al., 2013c).

12.8 POTENTIAL IMPLICATIONS

Potential applications of nanotechnology and utilization of nanomaterials in the textile fields for producing highly functional value-added textiles that are on the market today with properties such as antimicrobial function, UV protection, self-cleaning ability, flame protection, water repellency/oil repellency, insect repellency, super hydrophobicity, as well as functionalized nanofibers for advanced applications are now growing. However, the growing public concerns about the possible negative impacts and potential risks of preparation and application of nanomaterials on the environment, human health and safety, and sustainability (EHS/S) aspects have dramatically increased. It is suspected that engineered nanoparticles (ENPs) could be unintentionally liberated from nanoproducts because of external impacts during the life cycle and then could adversely affect EHS/S requirements (Morose, 2010; Chen et al., 2012; Köhler and Som, 2013). The EHS/S negative impacts of nanomaterials can derive from selection of starting materials, NP preparation route, textile treatment processes, extent of fixation, and finally subsequent use and disposal of the textile materials. The scientific knowledge regarding ENP toxicity, extent, and mechanism of their liberation during the life cycle of the product, the possible side effects, the potential risks, and their fate after release are insufficient and scarce. Köhler and Som (2013) have proposed the following precautionary and preventative measures to avoid or minimize EHS/S risks during the life cycle of nano-textiles: (i) *in situ* generation of ENPs or using ENPs in liquid or encapsulated form during the production phase; (ii) search for ENP-free alternatives (i.e., source reduction to minimize the exposure to the textile user during the use phase); (iii) make sure that the loaded ENPs are permanently fixed onto the textile product by functionalization to avoid recycling and disposal problems without adversely affecting the desired properties; (iv) reduce resource consumption as well as waste generation (Morose, 2010); and (v) highlight principles and favorable conditions for responsible manufacturing, handling, and use of ENPs, as well as disposal of nanomaterials containing textiles.

12.9 EVALUATION OF ANTIBACTERIAL EFFICACY

Various test methods have been developed to demonstrate the antibacterial efficacy of functionalized textiles against *E. coli* (G −ve) and *S. aureus* (G + ve) bacteria as test microorganisms. These tests broadly fall into two categories (Höfer, 2006; Gao and Cranston, 2008; Lam et al., 2012): (i) the agar diffusion test, expressed as zone of inhibition, as a preliminary qualitative methods (standard methods: AATCC 147-2004 and JIS L1902-2002 and SN 195920-1992) and (ii) suspension tests, expressed as bacteria counting, give a quantitative assessment of efficacy antibacterial treatment after incubation (standard methods: AATCC 100-2004 and JIS L1902-2002 and SN 195924-1992).

12.10 FUTURE SCOPE

Application of green chemistry principles to nanoscience is highly demanded and still a challenge for future development of environmentally friendly antimicrobial textiles. The development and production of novel, green, cost-effective functionalized nanomaterials, as well as their innovative diverse application for large-scale production, taking into consideration EHS/S concerns, are still a challenge for further development in the near future.

The future development of the textile and apparel industry, especially at the level of SME, will strongly depend on adoption and implementation of emerging technologies, building new technology capabilities, and transferring these high technologies into innovative/multifunctional/value-added textile products that meet the emerging eco-labeling standards. Additionally, there is an urgent need for sharing knowledge, exchanging best practices, engaging with the textile industry, facilitating the take-up of new technologies and innovations to make the emerging technologies more affordable, and supporting collaboration between international and national standardization bodies to promote synergy in handling EHS/S risks throughout the product life cycle and to help the transition to more sustainable, resource-efficient, and green markets for nano and smart-textiles.

Acknowledgment

The author gratefully acknowledges the kind and continued assistance of Associate Professor Basma M. Eid, Textile Research Division, National Research Centre, during preparation of this chapter and for generously providing relevant materials that made it possible.

References

Abdel-Aziz, M.S., Eid, B.M., Ibrahim, N.A., 2014. Biosynthesis silver nanoparticles for antibacterial treatment of cellulosic fabrics using O_2-plasma. AATCC J. Res. 1 (1), 6–12.

Abid, J.P., Wark, A.W., Brevet, P.F., Girault, H.H., 2002. Preparation of silver nanoparticles in solution from a silver salt by laser irradiation. Chem. Commun. 7, 792–793.

Ahmad, A., Senapati, S., Khan, M.I., Kumar, R., Sastry, M. J., 2005. Extra/intracellular biosynthesis of gold nanoparticles by an alkalotolerant fungus, *Trichothecium* sp. Biomed. Nanotechnol. 1, 47–53.

Ahmed, M., Khan, M.A.M., Siddiqui, M.K.J., Alshali, M.S., Alrokayan, S.A., 2011. Green synthesis, characterization and evaluation of biocompatibility of silver-nanoparticles. Physica E 43, 1266–1271.

Ali, D.M., Thajuddin, N., Jeganathan, K., Gunasekaran, M., 2011. Plant extract mediated synthesis of silver and gold nanoparticles and its antibacterial activity against clinically isolated pathogens. Colloids Surf. B Biointerfaces 85, 360–365.

Antonietti, M., Kuang, D., Smarsly, B., Zhou, Y., 2004. Ionic liquids for the convenient synthesis of functional nanoparticles and other inorganic nanostructures. Angew. Chem. Int. Ed. 43, 4988–4992.

Baldassari, S., Komarneni, S., Mariani, E., Villa, C., 2005. Microwave-hydrothermal process for the synthesis of rutile. Mater. Res. Bull. 40, 2014–2020.

Bankar, A., Joshi, B., Kumar, A.R., Zinjarde, S., 2010. Banana peel extract mediated novel route for the synthesis of silver nanoparticles. Colloids Surf. A Physicochem. Eng. Asp. 368, 58–63.

Baruwati, B., Polshettiwar, V., Varma, R.S., 2009. Glutathione promoted expeditious green synthesis of silver nanoparticles in water using microwaves. Green Chem. 11, 926–930.

Beveridge, T.J., Murray, R.G., 1980. Sites of metal deposition in the cell wall of *Bacillus subtilis*. J. Bacteriol. 141, 876–887.

Bhattacharya, D., Gupta, R.K., 2005. Nanotechnology and potential of microorganisms. Crit. Rev. Biotechnol. 25, 199–204.

Bide, M., 2009. Fiber sustainability: green chemistry is not black & white. AATCC Rev. 9, 34–37.

Black, S., 2009. The role of nanotechnology in sustainable textiles. In: Blakburn, R.S. (Ed.), Sustainable Textiles—Life Cycle and Environmental Impact. Woodhead Publisher, Cambridge, pp. 312–328, Chapter 12.

Chairam, S., Poolperm, C., Somsook, E., 2009. Starch vermicelli template-assisted synthesis of size/shape-controlled nanoparticles. Carbohydr. Polym. 75, 694–704.

Chen, J., Wang, J., Zhang, X., Jin, Y., 2008. Microwave-assisted green synthesis of silver nanoparticles by carboxymethyl cellulose sodium and silver nitrate. Mater. Chem. Phys. 108, 421–424.

Chen, J., Lu, Y.G., Sun, C., 2012. Safety and health assessment of manufactured nanoparticles in nano-coated textile products. Text. Light Ind. Sci. Technol. 1, 37–42.

Choi, K.-S., Lichtenegger, H.C., Stucky, G.D., McFarland, E. W., 2002. Electrochemical synthesis of nanostructured ZnO films utilizing self-assembly of surfactant molecules at solid – liquid interfaces. J. Am. Chem. Soc. 124, 12402–12403.

Choudhury, A.K.R., 2006a. Dyes and dyeing. Textile Preparation and Dyeing. Science Publishers, Enfield, N.H., USA, Chapters 6, pp. 326–399.

Choudhury, A.K.R., 2006b. Preparatory processes I: cleaning and preparatory. Textile Preparation and Dyeing. Science Publishers, Enfield, N.H., USA, Chapters 4, pp. 131–248.

Choudhury, A.K.R., 2011. Dyeing of synthetic fibres. In: Clark, M. (Ed.), Handbook of Textile and Industrial Dyeing, vol.2: Applications of Dyes. Woodhead Publishers, Cambridge, pp. 40–125.

Compagnini, G., Scalisi, A.A., Puglisi, O., 2002. Ablation of noble metals in liquids: a method to obtain nanoparticles in a thin polymeric film. Phys. Chem. Chem. Phys. 4, 2787–2791.

Compagnini, G., Scalisi, A.A., Puglisi, O., 2003. Production of gold nanoparticles by laser ablation in liquid alkanes. J. Appl. Physiol. 94, 7874–7877.

Cook, J.G., 2001. Handbook of Textile Fibers, vol.1: Natural Fibers. Woodhead, Cambridge.

Dahl, J., Maddux, B.L.S., Hutchison, J.E., 2007. Towards greener nanosynthesis. Chem. Rev. 107, 2228–2269.

Darroudi, M., Ahmed, M.B., Abdullah, A.H., Ibrahim, N. A., Shameli, K., 2010. Effect of accelerator in green synthesis of silver nanoparticles. Int. J. Mol. Sci. 11, 3898–3905.

Darroudi, M., Ahmed, M.B., Abdullah, A.H., Ibrahim, N. A., 2011a. Green synthesis and characterization of gelatin-based and sugar-reduced silver nanoparticles. Int. J. Nanomed. 6, 569–574.

Darroudi, M., Ahmad, M.B., Zamiri, R., Abdullah, A., Ibrahim, N.A., Shmeli, K., et al., 2011b. Preparation and characterization of gelatin mediated silver nanoparticles by laser ablation. J. Alloys Compd. 509, 1301–1304.

Dastjeradi, R., Montazer, M., Shahsavan, S., 2009. A new method to stabilize nanoparticles on textile surfaces. Colloids Surf. A Physicochem. Eng. Asp. 345, 202–210.

Dastjerdi, R., Montazer, M., 2010. A review on the application of inorganic nano-structured materials in the modification of textiles: focus on anti-microbial properties. Colloids Surf. B Biointerfaces 79, 5–18.

Doakhan, S., Montazer, M., Rashidi, A., Moniri, R., Moghadam, M.B., 2013. Influence of sericin/TiO_2 nanoparticles on cotton fabric: part 1. Enhanced antibacterial effect. Carbohydr. Polym. 94, 737–748.

El-Moghazy, Y., 2004. An integrated approach to analyzing the nature of multicomponent fiber blending. Text. Res. J. 74, 701–712.

Emam, H.E., Manian, A.P., Siroka, B., Duelli, H., Redl, B., Pipal, A., et al., 2013. Treatments to impart antimicrobial activity to clothing and household cellulosic-textiles-why "nano"-silver? J. Clean. Prod. 39, 17–23.

Erem, A.D., Ozcan, G., Skrifvars, M., 2011. Antibacterial activity of PA6/ZnO nanocomposite fibers. Text. Res. J. 81, 1638–1646.

Evans, F.N., Peters, S., Stingelin, N., 2012. Nanotechnology innovation for future development in the textile industry. In: Horne, L. (Ed.), New Product Development in Textiles—Innovation and Production. Woodhead Publisher, Cambridge, pp. 109–131, Chapter 7.

Gao, X., Wei, L., Yan, H., Xu, B., 2011. Green synthesis and characteristics of core-shell structure silver/starch nanoparticles. Mater. Lett. 65, 2963–2965.

Gao, Y., Cranston, R., 2008. Recent advances in antimicrobial treatments of textiles. Text. Res. J. 78, 60–72.

Goesmann, H., Feldmann, C., 2010. Nanoparticulate functional materials. Angew. Chem. Int. Ed. 49, 1362–1395.

Gorensek, M., Gorjanc, M., Bukosek, V., Kovac, J., Petrovic, Z., Puac, N., 2010. Functionalization of polyester fabric by Ar/N_2 plasma and silver. Text. Res. J. 80, 1633–1642.

Gouda, M., Ibrahim, N.A., 2008. New approach for improving antibacterial function to cotton fabric. J. Ind. Text. 37, 327–339.

Gowri, S., Almeida, L., Amorim, T., Carneiro, N., Souto, A. P., Esteves, M.F., 2010. Polymer nanocomposites for multifunctional finishing of textiles—a review. Text. Res. J. 80, 1290–1306.

Gulrajani, M.L., 2006. Nano finishes. Indian J. Fibre Test. Res. 31, 187–201.

Gulrajani, M.L., Gupta, D., 2011. Emerging techniques for functional finishing of textiles. Indian J. Fiber Text. Res. 36, 388–397.

Harifi, T., Montazer, M., 2012. Past, present and future prospects of cotton cross-linking: new insight into nano particles. Carbohyd. Polym. 88, 1125–1140.

Hashemikia, S., Montazer, M., 2012. Sodium hypophosphite and nano TiO_2 inorganic catalysts along with citric acid on textile producing multi-functional properties. Appl. Catalysis A General. 417-418, 200–208.

He, S., Guo, Z., Zhang, Y., Zhang, S., Wang, J., Gu, N., 2007. Biosynthesis of gold nanoparticles using the bacteria *Rhodopseudomonas capsulata*. Mater. Lett. 61, 3984–3987.

Hebeish, A., Ibrahim, N.A., 2007. The impact of frontier sciences on textile industry—a review. Colourage 54, 53–62.

Höfer, D., 2006. Antimicrobial textiles—evaluation of their effectiveness and safety. Curr. Probl. Dermatol. 33, 42–50.

Holme, I., 2007. Innovative technologies for high performance textiles. Color. Technol. 123, 59–73.

Hu, B., Wang, S.B., Wang, K., Zhang, M., Yu, S.H., 2008. Microwave assisted rapid facile "green" synthesis of uniform silver-nanoparticles: self assembly into multilayered films and their optical properties. J. Phys. Chem. C 112, 11169–11174.

Hu, Z., Chan, W.L., Szeto, Y.S., 2008. Nanocomposite of chitosan and silver oxide and its antibacterial activity. J. Appl. Polym. Sci. 108, 52–56.

Huang, H., Yang, X., 2004. Synthesis of polysaccharide stabilized gold and silver nanoparticles: a green method. Carbohyd. Res. 239, 2627–2631.

Hudlikar, M., Joglekar, S., Dhaygude, M., Kodam, K., 2012. Green synthesis of TiO_2 nanoparticles by using aqueous extract of *Jatropha curcas* L. latex. Mater. Lett. 75, 196–199.

Ibrahim, N.A., Gouda, M., El-shafei, A.M., Abdel-Fatah, O. M., 2007. Antimicrobial activity of cotton fabrics containing immobilized enzymes. J. Appl. Polym. Sci. 104, 1754–1761.

Ibrahim, N.A., Aly, A.A., Gouda, M., 2008. Enhancing the antibacterial properties of cotton fabrics. J. Ind. Text. 37, 203–212.

Ibrahim, N.A., Gouda, M., Husseiny, Sh. M., El-Gamal, A. R., Mahrous, F., 2009. UV-protecting and antibacterial finishing of cotton knits. J. Appl. Polym. Sci. 112, 3589–3596.

Ibrahim, N.A., Amr, A., Eid, B.M., El Sayed, Z.M., 2010a. Innovative Multifunctional treatments of lignocellulosic jute fabric. Carbohyd. Polym. 82, 1198–1204.

Ibrahim, N.A., Fahmy, H.M., Abdel Rehim, M., Sharaf, S.S., Abo-Shosha, M.H., 2010b. Finishing of cotton fabrics with hyperbranched poly (ester-amine) to enhance their antibacterial properties and UV protection. Polym. Plastic Technol. Eng. 49, 1279–1304.

Ibrahim, N.A., Mahrous, F., El–Gamal, A.R., Gouda, M., Husseiny, S.M., 2010c. Multifunctional anionic cotton dyeings. J. Appl. Polym. Sci. 115, 3249–3255.

Ibrahim, N.A., 2011. Dyeing of textile blends. In: Clark, M. (Ed.), Handbook of Textile and Industrial Dyeing, vol.2: Applications of Dyes. Woodhead Publishers, Cambridge, pp. 147–172.

Ibrahim, N.A., Eid, B.M., El-Zairy, E.R., 2011. Antibacterial functionalization of reactive-cellulosic prints via inclusion of bioactive neem oil/βCD complex. Carbohyd. Polym. 86, 1313–1319.

Ibrahim, N.A., Eid, B.M., El-Batal, H., 2012a. A Novel approach for adding smart functions to cellulosic fabrics. Carbohyd. Polym. 87, 744–751.

Ibrahim, N.A., Eid, B.M., Youssef, M.A., El-Sayed, S.A., Salah, A.M., 2012b. Functionalization of cellulose-containing fabrics by plasma and subsequent metal salt treatments. Carbohyd. Polym. 90, 908–914.

Ibrahim, N.A., Amr, A., Eid, B.M., Mohammed, Z.E., Fahmy, H.M., 2012c. Poly (acrylic acid)/poly (ethylene glycol) adduct for attaining multifunctional cellulosic fabrics. Carbohyd. Polym. 89, 684–660.

Ibrahim, N.A., Eid, B.M., Abou Elmaaty, T.M., Abd El-Aziz, E., 2013a. A smart approach to add antibacterial functionality to cellulosic pigment prints. Carbohyd. Polym. 94, 612–618.

Ibrahim, N.A., Eid, B.M., Youssef, M.A., Ameen, H.A., Salah, A.M., 2013b. Surface modification and smart functionalization of polyester containing fabrics. J. Ind. Text. 42, 353–357.

Ibrahim, N.A., Abou Elmaaty, T.M., Eid, B.M., Abd El-Aziz, E., 2013c. Combined antimicrobial finishing and pigment printing of cotton/polyester blends. Carbohyd. Polym. 95, 379–388.

Ibrahim, N.A., Khalil, H.M., El-Zairy, E.M.R., Abdalla, W. A., 2013d. Smart options for simultaneous functionalization and pigment coloration of cellulosic/wool blends. Carbohyd. Polym. 96, 200–210.

Ibrahim, N.A., Eid, B.M., Youssef, M.A., Ibrahim, H.A., Ameen, H.A., Salah, A.M., 2013e. Multifunctional finishing of cellulosic/polyester blended fabrics. Carbohyd. Polym. 97, 783–793.

Ibrahim, N.A., El-Zairy, M.R., El-Zairy, W.M., Ghazal, H. A., 2013f. Enhancing the UV-protection and antibacterial properties of polyamide-6 fabric by natural dyeing. Text. Light Ind. Sci. Technol. 2, 36–41.

Ibrahim, N.A., Amr, A., Eid, B.M., Almetwally, A.A., Mourad, M.M., 2013g. Functional finishes of stretch cotton fabrics. Carbohyd. Polym. 98, 1603–1609.

Ibrahim, N.A., Khalil, H.M., El-Zairy, E.M.R., Abdalla, W. A., 2013h. Smart options for simultaneous functionalization and pigment coloration of cellulosic/wool blends. Carbohyd. Polym. 96, 200–210.

Ibrahim, N.A., Eid, B.M., Abd El-Aziz, E., Abou Elmaaty, T.M., 2013i. Functionalization of linen/cotton pigment prints using nano-structure materials. Carbohyd. Polym. 97, 537–545.

Ibrahim, N.A., Abou Elmaaty, T.M., Eid, B.M., Abd El-Aziz, E., 2013j. Combined antimicrobial finishing and pigment printing of cotton/polyester blends. Carbohyd. Polym. 95, 379–388.

Jha, A.K., Prasad, K., Kulkarni, A.R., 2009. Synthesis of TiO_2 nanoparticles using microorganisms. Colloid Surf. B-Biointerfaces 71, 226–229.

Josh, M., Ali, S.W., Rajendran, S., 2007. Antibacterial finishing of polyester/cotton blend fabrics using neem (*Azadirachta indica*): a natural bioactive agent. J. Appl. Polym. Sci. 106, 793–800.

Joshi, M., Bhattcharyya, A., 2011. Nanotechnology—a new route to high performance functional textiles. Text. Prog. 34, 155–233.

Kawabata, A., Tylor, J.A., 2004. Effect of reactive dyes upon the uptake and antibacterial action of poly (hexamethylene biguanide) on cotton, part I: effect of bis (monochlorotriazinyl) dyes. Color Technol. 120, 213–219.

Kaylon, B.D., Olgun, U., 2001. Antibacterial efficacy of triclosan incorporated polymer. Am. J. Infect. Control. 29, 124–125.

Khalil-Abed, M.S., Yazdanshenas, M.E., Nateghi, M.R., 2009. Effect of cationization of silver nanoparticles on cotton surface and its antibacterial activity. Cellulose. 16, 1147–1157.

Kharissova, O.V., Dias, H.V.R., Kharisov, B.I., Pérez, B.O., Pérez, V.M.J., 2013. The greener synthesis of nanoparticles. Trends Biotechnol. 31, 240–248.

Khorsand, A.Z., Abd Majid, W.H., Darroudi, M., Yousefi, R, 2011. Synthesis and characterization of ZnO nanoparticles prepared in gelatin media. Mater. Lett. 65, 70–73.

Kim, K.-S., Choi, S., Cha, J.-H., Yeon, S.-H., Lee, H., 2006. Facile one-pot synthesis of gold nanoparticles using alcohol ionic liquids. J. Mater. Chem. 16, 1315–1317.

Kim, Y.H., Sun, G., 2001. Durable antimicrobial finishing of nylon fabrics with acid dyes and a quaternary ammonium salt. Text. Res. J. 71, 318–323.

Koh, J., 2011. Dyeing of cellulosic fibres. In: Clark, M. (Ed.), Handbook of Textile and Industrial Dyeing, vol.2: Applications of Dyes. Woodhead Publishers, Cambridge, pp. 129–146.

Köhler, A.R., Som, C., 2013. Risk preventive innovation strategies for emerging technologies the cases of nano-textiles and smart textiles. Technovation. Available from: <http://dx.doi.org/10.1016/j.technovation.2013.07.002>.

Kumar, A., Choudhury, R., 2013. Green chemistry and the textile industry. Text. Prog. 45, 3–143.

Lam, Y.L., Kan, C.W., Yuen, C.W.M., 2012. Development in functional finishing of cotton fibers-wrinkle resistance, flame–retardant and antimicrobial treatments. Text. Prog. 44, 175–249.

Lee, H.J., Yeo, S.Y., Jong, S.H., 2003. Antibacterial effect of nanosize silver colloidal solution in textile fabrics. J. Mater. Sci. 38, 2199–2204.

Lewis, D.M., 2011. The colouration of wool. In: Clark, M. (Ed.), Handbook of Textile and Industrial Dyeing, vol.2: Applications of Dyes. Woodhead Publishers, Cambridge, pp. 3–35.

Lim, S.H., Hudson, S.M., 2003. Review of chitosan and its derivatives as antimicrobial agents and their uses as textile chemicals. J. Macromol. Sci. Polymer Rev. 43, 223–269.

Lim, S.H., Hudson, S.M., 2004. Application of a fiber-reactive chitosan derivative to cotton fabric as an antimicrobial textile finish. Carbohyd. Polym. 56, 227–234.

Liu, F.-K., Ker, C.-J., Chang, Y.-C., Ko, F.-H., Chu, T.-C., Dai, B.-T., 2003. Microwave heating for the preparation of nanometer gold particles. Jpn. J. Appl. Phys. 42, 4152–4158.

Liu, F.-K., Huang, P.-W., Chang, Y.-C., Ko, C.-J., Ko, F.-H., Chu, T.-C., 2005. Formation of silver nanorods by microwave heating in the presence of gold seeds. J. Cryst. Growth 273, 439–445.

Liu, J., Anand, M., Roberts, C.B., 2006. Synthesis and extraction of β-d-glucose-stabilized Au nanoparticles processed into low-defect, wide-area thin films and ordered arrays using CO_2-expanded liquids. Langmuir 22, 3964–3971.

Liu, Y.-C., Chuang, T.C., 2003. Synthesis and characterization of gold/polypyrrole core – shell nanocomposites and elemental gold nanoparticles based on the gold-containing nanocomplexes prepared by electrochemical methods in aqueous solutions. J. Phys. Chem. B. 107, 12383–12386.

Liu, Y.-C., Juang, L.-C., 2004. Electrochemical methods for the preparation of gold-coated TiO_2 nanoparticles with variable coverages. Langmuir 20, 6951–6955.

Liu, Y.-C., Lin, L.-H., Chiu, W.-H., 2004. Size-controlled synthesis of gold nanoparticles from bulk gold substrates by sonoelectrochemical methods. J. Phys. Chem. B. 108, 19237–19240.

Lkhavajav, N., Yasa, I., Celik, E., Koizhaiganova, M., Sari, O., 2011. Antimicrobial activity of colloidal silver nanoparticles prepared by sol-gel method. Dig. J. Nanomater. Biostruct. 6, 149–154.

Ma, M.H., Sun, Y.Y., Sun, G., 2003. Antimicrobial cationic dyes. part I: synthesis and characterization. Dyes Pigments 58, 27–35.

Mahltig, R., Fiedler, D., Böttcher, H., 2004. Antimicrobial sol-gel coating. J. Sol-Gel Sci. Technol. 32, 219–222.

Mahltig, R., Haufe, H., Böttcher, H., 2005. Functionalization of textiles by inorganic sol-gel coatings. J. Mater. Chem. 15, 4385–4398.

Maillard, J.Y., 2002. Bacterial target sites for biocide action. J. Appl. Microbiol. Symp. 92 (Suppl.), 16S–27S.

Maity, D., Mollick, M.M.R., Mondal, D., Bhowmick, B., Bain, M.K., Bankura, K., et al., 2012. Synthesis of methycellulose-silver nanocomposite and investigation of mechanical and antimicrobial properties. Carbohyd. Polym. 90, 1818–1825.

Mao, J.W., Murphy, L., 2001. Durable freshness for textiles. AATCC Rev. 1, 28–31.

Marambio-Jones, C., Hoek, E.M., 2010. A review of antibacterial effects of silver nanomaterials and potential implications for human health and the environment. J. Nanopart. Res. 12, 1531–1551.

Montazer, M., Pakdel, E., 2011. Functionality of nanotitanium oxide on textiles with future aspects: focus on wool. J. Photochem. Photobiol. C-Photochem. Rev. 12, 293–303.

Montazer, M., Seifollahzadeh, S., 2011. Enhanced self-cleaning, antibacterial and UV-protection properties of nano TiO_2 treated textile through enzymatic pretreatment. Photochem. Photobiol. 87, 877–883.

Montazer, M., Shamei, A., Alimohammadi, F., 2012. Stabilized nano-silver loaded nylon knitted fabric using BTCA without yellowing. Prog. Org. Coat. 74, 270–276.

Morones, J.R., Elechiguerra, J.L., Camacho, A., Holt, K., Kouri, J.B., Ramírez, J.T., et al., 2005. The bactericidal effect of silver nanoparticles. Nanotechnology 16, 2346–2353.

Morose, G., 2010. The principle of "design" for safer nanotechnology. J. Clean. Prod. 18, 285–289.

Mukherjee, P., Ahmad, A., Mandal, D., Senapati, S., Sainkar, S.R., Khan, M.I., et al., 2001. Bioreduction of $AuCl_4-$ ions by the fungus, *Verticillium* sp. and surface trapping of the gold nanoparticles formed. Angew. Chem. Int. Ed. 40, 3585–3588.

Murugan, A.-V., Samuel, V., Ravi, V., 2006. Synthesis of nanocrystalline anatase TiO_2 by microwave hydrothermal method. Mater. Lett. 60, 479–480.

Narayanan, K.B., Sakthivel, N., 2010. Biological synthesis of metal nanoparticles by microbes. Adv. Colloid Interface Sci. 156, 1–13.

Onar, N., Akist, A.C., Sen, Y., Mutlu, M., 2011. Antibacterial, UV-protective and self-cleaning properties of cotton fabrics coated by dip-coating and solvothermal coating methods. Fiber Polym. 12, 461–470.

Pal, A., Esumi, K., Pal, T., 2005. Preparation of nanosized gold particles in a biopolymer using UV photoactivation. J. Colloid Interface Sci. 288, 396–401.

Panacek, A., Kvitek, L., Prucek, R., Kolar, M., Vecerova, R., Pizúrova, N., et al., 2006. Silver colloid nanoparticles: synthesis, characterization, and their antibacterial activity. J. Phys. Chem. B. 110, 16248–16253.

Panigrahi, S., Kundu, S., Ghosh, S.K., Nath, S., Pal, T., 2004. General method of synthesis for metal nanoparticles. J. Nanopart. Res. 6, 411–414.

Parikh, R.Y., Singh, S., Prasad, B.L.V., Patole, M.S., Sastry, M., Shouche, Y.S., 2008. Extracellular synthesis of crystalline silver nanoparticles and molecular evidence of silver resistance from *Morganella* sp.: towards understanding biochemical synthesis mechanism. ChemBioChem. 9, 1415–1422.

Perelshtein, I., Applerot, G., Perkas, N., Wehrschuetz-Sigl, E., Hasmann, A., Guebitz, G., et al., 2009. CuO–cotton nanocomposite: formation, morphology, and antibacterial activity. Surface Coating Technol. 204, 54–57.

Perelshtein, I., Applerot, G., Perkas, N., Grinblat, J., Hulla, E., Wehrschuetz-Sigl, E., et al., 2010. Ultrasound radiation as a "throwing stones" technique for the production of antibacterial nanocomposite textiles. ACS Appl. Mater. Interfaces 2, 1999–2004.

Perelshtein, I., Applerot, G., Perkas, N., Grinblat, J., Gedanken, A., 2012. A one-step process for the antimicrobial finishing of textiles with crystalline TiO_2 nanoparticles. Chem. Eur. J. 18, 4575–4582.

Philip, D., 2009. Biosynthesis of Au, Ag and Au–Ag nanoparticles using edible mushroom extract. Spectrochim. Acta A Mol. Biomol. Spectrosc. 73, 374–381.

Pugazhenthiran, N., Anandan, S., Kathiravan, G., Prakash, N.K.U., Crawford, S., Ashokkumar, M., 2009. Microbial synthesis of silver nanoparticles by *Bacillus* sp. J. Nanopart. Res. 11, 1811–1815.

Purwar, R., Joshi, M., 2004. Recent development in antimicrobial finishing of textiles—a review. AATCC Rev. 4, 22–26.

Qian, L., Sun, G., 2004. Durable and regenerable antimicrobial textiles: improving efficacy and durability of biocidal functions. J. Appl. Polym. Sci. 91, 2588–2593.

Radetic, M., 2013. Functionalization of textile methods with silver nanoparticles. J. Mater. Sci. 48, 95–107.

Rai, M., Yadav, A., Gade, A., 2009. Silver nanoparticles as a new generation of antimicrobials. Biotechnol. Adv. 27, 76–83.

Ramirez, I.M., Bashir, S., Luo, Z., Liu, J.L., 2009. Green synthesis and characterization of polymer stabilized silver nanoparticles. Colloid Surf. B-Biointerfaces 73, 185–191.

Raveendran, P., Fu, J., Wallen, S.L., 2003. Completely "green" synthesis and stabilization of metal nanoparticles. J. Am. Chem. Soc. 125, 13940–13941.

Ravindra, S., Mohan, Y.M., Reddy, N.N., Raju, K.M., 2010. Fabrication of antibacterial cotton fibres loaded with silver nanoparticles via "green approach". Colloid Surf. A-Physicochem. Eng. Asp. 367, 31–40.

Rinaudo, M., 2006. Chitin and chitosan: properties and applications. Prog. Polym. Sci. 31, 603–632.

Rizzello, L., Cingolani, R., Pompa, P.P., 2013. Nanotechnology tools for antibacterial materials. Nanomedicine 8, 807–821.

Sangeetha, G., Rajeshwari, S., Venckatesh, R., 2011. Green synthesis of zinc oxide nanoparticles by *Aloe barbadensis miller* leaf extract: structure and optical properties. Mater. Res. Bull. 46, 2560–2566.

Sawhney, A.P.S., Condon, B., Singth, K.V., Pang, S.S., Li, G., Hui, D., 2008. Modern applications of nanotechnology in textiles. Text. Res. J. 78, 731–739.

Schindler, W.D., Hauser, P., 2004. Antimicrobial finishes. Chemical Finishing of Textiles. Woodhead Publisher, Cambridge, Chapter 15, pp. 165–174.

Selvam, S., Gandhi, R.R., Suresh, J., Gowri, S., Ravikumar, S., Sundrarajan, M., 2012. Antibacterial effect of novel synthesized sulfate β-cyclodextrin crosslinked cotton fabric and its improved antibacterial activities with ZnO, TiO_2, and Ag nanoparticles coating. Inter. J. Pharm. 434, 366–374.

Shahid, M., Isalam, S., Mohammad, F., 2013. Recent advancements in natural dye applications: a review. J. Clean. Prod. 53, 310–331.

Shankar, S.S., Rai, A., Ahmad, A., Sastry, M., 2004. Rapid synthesis of Au, Ag, and bimetallic Au core–Ag shell nanoparticles using neem (*Azadirachta indica*) leaf broth. J. Colloid Interface Sci. 275, 496–502.

Sharma, V.K., Yngard, R.A., Lin, Y., 2009. Silver nanoparticles: green synthesis and their antimicrobial activities. Adv. Colloid Interface Sci. 145, 83–96.

Shore, J., 1998. Blends Dyeing. Society of Dyers and Colorists, Bradford.

Sileikaite, A., Prosycevas, I., Puiso, J., Juraitis, A., Guobiene, A., 2006. Analysis of silver nanoparticles produced by chemical reduction of silver salt solution. Mater. Sci. (Medziagotyra). 12, 287–291.

Simoncic, B., Tomsic, B., 2010. Structure of novel antimicrobial agents for textiles—a review. Text. Res. J. 80, 1721–1737.

Singh, M., Sinha, I., Mandal, R.K., 2009. Role of pH in the green synthesis of silver nanoparticles. Mater. Lett. 63, 425–427.

Son, Y.A., Kim, B.S., Ravikumar, K., Lee, S.G., 2006. Imparting durable antibacterial properties of cotton fabrics using quaternary ammonium salts through 4-aminobenzenesulfonic acid-chloro-triazine adduct. Eur. Polym. J. 42, 3059–3067.

Sondi, I., Goia, D.V., Matijevic, E., 2003. Preparation of highly concentrated stable dispersions of uniform silver nanoparticles. J. Colloid Interface Sci. 260, 75–81.

Su, C.-H., Wu, P.-L., Yeh, C.-S., 2003. Sonochemical synthesis of well-dispersed gold nanoparticles at the ice temperature. J. Phys. Chem. B. 107, 14240–14243.

Sun, Y.Y., Sun, G., 2001. Novel regenerable *N*-halamine polymeric biocide. III. Grafting hydantoin-containing monomers onto synthetic fabrics. J. Appl. Polym. Sci. 81, 1517–1525.

Thomas, V., Bajpai, M., Bajpai, S.K., 2011. *In situ* formation of silver nanoparticles with chitosan-attached cotton fabric for antibacterial property. J. Ind. Text. 40, 229–245.

Twu, Y.K., Chen, Y.W., Shih, C.M., 2008. Preparation of silver nanoparticles using chitosan suspension. Powder Technol. 185, 251–257.

Usha, R., Prabu, E., Palaniswamy, M., Venil, C.K., Rajendran, R., 2010. Synthesis of metal oxide nano particles by *Streptomyces* sp. for development of antimicrobial textiles. Global J. Biotechnol. Biochem. 5, 153–160.

Vigneshwaran, N., Nachane, R.P., Balasubramanya, R.H., Varadarajan, P.V., 2006a. A novel one-pot "green" synthesis of stable nanoparticles using soluble starch. Carbohyd. Res. 341, 2012–2018.

Vigenshwaran, N., Kumar, S., Kathe, A.A., Varadarajan, P. V., Prasad, V., 2006b. Functional finishing of cotton fabrics using zinc-oxide soluble starch nanocomposites. Nanotechnology 17, 5087–5095.

Vigneshwaran, N., Ashtaputre, N.M., Varadarajan, P.V., Nachane, R.P., Paralikar, K.M., Balasubramanya, R.H., 2007. Biological synthesis of silver nanoparticles using the fungus *Aspergillus flavus*. Mater. Lett. 61, 1413–1418.

Vigo, T.L., 1994. Fabrics with improved aesthetic and functional properties. Textile Processing and Properties: Preparation, Dyeing, Finishing and Performance. Elsevier, Netherlands, Chapter 4, pp. 206–272.

Wallace, M.L., 2001. Testing of efficacy of polyhexamrthylene biguanide as an antimicrobial treatment for cotton fabric. AATCC Rev. 1, 18–20.

Wang, H., Qiao, X., Chen, J., Ding, S., 2005. Preparation of silver nanoparticles by chemical reduction method. Colloid Surf. A-Physicochem. Eng. Asp. 256, 111–115.

Williams, J.F., HaloSource, V., Cho, U., 2005. Antimicrobial functions of synthetic fibers: recent developments. AATCC Rev. 5, 17–21.

Windler, L., Height, M., Nowack, B., 2013. Comparative evaluation of antimicrobials for textile applications. Environ. Inter. 52, 62–73.

Wu, C.C., Chen, D.H., 2007. A facile and a completely green route for synthesis gold nanoparticles by the use of drink additives. Gold Bull. 40, 206–212.

Xiong, J., Wang, Y., Xue, Q., Wu, X., 2011. Synthesis of highly stable dispersions of nanosized copper particles using L-ascorbic acid. Green Chem. 13, 900–904.

Yao, S.Y., Lee, H.J., Jeong, S.H., 2003. Preparation of nano composite fibers for permanent antibacterial effects. J. Mater. Sci. 38, 2143–2147.

Zhang, J.J., Gu, M.M., Zheng, T.T., Zhu, J.J., 2009. Synthesis of gelatin-stabilized gold nanoparticles and assembly of carboxylic single-walled carbon nanotubes/Au composites for cytosensing and drug uptake. Anal. Chem. 81, 6641–6648.

Zhang, Z.T., Chen, L., Ji, J.M., Huang, Y.L., Chen, D.H., 2003. Antibacterial properties of cotton fabrics treated with chitosan. Text. Res. J. 73, 1103–1106.

Zhao, T., Sun, G., 2006. Antimicrobial finishing of wool fabrics with quaternary aminopyridinium salts. J. Appl. Polym. Sci. 103, 482–486.

Zhou, X., Xie, X.-X., Jiang, Z.-Y., Kuang, Q., Zhang, S.-H., Xu, T., et al., 2005. Formation of ZnO hexagonal micropyramids: a successful control of the exposed polar surfaces with the assistance of an ionic liquid. Chem. Commun. 2005 (44), 5572–5574.

CHAPTER

13

Complexes of Metal-Based Nanoparticles with Chitosan Suppressing the Risk of *Staphylococcus aureus* and *Escherichia coli* Infections

Dagmar Chudobova[1], *Kristyna Cihalova*[1], *Pavel Kopel*[1,2], *Lukas Melichar*[1], *Branislav Ruttkay-Nedecky*[1,2], *Marketa Vaculovicova*[1,2], *Vojtech Adam*[1,2] *and Rene Kizek*[1,2]

[1]Department of Chemistry and Biochemistry, Faculty of Agronomy, Mendel University in Brno, Zemedelska, Brno, Czech Republic, European Union [2]Central European Institute of Technology, Brno University of Technology, Technicka, Brno, Czech Republic, European Union

13.1 INTRODUCTION

Nanoscale science shows growing potential not only in industry but also in biological and medicinal applications. Nanomedicine is a relatively new part of medicine dealing with nanoparticles that can be used for disease detection, prevention, and treatment of dangerous diseases such as cancer (Yang et al., 2012). Inorganic nanoparticles have many different applications in medicine, including their use as drug (Hajipour et al., 2012) or gene delivery complexes, therapeutic hyperthermia agents, and contrast agents in diagnostic systems. Modern imaging techniques based on nanoparticles allow detection of changes in tissues and organs. For example, magnetic particles are an appropriate approach in both imaging and targeted transport of drugs. In addition, a combination of magnetic particles and quantum dots (QDs) facilitates accurate localization of a tumor.

In vivo imaging is used for noninvasive evaluation of structure and body organ function in a living organism. There are many techniques including optical fluorescence, positron emission tomography (PET), single photon emission computed tomography (SPECT), magnetic resonance imaging (MRI), magnetic resonance

Nanotechnology in Diagnosis, Treatment and Prophylaxis of Infectious Diseases.
DOI: http://dx.doi.org/10.1016/B978-0-12-801317-5.00013-X

spectroscopy (MRS), and computed tomography (CT) to investigate the anatomical or functional changes in tissues. In these techniques, contrast agents provide the imaging signal, for example, fluorophores for optical imaging, radioisotopes for SPECT, and paramagnetic particles for MRI and high-density molecules for CT.

MRI is one of the most powerful and noninvasive diagnostic techniques for living organisms based on the interaction of protons with the surrounding molecules of tissues. The first MRI contrast agents were paramagnetic complexes of gadolinium (Gd^{3+}) and manganese (Mn^{2+}) (Kim et al., 2009; De et al., 2011). Recently, magnetic nanoparticles (MNPs) were also considered as contrast agents for MRI because they take advantage of property changes that occur when materials are nano-scaled (Cormode et al., 2013). *In vivo* investigations of bacterial infections previously relied mostly on bioluminescence imaging; however, *in vivo* bacteria tracking (Hoerr et al., 2013) or monitoring of progress of bacterial diseases (Ali et al., 2014) by MRI has been utilized. The fluorescence microscopy and flow cytometry are based on fluorescence marks utilizing externally added fluorophores that selectively bind to specific targets in tissues. The near-infrared region of the spectrum especially offers certain advantages for photon penetration and semiconductor QDs are now used for conjugation to targeting molecules (Leblond et al., 2010) as well as bacteria imaging (Wang et al., 2011; Zhu et al., 2012; Ong et al., 2014). Very often, multimodal imaging is used because a combination of two or more techniques can bring better results reducing disadvantages, which one of the techniques can possess. The combination improves diagnostic and therapeutic monitoring abilities. Most popular nanostructured multimodal imaging probes are combinations of MRI and optical imaging modalities (Kim et al., 2009).

Nanoparticles can also be used for targeted transport of drugs in organisms with the aim to reduce side effects of drugs, improve effectiveness, and increase circulation in the blood stream. Liposomes are the oldest and the most common type of lipid-based nanostructures used for biomedical applications (Allen and Cullis, 2013). Liposomes have the advantage of being able to contain hydrophobic drugs at the lipid bilayer itself and hydrophilic drugs inside the lipid bilayer; furthermore, cationic liposomes electrostatically bind anionic nucleic acids to their surface (Namiki et al., 2011). Delivery devices made from polymers are an attractive option because they degrade and gradually disappear after delivery. Of these polymers, poly (ε-caprolactone), poly(lactic acid), poly(glycolic acid), and their copolymers have been among the most extensively researched due to their biocompatibility, biodegradability, and regulatory approval (Brewer et al., 2011). Other polymers based on biological polysaccharides have been extensively investigated, including chitosan, cyclodextrin, and dextrans (Malam et al., 2009). Carbon nanotubes have also attracted attention because of their unique properties as one of the most promising nanomaterial for a variety of biomedical applications. Application of carbon nanotubes for the delivery of drugs to their site of action has become one of the main areas of interest of many researchers (Madani et al., 2011; Zhang et al., 2013).

13.2 SYNTHESIS OF METAL NANOPARTICLES, CHARACTERIZATION, AND MODIFICATION

In the past few years, MNPs have become a very useful tool in a variety of research areas such as biotechnology, biomedicine, MRI, in waste water treatment, and in information technology. The superparamagnetic nanoparticles have a fast response to a magnetic field. They randomize their directions and become neutral again almost immediately after the field is turned off because the thermal energy will flip the dipoles in random directions (Jeong et al., 2007; Lu et al., 2007). Magnetic particle size, biocompatibility, and monodispersity are critical parameters determining their use in biomedical applications (Tartaj et al., 2003). There are different synthesis methods such as co-precipitation, thermal decomposition, micro-emulsion, sol–gel, or microwave synthesis. Co-precipitation is a convenient way to synthesize iron oxides (either Fe_3O_4 or γ-Fe_2O_3) from aqueous Fe^{2+}/Fe^{3+} salt solutions by the addition of a base under inert atmospheres at room temperature or at higher temperatures (Lu et al., 2007). For example, ultra-small magnetite particles (2–4 nm) can be prepared by reaction of $FeCl_3$ and $FeCl_2$ aqueous solutions with a concentration ratio of 2:1 and drop-by-drop addition into 200 mL of an alkali solution under vigorous stirring for 40 min at a controlled temperature (20°C, 40°C, 60°C, or 80°C). Particles show high crystallinity and superparamagnetism (Wu et al., 2008). Nigam et al. (2011) developed a single-step process for preparation of biocompatible citric acid–functionalized Fe_3O_4 aqueous colloidal MNPs. Citrate-stabilized MNPs could offer an effective tool for hyperthermia treatment. Positively charged drugs can be trapped onto the surface of negatively charged nanoparticles through electrostatic interactions. Drug-bound molecules can then be released in the acidic environment of the tumor (Nigam et al., 2011; Behdadfar et al., 2012).

Thermal decomposition is another synthesis route for smaller nanoparticle preparation. For example, with decomposition of metal acetylacetonates in boiling organic solvents in the presence of stabilizing surfactants, the uniform MNPs can be prepared (Roca et al., 2006; Behdadfar et al., 2012). MNPs approximately 9 nm in size have been synthesized in water using the hydrothermal reduction method in the presence of citric acid as a nonexpensive and nontoxic reducing agent and stabilizer (Behdadfar et al., 2012).

In the absence of any surface coating, MNPs have hydrophobic surfaces with a large surface area-to-volume ratio. Due to the hydrophobic interactions, the particles agglomerate and form large clusters, resulting in increased particle size and thus sometimes lose the superparamagnetic properties. For effective stabilization of MNPs, often a coating is desirable. The coating procedure can be performed either during or after the synthesis of the particle. The coating also provides required properties for subsequent bioconjugation and further functionalization. Various biological molecules such as antibodies, proteins, targeting ligands, and others may be bound to the surfaces of nanoparticles by chemically coupling via amide or ester bonds to make the particles target-specific (Akbarzadeh et al., 2012).

In general, synthesis of different kinds of nanoparticles is as follows: any synthesis starts with a mixture of metal salts and usually capping ligands, after which the addition of a reducing agent or a change in pH causes formation of nanoparticles. Sometimes light irradiation, heating, or microwaves are necessary for synthesis. For example, syntheses of gold nanoparticles in water solution are based on the Turkevich method (Turkevich et al., 1951). A solution of chloroauric acid is heated to boiling and sodium citrate is added, reducing the gold ions and forming nanoparticles in a few minutes. The resulting citrate-capped nanoparticles are in a narrow size range and, according to the conditions, particles in the range of 15–150 nm can be produced. The citrate-capped gold nanoparticles are not too stable, but the capping citrate can be easily displaced with the desired coating material to make stable, biocompatible nanoparticles. The Turkevich method can be modified and thus nanoparticles of different shapes, such as nanorods or stars, can be prepared (Jana et al., 2001; Nehl et al., 2006). Gold nanoparticles can be covered with a gadolinium complex of cysteine diethylenetriaminepentaacetic acid because sulfur binds to gold very strongly (Park et al., 2010). MRI-active gold nanoparticles were also prepared directly by reduction of chloroauric acid in water in the presence of a thiol-containing gadolinium complex (Alric et al., 2008; Park et al., 2008; Cormode et al., 2013).

Nanoparticles can be prepared using various physicochemical methods (Mahl et al., 2012), but their synthesis using nontoxic and environmentally friendly methods is attractive, especially if they are intended for invasive applications in medicine. Several methods have been developed for biological or biogenic synthesis of nanoparticles from salts of the corresponding metals. Thus, microorganisms, plants, plant tissue, fruits, plant extracts, and marine algae can be used to produce nanoparticles. The reducing agents involved include the various water-soluble plant metabolites (e.g., alkaloids, phenolic compounds, terpenoids) and coenzymes. Silver and gold nanoparticles have been the particular focus of plant-based syntheses. Extracts of a diverse range of plant species have been successfully used in making nanoparticles (Mittal et al., 2013).

The optical properties of QDs depend on their size and their composition, which allows the tuning of the emission wavelength. Organic solvents or water can be used for synthesis of QDs. In the case of organic solvents, QDs are covered with hydrophobic ligands and ligand substitution is then necessary to prepare biocompatible QDs for medicinal purposes. As an example of organic solvent preparation is the hot injection method (Park et al., 2007); dimethyl cadmium and trioctylphosphine selenide are injected into the hot (300°C) trioctylphosphine oxide (TOPO) solvent. The resulting QDs are capped with TOPO and are dispersible in organic solvents. QDs can be prepared under reflux that requires a long reaction time and there are many surface defects on synthesized QDs that result in lower photoluminescence quantum yield. That is why the microwave synthesis is very often used for production of high-quality QDs in a much shorter time (Zhu et al., 2002; Huang and Han, 2010). The size of QDs can be easily tuned by changing the parameters of the microwave heating. The growth of the QDs stops when the microwave irradiation is turned off and the product is cooled down. The most frequent types of QDs synthesized using microwave irradiation are CdTe, CdSe, $Zn_{1-x}Cd_xSe$, and ZnSe. These QDs are mostly covered with thiol ligands such as mercaptopropionic acid, mercaptosuccinic acid, or glutathion (Qian et al., 2006; Ryvolova et al., 2012; Krejcova et al., 2013).

13.3 INTERACTION OF METAL NANOPARTICLES WITH CELL COMPONENTS AFFECTING CELLULAR PROCESSES

Sondi and Salopek-Sondi (2004) were among the first to describe silver nanoparticles (AgNPs) activity against *Escherichia coli* and to propose a likely explanation of observed effects. The authors revealed formation of "pits" in the bacterial cell wall and accumulation of AgNPs in the cellular membrane that led to an increase of its permeability and eventually to the death of bacterial cells. Also, they attempted to understand their mechanism of action. Presently, there are three main explanations that have been proposed to describe the antibacterial activity: (i) direct interaction of AgNPs with the bacterial cell membrane, causing subsequent membrane damage and complexation with components located inside the cells; (ii) interaction with thiol (−SH) groups and production of reactive oxygen species (ROS) (Banerjee et al., 2010); and (iii) release of silver ions that inhibit respiratory enzymes and also generate ROS (Pal et al., 2007).

Size-dependent activity of AgNPs was reported in many articles with smaller NPs having higher activity; this result may be explained by a relative increase in contact surface area (Liu et al., 2010). Shameli et al. (2012) demonstrated the antibacterial activity of different sizes of AgNPs in polyethylene glycol against Gram-positive (*Staphylococcus aureus*) and Gram-negative (*Salmonella typhimurium*) bacteria by the disc diffusion method. They reported the significant inhibition in growth of both these pathogens and concluded that the antibacterial activities of AgNPs in polyethylene glycol can be modified by controlling the size of nanoparticles because the activity of AgNPs decreases with the increase in the particle size.

Similarly, in bacteria, disturbances in membrane penetration may also lead to internalization of nanoparticles and subsequent intracellular effects, including ROS generation, interaction with –SH groups, inhibition of protein synthesis, and interaction with phosphorus-containing molecules, such as DNA.

The general understanding of the mode of action of AgNPs is that the NPs become attached to the sulfur-containing proteins on the bacterial cell wall, leading to increased permeability of the membrane, and finally causing cell death (Banerjee et al., 2010). In addition, Ag^{+} ions have been reported to induce generation of intracellular ROS in bacterial cells (Morones et al., 2005). A few reports have suggested Ag^{+} ions as the actual biocidal species, which are provided by active surfaces of AgNPs and silver oxide present on the surfaces of these nanoparticles. The silver ions enter the bacterial cells, where they are reduced as the cell attempts to remove them from the cell interior, eventually leading to cell destruction (Smetana et al., 2008). However, Banerjee et al. (2010) have recently demonstrated that AgNPs less than 10 nm in diameter make pores in the bacterial cells walls, thereby releasing the cytoplasmic content to the medium and leading to the cell death without affecting the intracellular and extracellular proteins and nucleic acids of the bacterium (Banerjee et al., 2010). A schematic summary of multiple bacterial actions of Ag^{+} ions or AgNPs is shown in Figure 13.1.

13.4 BIOCHEMICAL MECHANISM OF TOXICITY TO PROKARYOTIC CELLS

Use of metal nanoparticles represent a suitable tool for protection against bacterial infection, which is big threat for the human body and its immune system. It can lead to removal of extremities. Bacterial infections are caused by a wide range of pathogenic bacteria like *Staphylococcus* (its resistant varieties MRSA, VRSA), *E. coli*, and *Streptococcus*. *Staphylococci* dispose with the resistance to many antibiotic drugs like β-lactam antibiotics; especially dangerous varieties are methicillin-resistant *S. aureus* (MRSA) and vancomycin-resistant *S. aureus* (VRSA). Because of the increasing resistance ability of bacteria, it is necessary to focus research on the search for new antimicrobial drugs (Chudobova et al., 2014b). Substitutions for antibiotics must be bactericidal and safe for the human body at the same time. Mechanisms of inhibition must be different from that of antibiotics.

Toxicity of metal nanoparticles is caused by interaction of specific metals with cell structures (Rouch et al., 1995). Metals enter the cell in two different ways (Nies and Silver, 1995). Nonspecific transporters allow metals to enter the cell via chemoosmotic gradient; this way allows transporting of heavy metals as well (Schreurs and Rosenberg, 1982). This path provides fast transport of metal to the cells through the plasmatic membrane (Nieboer and Richardson, 1980). It is able to transport metal nanoparticles from the cell in case of excess. This system is called "open gate," and it explains the cause of the metal toxicity. Another type of metal transport is specific, slower than "open gate," consumes energy, and is used only when necessary (lack of nutrients) (Nies, 1999).

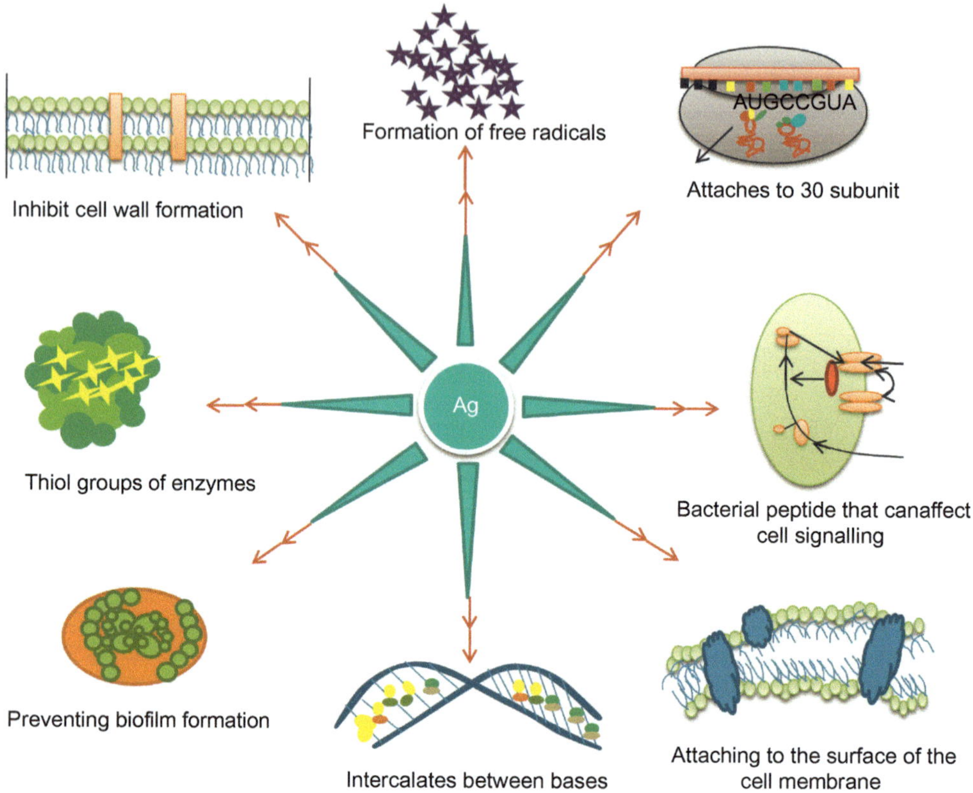

FIGURE 13.1 AgNPs showing multiple bactericidal actions. *Adapted from Rai et al. (2012).*

The main mechanism of metal toxicity—lipid peroxidation—involves the reaction of oxygen with polyunsaturated fatty acids (parts of cell membrane), which forms oxygen radicals. Other structures of the cells interact and are damaged by oxygen radicals.

13.5 OXIDATIVE STRESS AND FORMATION OF ROS BY METAL NANOPARTICLES

The most widely known bactericidal mechanism of the silver ion is its interaction with the thiol groups of the L-cysteine residues of proteins and consequent inactivation of their enzymatic functions (Liau et al., 1997). Although some other bactericidal mechanisms of silver ions such as the release of potassium (Russell and Hugo, 1994) and bonding to DNA (Arakawa et al., 2001) have also been reported, intracellular ROS generation by silver ions is strongly supported by the following reasons: (i) silver ions were found to induce an increase in the respiration rate because of the uncoupling of the respiratory chain, followed by cessation of respiration (Matsumura et al., 2003; Holt and Bard, 2005) and (ii) the interaction between silver ions and the thiol group could interrupt essential enzymes in the respiratory chain, such as NADH and succinate dehydrogenase, thereby obstructing adequate electron transfer to oxygen (Bragg and Rainnie, 1974; Messner and Imlay, 1999). Although these are plausible

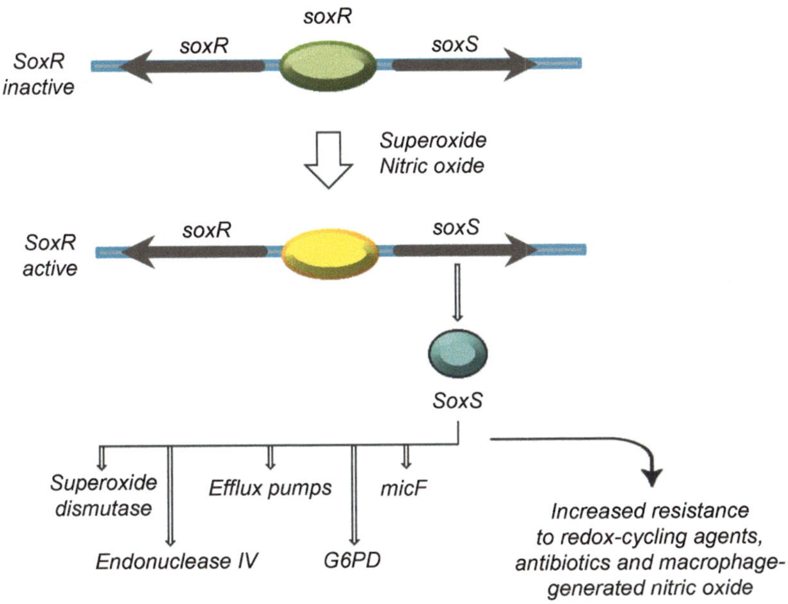

FIGURE 13.2 The soxRS regulon. The soxRS locus is composed of the divergently transcribed soxR and soxS genes. The SoxR protein is produced constitutively and is activated on exposure to superoxide-generating agents or nitric oxide (NO). The oxidized form of SoxR enhances the transcription of the soxS gene, the product of which is also a transcriptional activator. The SoxS protein activates transcription of genes that increase the resistance to oxidants. Additionally, activation of the SoxS-regulated genes increases the resistance to antibiotics and macrophage-generated NO. G6PD, glucose-6-Pdehydrogenase.

reasons for ROS generation, very little experimental evidence of silver-induced intracellular ROS generation and consequent bactericidal activity have been demonstrated. These mechanisms of the silver effect can be exerted in combination or otherwise, specifically in certain conditions where one mechanism occurs but the other does not, for example, thiol interaction but no ROS generation in anaerobic conditions. Therefore, to fully develop the antimicrobial activity of silver ions, more research to quantify the contribution of each mechanism and to determine the interrelations between the mechanisms is necessary. However, these attempts are still restricted by the limited understanding of silver–ion-mediated intracellular ROS generation.

ROS are a group of short-lived reactive oxidants that include the superoxide radical (O_2^-), hydroxyl radical (OH), hydrogen peroxide (H_2O_2), and singlet oxygen ($^1O_2$). ROS can be generated directly or indirectly inside cells, and oxidative stress results from an imbalance between ROS generation and cellular defensive functions, including those of antioxidant enzymes and antioxidants. Oxidative stress engenders many problems in cells, such as protein damage, DNA damage, and lipid peroxidation (Sies, 1997), and it is the major mechanism for bacterial killing by many drugs and antibiotics (Kohanski et al., 2007).

E. coli has been widely used as a model system to study the oxidative stress–related bactericidal phenomenon. *E. coli* has two distinct oxidative stress–sensory systems, namely, SoxR and OxyR, which are sensor-regulator proteins that respond to superoxide radicals, nitric oxide, and hydrogen peroxide by activating the defense regulon genes (Bauer et al., 1999; Storz and Imlay, 1999; Pomposiello and Demple, 2001). SoxR, a superoxide and nitric oxide radical sensor protein, induces only one gene, *soxS*, on exposure to superoxide radicals and nitric oxide. The SoxS protein acts as a transcription factor and then activates a set of superoxide-inducible genes, including *sodA* (superoxide dismutase), *acnA* (aconitase A), and *nfo* (endonuclease IV) (Figure 13.2).

However, OxyR is activated by hydrogen peroxide and alkyl hydroperoxide, and it induces the expression of genes, including *oxyS* (small regulatory RNA), *katG* (catalase), *ahpCF* (hydroperoxide reductase), *grxA* (glutaredoxin I), and *gorA* (glutathione reductase). The activities of these sensor proteins have been assayed, mostly by using the reporter fusions of their direct targets, that is, *soxS* and *oxyS* promoters, to monitor the intracellular ROS generation (Koo et al., 2003).

13.6 ANTIBACTERIAL EFFECT OF METAL NANOPARTICLES IN COMPLEX WITH CHITOSAN

Nanotechnology represents the ability to design, manipulate, and model functionalities on the nanometer scale. This discipline includes the study of nanoparticles, which can be classified as particles with a size less than 100 nm. Such particles with an antimicrobial function have received considerable attention within a range of diverse fields, including medicine and dentistry. For example, the antimicrobial properties of metal nanoparticles have been suggested to be due to their size and high surface-to-volume ratio. In theory, these characteristics should allow them to interact closely with microbial membranes and thus elicit an antimicrobial effect that is not solely due to the release of metal ions (Allaker and Memarzadeh, 2014).

Metal nanoparticles interact with cellular components (DNA, RNA, and ribosomes), deactivate, and effectively inhibit the cellular processes (Das et al., 2013). From the point of view of antimicrobial activity, the excellent results exhibit the metal nanoparticles at nanomolar concentrations. It is necessary to obtain the correct dimensions of the nanoparticles to avoid agglomeration, which significantly reduce the antimicrobial effect. The general mechanism of metal nanoparticle action has not been fully understood. However, several theories are known on the basis of damage of the microbial enzymes by the release of metal ions, change in the integrity of the membrane, penetration into the bacterial cytoplasm, and accumulation in the periplasmic space or the formation of ROS by the effect of metal nanoparticles. Metal nanoparticles display more antimicrobial effects on G^+ bacteria than on G^- bacteria. The effects are reflected by inhibition of cell division. The aim of this study is to compare the antimicrobial properties of different types of metal nanoparticles (silver, selenium, copper, or zinc nanoparticles) on the bacterial strains *S. aureus* and *E. coli*.

13.6.1 Chitosan

Chitosan is a cationic polysaccharide polymer composed of glucosamine and *N*-acetyl glucosamine linked with a β-1-4-glycosidic linkage obtained from chitin by deacetylation of chitin in the presence of alkali (Crini and Badot, 2008; Bonilla et al., 2013). Chitin is a biopolymer presented in the exoskeleton of crustaceans such as crabs, shrimp, and crawfish (Choi et al., 2001; Andrews et al., 2002). The positive charge of chitosan affords the polymer numerous physiological and biological properties (Honary et al., 2011). It is a biopolymer that is biocompatible and can be degraded by enzymes in the human body and the degradation products are nontoxic. Chitosan films and complexes have a great potential to be used as active material in medicine because of its antimicrobial activity, nontoxicity, and low permeability to oxygen (Kanatt et al., 2012). It has attracted considerable interest because of its antimicrobial and antitumor activities and its immune-enhancing effects (Suzuki et al., 1986; Jeon et al., 2001; No et al., 2002).

The exact mechanism for the antibacterial activity is not yet fully understood. Some studies have suggested that a positive charge on the NH_3^+ group of the glucosamine monomer at pH <6.3 allows interactions with negatively charged microbial cell membranes that lead to the leakage of intracellular constituents (Papineau et al., 1991; Sudarshan et al., 1992; Helander et al., 2001; Liu et al., 2004).

Other studies suggest that polymer exhibits antimicrobial activity by binding to the negatively charged bacterial cell wall, followed by attachment to the DNA, inhibiting its replication (Kurita et al., 1993). Also, the molecular weight and concentration of the used chitosan influences the antimicrobial effect (Honary et al., 2011; Sanchez-Gonzalez et al., 2011). For improvements in the bioactivity of chitosan, it is often combined with other bioactive materials such as drugs. However, there is no direct evidence for such effect.

In the study by Banerjee et al. (2010), synergy in antimicrobial activity of a chitosan–AgNP composite in the presence of molecular iodine was reported. Green fluorescent protein expressing recombinant *E. coli* bacteria have been used to test the efficacy and establish the mechanism of action. Experimental evidence indicated significantly high bactericidal activity of the nanocomposite in the presence of iodine due to the composite, chitosan, AgNPs, or iodine only. Transmission electron microscopy measurements revealed attachment of bacteria to the composite. In addition, flow cytometry results supported definite occurrence of cell wall damage of the bacteria treated with the composite in the presence of iodine. Further, the nanocomposite and iodine combination was found to exert ROS-generated oxidative stress in the cytoplasm of bacterial cells, leading to cell death. Elucidation of the mechanism of synergy due to three potential antibacterial components suggested that on the surface of AgNPs, molecular iodine possibly generated iodine atoms, thus contributing to free radical–induced oxidative stress, whereas chitosan and AgNPs facilitated the process of cell killing and thus collectively enhanced the potency of antimicrobial effects at the lowest concentrations of individual components (Figure 13.3).

In another study, Mathew and Kuriakose (2013) investigated the antibacterial effect of AgNP encapsulated in chitosan against *S. aureus* and *E. coli* and against fungal species such as *Aspergillus flavus* and *Aspergillus terreus*. This study showed that AgNP encapsulated in chitosan can be used for antibacterial and antifungal applications. In a further study, hybrid ZnO/chitosan nanoparticles were generated on cotton fabrics, and they showed higher antimicrobial activity against *S. aureus* and *E. coli* when compared with ZnO NPs alone (Petkova et al., 2014).

13.7 EFFECT OF METAL NANOPARTICLES IN SPECIFIC EXAMPLES

13.7.1 Silver Nanoparticles

More than 1,300 nanotechnology-enabled products have already entered the market. Silver is well-known for its significant antimicrobial activity and therapeutic potential. By far, the most commonly utilized commercial nanomaterials are antimicrobial AgNPs (Klasen, 2000; Silver, 2003; Ip et al., 2006). The release of silver ions by the nanoparticles has been proposed, and these ions can interact with many vital enzymes and inactivate them (Shi et al., 2004). Silver ions enter the bacterial cell; they inhibit several functions in the cell and damage the cell membrane. Even low concentrations of silver are very effective as a germicidal agent. The antimicrobial effect of silver ions or AgNPs was confirmed by several studies (Chudobova et al., 2013a, b, 2014a).

There is a wide range of microorganisms sensitive to silver. Silver has been identified to possess a good potential for treating cancer. Long-term exposure of AgNPs larger than 20 nm to eukaryotic cells may affect them negatively. A combination of AgNPs with a biopolymer like

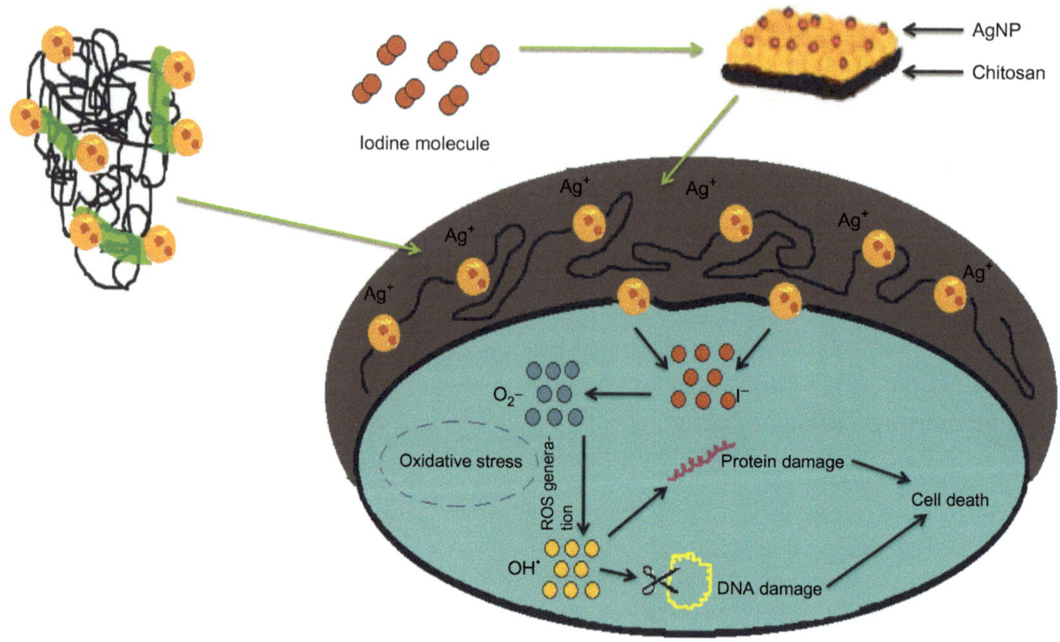

FIGURE 13.3 Schematic representation of the proposed mechanism of antibacterial activity of the iodinated chitosan–AgNP composite. *Adapted from Banerjee et al. (2010).*

chitosan is a possible solution for its eukaryotic or prokaryotic toxicity (Chudobova et al., 2013b).

It was observed that AgNPs can be used as an effective growth inhibitors for various microorganisms, and AgNPs possess high antibacterial activity against Gram-positive and Gram-negative bacteria (Hetrick et al., 2008; Chudobova et al., 2013a, b, 2014a). An effective biocidal concentration has a nanomolar level, in contrast to a micromolar level, of silver ions. Silver particles without coating are highly reactive and easily oxidized by air, leading to loss of dispersibility and antimicrobial effects (Kim et al., 2007; Sharma et al., 2009). It is possible to stabilize bare silver particles against degradation by bonding with other agents.

13.7.2 Selenium Nanoparticles

Selenium is an essential micro-mineral element that is relatively deficient in most areas of the world. Selenium has a large number of biological functions in the human organism. The most important and known action is its antioxidant effect because it forms selenocysteine, part of the active center of the glutathione peroxidase enzyme (Kryukov et al., 2003). In these enzymes, selenium is a catalyst of the oxidation of glutathione and the reduction of hydrogen peroxide to water. Its functions include regulation of metabolism, improvement of immunity, enhancement of reproductive performance, cancer prevention, resistance to free radicals, detoxification of some heavy metals, and prevention of some local epidemic diseases (Navarro-Alarcon and Lopez-Martinez, 2000; Mater et al., 2005).

Selenium is a naturally occurring essential element that is needed for a healthy life; the FDA recommends daily intake of selenium. A deficit of selenium is associated with an increased risk of mortality, poor immune function, and cognitive decline. Selenium particles are able to damage bacterial biofilm and inhibit bacterial growth (Tran et al., 2009; Tran and Webster, 2011; Chudobova et al., 2014a). Selenium atoms or complex substances can generate a superoxide that is toxic to cancer cells, bacteria, and viruses (Spallholz et al., 2001).

Selenium-modified materials are able to show antimicrobial effects (Chudobova et al., 2014a). Selenium has been studied for various medical applications and as a potential material for orthopedic implants (Perla and Webster, 2005) and anticancer applications. Selenium as a dietary supplement has been demonstrated to reduce the risks of various types of cancers, including prostate cancer, lung cancer, and esophageal and gastric-cardiac cancers (Clark et al., 1996, 1998; Wei et al., 2004; Rayman, 2005; Yang et al., 2009). Selenium-enriched probiotics have been shown to strongly inhibit the growth of pathogenic *E. coli in vivo* and *in vitro*.

13.7.3 Copper Nanoparticles

Copper nanoparticles demonstrate extraordinary performance as antibacterial and antimicrobial agents (Majzlik et al., 2010, 2011). Nanoparticles are transferred to various substrates to be used in practical applications. The most commonly used substrates are polymeric beads prepared from methacrylic acid and divinylbenzene, silica-coated hollow polystyrene beads, polystyrene–divinylbenzene ion exchangers, activated carbon granules, carbon aerogels, and silica beads. These substrates have been impregnated *ex situ* with metal nanoparticles (Khare et al., 2014). Copper nanoparticles play a dual role: (i) they act as an antibacterial agent, although less effectively than other metal nanoparticles and (ii) they enhance the porosity (internal surface area and pore volume) in the beads such that the dispersed AgNPs in the beads are relatively more accessible to bacteria (Khare et al., 2014).

In comparison with silver, few studies have reported the antimicrobial properties of copper. Copper may have a similar principle of action as that of silver; however, the precise mechanism regarding how copper nanoparticles cause activity against microorganisms is unclear. As with silver, it is thought that copper acts by combining with the –SH groups of key microbial enzymes. Studies demonstrated superior antimicrobial activity of copper nanoparticles against *E. coli* and spore-forming *Bacillus subtilis* compared with AgNPs (Allaker and Memarzadeh, 2014).

In recent studies, AgNPs were also used to prepare a metal-doped TiO_2 photocatalyst for bacterial (*E. coli*) disinfection under visible light irradiation. However, AgNPs are expensive. The use of a cheaper metal, such as copper, instead of expensive silver in a metal-doped TiO_2 photocatalyst without a significant decrease in the disinfection activity could contribute to the development of cheaper photocatalysts for practical uses. Copper ions and copper nanoparticles have exhibited high antibacterial activity against a wide range of bacteria, including *Salmonella enterica*, *Campylobacter jejuni*, *E. coli*, *Listeria monocytogenes*, and *S. aureus*. Copper is also an essential metal element for human health and is considered a low-toxicity metal in human organisms and is used in intrauterine devices (Pham and Lee, 2014).

Copper nanoparticles can be easily mixed with polymers and are relatively stable in terms of both chemical and physical properties. They may be particularly valuable antimicrobial agents because they can be prepared with extremely high surface areas of unusual crystal morphologies (Nezhad et al., 2014).

13.7.4 Zinc Nanoparticles

Zinc is an important trace element in the human body. Zinc shows attractive antimicrobial properties that are utilized in several pharmaceutical and cosmetic products. Zinc is also an endogenous metal that is involved in many physiological phenomena. It is involved *in vivo* in more than 300 enzymatic reactions as a cofactor (Haase et al., 2008). The proposed mechanisms of antibacterial activity with respect to zinc nanoparticles include generation of ROS and damage to the cell membrane with subsequent interaction of the nanoparticle with the intracellular contents (Allaker and Memarzadeh, 2014). Zinc is commonly used in the pharmaceutical industry as carrying agents, smoothing agents, and a protective coating. It is commonly used to treat diaper rash, acne, and minor burns (Arad et al., 1999). An advantage of zinc is its low price and easy clearance (Sawai et al., 1998). The mechanism of antimicrobial toxicity of zinc is caused by the generation of hydrogen peroxide (Sawai et al., 1995). It was found that this method could be applied as a simple means of inferring and classifying the influences and damage induced by physical and chemical stresses, because each stress induces different changes in the sensitivities of bacterial cells to antibiotics (Tapiero and Tew, 2003).

Recent studies confirm the higher concentration of zinc nanoparticles for growth inhibition and the killing effect against a range of pathogens including *E. coli* and MRSA.

13.8 CONCLUSION

The demand for antimicrobial agents with the same effect as antibiotic drugs is rapidly increasing. Metal nanoparticles dispose antimicrobial properties, leading to reduction in the risk of bacterial infection. Currently, testing for their antimicrobial efficacy is incorporated into a range of studies performed by research groups worldwide. Potential toxicity of metal nanoparticles to eukaryotic cells can be solved by the formation of a complex with a biopolymer substance like chitosan. The future of nanotechnology and the use of metal nanoparticles in practice will be directed to ensure maximal antimicrobial effects with minimal toxicity for the host organism.

Acknowledgment

Financial support from the CEITEC CZ.1.05/1.1.00/02.0068 project is gratefully acknowledged.

References

Akbarzadeh, A., Samiei, M., Davaran, S., 2012. Magnetic nanoparticles: preparation, physical properties, and applications in biomedicine. Nanoscale Res. Lett. 7, 1–13.

Ali, S.Z., Srinivasan, S., Peh, W.C.G., 2014. MRI in necrotizing fasciitis of the extremities. Br. J. Radiol. 87. Available from: <http://dx.doi.org/10.1186/1556-276X-7-144>.

Allaker, R.P., Memarzadeh, K., 2014. Nanoparticles and the control of oral infections. Int. J. Antimicrob. Agents 43, 95–104.

Allen, T.M., Cullis, P.R., 2013. Liposomal drug delivery systems: from concept to clinical applications. Adv. Drug Deliv. Rev. 65, 36–48.

Alric, C., Taleb, J., Le Duc, G., Mandon, C., Billotey, C., Le Meur-Herland, A., et al., 2008. Gadolinium chelate coated gold nanoparticles as contrast agents for both X-ray computed tomography and magnetic resonance imaging. J. Am. Chem. Soc. 130, 5908–5915.

Andrews, L.S., Keys, A.M., Martin, R.L., Grodner, R., Park, D.L., 2002. Chlorine dioxide wash of shrimp and crawfish an alternative to aqueous chlorine. Food Microbiol. 19, 261–267.

Arad, A., Mimouni, D., Ben-Amitai, D., Zeharia, A., Mimouni, M., 1999. Efficacy of topical application of eosin compared with zinc oxide paste and corticosteroid cream for diaper dermatitis. Dermatology 199, 319–322.

Arakawa, H., Neault, J.F., Tajmir-Riahi, H.A., 2001. Silver(I) complexes with DNA and RNA studied by Fourier transform infrared spectroscopy and capillary electrophoresis. Biophys. J. 81, 1580–1587.

Banerjee, M., Mallick, S., Paul, A., Chattopadhyay, A., Ghosh, S.S., 2010. Heightened reactive oxygen species generation in the antimicrobial activity of a three component iodinated chitosan–silver nanoparticle composite. Langmuir 26, 5901–5908.

Bauer, C.E., Elsen, S., Bird, T.H., 1999. Mechanisms for redox control of gene expression. Annu. Rev. Microbiol. 53, 495–523.

Behdadfar, B., Kermanpur, A., Sadeghi-Aliabadi, H., Morales, M.D., Mozaffari, M., 2012. Synthesis of high intrinsic loss power aqueous ferrofluids of iron oxide nanoparticles by citric acid-assisted hydrothermal-reduction route. J. Solid State Chem. 187, 20–26.

Bonilla, J., Atares, L., Vargas, M., Chiralt, A., 2013. Properties of wheat starch film-forming dispersions and films as affected by chitosan addition. J. Food Eng. 114, 303–312.

Bragg, P.D., Rainnie, D.J., 1974. Effect of silver ions on respiratory-chain of *Escherichia coli*. Can. J. Microbiol. 20, 883–889.

Brewer, E., Coleman, J., Lowman, A., 2011. Emerging technologies of polymeric nanoparticles in cancer drug delivery. J. Nanomater. Brewer. Available from: <http://dx.doi.org/10.1155/2011/408675> article ID: 408675.

Choi, B.K., Kim, K.Y., Yoo, Y.J., Oh, S.J., Choi, J.H., Kim, C. Y., 2001. *In vitro* antimicrobial activity of a chitooligosaccharide mixture against *Actinobacillus* actinomycetemcomitans and *Streptococcus mutans*. Int. J. Antimicrob. Agents 18, 553–557.

Chudobova, D., Dobes, J., Nejdl, L., Maskova, D., Merlos, M.A.R., Ruttkay-Nedecky, B., et al., 2013a. Oxidative stress in *Staphylococcus aureus* treated with silver(I) ions revealed by spectrometric and voltammetric assays. Int. J. Electrochem. Sci. 8, 4422–4440.

Chudobova, D., Nejdl, L., Gumulec, J., Krystofova, O., Merlos, M.A.R., Kynicky, J., et al., 2013b. Complexes of silver(I) ions and silver phosphate nanoparticles with hyaluronic acid and/or chitosan as promising antimicrobial agents for vascular grafts. Int. J. Mol. Sci. 14, 13592–13614.

Chudobova, D., Cihalova, K., Dostalova, S., Ruttkay-Nedecky, B., Merlos, M.A.R., Tmejova, K., et al., 2014a. Comparison of the effects of silver phosphate and selenium nanoparticles on *Staphylococcus aureus* growth reveals potential for selenium particles to prevent infection. FEMS Microbiol. Lett. 351, 195–201.

Chudobova, D., Dostalova, S., Blazkova, I., Michalek, P., Ruttkay-Nedecky, B., Sklenar, M., et al., 2014b. Effect of ampicillin, streptomycin, penicillin and tetracycline on metal resistant and non-resistant *Staphylococcus aureus*. Int. J. Environ. Res. Public Health 11, 3233–3255.

Clark, L.C., Combs, G.F., Turnbull, B.W., Slate, E.H., Chalker, D.K., Chow, J., et al., 1996. Effects of selenium supplementation for cancer prevention in patients with carcinoma of the skin a randomized controlled trial—a randomized controlled trial. J. Am. Med. Assoc. 276, 1957–1963.

Clark, L.C., Dalkin, B., Krongrad, A., Combs, G.F., Turnbull, B.W., Slate, E.H., et al., 1998. Decreased incidence of prostate cancer with selenium supplementation: results of a double-blind cancer prevention trial. Br. J. Urol. 81, 730–734.

Cormode, D.P., Sanchez-Gaytan, B.L., Mieszawska, A.J., Fayad, Z.A., Mulder, W.J.M., 2013. Inorganic nanocrystals as contrast agents in MRI: synthesis, coating and introduction of multifunctionality. NMR Biomed. 26, 766–780.

Crini, G., Badot, P.M., 2008. Application of chitosan, a natural aminopolysaccharide, for dye removal from aqueous solutions by adsorption processes using batch studies: a review of recent literature. Prog. Polym. Sci. 33, 399–447.

Das, M.R., Sarma, R.K., Borah, S.C., Kumari, R., Saikia, R., Deshmukh, A.B., et al., 2013. The synthesis of citrate-modified silver nanoparticles in an aqueous suspension of graphene oxide nanosheets and their antibacterial activity. Colloids Surf. B Biointerfaces 105, 128–136.

De, M., Chou, S.S., Joshi, H.M., Dravid, V.P., 2011. Hybrid magnetic nanostructures (MNS) for magnetic resonance imaging applications. Adv. Drug Deliv. Rev. 63, 1282–1299.

Haase, H., Overbeck, S., Rink, L., 2008. Zinc supplementation for the treatment or prevention of disease: current status and future perspectives. Exp. Gerontol. 43, 394–408.

Hajipour, M.J., Fromm, K.M., Ashkarran, A.A., de Aberasturi, D.J., de Larramendi, I.R., Rojo, T., et al., 2012. Antibacterial properties of nanoparticles. Trends Biotechnol. 30, 499–511.

Helander, I.M., Nurmiaho-Lassila, E.L., Ahvenainen, R., Rhoades, J., Roller, S., 2001. Chitosan disrupts the barrier properties of the outer membrane of Gram-negative bacteria. Int. J. Food Microbiol. 71, 235–244.

Hetrick, E.M., Shin, J.H., Stasko, N.A., Johnson, C.B., Wespe, D.A., Holmuhamedov, E., et al., 2008. Bactericidal efficacy of nitric oxide-releasing silica nanoparticles. ACS Nano. 2, 235–246.

Hoerr, V., Tuchscherr, L., Huve, J., Nippe, N., Loser, K., Glyvuk, N., et al., 2013. Bacteria tracking by *in vivo* magnetic resonance imaging. BMC Biol. 11. Available from: <http://dx.doi.org/10.1186/1741-7007-11-63>.

Holt, K.B., Bard, A.J., 2005. Interaction of silver(I) ions with the respiratory chain of *Escherichia coli*: an electrochemical and scanning electrochemical microscopy study of the antimicrobial mechanism of micromolar Ag. Biochemistry 44, 13214–13223.

Honary, S., Ghajar, K., Khazaeli, P., Shalchian, P., 2011. Preparation, characterization and antibacterial properties of silver–chitosan nanocomposites using different molecular weight grades of chitosan. Trop. J. Pharm. Res. 10, 69–74.

Huang, L., Han, H.Y., 2010. One-step synthesis of water-soluble ZnSe quantum dots via microwave irradiation. Mater. Lett. 64, 1099–1101.

Ip, M., Lui, S.L., Poon, V.K.M., Lung, I., Burd, A., 2006. Antimicrobial activities of silver dressings: an *in vitro* comparison. J. Med. Microbiol. 55, 59–63.

Jana, N.R., Gearheart, L., Murphy, C.J., 2001. Wet chemical synthesis of high aspect ratio cylindrical gold nanorods. J. Phys. Chem. B. 105, 4065–4067.

Jeon, Y.J., Park, P.J., Kim, S.K., 2001. Antimicrobial effect of chitooligosaccharides produced by bioreactor. Carbohydr. Polym. 44, 71–76.

Jeong, U., Teng, X.W., Wang, Y., Yang, H., Xia, Y.N., 2007. Superparamagnetic colloids: controlled synthesis and niche applications. Adv. Mater. 19, 33–60.

Kanatt, S.R., Rao, M.S., Chawla, S.P., Sharma, A., 2012. Active chitosan–polyvinyl alcohol films with natural extracts. Food Hydrocolloids 29, 290–297.

Khare, P., Sharma, A., Verma, N., 2014. Synthesis of phenolic precursor-based porous carbon beads in situ dispersed with copper-silver bimetal nanoparticles for antibacterial applications. J. Colloid Interface Sci. 418, 216–224.

Kim, J., Piao, Y., Hyeon, T., 2009. Multifunctional nanostructured materials for multimodal imaging, and simultaneous imaging and therapy. Chem. Soc. Rev. 38, 372–390.

Kim, J.S., Kuk, E., Yu, K.N., Kim, J.H., Park, S.J., Lee, H.J., et al., 2007. Antimicrobial effects of silver nanoparticles. Nanomed. Nanotechnol. Biol. Med. 3, 95–101.

Klasen, H.J., 2000. A historical review of the use of silver in the treatment of burns. II. Renewed interest for silver. Burns 26, 131–138.

Kohanski, M.A., Dwyer, D.J., Hayete, B., Lawrence, C.A., Collins, J.J., 2007. A common mechanism of cellular death induced by bactericidal antibiotics. Cell 130, 797–810.

Koo, M.S., Lee, J.H., Rah, S.Y., Yeo, W.S., Lee, J.W., Lee, K. L., et al., 2003. A reducing system of the superoxide sensor SoxR in *Escherichia coli*. EMBO J. 22, 2614–2622.

Krejcova, L., Nejdl, L., Hynek, D., Krizkova, S., Kopel, P., Adam, V., et al., 2013. Beads-based electrochemical assay for the detection of influenza hemagglutinin labeled with CdTe quantum dots. Molecules 18, 15573–15586.

Kryukov, G.V., Castellano, S., Novoselov, S.V., Lobanov, A. V., Zehtab, O., Guigo, R., et al., 2003. Characterization of mammalian selenoproteomes. Science 300, 1439–1443.

Kurita, K., Tomita, K., Tada, T., Nishimura, S., Ishii, S., 1993. Reactivity characteristics of a new form of chitosan—facile *n*-phthaloylation of chitosan prepared from squid beta-chitin for effective solubilization. Polym. Bull. 30, 429–433.

Leblond, F., Davis, S.C., Valdes, P.A., Pogue, B.W., 2010. Pre-clinical whole-body fluorescence imaging: review of instruments, methods and applications. J. Photochem. Photobiol. B Biol. 98, 77–94.

Liau, S.Y., Read, D.C., Pugh, W.J., Furr, J.R., Russell, A.D., 1997. Interaction of silver nitrate with readily identifiable groups: relationship to the antibacterial action of silver ions. Lett. Appl. Microbiol. 25, 279–283.

Liu, H., Du, Y.M., Wang, X.H., Sun, L.P., 2004. Chitosan kills bacteria through cell membrane damage. Int. J. Food Microbiol. 95, 147–155.

Liu, H.-L., Dai, S.A., Fu, K.-Y., Hsu, S.-h., 2010. Antibacterial properties of silver nanoparticles in three different sizes and their nanocomposites with a new waterborne polyurethane. Int. J. Nanomed. 5, 1017–1028.

Lu, A.H., Salabas, E.L., Schuth, F., 2007. Magnetic nanoparticles: synthesis, protection, functionalization, and application. Angew. Chem. Int. Ed. Engl. 46, 1222–1244.

Madani, S.Y., Naderi, N., Dissanayake, O., Tan, A., Seifalian, A.M., 2011. A new era of cancer treatment: carbon nanotubes as drug delivery tools. Int. J. Nanomedicine 6, 2963–2979.

Mahl, D., Diendorf, J., Ristig, S., Greulich, C., Li, Z.A., Farle, M., et al., 2012. Silver, gold, and alloyed silver–gold nanoparticles: characterization and comparative cell-biologic action. J. Nanopart. Res. 14.

Majzlik, P., Strasky, A., Nemec, M., Trnkova, L., Havel, L., Zehnalek, J., et al., 2010. Remediation potential of bacteria for removing heavy metals from environment. Listy Cukrov. Reparske. 126, 414–415.

Majzlik, P., Strasky, A., Adam, V., Nemec, M., Trnkova, L., Zehnalek, J., et al., 2011. Influence of zinc(II) and copper (II) Ions on *Streptomyces* bacteria revealed by electrochemistry. Int. J. Electrochem. Sci. 6, 2171–2191.

Malam, Y., Loizidou, M., Seifalian, A.M., 2009. Liposomes and nanoparticles: nanosized vehicles for drug delivery in cancer. Trends Pharmacol. Sci. 30, 592–599.

Mater, D.D.G., Bretigny, L., Firmesse, O., Flores, M.J., Mogenet, A., Bresson, J.L., et al., 2005. *Streptococcus thermophilus* and *Lactobacillus delbrueckii* subsp bulgaricus survive gastrointestinal transit of healthy volunteers consuming yogurt. FEMS Microbiol. Lett. 250, 185–187.

Mathew, T.V., Kuriakose, S., 2013. Photochemical and antimicrobial properties of silver nanoparticle-encapsulated chitosan functionalized with photoactive groups. Mater. Sci. Eng. C Mater. Biol. Appl. 33, 4409–4415.

Matsumura, Y., Yoshikata, K., Kunisaki, S., Tsuchido, T., 2003. Mode of bactericidal action of silver zeolite and its comparison with that of silver nitrate. Appl. Environ. Microbiol. 69, 4278–4281.

Messner, K.R., Imlay, J.A., 1999. The identification of primary sites of superoxide and hydrogen peroxide formation in the aerobic respiratory chain and sulfite reductase complex of *Escherichia coli*. J. Biol. Chem. 274, 10119–10128.

Mittal, A.K., Chisti, Y., Banerjee, U.C., 2013. Synthesis of metallic nanoparticles using plant extracts. Biotechnol. Adv. 31, 346–356.

Morones, J.R., Elechiguerra, J.L., Camacho, A., Holt, K., Kouri, J.B., Ramirez, J.T., et al., 2005. The bactericidal effect of silver nanoparticles. Nanotechnology 16, 2346–2353.

Namiki, Y., Fuchigami, T., Tada, N., Kawamura, R., Matsunuma, S., Kitamoto, Y., et al., 2011. Nanomedicine for cancer: lipid-based nanostructures for drug delivery and monitoring. Acc. Chem. Res. 44, 1080–1093.

Navarro-Alarcon, M., Lopez-Martinez, M.C., 2000. Essentiality of selenium in the human body: relationship with different diseases. Sci. Total Environ. 249, 347–371.

Nehl, C.L., Liao, H.W., Hafner, J.H., 2006. Optical properties of star-shaped gold nanoparticles. Nano Lett. 6, 683–688.

Nezhad, S.S., Khorasgani, M.R., Emtiazi, G., Yaghoobi, M.M., Shakeri, S., 2014. Isolation of copper oxide (CuO) nanoparticles resistant *Pseudomonas* strains from soil and investigation on possible mechanism for resistance. World J. Microbiol. Biotechnol. 30, 809–817.

Nieboer, E., Richardson, D.H.S., 1980. The replacement of the non-descript term heavy-metals by a biologically and chemically significant classification of metal-ions. Environ. Pollut. B. 1, 3–26.

Nies, D.H., 1999. Microbial heavy-metal resistance. Appl. Microbiol. Biotechnol. 51, 730–750.

Nies, D.H., Silver, S., 1995. Ion efflux systems involved in bacterial metal resistances. J. Ind. Microbiol. Biot. 14, 186–199.

Nigam, S., Barick, K.C., Bahadur, D., 2011. Development of citrate-stabilized Fe_3O_4 nanoparticles: conjugation and release of doxorubicin for therapeutic applications. J. Magn. Magn. Mater. 323, 237–243.

No, H.K., Park, N.Y., Lee, S.H., Meyers, S.P., 2002. Antibacterial activity of chitosans and chitosan oligomers with different molecular weights. Int. J. Food Microbiol. 74, 65–72.

Ong, L.C., Ang, L.Y., Alonso, S., Zhang, Y., 2014. Bacterial imaging with photostable upconversion fluorescent nanoparticles. Biomaterials 35, 2987–2998.

Pal, S., Tak, Y.K., Song, J.M., 2007. Does the antibacterial activity of silver nanoparticles depend on the shape of the nanoparticle? A study of the gram-negative bacterium *Escherichia coli*. Appl. Environ. Microbiol. 73, 1712–1720.

Papineau, A.M., Hoover, D.G., Knorr, D., Farkas, D.F., 1991. Antimicrobial effect of water-soluble chitosans with high hydrostatic-pressure. Food Biotechnol. 5, 45–57.

Park, J., Joo, J., Kwon, S.G., Jang, Y., Hyeon, T., 2007. Synthesis of monodisperse spherical nanocrystals. Angew. Chem. Int. Ed. Engl. 46, 4630–4660.

Park, J.A., Reddy, P.A.N., Kim, H.K., Kim, I.S., Kim, G.C., Chang, Y., et al., 2008. Gold nanoparticles functionalised by Gd-complex of DTPA-bis(amide) conjugate of glutathione as an MRI contrast agent. Bioorg. Med. Chem. Lett. 18, 6135–6137.

Park, J.A., Kim, H.K., Kim, J.H., Jeong, S.W., Jung, J.C., Lee, G.H., et al., 2010. Gold nanoparticles functionalized by gadolinium–DTPA conjugate of cysteine as a multimodal bioimaging agent. Bioorg. Med. Chem. Lett. 20, 2287–2291.

Perla, V., Webster, T.J., 2005. Better osteoblast adhesion on nanoparticulate selenium—a promising orthopedic implant material. J. Biomed. Mater. Res. A. 75A, 356–364.

Petkova, P., Francesko, A., Fernandes, M.M., Mendoza, E., Perelshtein, I., Gedanken, A., et al., 2014. Sonochemical coating of textiles with hybrid ZnO/chitosan antimicrobial nanoparticles. ACS Appl. Mater. Interfaces 6, 1164–1172.

Pham, T.D., Lee, B.K., 2014. Cu doped TiO_2/GF for photocatalytic disinfection of *Escherichia coli* in bioaerosols under visible light irradiation: application and mechanism. Appl. Surf. Sci. 296, 15–23.

Pomposiello, P.J., Demple, B., 2001. Redox-operated genetic switches: the SoxR and OxyR transcription factors. Trends Biotechnol. 19, 109–114.

Qian, H.F., Dong, C.Q., Weng, J.F., Ren, J.C., 2006. Facile one-pot synthesis of luminescent, water-soluble, and biocompatible glutathione-coated CdTe nanocrystals. Small 2, 747–751.

Rai, M.K., Deshmukh, S.D., Ingle, A.P., Gade, A.K., 2012. Silver nanoparticles: the powerful nanoweapon against multidrug-resistant bacteria. J. Appl. Microbiol. 112, 841–852.

Rayman, M.P., 2005. Selenium in cancer prevention: a review of the evidence and mechanism of action. Proc. Nutr. Soc. 64, 527–542.

Roca, A.G., Morales, M.P., O'Grady, K., Serna, C.J., 2006. Structural and magnetic properties of uniform magnetite nanoparticles prepared by high temperature decomposition of organic precursors. Nanotechnology 17, 2783–2788.

Rouch, D.A., Lee, B.T.O., Morby, A.P., 1995. Understanding cellular-responses to toxic agents—a model for mechanism-choice in bacterial metal resistance. J. Ind. Microbiol. 14, 132–141.

Russell, A.D., Hugo, W.B., 1994. Antimicrobial activity and action of silver. Prog. Med. Chem. 31, 351–370.

Ryvolova, M., Chomoucka, J., Drbohlavova, J., Kopel, P., Babula, P., Hynek, D., et al., 2012. Modern micro and nanoparticle-based imaging techniques. Sensors. 12, 14792–14820.

Sanchez-Gonzalez, L., Chafer, M., Hernandez, M., Chiralt, A., Gonzalez-Martinez, C., 2011. Antimicrobial activity of polysaccharide films containing essential oils. Food Control. 22, 1302–1310.

Sawai, J., Igarashi, H., Hashimoto, A., Kokugan, T., Shimizu, M., 1995. Evaluation of growth-inhibitory effect of ceramics powder slurry on bacteria by conductance method. J. Chem. Eng. Jpn. 28, 288–293.

Sawai, J., Shoji, S., Igarashi, H., Hashimoto, A., Kokugan, T., Shimizu, M., et al., 1998. Hydrogen peroxide as an antibacterial factor in zinc oxide powder slurry. J. Ferment. Bioeng. 86, 521–522.

Schreurs, W.J.A., Rosenberg, H., 1982. Effect of silver ions on transport and retention of phosphate by *Escherichia coli*. J. Bacteriol. 152, 7–13.

Shameli, K., Bin Ahmad, M., Jazayeri, S.D., Shabanzadeh, P., Sangpour, P., Jahangirian, H., et al., 2012. Investigation of antibacterial properties silver nanoparticles prepared via green method. Chem. Cent. J. 6.

Sharma, V.K., Yngard, R.A., Lin, Y., 2009. Silver nanoparticles: green synthesis and their antimicrobial activities. Adv. Colloid Interface Sci. 145, 83–96.

Shi, Z.L., Neoh, K.G., Kang, E.T., 2004. Surface-grafted viologen for precipitation of silver nanoparticles and their combined bactericidal activities. Langmuir 20, 6847–6852.

Sies, H., 1997. Oxidative stress: oxidants and antioxidants. Exp. Physiol. 82, 291–295.

Silver, S., 2003. Bacterial silver resistance: molecular biology and uses and misuses of silver compounds. Fems Microbiol. Rev. 27, 341–353.

Smetana, A.B., Klabunde, K.J., Marchin, G.R., Sorensen, C. M., 2008. Biocidal activity of nanocrystalline silver powders and particles. Langmuir 24, 7457–7464.

Sondi, I., Salopek-Sondi, B., 2004. Silver nanoparticles as a antimicrobial agent: a case study on E. coli as a model for Gram-negative bacteria. J. Colloid Interf Sci. 275, 177–182.

Spallholz, J.E., Shriver, B.J., Reid, T.W., 2001. Dimethyldiselenide and methylseleninic acid generate superoxide in an *in vitro* chemiluminescence assay in the presence of glutathione: implications for the anticarcinogenic activity of L-selenomethionine and L-Se-methylselenocysteine. Nutr. Cancer 40, 34–41.

Storz, G., Imlay, J.A., 1999. Oxidative stress. Curr. Opin. Microbiol. 2, 188–194.

Sudarshan, N.R., Hoover, D.G., Knorr, D., 1992. Antibacterial action of chitosan. Food Biotechnol. 6, 257–272.

Suzuki, K., Mikami, T., Okawa, Y., Tokoro, A., Suzuki, S., Suzuki, M., 1986. Antitumor effect of hexa-*n*-acetylchitohexaose and chitohexaose. Carbohydr. Res. 151, 403–408.

Tapiero, H., Tew, K.D., 2003. Trace elements in human physiology and pathology: zinc and metallothioneins. Biomed. Pharmacother. 57, 399–411.

Tartaj, P., Morales, M.D., Veintemillas-Verdaguer, S., Gonzalez-Carreno, T., Serna, C.J., 2003. The preparation of magnetic nanoparticles for applications in biomedicine. J. Phys. D Appl. Phys. 36, R182–R197.

Tran, P.A., Webster, T.J., 2011. Selenium nanoparticles inhibit *Staphylococcus aureus* growth. Int. J. Nanomed. 6.

Tran, P.L., Hammond, A.A., Mosley, T., Cortez, J., Gray, T., Colmer-Hamood, J.A., et al., 2009. Organoselenium coating on cellulose inhibits the formation of biofilms by *Pseudomonas aeruginosa* and *Staphylococcus aureus*. Appl. Environ. Microbiol. 75, 3586–3592.

Turkevich, J., Stevenson, P.C., Hillier, J., 1951. A study of the nucleation and growth processes in the synthesis of colloidal gold. Discuss. Faraday Soc. 55–75.

Wang, X.H., Du, Y.M., Li, Y., Li, D., Sun, R.C., 2011. Fluorescent identification and detection of *staphylococcus aureus* with carboxymethyl chitosan/CdS quantum dots bioconjugates. J. Biomater. Sci. Polym. Ed. 22, 1881–1893.

Wei, W.Q., Abnet, C.C., Qiao, Y.L., Dawsey, S.M., Dong, Z. W., Sun, X.D., et al., 2004. Prospective study of serum selenium concentrations and esophageal and gastric cardia cancer, heart disease, stroke, and total death. Am. J. Clin. Nutr. 79, 80–85.

Wu, J.H., Ko, S.P., Liu, H.L., Jung, M.H., Lee, J.H., Ju, J.S., et al., 2008. Sub 5 nm Fe_3O_4 nanocrystals via coprecipitation method. Colloids Surf. A Physicochem. Eng. Asp. 313, 268–272.

Yang, F., Jin, C., Subedi, S., Lee, C.L., Wang, Q., Jiang, Y.J., et al., 2012. Emerging inorganic nanomaterials for pancreatic cancer diagnosis and treatment. Cancer Treat. Rev. 38, 566–579.

Yang, J.J., Huang, K.H., Qin, S.Y., Wu, X.S., Zhao, Z.P., Chen, F., 2009. Antibacterial action of selenium-enriched probiotics against pathogenic *Escherichia coli*. Dig. Dis. Sci. 54, 246–254.

Zhang, Y., Chan, H.F., Leong, K.W., 2013. Advanced materials and processing for drug delivery: the past and the future. Adv. Drug Deliv. Rev. 65, 104–120.

Zhu, H.Y., Sikora, U., Ozcan, A., 2012. Quantum dot enabled detection of *Escherichia coli* using a cell-phone. Analyst. 137, 2541–2544.

Zhu, J.J., Wang, H., Zhu, J.M., Wang, J., 2002. A rapid synthesis route for the preparation of CdS nanoribbons by microwave irradiation. Mater. Sci. Eng. B Solid State Mater. Adv. Technol. 94, 136–140.

CHAPTER

14

Nanotechnology—Is There Any Hope for Treatment of HIV Infections or Is It Simply Impossible?

Ranjita Shegokar

Freie Universität Berlin, Institute of Pharmacy Department of Pharmaceutics, Biopharmaceutics & NutriCosmetics, Kelchstraße, Berlin, Germany

14.1 INTRODUCTION

In 2011, UNAIDS announced a target of reaching 15 million people with antiretroviral treatment by 2015. If this promise is reached, then it is estimated that up to 12 million infections and more than 7 million deaths can be averted by 2020. The total number of new infections would be reduced to more than half by 2015 if this attempt is successful. The treatment is mainly based on the use of approved conventional oral solid dosage forms of antiretrovirals.

HIV mainly infects immune cells (e.g., T4-lymphocyte), which have CD4 receptors on their surface, because this protein helps viruses to bind to cells and replicate once HIV DNA is integrated in the cell's DNA. Other organs that are infected by HIV are lymph nodes, brain, kidney, lungs, genitals, and others. The infection cycle of HIV includes five steps: binding, reverse transcription, integration, transcription and translation, viral assembly and maturation. In the binding stage, HIV binds to a $CD4^{+}$/CC chemokine receptor CCR5 or the CXC chemokine receptor or CXCR4 surface receptor, and it activates other proteins on the cell's surface, which allows the HIV envelope to fuse to the cell. The entire entry process is completed within 1 h of virus contact with the cell. Antiretroviral drugs which stop this binding are called as entry inhibitors. In the reverse-transcription stage, the viral capsid that contains the RNA and important enzymes are released into the host cell. A viral enzyme called reverse transcriptase makes a DNA copy of the RNA. This new DNA is called "proviral DNA." Nucleoside reverse-transcriptase inhibitors (NRTIs) and non-nucleoside reverse-transcriptase inhibitors (NNRTIs) class drugs are used to stop this transcription, thereby avoiding further new infection. In the third

Nanotechnology in Diagnosis, Treatment and Prophylaxis of Infectious Diseases.
DOI: http://dx.doi.org/10.1016/B978-0-12-801317-5.00014-1

stage, called integration, the formed HIV pro-DNA is then transported to the cell's nucleus. The viral enzyme called integrase hides the proviral DNA in the cell's DNA. Integration can be blocked by integrase inhibitors. In the fourth stage of transcription and translation, generally occurs over the next 10–15 h of infection, the strands of viral pro DNA in the nucleus separate and special enzymes create a complementary strand of mRNA. Transcription can be blocked by new class of drugs called antisense antivirals or transcription inhibitors (TIs). In the translation part, each mRNA strand orders production of a corresponding string of new viral protein chains that are required to make a copy of new HIV virus. In the last stage of viral assembly and maturation, long strings of proteins are cut into smaller proteins of different functional group-like structural elements of new HIV by a viral enzyme called protease, whereas others become enzymes, such as reverse transcriptase. These new viral particles are assembled with all components; they bud off the host cell and create a new virus. The virus then enters the maturation stage, where the processing of viral proteins takes place. Viral assembly can be blocked by protease inhibitors (PIs). Few industries are involved in development of anti-HIV drugs called maturation inhibitors. The first approved antiretroviral was introduced in 1996; since then, almost 30 FDA-approved antiretroviral drugs are available (Figure 14.1).

In 2012, almost 9.7 million people received antiretroviral therapy. However, in 2013 only 34% (32–37%) of the 28.6 million people received treatment. According to 2012, an estimated 35.3 (32.2–38.8) million people were living with HIV

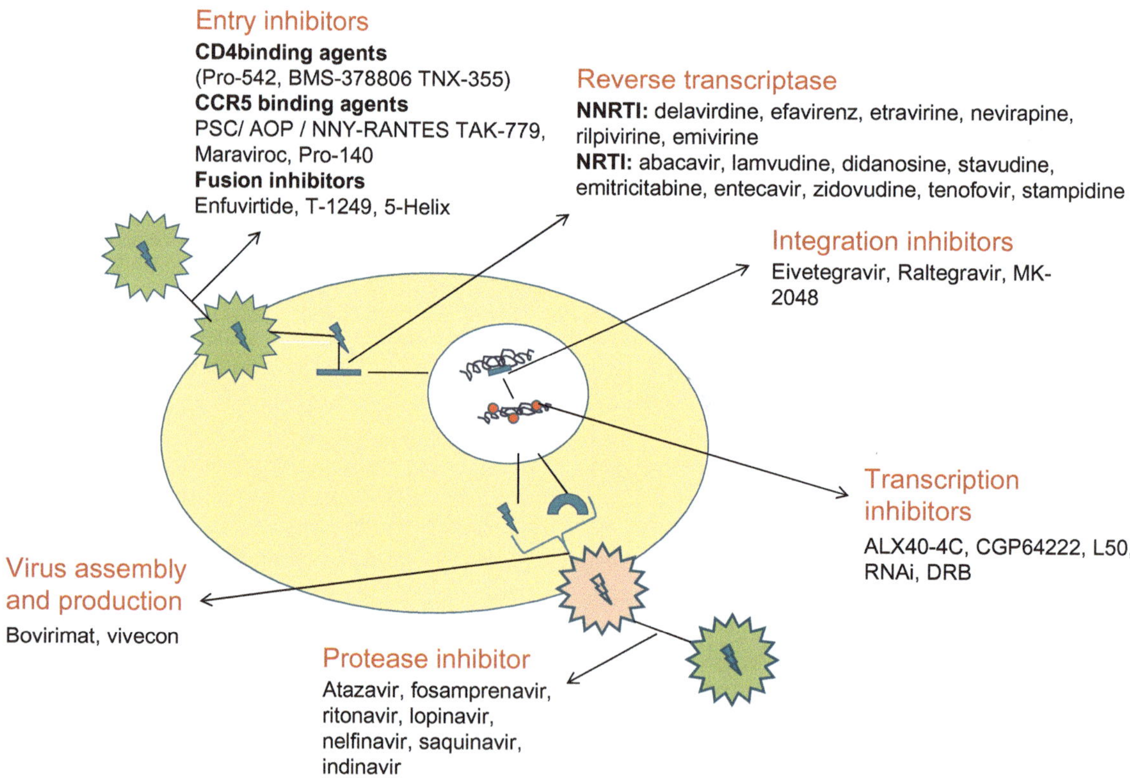

FIGURE 14.1 HIV replication cycle and categories of drugs developed to control the particular stage of cycle.

globally. However, there has been a 33% decline in new infections (2.3 million in 2012) globally compared with 2001 (3.4 million). A decline in the number of AIDS deaths have also seen, with 1.6 million AIDS deaths in 2012 (2.3 million in 2005). This is because of an increase in access to life-saving antiretroviral therapy compared with previous years using when various HIV/AIDS health programs were used. In 2012, an estimated US $18.9 billion was available for HIV programs in low-income and middle-income countries—a 10% increase since 2011 (da Costa et al., 2012; UNAIDS, 2013). The good news is that new infections and HIV/AIDS related deaths are declining, but there are still several unanswered questions regarding drug resistance in this population and whether HIV can be completely cured or if it remains in a latent stage. There are links between cured infections, new infections, and mother-to-child disease transmission. The duration of treatment and role of HIV programs have been debated.

14.2 CURRENT ANTIRETROVIRAL CHEMOTHERAPY

As discussed earlier, anti-retrovirals are distributed into six distinct classes based on their molecular mechanism and resistance profiles (Table 14.1). FDA-approved antiretrovirals include: (i) NRTIs; (ii) NNRTIs; (iii) integrase inhibitors; (iv) PIs; (v) fusion inhibitors; and (vi) co-receptor antagonists. A new class of antiretroviral called maturation inhibitors is under research. Antiretrovirals are prescribed as single drugs or as combinations depending on the disease stage and severity.

The development of reverse-transcriptase inhibitors and protease transcriptases began in the mid -1990s has revolutionized the overall efficacy and durability of antiretroviral therapy. Highly active antiretroviral therapy (HAART), a combination antiretroviral treatment approach, dramatically suppresses viral replication and reduces the plasma HIV-1 viral load to <50 RNA copies/mL, resulting in a significant reconstitution of the immune system. A combination of three drugs was found to be more effective than two-drug regimens. It is estimated that one mutation is introduced for every 1,000–10,000 nucleotides synthesized, resulting in 1–10 mutations in each viral genome with every replication cycle. The accumulation of latent virus in cellular compartments and organ reservoirs significantly contributes to the pathogenic disease state in a later stage of treatment.

The small-molecule and antibody-based attachment inhibitors like BMS-378806 and TNX-355 can target Gp120 and CD4. Both these compounds have been tested in clinical trials; however, they have not been approved for human use. Gp41 and CCR5 are targets for fuzeon and maraviroc, which belong to the fusion inhibitor and small-molecule CCR5 chemokine receptor antagonist class. HIV maturation is effectively blocked by inhibitor betulinic acid by attacking the viral plasma membrane, which is currently a widely studied area. At the molecular level, resistance mainly occurs by amino acid substitutions such as L100, K101, K103, E138, V179, Y181, and Y188 in the NNRTI-binding pocket of reverse transcriptase. HIV-1 protease initiates the cleavage of the viral gag and gag-pol polyprotein precursors during virion maturation. For PI resistance, mutations occur at sites like D30N, G48V, I50V, V82A, or I84V. These molecular targets can be used in developing new antiretrovirals. Three antagonists named VCV, MVC, and Aplaviroc showed promising results in humans by affecting virus replication (Arts and Hazuda, 2012). Gag proteins of HIV are held together by zinc finger and are involved in binding and assembly of viral RNA in new virus. These zinc fingers are located in cellular proteins and are functioned to capture zinc ions, thereby contributing to the capacity of the protein to bind to RNA or DNA. The special class of antiretroviral called zinc finger inhibitors is being developed. Zinc fingers can prevent Gag protein functioning, stop capturing, and packaging of new HIV genetic material. Thereby, zinc

TABLE 14.1 List of Antiretroviral Drugs Used in HIV/AIDS Chemotherapy

Drug	Brand name	Manufacturer
NRTIs		
Emtricitabine, FTC	Emtriva	Gilead Sciences
Tenofovir disoproxil fumarate and emtricitabine	Truvada	
Tenofovir disoproxil fumarate, TDF	Viread	
Lamivudine, 3TC	Epivir	GlaxoSmithKline
Lamivudine and zidovudine	Combivir	
Abacavir and lamivudine	Epzicom	
Abacavir sulfate, ABC	Ziagen	
Zidovudine/azidothymidine/AZT/ZDV	Retrovir	
Abacavir, zidovudine, and lamivudine	Trizivir	
Enteric coated didanosine/ddI EC	Videx EC	Bristol-Myers Squibb
Didanosine, dideoxyinosine, ddI	Videx	
Stavudine, d4T	Zerit	
Zalcitabine, dideoxycytidine, ddC	Hivid	Hoffmann-La Roche
NNRTIs		
Rilpivirine	Edurant	Tibotec Therapeutics
Etravirine	Intelence	
Delavirdine, DLV	Rescriptor	Pfizer
Efavirenz, EFV	Sustiva	Bristol-Myers Squibb
Nevirapine, NVP	Viramune (immediate release)	Boehringer Ingelheim
	Viramune XR (Extended Release)	
PIs		
Atazanavir sulfate, ATV	Reyataz	Bristol-Myers Squibb
Tipranavir, TPV	Aptivus	Boehringer Ingelheim
Indinavir, IDV,	Crixivan	Merck
Saquinavir mesylate, SQV	Invirase	Hoffmann-La Roche
Lopinavir and ritonavir, LPV/RTV	Kaletra	Abbott Laboratories
Ritonavir, RTV	Norvir	
Darunavir	Prezista	Tibotec, Inc.
Nelfinavir mesylate, NFV	Viracept	Agouron Pharmaceuticals
Fosamprenavir Calcium, FOS-APV	Lexiva	GlaxoSmithKline
Amprenavir, APV (no longer marketed)	Agenerase	

(*Continued*)

TABLE 14.1 (Continued)

Drug	Brand name	Manufacturer
FUSION INHIBITORS		
Enfuvirtide, T-20	Fuzeon	Hoffmann-La Roche and Trimeris
ENTRY INHIBITORS—CCR5 CORECEPTOR ANTAGONIST		
Maraviroc	Selzentry	Pfizer
HIV INTEGRASE STRAND TRANSFER INHIBITORS		
Raltegravir	Isentress	Merck & Co., Inc.
Dolutegravir	Tivicay	GlaxoSmithKline
HAART COMBINATION PRODUCT		
Efavirenz, emtricitabine, and tenofovir disoproxil fumarate	Atripla	Bristol-Myers Squibb and Gilead Sciences
Emtricitabine, rilpivirine, and tenofovir disoproxil fumarate	Complera	Gilead Sciences
Elvitegravir, cobicistat, emtricitabine, tenofovir disoproxil fumarate	Stribild	
abacavir sulfate, dolutegravir sodium, lamivudine (ABC/DTG/3TC)	Triumeq	ViiV Healthcare/GlaxoSmithKline

fingers help disruption of the nucleocapsid of HIV, which leads to the production of a dysfunctional virus that cannot infect new cells.

Gilead Sciences Inc. services the major portion of the world's HIV/AIDS drug needs and in 2010 achieved $6.3 billion in drug sales, representing almost 40% of the market. They have approximately 60 HIV drugs in various stages of development. GlaxoSmithKline is another big pharma company that is active in the antiretroviral field; the new drug dolutegravir, an integrase inhibitor also known as S/GSK1349572, recently entered the market. Tivicay® (dolutegravir) was approved by the US FDA and by Health Canada in 2013. In late 2013, based on the results of four pivotal phase III studies of dolutegravir, the Committee for Medicinal Products for Human Use (CHMP) of the European Medicines Agency (EMA) issued a positive opinion to regulatory authorities. Immunomedics Inc. has collaborations with Karolinska Institute in Sweden regarding a new class of antiretroviral drugs that prevent further viral infection. In developing countries, generic versions of antiretroviral drugs including zidovudine, didanosine, stavudine, and lamivudine are being produced and, according to GBI Research, they are estimated to have represented 46% of the global HIV market in 2010. The current market growth of 7% during 2011–2018 has been forecasted. Stribild is a multiclass combination drug with Elvitegravir that is boosted with cobicistat; it is a once-per-day pill made by Gliead Sciences. Selzentry UK-427857 developed by Pfizer comprises maraviroc and is the first CCR5 receptor antagonist on the market. This entry inhibitor blocks entry of HIV from entering white blood cells. Raltegravir (Isentress, formerly MK-0518), an integrase inhibitor, was launched by Merck & Co. In 2011, it received approval for pediatric use in a 2013 (phase II trial), its effects in multiple sclerosis were evaluated. New products are being launched and clinical trials are being performed; however, all of them are based on a conventional oral solid dosage form and none of them is based on a nanotechnology approach.

14.3 NANOTECHNOLOGY IN HIV CHEMOTHERAPY—WHY?

Looking at the major failures in clinical trials, it seems that vaccine research will take another one or two decades to find a safe and effective candidate that would reach the market. Currently, there is no ideal cure for HIV/AIDS. The complete eradication of the virus from the body is not possible with the currently available antiretroviral therapies. As stated, combination antiretroviral therapy can dramatically improve treatment, but it has certain limitations such as lifetime therapy, major side effects, and ineffectiveness in virus resistance cases. HAART is manly associated with major adverse effects and co-infections cause increased rates of heart disease, diabetes, liver disease, and cancer. Nanotechnology-based specific treatments can be used to cure HIV infections and co-infections. A virus in the latent state in $CD4^+$ T cells and cells of the macrophage–monocyte lineage cannot be killed by HAART and needs special delivery systems that can reach these cellular and organ sites. The other organs that harbor latent HIV are lymphoid tissue, testes, liver, kidney, lungs, gut, and the central nervous system (CNS). In the past two decades, rigorous research and development has been performed in the area of microbicides, resulting in many effective topical microbicides candidates that are currently in clinical trials.

Nanotechnology is a widely studied and evaluated platform in an emerging multidisciplinary field of science. It allows engineering and manufacturing of drug delivery systems within a nanometer range and sometimes allows flexibility to experiment at the atomic and molecular levels. It has been revolutionized over the decades in main disease areas such as photodynamic therapy, gene therapy, immunotherapy, vaccinology, and cancer therapy. Many contributions have been made by laboratory research regarding exploring the potential of nanotechnology for the treatment and prevention of HIV/AIDS. However, there are several challenges and questions concerning the potential of nanotechnology to provide more effective treatments for HIV/AIDS. Nanotechnology is being explored for antiretroviral therapy (oral/parenteral/transdermal route/pulmonary), vaccine delivery, and microbicides. Novel approaches for identifying new antigens and adjuvants for better HIV therapy are being studied; these approaches mainly suffer from drug delivery and formulation stability issues. Now is the time to explore nanotechnological platforms and their potential in HIV treatment and prevention. HIV/AIDS therapy desperately needs nontoxic, sustained drug release systems that are effective during viral latent stages to avoid lifetime antiretroviral therapy. Targeting at the cellular and organ levels could solve many issues related to current single-dose therapy and HAART. A number of research groups have evaluated and are evaluating the potential of nanoparticles (Figure 14.2) for HIV

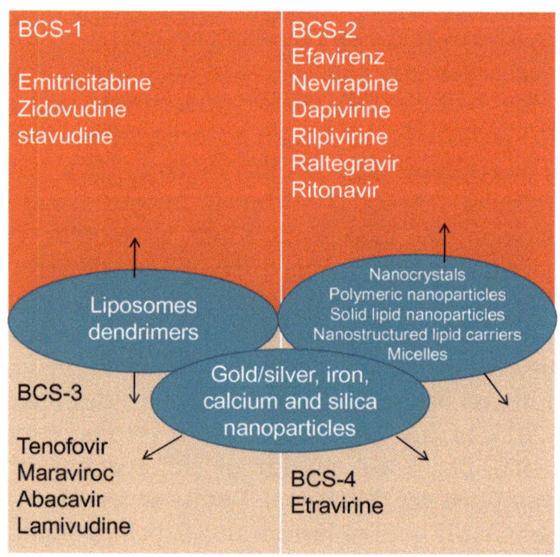

FIGURE 14.2 Overview of antiretroviral drugs classified as per the biopharmaceutical classification system (BCS) system and literature reported nanotechnology-based drug delivery systems developed for antiretrovirals.

therapy at preclinical stages; however, no efforts have been successful in reaching clinical trials. Overall, the nanoparticle platform shows promising results and it is worth exploring its potential at preclinical and clinical levels. The role of nanoparticles and their specific benefits have been evaluated individually.

14.4 NANOPARTICLE RESEARCH

14.4.1 Drug Targeting and Prolong Circulation

Drug targeting entails accumulation of a drug at a particular organ or cell. The prolonged residence of the drug in systemic circulation involves the stealth nature of the system (Figure 14.3). Both approaches provide unique benefits and have limitations. Several attempts are made to use this type of targeting or prolonged circulation for antiretroviral drugs. In a recent study, rilpivirine (TMC278) nanosuspensions, 200 nm, are prepared and are stabilized by polyethylene–polypropylene glycol (poloxamer 338) and polyethylene glycolylated (PEGylated) tocopheryl succinate ester (TPGS 1000) (van't Klooster et al., 2010). A single-dose administration of the drug in the form of nanosuspensions in dogs and mice resulted in release over 3 months in dogs and 3 weeks in mice. This formulation can offer lower dosing frequencies and can help to improve adherence. Nanosuspension is a widely studied platform for treatment of infectious and noninfectious diseases (Shegokar, 2013). Lipoid E80 stabilized indinavir nanosuspension could deliver the drug to various tissues when loaded into macrophages and injected intravenously to mice. An increased amount of drug was found in the lungs, liver, and spleen. However, the single dose of the nanoparticle-loaded macrophages showed significant antiviral activity up to 14 days in the brain when administered intravenously in an HIV-infected mouse model (Dou et al., 2007).

The mannose- and galactose-coated stavudine liposomes (size, 100–200 nm) showed improved cellular uptake compared with free drugs. A significantly higher amount of the drug is found in liver, spleen, and lungs (Garg et al., 2006). Similar attempts were also made for zidovudine. A mannose-targeted zidovudine liposome resulted in increased accumulation in lymph nodes and spleen (Jain et al., 2008; Kaur et al., 2008). An *in vitro* study show that a

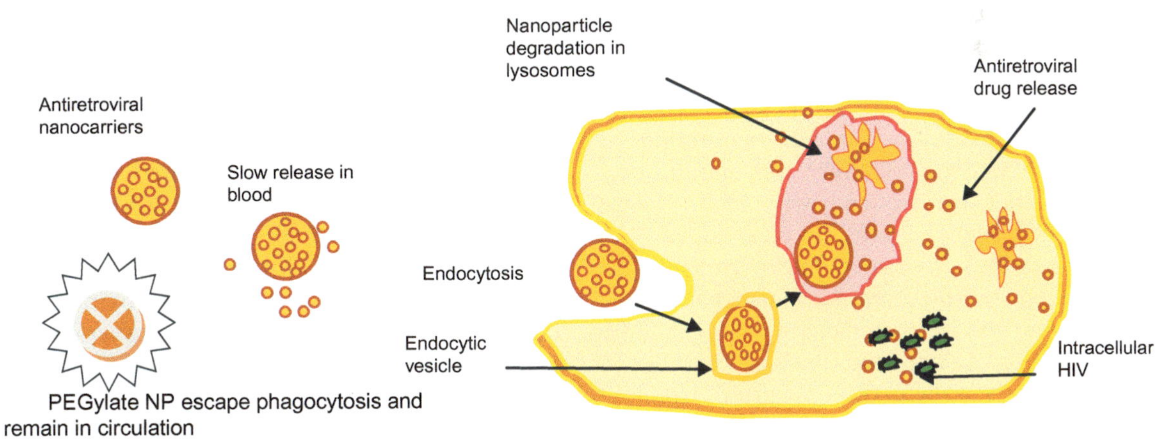

FIGURE 14.3 Mechanisms of drugs targeting 1) uncoated nanoparticles and 2) long circulation (PEGylation) of antiretrovirals using nanotechnology-based drug delivery systems.

mannose-targeted poly (propyleneimine) efavirenz dendrimer can effectively deliver the drug to human monocytes/macrophages and showed almost 12-fold increase in cellular uptake compared with free drugs (Dutta et al., 2007). The tetra-peptide tuftsin (Thr–Lys–Pro–Arg) decorated dendrimer was effective for targeting macrophages (Dutta et al., 2008). *In vitro* experiments showed sixfold prolonged release and 34-fold increase in cellular uptake. An anti-HIV effect of the surface-modified dendrimer was seven fold higher than that for free drugs.

Another attempt was made to target macrophages, which serve as a major reservoir for HIV. A peptide-based nanocarrier was developed where a drug is conjugated to the backbone of peptide–PEG and *N*-formyl-methionyl-leucyl-phenylalanine (fMLF). An fMLF is a bacterial peptide sequence for which macrophages express a receptor. Surface-modified, fMLF-targeted peptide–PEG nanoparticles resulted in increased cellular uptake and accumulation in liver, kidney, and spleen compared with uncoated nanoparticles (Wan et al., 2007).

14.4.2 Drug Solubility Enhancement

Wet milling technique is used to dimunate particles, thereby increasing the surface area and solubility. Ritonavir nanosuspension was produced by bead milling technique in the presence of poloxamer 407. Ritonavir drug particle size was reduced from 52 μm to 385 nm with PDI of 0.268 (−25.4 mV) after 2 h of milling at 3,000 rpm. The drug solubility was significantly increased after milling and was found to be 1.62 mg/mL. Nanosuspension was resistant to the dilution and freeze–thaw experiment. The drug release profile from nanosuspension was much faster (91% in 60 min) compared with (28% in 60 min) that of pure drug (Prakash et al., 2013).

Efavirenz is a BCS class II drug, has low solubility of 0.9 μg/mL, and has a low intrinsic dissolution rate of 0.037 $mg/cm^2/min$. It was co-processed with sodium lauryl sulfate (SLS) and polyvinylpyrrolidone (PVP) co-micronization to improve dissolution properties. The final particle size of drug processed with SLS was much smaller compared with PVP, and both sizes existed in the micron range. Both materials formed a hydrophilic layer around the insoluble drug particles, resulting in several times higher dissolution compared with unprocessed drug (da Costa et al., 2013). Madhavi et al. prepared solid dispersions of Efavirenz by using the solvent evaporation method using PEG 6000 at a different ratio. Drug release was significantly increased from 16% to 70% when converted to solid dispersions, and the proposed effect could be due to the encapsulation of amorphous particles into the hydrophilic matrix (Madhavi et al., 2011).

Natural biopolymer obtained from seeds of *Buchanania lanzan* was used as a stabilizer to form zidovudine nanosuspension. The solvent evaporation method was used to formulate nanosuspensions (Tyagi and Madhav, 2013). At the end of 8 h, up to 85–95% release was obtained depending upon the composition of formulation. Unfortunately, no information on solubility was discussed in this article. Another antiretroviral drug saquinavir possesses only 4% of bioavailability; to enhance its *in vivo* effects, it was encapsulated in a nanostructured lipid carriers (NLCs) (165 nm) by a high-pressure homogenization technique. The permeability of saquinavir was significantly increased when tested in an enterocyte-like model (Caco-2 monolayers) and the FAE monolayers (Caco-2/Raji cell co-culture) model. The size-dependent accumulation was observed in Caco-2 cells. According to the author, it is the first kind of mechanistic study on NLC transport across intestinal models *in vitro* (Beloqui et al., 2013). The antiretroviral drug nevirapine was processed as nanosuspensions (457 nm) by Shegokar and Singh (2011a) using high-pressure homogenization. The nanosuspension showed improved bioavailability in Wistar rats and improved accumulation in primary macrophages due to increased drug solubility (Shegokar and Singh, 2011b). Nanosuspensions were scaled-up to pilot scale using varied capacity homogenizers and were compared with milling technique (Shegokar et al., 2011a,b). Nanosuspensions were further converted into dry powder by fluid bed drying for oral application for parenteral application (after reconstitution with water for injection) by lyophilization. The stability of nanosuspensions upon sterilization was also studied (Shegokar and Singh, 2012a).

Other solubility enhancement attempts were made for ritonavir (Fukushima et al., 2008; Ilevbare et al., 2013; Rodriguez-Spong et al., 2008; Sinha et al., 2010), nevirapine (Raju et al., 2014), efavirenz (Alves et al., 2014; Chiappetta et al., 2011), 5-chloro-3-phenylsulfonylindole-2-carboxamide (Obitte et al., 2013), and lopinavir (Jain et al., 2013).

14.4.3 Transport Across Blood–Brain Barrier

SLNs of stavudine (175–393 nm) on oral administration of a single dose of 4 mg/kg body weight showed accumulation in tissues of heart, liver, spleen, lung, kidney, brain, GI track, muscle, bone marrow, and lymph node. The main organ includes kidney, liver, lymph nodes, and spleen. The levels detected for nanoparticles are much higher than that for pure drug. Nanoparticles crossed the blood–brain barrier (BBB) and showed drug levels (689 ng) 2 h after administration. The total amount of stavudine in liver, spleen, lungs, bone marrow, and lymph nodes was approximately 60% of the total dose after 2 h and was only 19% for pure drug. Lipid nanoparticles increased the cellular uptake and organ targeting when drug was encapsulated (Dandagi et al., 2012). Nevirapine nanosuspensions surface-modified with bovine serum albumin (BNS) showed superior accumulation in brain compared with uncoated drug. The drug levels lasted for more than 24 h in brain. The AUC(brain)-to-AUC(blood) ratio of BNS was found to be 9.33. The formulation BNS showed almost 21-fold enhancement in gamma count at the end of 1 h (with AUC^{0-24} of 194.74 counts h/g) with prolonged MRT of 7.043 h. The clearance rate was observed to be very slow for the nanosuspension coated with serum albumin. None of the other nanosuspension formulations (coated with PEG, dextran) showed levels in the brain. In another study, saquinavir was delivered to brain microvascular endothelial cells in the form of transferrin-conjugated quantum rods (Mahajan et al., 2010), nevirapine as transferrin decorated PLGA nanoparticles to the brain (Kuo et al., 2011), and methyl methacrylate–sulfopropyl methacrylate nanoparticles surface-modified with RMP-7 to deliver stavudine, delavirdine, and saquinavir across the BBB (Kuo and Lee, 2012).

14.4.4 Toxicity Reduction and Dose Reduction

Various attempts were made to encapsulate antiretroviral drugs into nanoparticles, thereby increasing the efficacy and targeting that resulted in an ultimate effective dose required to reduce the viral burden. Some of these studies include poly(lactide-co-glycolide) nanoparticles containing ritonavir, lopinavir, and efavirenz, amorphous dispersion of efavirenz prepared using spray-drying technology, solid dispersions by solvent evaporation, and physical mixture methods using PEG as the hydrophilic carrier (Maíra Assis da Costa et al., 2013).

Shegokar et al. encapsulated stavudine into solid lipid nanoparticles (SLNs) to target the viral reservoirs and to reduce toxicity associated with pure drug. Lipid nanoparticles were prepared by hot pressure homogenization at laboratory scale and scaled up successfully at pilot batch size (Shegokar et al., 2011a,b), and were stable for more than 3 years at room temperature and refrigeration conditions. The studies on cellular uptake and *in vivo* biodistribution showed enhanced cellular uptake and targeting potential for both coated (with albumin, dextran, and PEG) and uncoated solid lipid nanoparticles (Figure 14.4). An acute toxicity study showed no signs of abnormal behavioral or physiological changes after dose administration over the observation period of 14 days. The acute dose toxicity experiment confirmed that at 25-times higher dose than that of the therapeutic dose of the developed lipid formulations had no pharmacotoxic signs, rendering them pharmacologically safe. None of the animals from repeated dose toxicity studies showed abnormal behavioral or physiological changes on the day of dose administration and thereafter over the period of 28 days. The histological observations on day 28 confirmed complete recovery of the animals, as evident from absolutely normal conditions of the tissues. The developed uncoated and coated lipid nanoparticles were safe for administration over the required therapeutic schedule, with few mild and self-limiting pathological effects (Shegokar and Singh, 2012b).

Polylysine heparin–functionalized solid lipid nanoparticles (fSLNs; 153 nm, −51 mV) were developed as a vaginal microbicide by a layer-by-layer deposition method. At selected concentrations, more than 80% cell viability was observed compared with the positive control. The LDH membrane integrity assay showed very low levels, whereas positive control showed 15% LDH release. The developed nanoparticles were found to be less toxic compared with the positive control. This effect is due to the entrapment of the drug in the lipidic biodegradable matrix (Alukda et al., 2011). Kasongo et al. (2011b) developed NLC composed of Precirol® ATO 5 and Transcutol® HP for delivery of didanosine to the CNS. *In vitro* protein adsorption studies showed incorporation of ddI in NLC to facilitate brain targeting. Data obtained from 2D PAGE analysis revealed that DDI-loaded NLC preferentially adsorb proteins *in vitro* that are responsible for specific brain targeting *in vivo* (Kasongo et al., 2011a).

However, efavirenz nanosuspensions prepared by freeze-drying techniques showed improved bioavailability, thereby improving targeting and reducing resultant doses (Patel et al., 2014). da Costa et al. (2012) observed an improvement in dissolution of efavirence after co-micronization with polyvinyl pyrrolidone

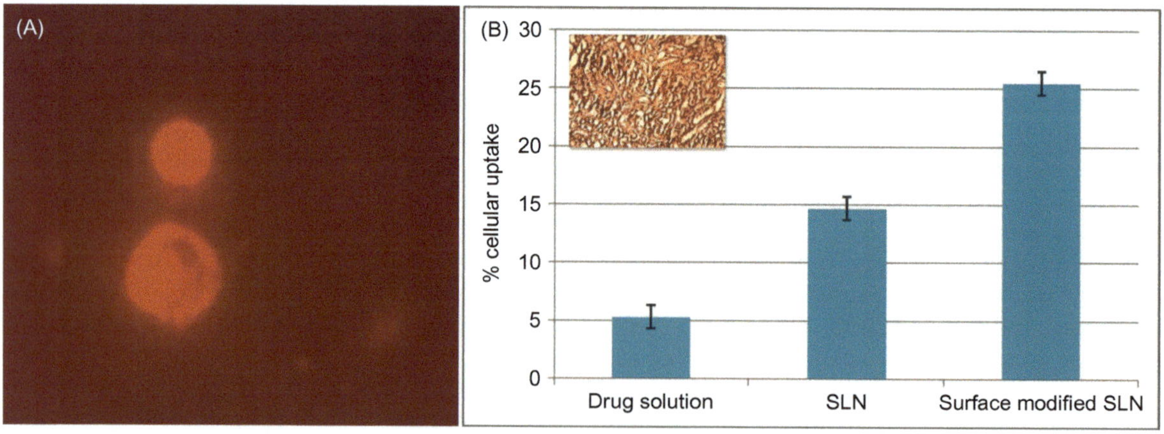

FIGURE 14.4 Photomicrograph of primary macrophages after uptake of stavudine lipid nanoparticles at 30 min (A) and percentage of cellular uptake of uncoated and surface-modified (coated) drug-loaded lipid nanoparticles (B). A histopathology photomicrograph of the spleen in the left corner of the graph shows no sign of pathology in the spleen after repeated administration of uncoated lipid nanoparticles for 28 days.

(PVP) and sodium lauryl sulfate. Efavirenz solid lipid nanoparticles of 124 nm showed fast absorption *in vivo* (Gaur et al., 2014). Nelfinavir mesylate (pK_a 14) nanocrystals 119 − 740 nm/23 mV in size were prepared by the ball milling (30 min) technique to improve bioavailability profile, which is associated with solubility enhancement. Nanosuspensions were stabilized by poloxamer 407, and could successfully stabilize drug particles as dispersions and also on redispersion when freeze-dried using mannitol as cryoprotectant. The solubility of the drug was increased to 8 mg/mL, a two-times higher increase (Naresh et al., 2013). Nevirapine nanosuspensions (772 nm) prepared by solvent evaporation method, showed improved drug release (Chowdary, 2013). Rit-NLC was prepared by sonication method and freeze-dried in the presence of mannitol (6%, w/v) for oral delivery system (Walimbe et al., 2012). Unfortunately, no information on *in vivo* data is reported; it is expected that the lipid carrier helps to reduce associated toxicity *in vivo*.

14.4.5 Metallic Nanoparticles in Chemotherapy

Different types of fullerene-based (C-60) structures, dendrimers, and inorganic nanoparticles, such as gold and silver, showed promising anti-HIV activity. However, these effects are limited to *in vitro* studies in inhibiting HIV replication. The reported metals for anti-HIV action are silver (Lara et al., 2010; Galdiero et al., 2011), gold (Vijayakumar and Ganesan, 2012), and copper (Mastro, 2010). Readers are requested to refer other chapters in this book which covers use of metallic nanoparticles in HIV therapy.

14.4.6 Gene Therapy

Gene therapy is another promising approach for treatment of HIV/AIDS therapy, using nucleic acid–based compounds like DNA, siRNA, RNA decoys, ribozymes, and aptamers or protein-based fusion inhibitors and zinc-finger nucleases. It can also be used to interfere with viral replication. Viral vectors and RNAi can target the various stages of the viral replication cycle or can be directed to the cellular targets like CD4, CCR5, and/or CXCR4. A fusion protein with a peptide transduction domain was used to deliver siRNA to T cells and resulted in CD4-specific and CD8-specific siRNA responses and showed no adverse effects. Single-walled carbon nanotubes showed 90% knockdown of CXCR4 and 60% knockdown of CD4-specific siRNA in human T cells and peripheral blood mononuclear cells. Amino-terminated carbosilane dendrimers (with interior carbon silicon bonds) was successfully tested for delivery of siRNA to HIV-infected lymphocytes and showed promising results. Gene therapy can exhibit synergistic effects if used in combination with chemotherapy or as an adjuvant therapy. Nanoparticles play important role in gene therapy.

14.4.7 Immunization

This approach involves use of immunomodulatory agents to modulate an immune response. The most common approach is by delivery of cytokines like IL-2, IL-7, and IL-15 or antigens. PEG stabilized poly(propylene sulfide) polymer nanoparticles and cross-linked polymer nanoparticles with a pH-responsive core and hydrophilic charged shell. Nanoparticles of the co-polymer poly(D,L-lacticide-co-glycolide) are used as a platform for immunization therapy. HIV p24 protein adsorbed on the surface of surfactant free anionic poly(D,L-lactide) (PLA) nanoparticles showed efficient delivery of antigens and enhanced cellular and mucosal immune responses. A successful example of nanoparticles that reached clinical trials is the DermaVir patch. It has targeted nanoparticles composed of polyethyleimine mannose, glucose, and HIV antigen coding DNA plasmid ~100 nm in size, and it has now entered phase II trials. Preclinical studies and phase I clinical trials confirmed safety and tolerability of the DermaVir patch.

14.4.8 Vaccines

The preventive HIV/AIDS vaccine suffers from various challenges like viral strain and sequence diversity, viral evasion of humoral and cellular immune responses, and lack of methods to elicit broadly reactive neutralizing antibodies and cytotoxic T cells. In addition to formulation and stability issues of the nanoparticulate drug delivery system in body fluids, nanoparticles present a major opportunity for delivery of antigen, initiation of immune responses, and can be adapted for various routes of administration. Currently, various lipid-based systems have been explored for HIV/AIDS vaccine delivery.

In one study, nasal immunization of mice with the HIV gp160 protein liposome composed of cholesterol, sphingomyelin, phosphatidylethanolamine, phosphatidylcholine, and phosphatidylserine resulted in high titers of gp160-specific neutralizing antibody responses. In another study, the HIV gp41 proteins were delivered in the form of liposomes (110–400 nm), which exhibited strong antibody responses in mice and rabbits. Immunization with CpG ODN delivered through liposomes (made of 1-(2-[oleoyloxy] ethyl)-2-oleyl-3-(2-hydroxy-ethyl) imidazolinium chloride (DOTIM) and cholesterol). Nanoemulsion (~350 nm) composed of cetylpyridinium chloride and soybean oil was used for delivery of gp120 protein through the nasal route. Both forms elicited strong immune responses. Another candidate, MF59, in the form of nanoemulsion is approved for its use in more than 20 countries as an adjuvant for an influenza vaccine. Polymeric nanoparticles based on PLA and PLGA were used to deliver protein and DNA showed strong cytotoxic T-cell responses. Surface modification with p24 and gp120 proteins on PLA nanoparticles exhibited high antibody titers. In a separate study, polystyrene nanospheres (~350 nm) coated with inactivated HIV particles were used for nasal immunization (Miyake et al., 2004). Figure 14.5 lists vaccine candidates presently being tested in various phases of clinical trials.

14.4.9 Microbicide

There are more than 40 microbial candidates in preclinical trials and almost 12 candidates in clinical trials. The first dendrimer-based microbicide gel tested clinically, SPL7013 (Vivagel, Starpharma Holdings Ltd., Melbourne, Australia), is a lysine-based dendrimer with naphthalene disulfonic acid surface groups and can be engineered with optimized potency against HIV and HSV89. In a phase I clinical trial, it was found to be safe and well tolerated in healthy women, with no evidence of systemic toxicity or absorption. PSC-RANTES encapsulated PLGA nanoparticles; ~260 nm is being explored for use as a microbicide. A similar type of polymeric nanoparticles is also used for delivery of siRNA and showed sustained gene silencing for up to 14 days. Beta-cyclodextrin-based (beta-CD) drug delivery system was used to improve solubility of UC781. Various antiretroviral drugs showed their presence in development of microbicide, including ritonavir, tenofovir, and stavudine (Cohly et al., 2003).

Chinsembu and Hedimbi (2009) gathered information on 36 plant families containing 46 plant species with known anti-HIV active compounds and known modes of action. Topical microbicides that can be self-administered are being developed as a subset of pre-exposure prophylaxis strategies that together with vaccines might significantly reduce HIV infection. Praneem is a combination of herbal extracts that

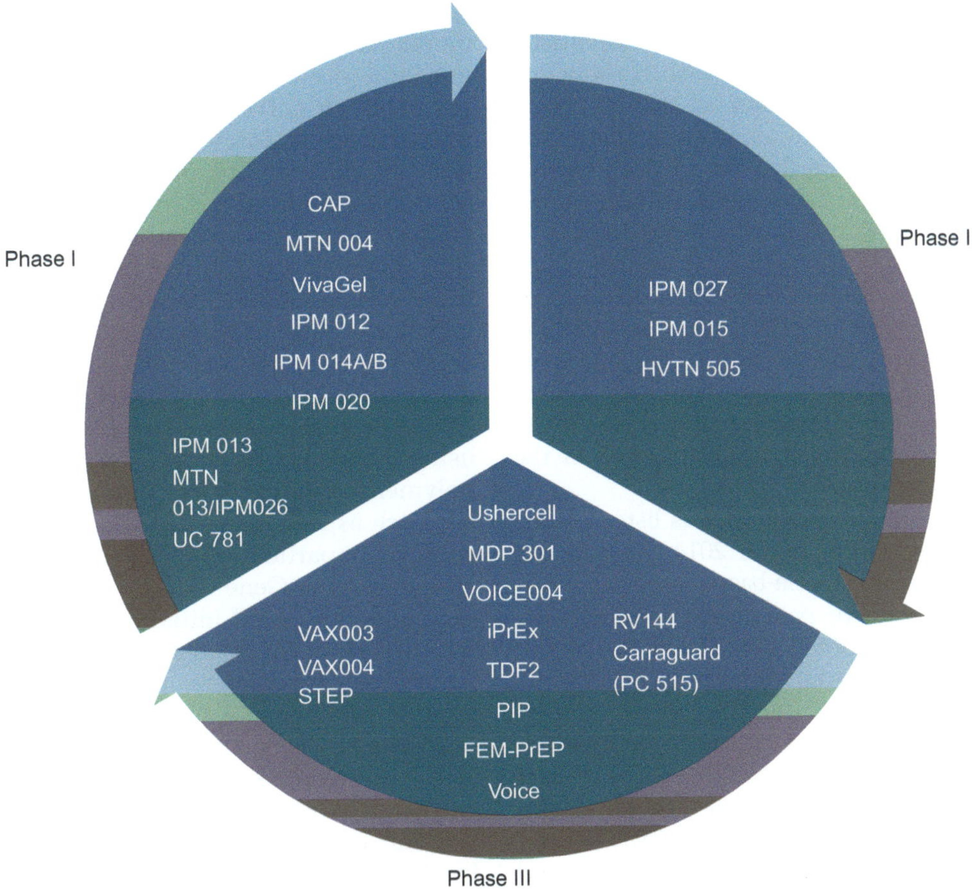

FIGURE 14.5 Overview of vaccine candidates in clinical trials for prophylaxis of HIV/AIDS.

has undergone phase I and phase II safety and acceptability studies (Weblink1). Similarly, another polyherbal cream and stavudine microbicidal gel showed promising HIV activity (Talwar et al., 2008; Shegokar and Singh, 2015). All of the studied candidates are conventional. Nanotechnology can offer enhanced activity if encapsulated in polymeric or lipidic form. Herbal candidates are no exceptions if formulation issues are solved. Use of nanotechnology in developing herbal microbicides is in the preliminary stages and is expected to increase in coming years.

14.5 INDUSTRY APPROACH AND COMMERCIALIZATION SUCCESS

In the past decades, nanotechnology has been rigorously explored for its effectiveness in diagnosis, treatment, and monitoring of disease. Different types of nanoparticle delivery systems are in their infancy, whereas liposomes and polymer-based nanoparticles are being used in various approved therapies including cancer. At Liverpool University, SLNs are developed using freeze-drying or spray-drying of emulsions. These antiretroviral drug nanoparticles can reduce overall drug dose and offer benefits for pediatric administration. These nanoparticles are robust during large-scale GMP production and stability testing at different storage conditions to support clinical documentation in 2014 (Weblink2). Recently, two global pharmaceutical companies have developed once-per-month long-acting nanoformulations for use in chronic diseases like HIV. Two of such HIV drugs are now in later stages of development. Many active firms, such as AbbVie, ViiV Healthcare, Merck & Co., Inc., Janssen Therapeutics, Bristol-Myers Squibb, Gilead Sciences, Inc., Salix Pharmaceuticals, Inc., and Vertex Pharmaceuticals, Inc., are active in the area of antiretrovirals.

One phase I/II clinical trial was listed on the NCT database during April 2014; it evaluates the effect of nanoparticle-based vaccine candidates in seropositive patients by targeting the HIV Tat Protein (EVA TAT). The trial is sponsored by Bioantech (trial number NCT01793818). This trial is conducted using volunteers undergoing antiretroviral treatment at three different doses and placebo; the volunteers have had undetectable viremia (<40 copies/mL) and CD4 cell levels $>350/mm^3$ for at least 1 year. In early HIV infection, HIV-1 Tat protein is secreted and is assumed to protect HIV-1-infected cells from the cellular immune response. Developing a synthetic nanovaccine called Tat Oyi could neutralize the antibodies against Tat variants representative of the five main HIV-1 subtypes and can eliminate virus-infected cells.

The research phase generally costs $30 million; this cost is double in the early development phase and seven- to eight-times higher in the late development phase. Nanotechnological systems need a one-time investment during the early research phase because of the use of sophisticated and advanced equipment and techniques. Major players in the vaccine market are Sanofi, GSK, Merck & Co. Inc., Wyeth, and Novartis, followed by Crucell, Medimmune, Sinovac, and Solvay.

HIV latent reservoirs at anatomical (lymphoid tissue, testes, liver, kidney, lungs, the gut, and the brain) and organ sites are not reachable by current therapies. Among them, microphages mainly contribute to the generation of elusive mutant viral genotypes by serving as the host for viral genetic recombination. Current therapy suffers challenges such as poor eradication of the virus from these reservoirs and longer duration of treatments, resulting in low patient adherence and, most importantly, resistance against certain drugs. Literature data showed successful targeting of anatomical and organ reservoirs by using nanocrystals, lipid nanoparticles, and polymeric nanoparticles. Various nanomaterials such as fullerenes, dendrimers, silver, and gold nanoparticles have shown anti-HIV effects *in vitro*. Gene therapy based on siRNA is in its infancy and initial results are promising for HIV/AIDS treatment. Single-walled nanotubes, dendrimers, fusion proteins, and peptide–antibody conjugates have all been used for delivery of siRNA to HIV-specific cells. Immunotherapy using DNA plasmid encapsulated in nanoparticles delivered through mannose-targeted polyethyleimine is in phase II clinical trials. Local therapy using microbicides is explored with the use of nanoparticles to enhance prophylactic effects, for example, VivaGel, dendrimer-based microbicide gel, and polymeric nanoparticles for delivery of CCR5 inhibitor PSC-RANTES.

14.6 CONCLUSION AND PERSPECTIVES

Vaccination is the most effective way to control HIV/AIDS, but it would take another decade or more for a suitable vaccine candidate to reach the market. Until then, the only available single drug or multiple drug therapy has to play an important role. It takes more than 8 years for only 1 out of 1,000 compounds to cross the gate of the laboratory to clinical trials (toxicity and efficacy) in humans, and only one out of few selected compounds gets final approval. These new patented compounds come at a much higher cost to the patient. In some countries like South Africa, India, and Brazil, governments gave permission to manufacturers to infringe on foreign patents and produce low-cost versions of antiretrovirals without affecting functionality. Recently, many major pharmaceutical companies have lowered the prices of their drugs in selected countries. This ensures that the patient at least has access to affordable antiretroviral therapy. Issues like large dose, poor patient adherence and efficacy, and high toxicity after oral administration of antiretrovirals remain and can only be remedied by intelligent nanoparticle design. Research performed until now shows that nanoparticles have the potential to reduce dosing frequency and toxicity, and can target particular organs or cells and eradicate viral reservoirs if administered parenterally, which is debated all the time. The advantages offered by the nanoparticle platform are worth to explore with a limited production cost increase. Nanoparticles might serve as the best way to improve patient compliance, for example, sustained delivery systems by reducing dosing frequency and maybe duration of treatment (not long) and improvement in efficacy of antiretrovirals. However, the fear of HIV transmission after parenteral administration of nanoparticles needs to be judged by experts from all disciplines and must be discussed from all aspects. Furthermore, exhaustive preclinical and clinical data are required to justify the arguments regarding the administration of nanoparticles in HIV/AIDS patients. Future collaborations of clinicians, scientists, and regulatory bodies might bring hope for using nanoparticles in HIV therapy. The research outcome is still very positive and hope still exists regarding the entry of nanoparticle-based formulations in clinical trials soon.

References

Alukda, D., Sturgis, T., Youan, B.B., 2011. Formulation of tenofovir-loaded functionalized solid lipid nanoparticles intended for HIV prevention. J. Pharm. Sci. 100, 3345–3356.

Alves, L.D., de La Roca Soares, M.F., de Albuquerque, C.T., da Silva, E.R., Vieira, A.C., Fontes, D.A., et al., 2014. Solid dispersion of efavirenz in PVP K-30 by conventional solvent and kneading methods. Carbohydr. Polym. 104, 166–174.

Arts, E.J., Hazuda, D.J., 2012. HIV-1 antiretroviral drug therapy. Cold Spring Harb. Perspect. Med. 2, a007161.

Beloqui, A., Solinis, M.A., Gascon, A.R., del Pozo-Rodriguez, A., des Rieux, A., Preat, V., 2013. Mechanism of transport of saquinavir-loaded nanostructured lipid carriers across the intestinal barrier. J. Control Release 166, 115–123.

Chiappetta, D.A., Hocht, C., Taira, C., Sosnik, A., 2011. Oral pharmacokinetics of the anti-HIV efavirenz encapsulated within polymeric micelles. Biomaterials 32, 2379–2387.

Chinsembu, K.C., Hedimbi, M., 2009. A survey of plants with anti-HIV active compounds and their modes of action. Med. J. Zambia. 36, 178–186.

Chowdary, P.M., 2013. Preparation characterization and *in vitro* evaluation of antoviral nanoparticles. Thesis. KLE University, Department of Pharmaceutical Technology, Bangalore, India.

Cohly, H.H.P., Asad, S., Das, S.K., Angel, M.F., Rao, M., 2003. Effect of antioxidant (turmeric, turmerin and curcumin) on human immunodeficiency virus. Int. J. Mol. Sci. 4.

da Costa, M.A., Seiceira, R.C., Rodrigues, C.R., Hoffmeister, C.R., Cabral, L.M., Rocha, H.V., 2012. Efavirenz dissolution enhancement I: co-micronization. Pharmaceutics 5, 1–22.

da Costa, M.A., Seiceira, R.C., Rodrigues, C.R., Hoffmeister, C.R.D., Cabral, L.M., Rocha, H.V.A., 2013. Efavirenz dissolution enhancement I: co-micronization. Pharmaceutics 5, 1–22.

Dandagi, P.M., Patel, P.D., Gadad, A.P., Aravapalli, A.K., 2012. RES and brain targeting stavudine-loaded solid lipid nanoparticles for AIDS therapy. Asian J. Pharm. 6, 116–123.

Dou, H., Morehead, J., Destache, C.J., Kingsley, J.D., Shlyakhtenko, L., Zhou, Y., et al., 2007. Laboratory investigations for the morphologic, pharmacokinetic, and anti-retroviral properties of indinavir nanoparticles in human monocyte-derived macrophages. Virology 358, 148–158.

Dutta, T., Agashe, H.B., Garg, M., Balakrishnan, P., Kabra, M., Jain, N.K., 2007. Poly (propyleneimine) dendrimer based nanocontainers for targeting of efavirenz to human monocytes/macrophages *in vitro*. J. Drug Target 15, 89–98.

Dutta, T., Garg, M., Jain, N.K., 2008. Targeting of efavirenz loaded tuftsin conjugated poly(propyleneimine) dendrimers to HIV infected macrophages *in vitro*. Eur. J. Pharm. Sci. 34, 181–189.

Fukushima, K., Haraya, K., Terasaka, S., Ito, Y., Sugioka, N., Takada, K., 2008. Long-term pharmacokinetic efficacy and safety of low-dose ritonavir as a booster and atazanavir pharmaceutical formulation based on solid dispersion system in rats. Biol. Pharm. Bull. 31, 1209–1214.

Galdiero, S., et al., 2011. Silver nanoparticles as potential antiviral agents. Molecules 16, 8894–8918.

Garg, M., Asthana, A., Agashe, H.B., Agrawal, G.P., Jain, N.K., 2006. Stavudine-loaded mannosylated liposomes: *in-vitro* anti-HIV-I activity, tissue distribution and pharmacokinetics. J. Pharm. Pharmacol. 58, 605–616.

Gaur, P.K., Mishra, S., Bajpai, M., Mishra, A., 2014. Enhanced oral bioavailability of efavirenz by solid lipid nanoparticles: *in vitro* drug release and pharmacokinetics studies. Biomed. Res. Int. 2014, 363404.

Ilevbare, G.A., Liu, H., Pereira, J., Edgar, K.J., Taylor, L.S., 2013. Influence of additives on the properties of nanodroplets formed in highly supersaturated aqueous solutions of ritonavir. Mol. Pharm. 10, 3392–3403.

Jain, S., Sharma, J.M., Jain, A.K., Mahajan, R.R., 2013. Surface-stabilized lopinavir nanoparticles enhance oral bioavailability without coadministration of ritonavir. Nanomedicine 8, 1639–1655.

Jain, S., Tiwary, A.K., Jain, N.K., 2008. PEGylated elastic liposomal formulation for lymphatic targeting of zidovudine. Curr. Drug Deliv. 5, 275–281.

Kasongo, K.W., Jansch, M., Muller, R.H., Walker, R.B., 2011a. Evaluation of the *in vitro* differential protein adsorption patterns of didanosine-loaded nanostructured lipid carriers (NLCs) for potential targeting to the brain. J. Liposome Res. 21, 245–254.

Kasongo, K.W., Shegokar, R., Muller, R.H., Walker, R.B., 2011b. Formulation development and *in vitro* evaluation of didanosine-loaded nanostructured lipid carriers for the potential treatment of AIDS dementia complex. Drug Dev. Ind. Pharm. 37, 396–407.

Kaur, C.D., Nahar, M., Jain, N.K., 2008. Lymphatic targeting of zidovudine using surface-engineered liposomes. J. Drug Target 16, 798–805.

Kuo, Y.C., Lee, C.L., 2012. Methylmethacrylate-sulfopropylmethacrylate nanoparticles with surface RMP-7 for targeting delivery of antiretroviral drugs across the blood–brain barrier. Colloids Surf. B Biointerfaces 90, 75–82.

Kuo, Y.-C., Lin, P.-I., Wang, C.-C., 2011. Targeting nevirapine delivery across human brain microvascular endothelial cells using transferrin-grafted poly(lactide-co-glycolide) nanoparticles. Nanomedicine 6, 1011–1026.

Lara, H.H., Ayala-Nunez, N.V., Ixtepan-Turrent, L., Rodriguez-Padilla, C., 2010. Mode of antiviral action of silver nanoparticles against HIV-1. J. Nanobiotechnology 8, 1.

Madhavi, B.B., Kusum, B., Krishna Chatanya, C.H., Madhu, M.N., Sri Harsha, V., Banji, D., 2011. Dissolution enhancement of efavirenz by solid dispersion and PEGylation techniques. Int. J. Pharm. Investig. 1, 29–34.

Mahajan, S.D., Roy, I., Xu, G., Yong, K.T., Ding, H., Aalinkeel, R., et al., 2010. Enhancing the delivery of anti retroviral drug "Saquinavir" across the blood brain barrier using nanoparticles. Curr. HIV Res. 8, 396–404.

Mastro, E.A., 2010. Non-toxic inhibition of HIV-1 replication with silver–copper nanoparticles. Med. Chem. Res. 19, 1074–1081.

Miyake, A., Akagi, T., Enose, Y., Ueno, M., Kawamura, M., Horiuchi, R., et al., 2004. Induction of HIV-specific antibody response and protection against vaginal SHIV transmission by intranasal immunization with inactivated SHIV-capturing nanospheres in macaques. J. Med. Virol. 73, 368–377.

Naresh, D.N., Nayak, U.Y., Ranjan, O.P., 2013. Preparation and characterization of nelfinavir mesylate nanocrystals by ball milling. Int. J. Res. Pharm. Sci. 4, 586–592.

Obitte, N.C., Rohan, L.C., Adeyeye, C.M., Parniak, M.A., Esimone, C.O., 2013. The utility of self-emulsifying oil formulation to improve the poor solubility of the anti HIV drug CSIC. AIDS Res. Ther. 10, 14.

Patel, G.V., Patel, V.B., Pathak, A., Rajput, S.J., 2014. Nanosuspension of efavirenz for improved oral bioavailability: formulation optimization, *in vitro*, *in situ* and *in vivo* evaluation. Drug Dev. Ind. Pharm. 40, 80–91.

Prakash, S., Vidyadhara, S., Sasidhar, R.L.C., Abhijit, D., Akhilesh, D., 2013. Development and characterization of Ritonavir nanosuspension for oral use. Der Pharmacia Lettre. 5, 48–55.

Raju, A., Reddy, A.J., Satheesh, J., Jithan, A.V., 2014. Preparation and characterisation of nevirapine oral nanosuspensions. Indian J. Pharm. Sci. 76, 62–71.

Rodriguez-Spong, B., Acciacca, A., Fleisher, D., Rodriguez-Hornedo, N., 2008. PH-induced nanosegregation of ritonavir to lyotropic liquid crystal of higher solubility than crystalline polymorphs. Mol. Pharm. 5, 956–967.

Shegokar, R., 2013. Nanosuspensions: a new approach for organ and cellular targeting in infectious diseases. J. Pharm. Invest. 43, 1–26.

Shegokar, R., Singh, K.K., 2011a. Nevirapine nanosuspensions for HIV reservoir targeting. Pharmazie 66 (6), 408–415.

Shegokar, R., Singh, K.K., 2011b. Surface modified nevirapine nanosuspensions for viral reservoir targeting: *in vitro* and *in vivo* evaluation. Int. J. Pharm. 421, 341–352.

Shegokar, R., Singh, K.K., 2012a. Nevirapine nanosuspensions: stability, plasma compatibility and sterilization. J. Pharm. Invest. 42, 257–269.

Shegokar, R., Singh, K.K., 2012b. Preparation, characterization and cell based delivery of stavudine surface modified lipid nanoparticles. J. Nanomed. Biotherapeutic Discov. 2, 1–9.

Shegokar, R., Singh, K.K., 2015. Development of mucoadhesive gel microbicide to target mucosal HIV reservoirs. J. HIV Clin. Scientific Res. 2 (1), 106.

Shegokar, R., Singh, K.K., Muller, R.H., 2011a. Production & stability of stavudine solid lipid nanoparticles—from lab to industrial scale. Int. J. Pharm. 416, 461–470.

Shegokar, R., Singh, K.K., Müller, R.H., 2011b. Nevirapine nanosuspension: comparative investigation of production methods. Nanotechnol. Dev. e4.

Sinha, S., Ali, M., Baboota, S., Ahuja, A., Kumar, A., Ali, J., 2010. Solid dispersion as an approach for bioavailability enhancement of poorly water-soluble drug ritonavir. AAPS PharmSciTech. 11, 518–527.

Talwar, G.P., Dar, S.A., Rai, M.K., Reddy, K.V., Mitra, D., Kulkarni, S.V., et al., 2008. A novel polyherbal microbicide with inhibitory effect on bacterial, fungal and viral genital pathogens. Int. J. Antimicrob. Agents 32, 180–185.

Tyagi, N., Madhav, N.V.S., 2013. Formulation and evaluation of Zidovudine nanosuspensions using a novel biopolymer from the seeds of *Buchanania lanzan*. J. Drug Deliv. Ther. 3, 85–88.

UNAIDS, 2013. Global report—UNAIDS report on the Global AIDS Epidemic 2013.

van't Klooster, G., Hoeben, E., Borghys, H., Looszova, A., Bouche, M.P., van Velsen, F., et al., 2010. Nanosuspension of rilpivirine (TMC278) as a long-acting injectable antiretroviral formulation: pharmacokinetics and disposition. Antimicrob. Agents Chemother. 54 (5), 2042–2050.

Vijayakumar, S., Ganesan, S., 2012. Gold nanoparticles as an HIV entry inhibitor. Curr. HIV Res. 10, 643–646.

Walimbe, C.A., More, S.S., Walawalkar, R.U., Shah, R.R., Ghodake, D., 2012. Optimisation of nanostructured lipid carriers of Ritonavir. NDDS. 4, 1–8.

Wan, L., Pooyan, S., Hu, P., Leibowitz, M.J., Stein, S., Sinko, P.J., 2007. Peritoneal macrophage uptake, pharmacokinetics and biodistribution of macrophage-targeted PEG-fMLF (*N*-formyl-methionyl-leucyl-phenylalanine) nanocarriers for improving HIV drug delivery. Pharm. Res. 24, 2110–2119.

Weblink1, <http://www.pubfacts.com/search/Praneem>.

Weblink2, <http://nanomednorth.com/2014/02/24/interview-with-the-conference-chair-dr-andrew-owen-nanomedicine-march-2014-edinburgh-scotland/>.

CHAPTER

15

Nanotherapeutic Approach to Targeting HIV-1 in the CNS: Role of Tight Junction Permeability and Blood–Brain Barrier Integrity

Supriya D. Mahajan[2], *Ravikumar Aalinkeel*[2], *Jessica L. Reynolds*[2], *Bindukumar B. Nair*[2], *Manoj J. Mammen*[2], *Lili Dai*[2], *Paras N. Prasad*[1] and *Stanley A. Schwartz*[2]

[1]Institute for Laser, Photonics and Biophotonics, State University of New York at Buffalo, Buffalo, NY, USA [2]Department of Medicine, Division of Allergy, Immunology, and Rheumatology, State University of New York at Buffalo, Clinical Translational Research Center, Buffalo, NY, USA

15.1 INTRODUCTION

Neurological illness is the initial manifestation of AIDS in up to 20% of patients, and more than 50% of all AIDS patients experience some form of neurological abnormality at some point during the disease (Gendelman et al., 1997; Ozdener, 2005, Trujillo et al., 2005; Hult et al., 2008; Vivithanaporn et al., 2011; Ganau et al., 2012; Sagar et al., 2014). The most prominent form of neurological complication associated with HIV-1 infection, excluding opportunistic infections, is a syndrome of combined cognitive and motor dysfunction, referred to as either AIDS/HIV encephalopathy or AIDS dementia complex (ADC) or HIV-associated neurological disorders (Price and Spudich, 2008; Vivithanaporn et al., 2011; Ganau et al., 2012; Sagar et al., 2014). HIV-associated neurological disorders (HAND) are common among HIV patients, and although the clinical and pathological manifestations of HAND have been well-characterized, the pathogenesis of this progressive central nervous system (CNS) disorder remains undefined. The pathogenesis of HAND may be partly attributed to infiltration of the CNS by HIV-1-infected mononuclear cells because the transmigration of HIV-1-infected monocytes and lymphocytes

Nanotechnology in Diagnosis, Treatment and Prophylaxis of Infectious Diseases.
DOI: http://dx.doi.org/10.1016/B978-0-12-801317-5.00015-3

across the blood–brain barrier (BBB) into the CNS is a prerequisite for the development of HIV-associated encephalitis (HIVE) (Ricardo-Dukelow et al., 2007; Wu et al., 2000; Worthylake and Burridge, 2001). The passage of HIV-infected monocytes or neurotoxins into the brain is a highly regulated process involving a complex network of cells of the CNS including brain microvascular endothelial cells (BMVECs) that form the microvasculature of the BBB, astrocytes, perivascular cells, and microglia. Interactions between the activated microglia and macrophages, endothelial cells, and astrocytes enhance neuroinflammation through signaling events that modulate TJ and affect BBB permeability and, consequently, viral load in the CNS. Both chronic inflammation and systemic HIV infection contribute to the disruption of the BBB, further promoting neurotoxicity.

Despite the success of highly active antiretroviral therapy (HAART) that has resulted in increased life expectancy, there continues to be a high incidence of the more common subsyndromal neuropsychological and neurocognitive disorders (HAND) in HIV-1-infected patients (Valcour et al., 2004; Wojna and Nath, 2006; Ances and Ellis, 2007), HIV^+ patients who are relatively healthy and respond well to HAART by immunological suppression as reflected by the improvement in their virologic factors such as $CD4^+$ may continue to be at risk for HAND. HAART that has proven most effective in reducing systemic viral loads in HIV-infected subject has been unable to purge the HIV virus from quiescent reservoirs such as the brain, where the virus is believed to be sequestered for long periods of time. The reduced efficacy of HAART to treat HIV sequestered in the brain is because HAART does not have a direct effect on the HIV-associated inflammatory response, ARV drugs do not cross the BBB in significant amounts, thereby reducing its efficacy on latent viral reservoirs and additionally causing emergence of ARV drug-resistant viral strains, limited BBB permeability, and the short half-life and low bioavailability of ARV drugs. Thus, the structural and functional complexity of BBB is the main impediment to effective ART to the CNS.

Targeted drug delivery using multimodal nanotechnology promises a novel approach in treating neuro-AIDS. Multimodality allows combination of therapeutic and novel diagnostic agents, with the provision of controlled drug delivery and real-time monitoring of drug action. The versatile targeting ability of drug-doped nanoparticles can be exploited for drug delivery across the BBB in a sustained manner and specific targeting to viral sites within the CNS. Nanotechnology-mediated drug delivery across the BBB has been pioneered by Kreuter (2014), who demonstrated brain-specific targeting of surfactant-coated polymeric nanoparticles after their systemic administration. However, this approach often caused irreparable damage to the BBB, allowing hazardous agents such as neurotoxins to infiltrate the brain from general circulation. Pardridge (2003a,b) was first to demonstrate active diffusion of systemically delivered liposomal vehicles into the brain via targeting certain receptors (e.g., transferrin receptors, insulin receptors) that are overexpressed on the apical surface of the BBB. We and others have incorporated transferrin and folates and other peptides such as Angiopep-2 and TAT on the surface of nanoparticles, facilitating BBB permeability via binding of these ligands to receptors present on the BMVECs, which constitute the BBB (Rao et al., 2009; Mahajan et al., 2012a,b). Kanmogne et al. (2012) showed nanoconstructs containing antiretroviral drugs are packaged into mononuclear phagocytes (MP) and can transverse through the human brain microvascular endothelial cells (HBMEC) that constitute the BBB via cell-to-cell contact, and these nanoconstructs had minimal cytotoxicity and incorporation of folate coating on these nanoconstructs, further improving their BBB transversing ability.

Several studies have shown that the nanoparticles are able to transverse the BBB (Mahajan et al., 2008, 2010, 2012a; Bonoiu et al., 2009; Rao et al., 2009; Wong et al., 2010, 2012; Saxena et al., 2012; Pilakka-Kanthikeel et al., 2013; Gidwani and Singh, 2014). The BBB shields the brain from toxic substances in the blood, supplies brain tissues with nutrients, and filters harmful compounds from the brain back to the bloodstream. The close interaction between BMVEC and other components of the neurovascular unit (astrocytes, pericytes, neurons, and basement membrane) ensures proper function of the CNS. Transport across the BBB is strictly limited through physical (tight junctions [TJs]) and metabolic barriers (enzymes, diverse transport systems). The BBB is a limiting factor for the delivery of therapeutic agents into the CNS. BBB breakdown or alterations in transport systems play an important role in the pathogenesis of many CNS diseases, including neuro-AIDS. The TJs play an important role in the regulation of the structural and functional integrity of the BBB. Nanoparticles can effectively cross various *in vitro* and *in vivo* BBB models by endocytosis and/or transcytosis, and current nanotherapeutic strategies are based on increasing drug trafficking across the BBB, improving targeting specificity in the CNS by using novel targeting moieties, improving BBB permeability, and reducing neurotoxicity of these various nanoformulations. Despite these studies, not much information is available on TJ modulation by nanoparticles commonly used in HIV-1 nanotherapeutics. This chapter highlights the effect of nanoparticle modulation of the BBB and its implication for CNS nanotherapeutics, discussed in the context of neuro-AIDS. A better understanding of how these commonly used nanoparticles modulate TJ expression, thereby affecting the transport of the drug nanoformulation across the BBB and allowing the development of nanotherapeutics to improve the BBB permeability and targeted therapy to the CNS.

15.2 HIV-1 RESERVOIR IN THE BRAIN

A latent HIV-1 virus reservoir exists in the brain, mainly in the form of resting $CD4^+$ T cells; however, subpopulations of monocytes, macrophages, dendritic cells, NK cells, and mast cells have been reported to harbor the virus (Chun et al., 2000). The latent HIV reservoirs in the CNS serve as major depot for HIV persistence; therefore, effective nanotechnology-based approaches to eradicate HIV could provide valuable steps toward HIV eradication. Latently infected cells are in cellular quiescence, and antigen stimulation or cytokine induction reactivates the latent provirus, leading to viral replication and reinfection. This reservoir prevents the total eradication of the virus. As a result, the infection is converted into a chronic disease; therefore, to achieve HIV eradication in infected patients, ART has to be combined with drugs that reactivate the dormant viruses. Histone deacetylase (HDAC) inhibitor and protein kinase C activator bryostatin-2 have recently been combined in a nanoformulation that also contains an ARV drug in an attempt to produce a particle capable of both activating latent virus and inhibiting viral spread (Choi et al., 2011; Kovochich et al., 2011). Thus, nanotechnological approaches that provide improved methods for activating latent HIV and simultaneously inhibit viral replication are an ideal therapeutic strategy to target HIV in sequestered sites. especially the brain.

15.3 CHARACTERISTICS OF NANOPARTICLES THAT ENHANCE THEIR APPLICABILITY TO BIOMEDICAL APPLICATION

The advent of nanomaterials has revolutionized the field of medicine (in both diagnosis and therapy) because of their robustness, safety, and, most importantly, multimodality, that is, the ability to perform multiple functions in the same agent. Advantages that nanoparticulate imaging agents have over traditional molecular imaging agents are as follows: (i) they provide a tunable, fluorescent chassis on which targeting agents (antibodies, peptides, or small molecules) can be added or changed to suit a specific need; (ii) they allow for multimodality (e.g., optical, MR, and radionuclide) imaging, thus permitting cross-evaluation for the same nanoparticle across different imaging platforms; (iii) they can be functionalized with both imaging and therapeutic abilities; (iv) they may be designed to provide salutary benefits related to pharmacokinetics by renal or fecal excretion; (v) they are of sufficient size to permit multivalency and, therefore, have the potential for higher affinity binding than standard agents; and (vi) they enable imaging from the single cell level to the entire intact organism *in vivo*.

Antiretrovirals either alone or in combination can be encapsulated or chemically linked to the surface of nanoparticles, which then offer more stability to encapsulated drugs in biological fluids and also protect the encapsulated drugs against enzymatic degradation. Because of their small size, they can be taken up by cells in systemic circulation or even cells such as BMVEC that line the BBB, whereas larger particles are often precluded by a barrier such as the BBB. Further, nanomaterials can provide improved drug delivery by virtue of their robustness, safety, and multimodality or multifuctionality. The multifunctional nanoparticle can simultaneously carry therapeutic agents, targeting molecules such as conjugated antibodies and imaging signal-contrast agents. Nanoparticles can be used for delivery of antivirals to targeted infected cells within reservoirs by virtue of targeting agents and molecules, and these nanoparticles can also act as imaging agents that allow for real-time tracking of the drug within cells. Nanotechnology-based delivery systems enhance the distribution of hydrophobic and hydrophilic drugs into and within tissue compartments; this ability renders them appropriate for clinical use in HIV-1 therapeutics. To have clinical applicability, nanoparticles should have certain characteristics. Nanoparticles used in biomedical applications are solid colloidal particles typically in the size range of 10–100 nm. Typically, if they are designed as nonspherical (rod like), this can dramatically extend the particle's circulation time *in vivo*. If particles are spherical and are in the range of 100–200 nm in size, then they have the highest potential for prolonged circulation because they are large enough to avoid uptake in the liver but small enough to avoid filtration by the spleen. To modulate the biodistribution profile of nanoparticles, that is, to increase their circulating time *in vivo*, particles must avoid being taken-up by the liver and the spleen. This is accomplished by engineering the particle such that the particle size is <200 nm while maintaining its two-dimensional structure; however, its accumulation in the liver must be prevented while still allowing the nanoparticles to navigate the sinusoids of the spleen. The surface chemistry of the nanoformulation heavily influences the process of opsonization, which allows the nanoparticle to evade capture by the reticuloendothelial system (RES). The use of specific cellular targeting ligands, which will then bind cell surface receptors of specific cells in the CNS, to which therapy is directed to, is also a necessary characteristic. Few types of nanoparticles have already been synthesized and used as drug carriers or targeted contrast agent for targeted drug delivery across the BBB, which are mainly lipid- or polymer-based (Table 15.1). However, poor efficiency of CNS-specific delivery, triggering of immunogenic reactions, and capture by nontarget sites such as liver and spleen have been some of the major limitations with such systems

TABLE 15.1 Nanocarriers Used in HIV-1 Therapeutics Directed to the CNS

Nanoparticle type	Key references
Natural biodegradable polymers: Chitosan	• Ramana et al. (2014) • Aghasadeghi et al. (2013) • Giacalone et al. (2013) • Belletti et al. (2012) • Al-Ghananeem et al. (2010) • Yang et al. (2010) • Föger et al. (2007) • Nayak et al. (2009)
Synthetic biodegradable polymers: Poly(L-lactide) or Poly(lactide-co-glycolide) based nanoparticles	• Chaowanachan et al. (2013) • Mainardes and Gremião (2012) • Shibata et al. (2013) • Kreuter (2013) • Rao (2008) • Mahajan et al. (2014)
Dendrimers	• Gajbhiye et al. (2013) • Jiménez et al (2010)
Solid polymeric nanoparticles	• Kuo and Lee (2012) • Kuo and Chung (2012) • Kuo and Chen (2006) • Kuo and Su (2007) • Kuo and Kuo (2008) • Lin (2012)
Inorganic nanoparticles	• Mahajan et al. (2010) • Bonoiu et al. (2009)
Magnetoelectric nanoparticles	• Nair et al. (2013) • Pilakka-Kanthikeel et al. (2013)
Liposomes	• Pinzón-Daza (2013)

that have significantly slowed the progress of nanodelivery to the brain. The use of high-quality inorganic nanoparticles as targeted delivery vehicles to facilitate the transmigration of specific molecules through BBB may open new avenues that can expedite CNS-specific drug delivery. The use of inorganic nanoparticles as high-resolution image-contrast agents as well as drug/gene delivery vehicles for numerous biological and biomedical applications has become an area of intense research focus over the past decade (Bruchez Jr. et al., 1998; Chan and Nie, 1998; Pardridge, 2003a,b; Vinogradov, 2004, Costantino et al., 2005; Tiwari and Amiji, 2006; Yong et al., 2007). Inorganic nanoparticles offer several advantages over polymeric and liposomal carriers, such as ultrasmall size, robust composition, nonantigenicity, and sustained release of therapeutics. The nanoparticle surface can be functionalized with a variety of biomolecules for the purpose of evading capture by the RES as well as delivery across biological barriers and target specificity.

Gold nanoparticles (GNPs) and nanorods (GNRs) have wide-ranging applications in the biomedical sciences because of their biocompatibility and ease of bioconjugation. The use of GNRs as multimodal probes for numerous biological and biomedical applications has become an area of intense research over the past decade (Shah M et al., 2014; Majdalawieh et al., 2014). GNPs and GNRs have valuable qualities such as low toxicity, low immunogenicity, and excellent biocompatibility (Majdalawieh et al., 2014). Their surfaces can be easily modified to incorporate a high cationic charge, which in turn is capable of forming electrostatic complexes with anionic polynucleotides (DNA and RNA).

We previously reported on the use of high-quality quantum rods (QRs) as targeted probes to study the transmigration of specific molecules through the BBB (Xu et al., 2008). The use of nanoparticles as luminescent probes for numerous biological and biomedical applications is an area of intense research focus (Bruchez Jr. et al., 1998; Chan and Nie, 1998). QRs are bright, photostable CdSe/ZnS fluorescent nanocrystals that exhibit tunable emission properties for a wide range of color possibilities. QRs offer several advantages over organic dyes, including increased brightness, stability against photobleaching, excitation within a broad spectral range, and a tunable and narrow emission spectrum. The QR surface can be functionalized with a variety of biomolecules and have shown the potential to dramatically outperform conventional organic dyes in the imaging of cellular and subcellular structures and in a variety of bioassays such as immunostaining and reporter gene expression.

Because of the apprehension in the pharmaceutical industry and the US Food and Drug Administration (FDA) regarding the use of inorganic metal-based nanoparticles in therapeutics, more research is now being focused on using biodegradable nanoparticle-based therapeutics. A similar trend is observed for HIV therapeutics as well. Biodegradable nanoparticles consist of those derived from natural and synthetic polymers. Chitosan is a fully biodegradable and biocompatible natural polymer. Chitosan occurs naturally in fungal cell walls and crustacean shells. Chitosan has been investigated extensively as a potential drug carrier because of its biocompatibility. The degree of deacetylation and the molecular weight of chitosan can be altered to obtain different physicomechanical properties, and studies have shown that they can be used as drug delivery systems (Föger et al., 2007; Nayak et al., 2009; Al-Ghananeem et al., 2010; Yang, L. et al., 2010, Belletti et al., 2012; Ramana et al., 2014; Aghasadeghi et al., 2013). Biodegradable synthetic polymers such as poly(L-lactide)-based or poly(lactide-co-glycolide)-based nanoparticles are derived via a synthetic route via ring opening polymerization (ROP) and click chemistries designed to achieve well-defined cationically modified poly-lactides (CPLAs). Typically, these well-defined CPLAs are successfully obtained through thiolene reaction of a tertiary amine-based thiol with an allyl-functionalized polylactide. The extent of cationic modification with tertiary amines can be effectively controlled by adjusting the feed ratio of reagents (Li et al., 2011). After cationic substitution, the remaining allyl groups will be modified using thiol-PEG. PEGylation will result in a "polymeric brush" that will shield the hydrophobic and cationic groups within the polymer to generate a stealth imaging and gene delivery system that can permeate across the BBB to reach the CNS (Li et al., 2011; Lu et al., 2005). These CPLAs demonstrated significant hydrolytic degradability and relatively low cytotoxicity. A biodegradable nanoplatform using CPLAs as scaffolds for drug delivery is a design that satisfies the essential requirements of safe and efficient nanocarriers. Moreover, additional ligands can be added to nanoplexes to target them to specific cells and tissues of interest. We previously reported that the efficiency of nanoplex-mediated gene silencing is significantly better than that achieved with commercial siRNA delivery agents (Bonoiu et al., 2009). The biodegradable nanoplex protects the drug from degradation by virtue of its attachment to the polymeric scaffold that prevent its degradation by the extracellular matrix (ECM), intracellular enzymatic degradation, and the acidic endosomal environment. Cellular uptake of this nanoplex is enhanced by a net positive surface charge on the protonable amine in the nanocarrier that can act as a weak base to absorb protons generated by ATPase, thereby facilitating endosomal escape and improvement of the transfection efficiency of the nanoplex. Aerosolization of this biodegradable nanocarrier for intranasal delivery to the CNS to circumvent the BBB is an ARV drug delivery strategy that remains to be explored.

Multifunctional NPs can be fabricated using the biodegradable polymer poly(lactic-co-glycolic acid) (PLGA). The PLGA-based NPs can increase the efficacy of ARV treatments because of the sustained release of the therapeutic agent from stable nanoparticles. They can improve pharmacokinetic and pharmacodynamic profiles. Important features of PLGA-based nanoformulation for antiretroviral drug delivery are as follows: (i) its biodegradability and biocompatibility are well-established in several preclinical studies; (ii) the FDA has approved its use in drug delivery systems for parenteral administration; therefore, it can quickly move to phase III clinical trials; (iii) PLGA has well-described formulations and methods of production that can be adapted to drugs that are hydrophilic or hydrophobic small molecules or macromolecules; (iv) PLGA protects the antiretroviral drug from degradation; (v) it allows sustained release of the ARV drug from the nanoformulation; and (vi) PLGA also allows the modification of its surface properties to allow better interaction with biological ligands such as antibodies and other targeting peptides. PLGA NPs have been used as vaccine adjuvants in HIV vaccines and in conjunction with other HIV drugs (Pawar et al., 2013).

15.4 NANOTECHNOLOGY-BASED HIV THERAPEUTICS

Several nanotechnology-based systems have been explored for HIV therapeutics. Nanotechnology-based HIV therapy offers unique advantages like enhancement of bioavailability, water solubility, stability, and targeting ability of ARV drugs. Nanoparticles used in HIV therapeutics mainly include liposomes, organic and inorganic nanoparticles, polymeric micelles, and dendrimers (Sharma et al., 2012; Parboosing et al., 2012; Mahajan et al., 2012a,b; Ramana et al., 2014; Date and Destache, 2013; Martinez-Skinner et al., 2013; Siccardi et al., 2013; Giacalone et al., 2013; Kumar et al., 2014). We have shown significant uptake of a quantum rod-transferin-Saquinavir (QR-Tf-Saquinavir) conjugate by BMVECs, along with a marked decrease in HIV-1 viral replication in the PBMCs treated with this (QR-Tf-Saquinavir) nanoconjugate (Mahajan et al., 2010). Additionally, we have stably incorporated the antiretroviral drug, Amprenavir, within a transferrin (Tf)-conjugated quantum dot (QD), and evaluated its BBB transversing ability and analyzed its antiviral efficacy in HIV-1-infected monocytes (Mahajan et al., 2012a,b). Although these antiretroviral delivery approaches using inorganic nanoparticles show promise, barriers to the clinical translation of these nanoformulations are rooted in the fact that these inorganic nanoparticles in larger sustained doses may cause significant cytotoxicity; therefore, current ARV nanotherapy approaches are focused on using biodegradable nanoparticles that have no significant toxicity *in vivo*. Recently, magnetic nanoparticles have been used as carriers to improve the delivery of nanoformulated ARV drugs across the BBB (Nair et al., 2013). ARVs with different physicochemical properties can be encapsulated individually into nanoparticles to potently inhibit HIV. Dash et al. (2012) recently showed that combination therapy or nano-ART, wherein ART such as Atazanavir and ritonavir was formulated in a poloxamer-188, which is a nonionic triblock copolymers that has surfactant properties that make them useful in increasing the miscibility of two substances with different hydrophobicities, has significant antiretroviral and neuroprotective activities *in vivo* in humanized NOD/scid-IL-2Rgc (NSG) mice, concluding that nano-ART has translational potential with sustained and targeted efficacy with limited toxicity. Dual targeting of the anti-HIV drug zidovudine (ZDV) via sialic acid–conjugated mannosylated poly (propyleneimine) (PPI) dendritic nanoconstructs also showed enhanced biocompatibility and site-specific delivery of the antiretroviral drug ZDV (Gajbhiye et al., 2013). Recent studies by Destache et al. (2009) showed that three antiretroviral drugs, ritonavir, lopinavir, and efavirenz, could be conjugated to PLGA (cART-NPs), and that sustained release of these antiretrovirals was observed from PLGA NP for 28 days without any noticeable toxicity (Destache et al., 2009). The same group demonstrated that in HIV-1-infected H9 monocytic cells treated with cART-NPs subcellular fractionation, these cells contained significantly higher nuclear, cytoskeleton, and membrane antiretroviral drug levels compared with cells treated with ARV drug alone. cART-NPs efficiently inhibited HIV-1 infection and transduction, thus demonstrating the efficacy of the PLGA NP formulation in inhibiting HIV-1 replication (Shibata et al., 2013). Chaowanachan et al. (2013) showed that antiretroviral drugs with different physicochemical properties such as Saquinavir, Efavirenz, and Tenofovir can be encapsulated individually into PLGA nanoparticles to potently inhibit HIV. Kuo and Su (2007) demonstrated that nanoparticles synthesized using a combination of three biodegradable components and the antiretroviral drug saquinavir (SQV) are encapsulated within the particle core, which is composed of PLGA to form SQV-PLGA NPs, and the surface of SQV-PLGA NPs was grafted successively with hydrophilic polyethyleneimine (PEI) and poly-(γ-glutamic acid) (γ-PGA). They showed that this nanoparticulate carrier system proved efficacious in delivering antiretroviral drugs across the BBB (Kuo and Chen, 2006; Kuo and Su, 2007; Kuo and Kuo, 2008; Kuo and Yu, 2011; Kuo and Chung, 2012; Kuo and Lee, 2012). In addition to CNS delivery of a nanoformulation that encapsulates the ARV drug, other nanotechnology-based strategies that can modulate the permeability of the BBB can facilitate the crossing-over of the ARV drugs from systemic circulation to the CNS. We have previously shown that a stable complex between MMP-9-specific siRNA and a QD can be formed that protects the MMP-9 siRNA from biological degradation. Treatment of BMVEC with this QD-MMP-9-siRNA (nanoplex) downregulated the expression of the MMP-9 gene, and this silencing of MMP-9 gene expression resulted in the upregulation of ECM proteins like collagen I, IV, V, and a decrease in endothelial permeability, as reflected by reduction of transendothelial resistance across the BBB, demonstrating that such strategies could be used to modulate the permeability of the BBB (Bonoiu et al., 2009).

Thus, nanotechnology can revolutionize the field of HIV medicine not only by improving diagnosis but also by improving delivery of antiretrovirals to targeted regions in the body, and by significantly enhancing the efficacy of the currently available antiretroviral medications.

15.5 THE BLOOD–BRAIN BARRIER

The BBB is a critical interface. It acts as a physical and metabolic barrier between the CNS and the peripheral circulation that serves to regulate and protect the microenvironment of the brain. The primary function of the normal BBB is to establish and maintain homeostasis in the CNS (Bradbury, 1993). The BBB is not rigid and comprises dynamic vessels that are capable of responding to rapid changes in the brain or blood. The BBB is composed of specialized brain capillary endothelial cells and astrocytic end feet that enhance the differentiation of the BBB endothelium. The BBB is composed of at least three types of cell-to-cell junctional structures between adjacent endothelial cells and/or between endothelial cells and astrocytes, the gap junctions, adherens junction, and the TJs. The high expression of TJ proteins is a special characteristic of the BBB.

15.6 *IN VITRO* MODEL OF THE HUMAN BBB

Several *in vitro* tissue culture systems have been developed to reproduce the physical and biochemical properties of the intact BBB. However, most of these systems lack essential features that are characteristic of the *in vivo* BBB. There is a considerable interest in establishing *in vitro* BBB cell culture models for several reasons, including studying drugs that penetrate the BBB and understanding how dysfunction of the BBB is involved in the pathogenesis of various neurological diseases. A good *in vitro* BBB model used to study neuroAIDS must reproduce the salient features of the *in situ* BBB and also must allow for manipulations to enable the researcher to mimic neuropathogenic process. In our laboratory, we have validated the transwell co-culture models (Persidsky and Gendelman, 1997; Persidsky et al., 2000, 2006a,b; Mukhtar and Pomerantz, 2000) that use primary normal human astrocytes (NHAs) and BMVECs, both of which are cell types that constitute the *in vivo* BBB and are grown to confluence on a PET membrane insert, thus providing an opportunity to examine the expression of TJ proteins, permeability, and transmigration of different leukocyte populations under different experimental stimuli. The *in vitro* BBB system we use has been extensively used and well-validated by several researchers (Persidsky and Gendelman, 1997; Persidsky et al., 2000, 2006a; Mukhtar and Pomerantz, 2000). This *in vivo* BBB model uses primary cultures of both BMVECs (Cat# ACBRI-376) and NHAs (Cat# ACBRI-371), which were obtained from Applied Cell Biology Research Institute (ACBRI, Kirkland, WA). Characterization of BMVECs demonstrated that >95% cells were positive for cytoplasmic von Willibrand factor/factor VIII. BMVECs were cultured in CS-C complete serum-free medium (ABCRI, Cat# SF-4Z0-500) with attachment factors (ABCRI, Cat# 4Z0-210) and Passage Reagent Group™ (ABCRI, Cat# 4Z0-800). NHAs were cultured in the CS-C medium, supplemented with 10 μg/mL human epidermal growth factor, 10 mg/mL insulin, 25 μg/mL progesterone, 50 mg/mL transferrin, 50 mg/mL gentamicin, 50 μg/mL amphotericin-B, and 10% FBS. NHAs were characterized on the basis of >99% of these cells being positive for glial fibrillary acidic protein (GFAP). Both BMVECs and NHAs were obtained at passage 2 for each experiment and were used for all experiments between 2 and 8 passages within the 6 to 27 cumulative population doublings. The BBB model used consists of two-compartment wells in a six-well culture plate, with the upper compartment separated from the lower by a 3-μM polyethylene terephthalate (PET) insert (surface area = 4.67 cm^2). The BMVECs were grown to confluency on the upper side of the insert, and a confluent layer of NHAs was grown on the underside. The formation of a functional and intact BBB takes a minimum of 5 days, which can be confirmed by determining the transendothelial electrical resistance (TEER) value. TEER across the *in vitro* BBB is measured using an Ohm meter (Millicell ERS system; Millipore, Bedford, MA, Cat# MERS 000−01). Electrodes were sterilized using 95% alcohol and rinsed in distilled water before measurement. A constant distance of 0.6 cm was maintained between the electrodes at all times during TEER measurement.

15.7 ROLE OF TJ PROTEIN IN BBB PRESERVATION

Endothelial TJs that are present between BMVECs that constitute the BBB form a diffusion barrier that selectively exclude most blood-borne substances from entering the brain. Astrocytic end feet tightly sheath the vessel wall and are critical for the induction and maintenance of the TJ barrier (Ballabh et al., 2004; Persidsky et al., 2006b). TJs of the BBB play a critical role in controlling cellular traffic into the CNS (Annunziata et al., 1998; Aurrand-Lions et al., 2000, 2001; Andras et al., 2003, 2005). Neuroinflammation as a result of viral−host interactions leads to breakdown of the BBB and subsequent dysfunction of the CNS. In HIV-1-infected patients, there is an initial breach in the BBB that also permits the influx of soluble, circulating HIV-1 proteins that may cause neurotoxicity (Kanmogne et al., 2002, 2005, 2007). The major molecular components of the TJ include the transmembranous and structural proteins, occludin, JAM, and claudins, and the submembranous peripheral ZO proteins (Annunziata et al., 1998; Furuse et al., 1998, 1999; Andras et al., 2003, 2005). ZO proteins are essential for targeting TJ structures, and they are linked to the actin cytoskeleton and related signal transducing mechanisms critical for TJ function (Tsukita et al., 1997). JAM proteins are not connected to the ancillary proteins of the cytoplasm, but they affect passage of cells when endothelial or mononuclear cells are activated (Andras et al., 2003, 2005). TJ and ZO are highly sensitive to the microenvironment and respond to inflammatory cytokines *in vitro*, resulting in an alteration in the subcellular localization and dissociation of the occludin/ZO complex, which are associated with an impaired BBB.

15.8 KEY TJ PROTEINS—JAM-2, ZO-1, CLAUDIN-5, AND OCCLUDIN

TJ formation and disruption are processes that involve a complex interaction between different TJ proteins and can be modulated by HIV-1 viral proteins (Andras et al., 2003, 2005). Occludin is a 60-kDa integral membrane protein shown to be functionally important for barrier function; specifically, occludin plays a key role in the redox regulation of TJs (Balda et al., 1996). Claudins constitute a large family of 20- to 27-kDa membrane proteins (with four transmembrane domains) expressed in TJs. Brain endothelial cells predominantly express claudin-3 and claudin-5 and significantly contribute to TJ formation and BBB integrity (Morita et al., 1999; Wolburg et al., 2003). In addition to occludin and claudins, JAM proteins, although not essential to TJ formation in endothelial cells, are involved in the facilitation of assembly of TJ components and in the establishment of cell polarity. Increased transmigration *in vitro* across endothelial cells overexpressing JAM-2 at intercellular contacts and regulation of paracellular permeability by JAM-2 were observed by Aurrand-Lions et al. in 2000. Under physiological conditions, phosphorylation regulates the maintenance and assembly of TJs. Excessive or unregulated phosphorylation leads to decreased expression of occludin at the TJ, resulting in increased BBB permeability. However, excessive phosphorylation of ZO-1 does not decrease ZO-1 expression; rather, inflammation results in an increase in ZO-1 expression, which is attributed to an increase in ZO-1 transcription (Wong, 1997). Claudin expression remains unchanged despite an inflammation stimulus, suggesting its involvement as a major protein in maintaining TJ integrity (Huber et al., 2001). Several claudin variants exist; however, claudin-5 plays a key role in TJ integrity at the BBB (Huber et al., 2001). Several studies highlight the complex yet variable role of these TJ proteins in maintaining BBB integrity under neuroinflammatory conditions (Balda et al., 1996; Morita et al., 1999; Wolburg et al., 2003).

15.9 MECHANISMS OF TJ MODULATION

The function of TJs and their regulation may involve specific expression of junctional proteins or distinct activation events, which regulate the stability of these TJs. Important cell–cell adhesive events between transmigrating leukocytes and the apical junction complex and the subsequent signaling events result in transmigration across the BBB. Disruptions of the TJ involve the phosphorylation of myosin light chain (MLC) and the regulation of TJ proteins (Huber et al., 2001). MLCK is a Ca^{2+}-calmodulin–dependent multifunctional enzyme that plays a critical role in cellular migration. It regulates the contractile interaction between actin microfilaments and myosin by phosphorylating the MLC during nonmuscle cell contraction, cytokinesis, stress fiber formation, and cell motility. Phosphorylation of MLC results in cytoskeletal alterations, and these cytoskeletal alterations and modulation of TJ proteins result in BBB impairment. We have shown that morphine and HIV-1 viral protein tat alone and in combination result in the initiation of an inflammatory response that releases pro-inflammatory cytokines, which stimulate intracellular calcium release and MLCK activation, and modulates changes in the expression of cytoskeletal/TJ proteins, resulting in alterations in BBB integrity, thereby enhancing leukocyte migration across the *in vitro* BBB (Mahajan et al., 2008). BBB integrity is sensitive to the phosphorylation state of specific residues within myosin regulatory light chain (RLC), and its activation can lead to BBB dysfunction, thereby promoting leukocyte migration across the BBB. An increase in MMP-9 activity at sites of BBB disruption exhibiting leukocyte infiltration has been reported (Yang, Y. et al., 2007). A recent study has shown that myosin light chain kinase (MLCK) significantly reduced the inductive effect of TGF-β1 on MMP-9 synthesis (Sinpitaksakul et al., 2008), demonstrating that TGF-β1 induced MMP-9 through MLCK activation. Additionally, Rho GTPases have been reported to play an important role in the adhesion of transmigrating leukocytes to the BBB and their subsequent diapedesis across the BBB (Persidsky et al., 2006b).

15.10 EFFECT OF NANOPARTICLES ON TJ PROTEINS IN BMVEC CULTURES

Transport of drugs across the BMVEC occurs through traversing occludins and claudins in the TJs or by attaching drugs to the existing BBB transporter systems. The BBB allows small molecules to enter the CNS and, typically, potential CNS drugs have large structures; therefore, utilization of nanocarriers for transport of CNS drugs that use the existing BBB transporter system such as receptor endocytosis or transcytosis is most common. Recently, Alyautdin et al. (2014) showed that the inulin spaces after intravenous injection of polysorbate 80-coated nanoparticles were increased by 1% compared with controls, indicating that there is no large-scale opening of the TJs of the brain endothelium by the polysorbate 80-coated nanoparticles. Limited information is available regarding the effect of nanoparticles on TJ modulation. Although many studies have reported the role of TJs in BBB integrity, not much information is available with respect to how nanoparticles affect TJ expression and related signaling mechanisms involved in TJ modulation, cytoskeletal reorganization, and the complex protein–protein interactions necessary for the maintenance of TJ integrity. We evaluated the effects of four different types of nanoparticles (chitosan, PLGA, CPLA, GNR, and QR) commonly used in neurotherapeutics on ZO-1, JAM-2, claudin-5, and occludin expression. Further, we also examined if these nanoparticles altered BBB integrity by measuring the TEER across the BBB using a well-established *in vitro* BBB model. We evaluated the effect of four different nanoparticles, chitosan, PLGA, CPLA, GNR, and QR, on TJ expression. BMVECs were treated with 3–5 mg/mL concentrations of chitosan, PLGA, CPLA, GNR, and QR for 48 h, and the gene expression levels of TJ proteins such as claudin-5, JAM-2, ZO-1, and occludin were quantitated using real-time quantitative polymerase chain reaction (QPCR). Figure 15.1A–D shows that there was no significant change in the gene expression levels of TJ proteins claudin-5, JAM-2, ZO-1, and occludin as compared with the untreated control. Further, we treated the *in vitro* BBB model with the same nanoparticles for a period of 48 h and did not observe any change in BBB permeability as assessed by measuring the TEER across

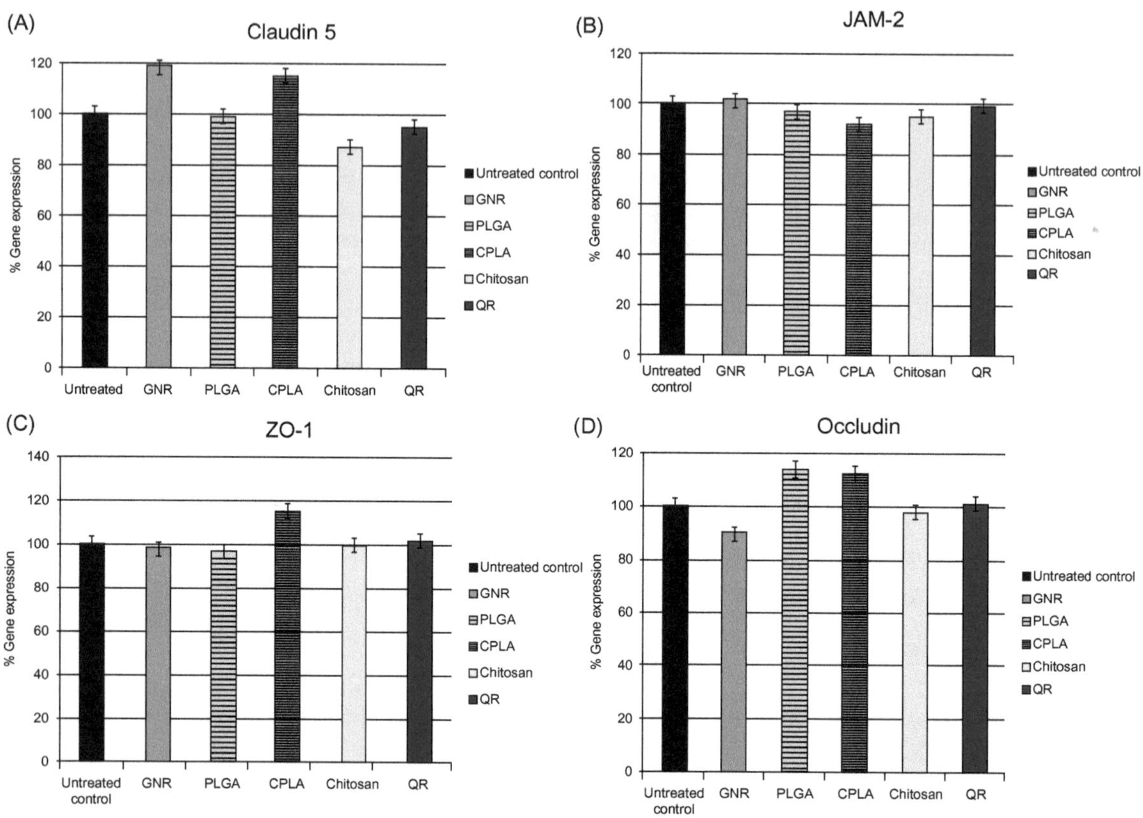

FIGURE 15.1 Effect of nanoparticles on TJ expression. BMVECs were treated with 3–5 mg/mL concentrations of chitosan, PLGA, CPLA GNR, and QR for 48 h, and the gene expression levels of TJ proteins such as claudin-5, JAM-2, ZO-1, and occludin were quantitated using real-time QPCR. Figure 15.1A–D: There was no significant change in the gene expression levels of TJ proteins claudin-5, JAM-2, ZO-1, and occludin as compared with the untreated control.

the BBB (Figure 15.2). We also determined the cell viability of the BMVECs before and after treatment with the nanoparticles to determine if the nanoparticles had any cytotoxic effects. Typically, 10,000 BMVECs suspended in 100 μL of media were incubated for 48 h with the chitosan, PLGA, CPLA, GNR, and QR nanoparticles in a 96-well plate at doses (~3–5 mg/mL) that are typically used for preparing the nanoformulation. At the end of the incubation period, 10 μL of MTT reagent (Cat# 30–1010 K; ATCC) was added to the cells, followed by incubation for ~3 h, which was followed by addition of a detergent solution to lyse the cells and solubilize the colored crystals. The amount of color produced was directly proportional to the number of viable cells. Colorimetric detection was performed at a wavelength of 570 nm. MTT cell proliferation assay measures the reduction of a tetrazolium component [MTT-3-(4,5-dimethylthiazol-2-yl)-2,5-diphenyltetrazolium bromide, a tetrazole] into an insoluble formazan product by the mitochondria of viable cells. The results of our MTT assay showed that BMVECs treated with chitosan, PLGA, CPLA, GNR, and QR nanoparticles for 48 h showed no significant cytotoxicity. All these data indicate that these

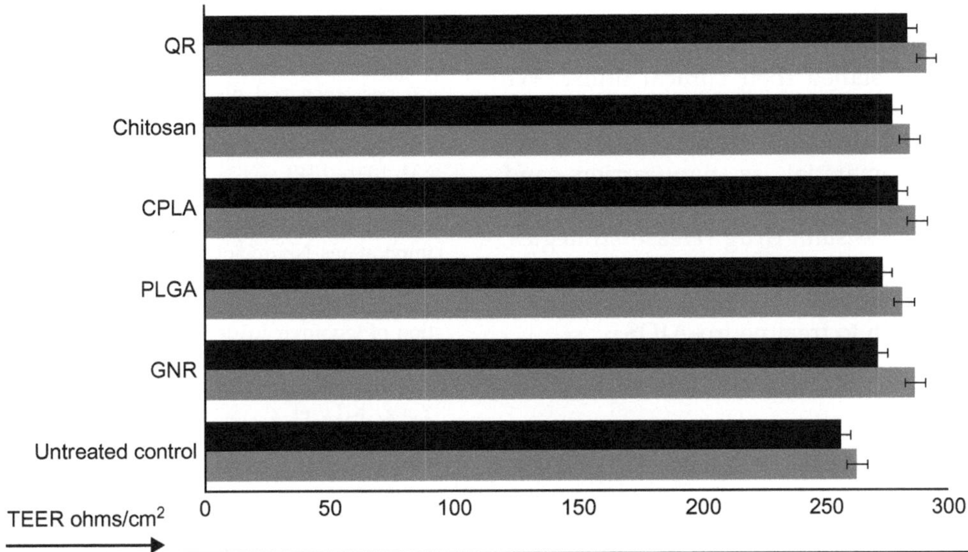

	Untreated control	GNR	PLGA	CPLA	Chitosan	QR
■ After	256	271	273	279	277	283
■ Before	262.66	286	281	287	284	291

FIGURE 15.2 Effect of nanoparticles on BBB integrity. Using a well-validated *in vitro* BBB model, we evaluated the effect of chitosan, PLGA, CPLA GNR, and QR nanoparticles on BBB integrity as assessed by measuring the TEER across the BBB. Our data show no significant change in TEER values between the nanoparticle-treated BBB as compared with the untreated BBB.

nanoparticles by themselves are nontoxic and do not affect BBB integrity at the concentrations tested; these concentrations are typically used in drug nanoformulations and therefore are ideal nanocarriers for antiretroviral drugs.

15.11 CONCLUSION

Targeting HIV sequestered in sanctuary sites such as the brain remains a formidable challenge in HIV therapeutics. However, nanotechnology approaches have shown tremendous promise because of their ability to prolong systemic circulation, cross the BBB, provide specific cell targeting, enhance intracellular drug levels in the CNS, and reduce the inherent toxicity of many antiretroviral drugs. Design and development of an ideal nanoformulation for CNS therapeutics require an interdisciplinary effort with expertise from nanochemists, biomedical engineers, biologists, biomedical scientists, and clinicians. Expertise in identification of biological targets, colloidal chemistry, drug–drug interactions, and associated renal-hepato toxicity are key to developing a new nanotherapeutic drug. Synthesis strategies should be focused on developing nontoxic, biocompatible, and biodegradable nanoparticles that overcome nonspecific organ uptake and RES. Extensive safety and toxicological studies need to be performed in parallel with biodistribution studies. Several antiretroviral drug–based nanoformulations are at various preclinical stages of study; however, no successful clinical trials that use nanoformulations have been reported. Rigorous pharmacodynamics and pharmacokinetic studies are needed before using these antiretroviral drug-based nanoformulations. Additionally, issues regarding HIV latency and the effects of long-term accumulation of these nanoparticles in the CNS have stalled their clinical utility. We believe that an understanding of the structure and physiology of the BBB, utilization of biodegradable nanomaterial as nanocarriers and development of specific strategies to target specific CNS cells, sustain drug release strategies, and create ease of administration of nanoformulations will be key to the development of a nanoformulation to treat neuro-AIDS.

References

Aghasadeghi, M.R., Heidari, H., Sadat, S.M., Irani, S., Amini, S., Siadat, S.D., et al., 2013. Lamivudine-PEGylated chitosan: a novel effective nanosized antiretroviral agent. Curr. HIV Res. 11 (4), 309–320.

Al-Ghananeem, A.M., Saeed, H., Florence, R., Yokel, R.A., Malkawi, A.H., 2010. Intranasal drug delivery of didanosine-loaded chitosan nanoparticles for brain targeting: an attractive route against infections caused by AIDS viruses. J. Drug Target 18 (5), 381–388.

Alyautdin, R., Khalin, I., Nafeeza, M.I., Haron, M.H., Kuznetsov, D., 2014. Nanoscale drug delivery systems and the blood–brain barrier. Int. J. Nanomed. 9, 795–811.

Ances, B.M., Ellis, R.J., 2007. Dementia and neurocognitive disorders due to HIV-1 infection. Semin. Neurol. 27 (1), 86–92.

Andras, I.E., Pu, H., Deli, M.A., Nath, A., Hennig, B., Toborek, M., 2003. HIV-1 Tat protein alters tight junction protein expression and distribution in cultured brain endothelial cells. J. Neurosci. Res. 74, 255.

Andras, I.E., Pu, H., Tian, J., Deli, M.A., Nath, A., Hennig, B., et al., 2005. Signaling mechanisms of HIV-1 Tat-induced alterations of claudin-5 expression in brain endothelial cells. J. Cereb. Blood Flow Metab. 25 (9), 1159–1170.

Annunziata, P., Cioni, C., Toneatto, S., Paccagnini, E., 1998. HIV-1 gp120 increases the permeability of rat brain endothelium cultures by a mechanism involving substance P. AIDS 12, 2377–2385.

Aurrand-Lions, M., Duncan, L., Ballestrem, C., Imhof, B.A., 2001. JAM-2, a novel immunoglobulin superfamily molecule, expressed by endothelial and lymphatic cell. J. Biol. Chem. 276 (4), 2733–2741.

Aurrand-Lions, M.A., Duncan, L., Du Pasquier, L., Imhof, B.A., 2000. Cloning of JAM-2 and JAM-3: an emerging junctional adhesion molecular family?. Curr. Top. Microbiol. Immunol. 251, 91–98.

Balda, M.S., Whitney, J.A., Flores, C., Gonzalez, S., Cereijido, M., Matter, K., 1996. Functional dissociation of paracellular permeability and transepithelial electrical resistance and disruption of the apical-basolateral intramembrane diffusion barrier by expression of a mutant tight junction membrane protein. J. Cell Biol. 134, 1031–1049.

Ballabh, P., Braun, A., Nedergaard, M., 2004. The blood–brain barrier: an overview: structure, regulation, and clinical implications. Neurobiol. Dis. 16 (1), 1–13.

Belletti, D., Tosi, G., Forni, F., Gamberini, M.C., Baraldi, C., Vandelli, M.A., et al., 2012. Chemico-physical investigation of tenofovir loaded polymeric nanoparticles. Int. J. Pharm. 436 (1–2), 753–763.

Bonoiu, A., Mahajan, S.D., Ye, L., Kumar, R., Ding, H., Yong, K.T., et al., 2009. MMP-9 gene silencing by a quantum dot-siRNA nanoplex delivery to maintain the integrity of the blood brain barrier. Brain Res. 1282, 142–155.

Bradbury, M.W., 1993. The blood–brain barrier. Exp. Physiol. 78 (4), 453–472.

Bruchez Jr., M., Moronne, M., Gin, P., Weiss, S., Alivisatos, A.P., 1998. Semiconductor nanocrystals as fluorescent biological labels. Science 281 (5385), 2013–2016.

Chan, W.C., Nie, S., 1998. Quantum dot bioconjugates for ultrasensitive nonisotopic detection. Science 281 (5385), 2016–2018.

Chaowanachan, T., Krogstad, E., Ball, C., Woodrow, K.A., 2013. Drug synergy of tenofovir and nanoparticle-based antiretrovirals for HIV prophylaxis. PLoS One 8, e61416.

Choi, S., Lee, J., Kumar, P., Lee, K.Y., Lee, S.K., 2011. Single chain variable fragment CD7 antibody conjugated PLGA/HDAC inhibitor immuno-nanoparticles: developing human T cell-specific nano-technology for delivery of therapeutic drugs targeting latent HIV. J. Control Release 152 (Suppl. 1), e9–10.

Chun, T.W., Davey Jr., R.T., Ostrowski, M., Shawn Justement, J., Engel, D., Mullins, J.I., et al., 2000. Relationship between pre-existing viral reservoirs and the re-emergence of plasma viremia after discontinuation of highly active anti-retroviral therapy. Nat. Med. 6, 757–761.

Costantino, L., Gandolfi, F., Tosi, G., Rivasi, F., Vandelli, M.A., Forni, F., 2005. Peptide-derivatized biodegradable nanoparticles able to cross the blood–brain barrier. J. Control Release 108 (1), 84–96.

Dash, P.K., Gendelman, H.E., Roy, U., Balkundi, S., Alnouti, Y., Mosley, R.L., et al., 2012. Long-acting nanoformulated antiretroviral therapy elicits potent antiretroviral and neuroprotective responses in HIV-1-infected humanized mice. AIDS 26 (17), 2135–2144.

Date, A.A., Destache, C.J., 2013. A review of nanotechnological approaches for the prophylaxis of HIV/AIDS. Biomaterials 34 (26), 6202–6228.

Destache, C.J., Belgum, T., Christensen, K., Shibata, A., Sharma, A., Dash, A., 2009. Combination antiretroviral drugs in PLGA nanoparticle for HIV-1. BMC Infect. Dis. 9, 198.

Föger, F., Kafedjiiski, K., Hoyer, H., Loretz, B., Bernkop-Schnürch, A., 2007. Enhanced transport of P-glycoprotein substrate saquinavir in presence of thiolated chitosan. J. Drug Target 15 (2), 132–139.

Furuse, M., Fujita, K., Hiiragi, T., Fujimoto, K., Tsukita, S., 1998. Claudin-1 and -2: novel integral membrane proteins localizing at tight junctions with no sequence similarity to occludin. J. Cell. Biol. 141 (7), 1539–1550.

Furuse, M., Sasaki, H., Tsukita, S., 1999. Manner of interaction of heterogeneous claudin species within and between tight junction strands. J. Cell Biol. 147, 891.

Gajbhiye, V., Ganesh, N., Barve, J., Jain, N.K., 2013. Synthesis, characterization and targeting potential of zidovudine loaded sialic acid conjugated-mannosylated poly(propyleneimine) dendrimers. Eur. J. Pharm. Sci. 48 (4–5), 668–679.

Ganau, M., Prisco, L., Pescador, D., Ganau, L., 2012. Challenging new targets for CNS-HIV infection. Front Neurol. 3, 43.

Gendelman, H.E., Persidsky, Y., Ghorpade, A., Limoges, J., Stins, M., Fiala, M., et al., 1997. The neuropathogenesis of the AIDS dementia complex. AIDS 11 (Suppl. A), S35–S45.

Giacalone, G., Bochot, A., Fattal, E., Hillaireau, H., 2013. Drug-induced nanocarrier assembly as a strategy for the cellular delivery of nucleotides and nucleotide analogues. Biomacromolecules 14 (3), 737–742.

Gidwani, M., Singh, A.V., 2014. Nanoparticle enabled drug delivery across the blood brainbarrier: *in vivo* and *in vitro* models, opportunities and challenges. Curr. Pharm. Biotechnol. 14 (14), 1201–1212.

Huber, J.D., Egleton, R.D., Davis, T.P., 2001. Molecular physiology and pathophysiology of tight junctions in the blood–brain barrier. Trends Neurosci. 24, 719.

Hult, B., Chana, G., Masliah, E., Everall, I., 2008. Neurobiology of HIV. Int. Rev. Psychiatry. 20 (1), 3–13.

Jiménez, J.L., Clemente, M.I., Weber, N.D., Sanchez, J., Ortega, P., de la Mata, F.J., et al., 2010. Carbosilane dendrimers to transfect human astrocytes with small interfering RNA targeting human immunodeficiency virus. BioDrugs. 24 (5), 331–343.

Kanmogne, G.D., Kennedy, R.C., Grammas, P., 2002. HIV-1 gp120 proteins and gp160 peptides are toxic to brain endothelial cells and neurons: possible pathway for HIV entry into the brain and HIV-associated dementia. J. Neuropathol. Exp. Neurol. 61 (11), 992–1000.

Kanmogne, G.D., Primeaux, C., Grammas, P., 2005. HIV-1 gp120 proteins alter tight junction protein expression and brain endothelial cell permeability: implications for the pathogenesis of HIV-associated dementia. J. Neuropathol. Exp. Neurol. 64 (6), 498–505.

Kanmogne, G.D., Schall, K., Leibhart, J., Knipe, B., Gendelman, H.E., Persidsky, Y., 2007. HIV-1 gp120 compromises blood–brain barrier integrity and enhances monocyte migration across blood–brain barrier: implication for viral neuropathogenesis. J. Cereb. Blood Flow Metab. 27 (1), 123–134.

Kanmogne, G.D., Singh, S., Roy, U., Liu, X., McMillan, J., Gorantla, S., et al., 2012. Mononuclear phagocyte intercellular crosstalk facilitates transmission of cell-targeted nanoformulated antiretroviral drugs to human brain endothelial cells. Int. J. Nanomed. 7, 2373–2388.

Kovochich, M., Marsden, M.D., Zack, J.A., 2011. Activation of latent HIV using drug-loaded nanoparticles. PLoS One 6 (4), e18270.

Kreuter, J., 2013. Mechanism of polymeric nanoparticle-based drug transport across the blood–brain barrier (BBB). J. Microencapsul. 30 (1), 49–54.

Kreuter, J., 2014. Drug delivery to the central nervous system by polymeric nanoparticles: what do we know? Adv. Drug Deliv. Rev. 71C, 2–14.

Kumar, L., Verma, S., Prasad, D.N., Bhardwaj, A., Vaidya, B., Jain, A.K., 2014. Nanotechnology: a magic bullet for HIV AIDS treatment. Artif. Cells Nanomed. Biotechnol. Available from: <http://dx.doi.org/10.3109/21691401.2014.883400>.

Kuo, Y.C., Chen, H.H., 2006. Effect of nanoparticulate polybutylcyanoacrylate and methylmethacrylate–sulfopropylmethacrylate on the permeability of zidovudine and lamivudine across the *in vitro* blood–brain barrier. Int. J. Pharm. 327 (1–2), 160–169.

Kuo, Y.C., Chung, C.Y., 2012. Transcytosis of CRM197-grafted polybutylcyanoacrylate nanoparticles for delivering zidovudine across human brain-microvascular endothelial cells. Colloids Surf. B Biointerfaces 91, 242–249.

Kuo, Y.C., Kuo, C.Y., 2008. Electromagnetic interference in the permeability of saquinavir across the blood–brain barrier using nanoparticulate carriers. Int. J. Pharm. 351 (1–2), 271–281.

Kuo, Y.C., Lee, C.L., 2012. Methylmethacrylate–sulfopropylmethacrylate nanoparticles with surface RMP-7 for targeting delivery of antiretroviral drugs across the blood–brain barrier. Colloids Surf. B Biointerfaces 90, 75–82.

Kuo, Y.C., Su, F.L., 2007. Transport of stavudine, delavirdine, and saquinavir across the blood–brain barrier by polybutylcyanoacrylate, methylmethacrylate-sulfopropylmethacrylate, and solid lipid nanoparticles. Int. J. Pharm. 340 (1–2), 143–152.

Kuo, Y.C., Yu, H.W., 2011. Transport of saquinavir across human brain-microvascular endothelial cells by poly (lactide-co-glycolide) nanoparticles with surface poly-(γ-glutamic acid). Int. J. Pharm. 416, 365–375.

Li, J., Feng, L., Fan, L., Zha, Y., Guo, L., Zhang, Q., et al., 2011. Targeting the brain with PEG-PLGA nanoparticles modified with phage-displayed peptides. Biomaterials 32 (21), 4943–4950.

Lin, C.M., Lu, T.Y., 2012. C60 fullerene derivatized nanoparticles and their application to therapeutics. Recent Pat. Nanotechnol. 6 (2), 105–113.

Lu, W., Tan, Y.Z., Hu, K.L., Jiang, X.G., 2005. Cationic albumin conjugated PEGylated nanoparticle with its transcytosis ability and little toxicity against blood–brain barrier. Int. J. Pharm. 295 (1–2), 247–260.

Mahajan, S.D., Aalinkeel, R., Sykes, D.E., Reynolds, J.L., Nair, B.B., Fernandez, S.F., et al., 2008. Tight junction regulation by morphine and HIV-1 tat modulates blood brain barrier permeability. J. Clin. Immunol. 28 (5), 528–541.

Mahajan, S.D., Roy, I., Xu, G., Yong, K.T., Ding, H., Aalinkeel, R., et al., 2010. Enhancing the delivery of antiretroviral drug "Saquinavir" across the blood brain barrier using nanoparticles. Curr. HIV Res. 8 (5), 396–404.

Mahajan, S.D., Law, W.C., Aalinkeel, R., Reynolds, J.L., Nair, B.B., Yong, K.T., et al., 2012a. Nanoparticle-mediated targeted delivery of antiretrovirals to the brain. Methods Enzymol. 509, 41–60.

Mahajan, S.D., Aalinkeel, R., Law, W.C., Reynolds, J.L., Nair, B.B., Sykes, D.E., et al., 2012b. Anti-HIV-1 nanotherapeutics: promises and challenges for the future. Int. J. Nanomed. 7, 5301–5314.

Mahajan, S.D., Yun, Y., Aalinkeel, R., Reynolds, J.L., Nair, B. B., Mammen, M.J., et al., 2014. Biodegradable nanoparticle based antiretroviral therapy across the blood brain barrier. In: Bawa, R., Audette, Gerald F, Rubinstein, I. (Eds.), Handbook of Clinical Nanomedicine-From Benchto Bedside, Pan Stanford Series in Nanomedicine, vol. 1. Pan Stanford Publishing/CRC Press, Singapore.

Mainardes, R.M., Gremião, M.P., 2012. Nanoencapsulation and characterization of zidovudine on poly(L-lactide) and poly(L-lactide)-poly(ethylene glycol)-blend nanoparticles. J. Nanosci. Nanotechnol. 12 (11), 8513–8521.

Majdalawieh, A., Kanan, M.C., El-Kadri, O., Kanan, S.M., 2014. Recent advances in gold and silver nanoparticles: synthesis and applications. J. Nanosci. Nanotechnol. 14 (7), 4757–4780.

Martinez-Skinner, A.L., Veerubhotla, R.S., Liu, H., Xiong, H., Yu, F., McMillan, J.M., et al., 2013. Functional proteome of macrophage carried nanoformulated antiretroviral therapy demonstrates enhanced particle carrying capacity. J. Proteome Res. 12 (5), 2282–2294.

Morita, K., Furuse, M., Fujimoto, K., Tsukita, S., 1999. Claudin multigene family encoding four-transmembrane domain protein components of tight junction strands. Proc. Natl. Acad. Sci. U.S.A. 96, 511–516.

Mukhtar, M., Pomerantz, R.J., 2000. Development of an *in vitro* blood–brain barrier model to study molecular neuropathogenesis and neurovirologic disorders induced by human immunodeficiency virus type 1 infection. J. Hum. Virol. 3 (6), 324–334.

Nair, M., Guduru, R., Liang, P., Hong, J., Sagar, V., Khizroev, S., 2013. Externally controlled on-demand release of anti-HIV drug using magneto-electric nanoparticles as carriers. Nat. Commun. 4, 1707.

Nayak, U.Y., Gopal, S., Mutalik, S., Ranjith, A.K., Reddy, M.S., Gupta, P., et al., 2009. Glutaraldehyde cross-linked chitosan microspheres for controlled delivery of zidovudine. J. Microencapsul. 26 (3), 214–222.

Ozdener, H., 2005. Molecular mechanisms of HIV-1 associated neurodegeneration. J. Biosci. 30 (3), 391–405.

Pardridge, W.M., 2003a. Blood–brain barrier genomics and the use of endogenous transporters to cause drug penetration into the brain. Curr. Opin. Drug Discov. Devel. 6 (5), 683–691.

Parboosing, R., Maguire, G.E., Govender, P., Kruger, H.G., 2012. Nanotechnology and the treatment of HIV infection. Viruses 4 (4), 488–520.

Pardridge, W.M., 2003b. Blood–brain barrier drug targeting: the future of brain drug development. Mol. Interv. 3 (2), 90–105, 51.

Pawar, D., Mangal, S., Goswami, R., Jaganathan, K.S., 2013. Development and characterization of surface modified PLGA nanoparticles for nasal vaccine delivery: effect of mucoadhesive coating on antigen uptake and immune adjuvant activity. Eur. J. Pharm. Biopharm. 85 (3 Pt A), 550–559.

Persidsky, Y., Gendelman, H.E., 1997. Development of laboratory and animal model systems for HIV-1 encephalitis and its associated dementia. J. Leukoc. Biol. 62 (1), 100–106.

Persidsky, Y., Zheng, J., Miller, D., Gendelman, H.E., 2000. Mononuclear phagocytes mediate blood–brain barrier compromise and neuronal injury during HIV-1-associated dementia. J. Leukoc. Biol. 68 (3), 413–422.

Persidsky, Y., Ramirez, S.H., Haorah, J., Kanmogne, G.D., 2006a. Blood–brain barrier: structural components and function under physiologic and pathologic conditions. J. Neuroimmune Pharmacol. 1 (3), 223–236.

Persidsky, Y., Heilman, D., Haorah, J., Zelivyanskaya, M., Persidsky, R., Weber, G.A., et al., 2006b. Rho-mediated regulation of tight junctions during monocyte migration across blood–brain barrier in HIV-1 encephalitis (HIVE). Blood 107 (12), 4770–4780.

Pilakka-Kanthikeel, S., Atluri, V.S., Sagar, V., Saxena, S.K., Nair, M., 2013. Targeted brain derived neurotropic factors (BDNF) delivery across the blood–brain barrier for neuro-protection using magnetic nano carriers: an *in-vitro* study. PLoS One 8 (4), e62241.

Price, R.W., Spudich, S., 2008. Antiretroviral therapy and central nervous system HIV type 1 infection. J. Infect. Dis. 197 (Suppl. 3), S294–S306.

Pinzón-Daza, M.L., Campia, I., Kopecka, J., Garzón, R., Ghigo, D., Riganti, C., 2013. Nanoparticle- and liposome-carried drugs: new strategies for active targeting and drug delivery across blood–brain barrier. Curr. Drug Metab. 14 (6), 625–640.

Ramana, L.N., Sharma, S., Sethuraman, S., Ranga, U., Krishnan, U.M., 2014. Evaluation of chitosan nanoformulations as potent anti-HIV therapeutic systems. Biochim. Biophys. Acta 1840 (1), 476–484.

Rao, K.S., Reddy, M.K., Horning, J.L., Labhasetwar, V., 2008. TAT-conjugated nanoparticles for the CNS delivery of anti-HIV drugs. Biomaterials 29 (33), 4429–4438.

Rao, K.S., Ghorpade, A., Labhasetwar, V., 2009. Targeting anti-HIV drugs to the CNS. Expert Opin. Drug Deliv. 6 (8), 771–784.

Ricardo-Dukelow, M., Kadiu, I., Rozek, W., Schlautman, J., Persidsky, Y., Ciborowski, P., et al., 2007. HIV-1 infected monocyte-derived macrophages affect the human brain microvascular endothelial cell proteome: new insights into blood–brain barrier dysfunction for HIV-1-associated dementia. J. Neuroimmunol. 185 (1–2), 37–46.

Sagar, V., Pilakka-Kanthikeel, S., Pottathil, R., Saxena, S.K., Nair, M., 2014. Towards nanomedicines for neuroAIDS. Rev. Med. Virol. 24 (2), 103–124.

Saxena, S.K., Tiwari, S., Nair, M.P., 2012. Nanotherapeutics: emerging competent technology in neuroAIDS and CNS drug delivery. Nanomedicine (Lond). 7 (7), 941–944.

Shah, M., Badwaik, V.D., Dakshinamurthy, R., 2014. Biological applications of gold nanoparticles. J. Nanosci. Nanotechnol. 14 (1), 344–362.

Sharma, P., Chawla, A., Arora, S., Pawar, P., 2012. Novel drug delivery approaches on antiviral and antiretroviral agents. J. Adv. Pharm. Technol. Res. 3 (3), 147–159.

Shibata, A., McMullen, E., Pham, A., Belshan, M., Sanford, B., Zhou, Y., et al., 2013. Polymeric nanoparticles containing combination antiretroviral drugs for HIV type 1 treatment. AIDS Res. Hum. Retroviruses 29, 746–754.

Siccardi, M., Martin, P., McDonald, T.O., Liptrott, N.J., Giardiello, M., Rannard, S., et al., 2013. Nanomedicines for HIV therapy. Ther. Deliv. 4 (2), 153–156.

Sinpitaksakul, S.N., Pimkhaokham, A., Sanchavanakit, N., Pavasant, P., 2008. TGF-beta1 induced MMP-9 expression in HNSCC cell lines via Smad/MLCK pathway. Biochem. Biophys. Res. Commun. 371 (4), 713–718.

Tiwari, S.B., Amiji, M.M., 2006. A review of nanocarrier-based CNS delivery systems. Curr. Drug Deliv. 3 (2), 219–232.

Trujillo, J.R., Jaramillo-Rangel, G., Ortega-Martinez, M., Penalva de Oliveira, A.C., Vidal, J.E., Bryant, J., et al., 2005. International NeuroAIDS: prospects of HIV-1 associated neurological complications. Cell Res. 15 (11–12), 962–969.

Tsukita, S., Furuse, M., Itoh, M., 1997. Molecular architecture of tight junctions: occludin and ZO-1. Soc. Gen. Physiol. Ser. 52, 69–76.

Valcour, V.G., Shikuma, C.M., Watters, M.R., Sacktor, N.C., 2004. Cognitive impairment in older HIV-1-seropositive individuals: prevalence and potential mechanisms. AIDS 18 (Suppl. 1), S79–S86.

Vinogradov, S., 2004. The second annual symposium on nanomedicine and drug delivery: exploring recent developments and assessing major advances. 19–20 August 2004, Polytechnic University, Brooklyn, NY. Expert Opin. Drug Deliv. 1 (1), 181–184.

Vivithanaporn, P., Gill, M.J., Power, C., 2011. Impact of current antiretroviral therapies on neuroAIDS. Expert Rev. Anti Infect. Ther. 9 (4), 371–374.

Wojna, V., Nath, A., 2006. Challenges to the diagnosis and management of HIV dementia. AIDS Read. 16 (11), 615–616, 621–624, 626, 629–632.

Wolburg, H., Wolburg-Buchholz, K., Kraus, J., Rascher-Eggstein, G., Liebner, S., Hamm, S., et al., 2003. Localization of claudin-3 in tight junctions of the blood–brain barrier is selectively lost during experimental autoimmune encephalomyelitis and human glioblastoma multiforme. Acta Neuropathol. 105, 586–592.

Wong, H.L., Chattopadhyay, N., Wu, X.Y., Bendayan, R., 2010. Nanotechnology applications for improved delivery of antiretroviral drugs to the brain. Adv. Drug Deliv. Rev. 62 (4–5), 503–517.

Wong, H.L., Wu, X.Y., Bendayan, R., 2012. Nanotechnological advances for the delivery of CNS therapeutics. Adv. Drug Deliv. Rev. 64 (7), 686–700.

Wong, V., 1997. Phosphorylation of occludin correlates with occludin localization and function at the tight junction. Am. J. Physiol. Cell Physiol. 273, C1859–C1867.

Worthylake, R.A., Burridge, K., 2001. Leukocyte transendothelial migration: orchestrating the underlying molecular machinery. Curr. Opin. Cell Biol. 13, 569.

Wu, D.T., Woodman, S.E., Weiss, J.M., McManu, C.M., D'Aversa, T.G., Hesselgesser, J., et al., 2000. Mechanisms of leukocyte trafficking into the CNS. J. Neurovirol. 6 (Suppl. 1), S82.

Xu, G., Yong, K.T., Roy, I., Mahajan, S.D., Ding, H., Schwartz, S.A., et al., 2008. Bioconjugated quantum rods as targeted probes for efficient transmigration across an *in vitro* blood–brain barrier. Bioconjug Chem. 19 (6), 1179–1185.

Yang, L., Chen, L., Zeng, R., Li, C., Qiao, R., Hu, L., et al., 2010. Synthesis, nanosizing and *in vitro* drug release of a novel anti-HIV polymeric prodrug: chitosan-*O*-isopropyl-5′-*O*-d4T monophosphate conjugate. Bioorg. Med. Chem. 18 (1), 117–123.

Yang, Y., Estrada, E.Y., Thompson, J., Liu, W., Rosenberg, G. A., 2007. Matrix metalloproteinase-mediated disruption of tight junction proteins in cerebral vessels is reversed by synthetic matrix metalloproteinase inhibitor in focal ischemia in rat. J. Cereb. Blood Flow Metab. 27 (4), 697–709.

Yong, K.T., Roy, I., Lee, H.H., Bergey, E.J., Tramposch, K. M., He, S., et al., 2007. Quantum rod bioconjugates as targeted probes for confocal and two-photon fluorescence imaging of cancer cells. Nano Lett. 7 (3), 761–765.

CHAPTER

16

A Novel Fungicidal Action of Silver Nanoparticles: Apoptosis Induction

Won Young Lee and Dong Gun Lee

School of Life Sciences and Biotechnology, College of Natural Sciences, Kyungpook National University, Daegu, Republic of Korea

16.1 INTRODUCTION

Because of the appearance of microorganisms that are insensitive to conventional drugs, the unique properties and wide application of silver and silver compounds have attracted researchers from all over the world to study their effective antimicrobial activities (Klasen, 2000; Sharma et al., 2009; Hwang et al., 2012b). Silver-based antiseptics that may have broad-spectrum activity against bacteria, fungi, and viruses are much less likely to induce microbial resistance than conventional antibiotics (Jones et al., 2004). In particular, silver nanoparticles (nano-Ag) are most promising because they show good antimicrobial properties due to their large ratio of surface area to volume; these properties are used as bactericide on burn wounds, fillers in dental cavities to prevent infection, thin coats on medical devices to prevent microbial biofilm formation, in air and water purification systems, in wastewater treatment plants, and in food processing for controlling microbial contamination (Jain and Pradeep, 2005; Kumar et al., 2008; Parashar et al., 2011; Soni et al., 2010). Silver is also known as being nontoxic and safe to the human body at lower concentrations, unlike other metal nanoparticles (Oberdörster et al., 2005; Gurunathan et al., 2009).

It is generally accepted that nano-Ag act in three main ways with the microorganism to produce antimicrobial effects: interacting with membrane or cell wall (Yamanaka et al., 2005; Kim et al., 2009); binding with DNA (Yang et al., 2009); and having an effect on microbial proteins (Yamanaka et al., 2005). Recently, nano-Ag have been reported to exert antimicrobial activity against yeast via a different mechanism known as apoptosis (Hwang et al., 2012b).

As important regulators of yeast apoptosis, the accumulation of intracellular reactive oxygen species (ROS) has been widely recognized as a crucial factor of cell death and has been connected to many of the known apoptotic pathways in yeast (Benaroudj et al., 2001; Carmona-Gutierrez et al., 2010). In this chapter, it is

Nanotechnology in Diagnosis, Treatment and Prophylaxis of Infectious Diseases.
DOI: http://dx.doi.org/10.1016/B978-0-12-801317-5.00016-5

introduced to address the relationship between ROS and apoptosis in *Candida albicans*, which is one of the most successful opportunistic pathogens in humans and causes life-threatening contagious infections with high mortality rates (Hwang et al., 2012b). Additionally, the cells undergoing nano-Ag-induced apoptosis experience particular phenomena, including ROS production, phosphatidyl serine exposure, cytochrome *c* release, mitochondrial membrane depolarization, caspase activation, DNA fragmentation, nuclear condensation, and cell cycle arrest (Figure 16.1) (Madeo et al., 1997; Bortner and Cidlowski, 2007). Furthermore, the synergistic effects of nano-Ag with conventional antibiotics were investigated (Hwang et al., 2012a) and it was confirmed that nano-Ag have notable antimicrobial effects for clinical application.

16.2 ROS ACCUMULATION

Yeast cells produce ROS from intracellular metabolites under normal physiological conditions, but cellular damage that these can cause is prevented by antioxidant machinery that neutralizes the ROS and repairs molecular damage or degrades oxidized molecules. However, when the cells are placed under specific stress conditions, ROS accumulates and consequently exceeds the antioxidant system of

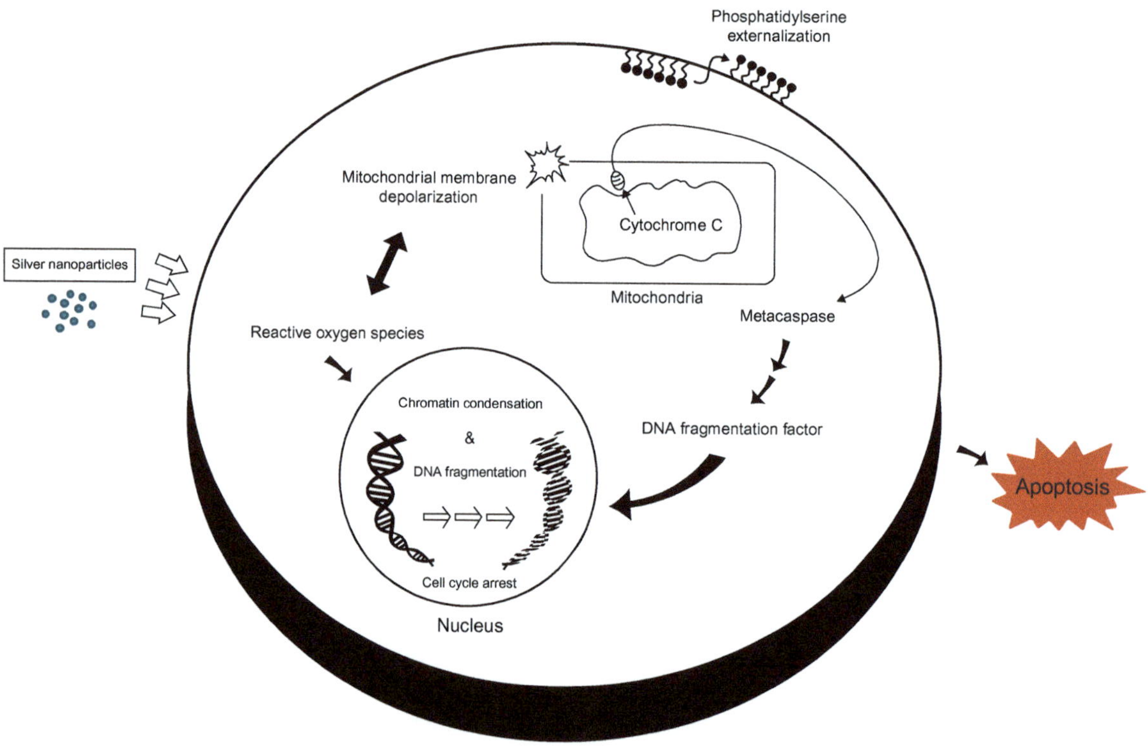

FIGURE 16.1 Apoptotic phenotypes in yeast cells (Cho et al., 2012). The •OH formation by nano-Ag in *C. albicans* triggers mitochondrial dysfunction and promotes the release of proapoptotic factors. Thereafter, metacaspase activation, an important regulator in the induction of apoptosis, induced many apoptotic phenotypes.

the cells (Costa and Moradas-Ferreira, 2001). The superfluity of ROS, such as hydrogen peroxide (H_2O_2), nitric oxide (NO), superoxide anion (O_2^-), and hydroxyl radicals (•OH), is observed in almost every apoptotic scenario and regulates the apoptosis as early signal mediators (Madeo et al., 1999; Fröhlich and Madeo, 2000).

Apoptosis induced by nano-Ag triggered accumulation of ROS (Hwang et al., 2012b), and intracellular levels of ROS could be detected by dihydrorhodamine-123 (DHR-123). DHR-123 is a ROS-sensitive dye that easily enters a cell with lipophilic property, localizes to the mitochondria, and oxidizes to the highly stable, fluorescent derivative rhodamine-123 by mitochondrial ROS (Sakurada et al., 1992; Ranganathan et al., 2009). Cellular ROS levels indicate a significant increase in the fluorescent signal that reflects ROS production in *C. albicans* cells treated with nano-Ag (Figure 16.2A) (Hwang et al., 2012b). This observation shows the activity of nano-Ag for chemically generated ROS.

Through Fenton and Habor–Weiss reactions, oxygen radicals convert to •OH, and •OH can be fatal when it reacts indiscriminately with cellular components such as unsaturated fatty acids, amino acid residues, and DNA (Rollet-Labelle et al., 1998; Perrone et al., 2008). Because •OH has been suggested to be a common mediator of apoptosis in many studies (Haruna et al., 2002), it is necessary to observe •OH formation during apoptosis (Lee et al., 2012). Using 3′-(*p*-hydroxyphenyl) fluorescein (HPF), a cell-permeable and fluorescent probe oxidized by •OH with high specificity (Setsukinai et al., 2003), reactive •OH formation was shown in

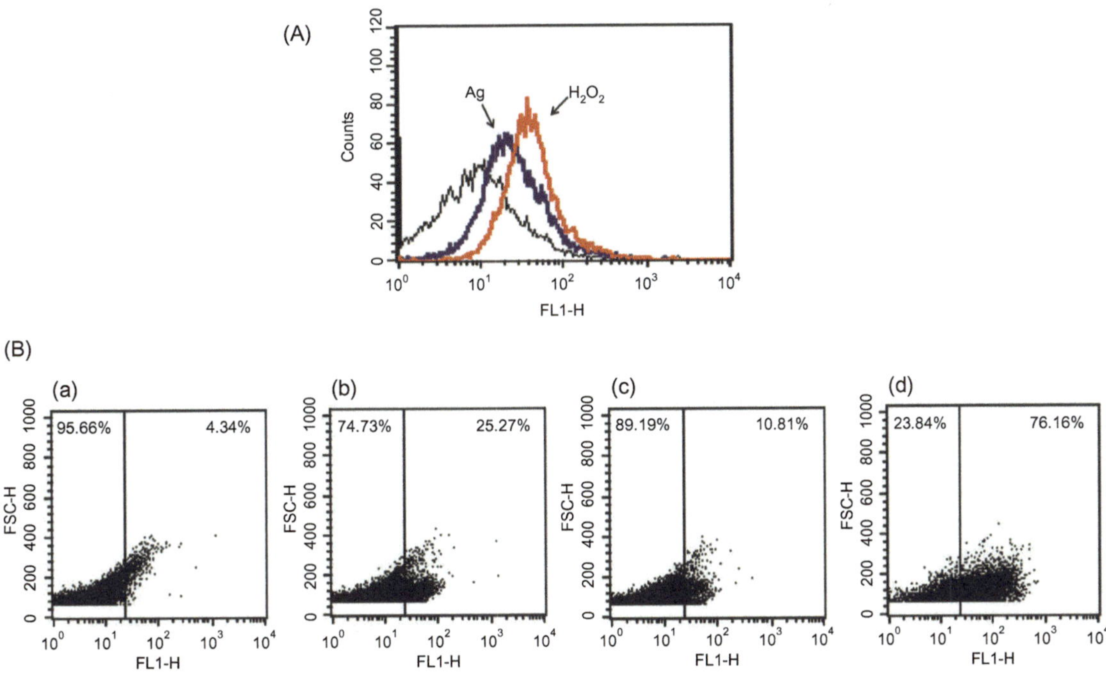

FIGURE 16.2 (A) Flow cytometric analysis of ROS accumulation in nano-Ag-treated (blue) and H_2O_2-treated (red solid line) *C. albicans* cells stained with DHR-123. (B) Increased intracellular •OH in *C. albicans* was detected by flow cytometry using HPF assay. (a) Untreated cells, (b) cells exposed to nano-Ag, (c) cells exposed to nano-Ag with thiourea, and (d) cells exposed to H_2O_2 (Hwang et al., 2012b).

C. albicans cells treated with nano-Ag (Figure 16.2B) (Hwang et al., 2012b). To determine whether •OH formation was essential for yeast apoptosis induced by nano-Ag, thiourea, which acts as a •OH scavenger, was also treated. The •OH seems to occupy most of the total ROS produced by nano-Ag, suggesting that •OH is a crucial component of ROS-induced apoptosis (Hwang et al., 2012b).

16.3 PHOSPHATIDYL SERINE EXPOSURE

Phosphatidyl serine is predominantly oriented toward the cytoplasm distributed on the compositional asymmetry of phospholipids within the cytoplasmic membrane, which is maintained by the ATP-binding cassette transporters in *C. albicans* (Cerbón and Calderón, 1991). The normal distribution of this lipid on the inner leaflet of the membrane bilayer is then disrupted because of stimulation of enzymes such as flippases or scramblase, which can move phosphatidyl serine in both directions across the membrane, and inhibition of aminophospholipid translocases, which returns the lipid to the inner side of the membrane (Verhoven et al., 1995). Because this phospholipid is translocated to the outer leaflet during early stages of apoptosis in response to particular calcium-dependent stimuli (Martin et al., 1995), the phosphatidyl serine exposure serves as a sensitive marker for early stages of yeast apoptosis. With a double staining method using fluorescein isothiocyanate (FITC)-Annexin V, which binds to outer leaflet phosphatidyl serine with high affinity in the presence of Ca^{2+}, and propidium iodide (PI), which is an intact membrane-impermeable dye, apoptotic cells can be distinguished from late apoptosis and necrotic cells (Smrz et al., 2007; Cho et al., 2012). Early apoptotic cells can only be stained with Annexin V-FITC, whereas late apoptotic and necrotic cells stain with both Annexin V-FITC and PI. As shown in Figure 16.3, the cell population in the lower right (LR) quadrant, which indicates the proportion of early apoptotic cells (Annexin V−positive/PI-negative), increased after treating the cells with nano-Ag but the fluorescent intensity of PI did not change. Curiously, the percentage of nano-Ag treated with thiourea did not increase as significantly as when treated solely with nano-Ag. From the distinct difference, it was confirmed that nano-Ag led to the translocation of the membrane phosphatidyl serine from the inner leaflet to the outer leaflet of the plasma membrane without damaging the plasma membrane permeability and that the generation and accumulation of intracellular ROS, specifically hydroxyl radicals, induced by nano-Ag was related to an apoptotic mechanism in *C. albicans* cells (Hwang et al., 2012b).

16.4 MITOCHONDRIAL DYSFUNCTION

Release of cytochrome *c* from the mitochondrial membrane to the cytosol is a crucial event in apoptotic cell death. Bax channel is formed on the outer mitochondrial membrane in response to various apoptotic signals. This mitochondrial apoptosis-induced channel facilitates release of cytochrome *c* from the mitochondria to the cytosol (Dejean et al., 2006). Cytochrome *c*, which is a small heme protein loosely associated with the inner membrane of the mitochondrion, is an essential component of the respiratory chain that acts as an electron carrier. This hemoprotein is also an intermediate in apoptosis. During the early phase of apoptosis, mitochondrial ROS production is stimulated, and cytochrome *c* is then detached from the mitochondrial inner membrane and can be extruded into the soluble cytoplasm through pores in the outer membrane (Liu et al., 1996; Orrenius and Zhivotovsky, 2005). When the cytochrome *c* is released into the

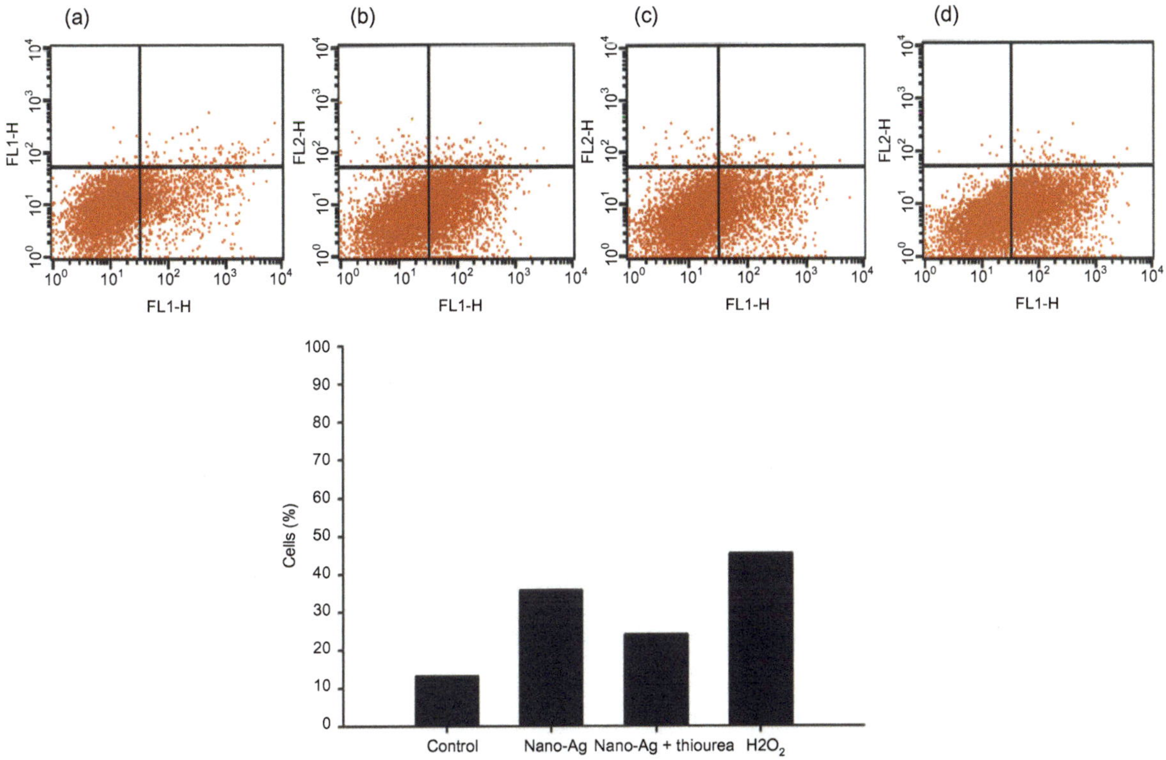

FIGURE 16.3 Phosphatidyl serine externalization, which is observed at an early stage of apoptosis, was shown by Annexin V-FITC and PI staining. (a) Untreated cells, (b) cells exposed to nano-Ag, (c) cells exposed to nano-Ag with thiourea, and (d) cells exposed to H_2O_2. The bottom bar graph shows the percentage of apoptotic cells (Hwang et al., 2012a,b).

cytosol, it is reduced because of the loss of the oxidase activity, and then caspase is activated as a representative of the other apoptotic proteases (Pereira et al., 2007). The amount of cytochrome *c* was detected in the cytosolic buffer medium after the cells were treated with nano-Ag, suggesting that nano-Ag may trigger cytochrome *c*–mediated intrinsic apoptosis (Figure 16.4A). Also, the addition of thiourea to yeast cells treated with nano-Ag, which do not usually produce hydroxyl radicals, exhibited reduced cytochrome *c* release compared with those treated with only nano-Ag (Figure 16.4A) (Hwang et al., 2012b).

In many cases, the opening of the transition pore of the mitochondrial membrane induces the collapse of the mitochondrial membrane potential, which can be detectable with JC-1 (5,5′,6,6′-tetra-chloro-1,1′,3,3′-tetraethylbenzimidazolylcarbocyanine chloride). JC-1 has advantages over other cationic dyes in that it can selectively enter into mitochondria and reversibly change color from red to green as the membrane potential decreases (Sampson et al., 2010). The ratio of green to red fluorescence is dependent only on the membrane potential and not other factors such as mitochondrial size, shape, and density, which may influence single-component fluorescence signals. As shown in Figure 16.4B, each nano-Ag or H_2O_2 treatment induced a significant decrease in mitochondrial membrane potential, whereas treatment with nano-Ag and thiourea appeared to have only a slight effect. This result suggests

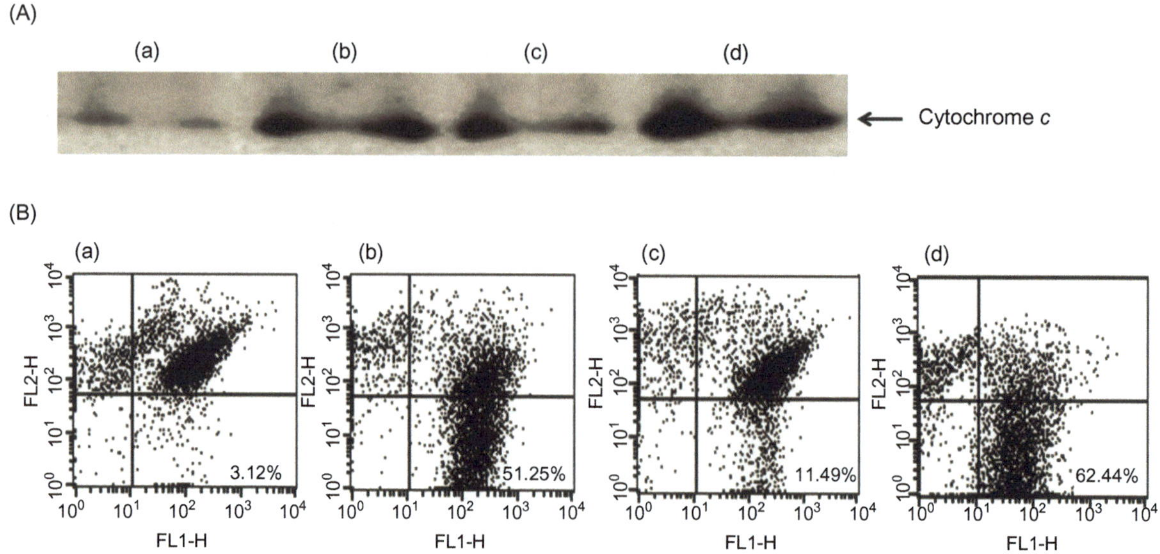

FIGURE 16.4 (A) Detection of cytochrome *c* release from nano-Ag-treated *C. albicans* mitochondria using Western blotting method. (B) JC-1 staining assay with *C. albicans* cells. (a) Untreated cells, (b) cells exposed to nano-Ag, (c) cells exposed to nano-Ag with thiourea, and (d) cells exposed to H_2O_2 (Hwang et al., 2012b).

that nano-Ag induce the breakdown of $\Delta\Psi_m$, which is a critical step in cells that are undergoing apoptosis and in the loss of mitochondrial permeability (Hwang et al., 2012b). In addition, it can be inferred that nano-Ag induce apoptosis through the formation of •OH and that •OH dissipates the balance of the mitochondrial membrane acting in an important role in the apoptotic process.

16.5 CASPASE ACTIVATION

Caspases (cysteine-dependent aspartate-specific proteases) are essential in cells for apoptosis and have been termed executioner proteins for their roles in the cell. They are regulated at a posttranslational level, ensuring that they can be rapidly activated. They are first synthesized as inactive pro-caspases that consist of a prodomain, a small subunit, and a large subunit. Initiator caspases possess a longer prodomain than the effector caspases, whose prodomain is very small. The prodomain of the initiator caspases contains domains that enable the caspases to interact with other molecules that regulate their activation. The activation of caspase is thought to be associated with cytochrome *c* release, and they play a major role in the apoptotic signaling pathway in the early stages of the apoptotic signaling network (Yuan and Horvitz, 2004; Zivna et al., 2010). There are metacaspases in fungi, that is orthologs of caspase in animals (Barroso et al., 2006), and that is considering involved in nano-Ag-induced cell death in *C. albicans*.

The accumulation of intracellular ROS caused by nano-Ag induces the activation of metacaspase, and metacaspase activation can inhibit various cellular fundamental processes, including DNA replication, mitochondrial function, and RNA and protein stability (Uren et al., 2000; Madeo et al., 2002). Metacaspase activity can be detected with FITC-VAD-FMK, which enters the cell and binds specifically to the activated

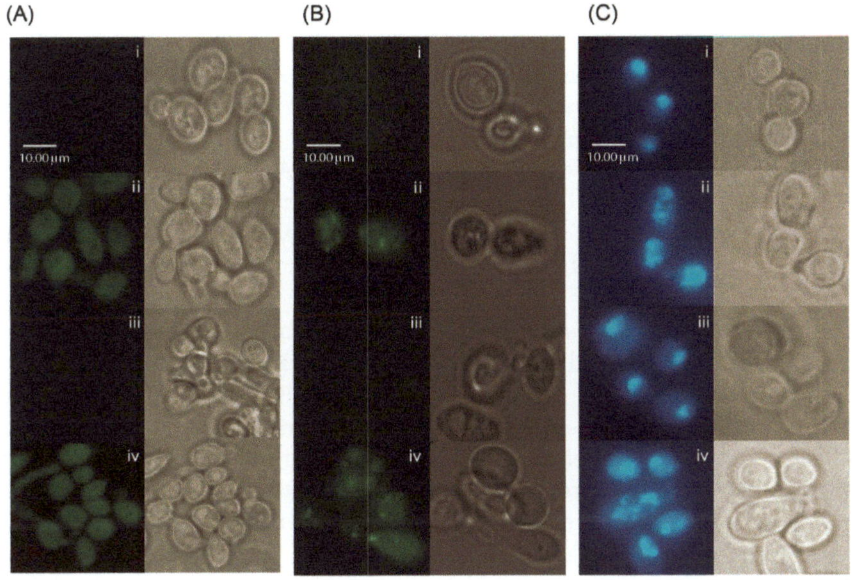

FIGURE 16.5 (A) Effect of nano-Ag on the activity of metacaspase in *C. albicans*. (B) DNA fragmentation was visualized by TUNEL assay. (C) Using DAPI staining, nuclear condensation was observed in nano-Ag-treated *C. albicans* (Hwang et al., 2012b).

caspases (Wu et al., 2010). Fluorescence analysis of the cells treated with nano-Ag shows significant green fluorescence in the FITC-VAD-FMK–loaded cells and the number of activated metacaspases decreased, which also reduced •OH formation in thiourea-treated cells (Figure 16.5A) (Hwang et al., 2012b).

16.6 DNA FRAGMENTATION AND CHROMOSOME CONDENSATION

DNA fragmentation and nuclear condensation can be features of late apoptosis (Wadskog et al., 2004). The activation of caspase protein vitalizes nuclease in yeast cells, such as DNA fragmentation factor DFF40 in mammalians (Liu et al., 1999). Chromatin DNA during apoptosis is broken into short fragments by activated endonucleases and is located in the margin of nuclei. As DNA cleavage exposes more abundant free 3′-OH termini, it can be detected by terminal deoxynucleotidyl transferase-mediated dUTP nick end labeling (TUNEL) assay, which labels modified nucleotides catalyzed by terminal deoxynucleotidyl transferase (Phillips et al., 2003). The TUNEL method is a fast and sensitive way to visualize DNA fragmentation in individual cells with fluorescence microscopy (Ribeiro et al., 2006). In yeast cells, apoptotic phenotypes (TUNEL-positive) were observed with a fluorescence microscope, which indicated DNA fragmentation and margination (Figure 16.5B) (Hwang et al., 2012b).

Chromatin condensation can be visualized by fluorescence microscopy after 4′-6-diamidino-2-phenylindole (DAPI) staining. DAPI, a cell-permeable fluorescence dye, is commonly used to analyze nuclear morphologic changes such as chromosome condensation and fragmentation at late-state apoptosis because of its strong binding ability to A–T-rich regions in DNA (Kapuscinski, 1995). DAPI staining of the yeast cells treated with nano-Ag showed more concentrated and split fluorescence intensity, indicating a greater degree of nuclear condensation of *C. albicans* cells compared with intact nuclei of normal control cells (Figure 16.5C) (Hwang et al., 2012b).

16.7 CELL-CYCLE ARREST

In some studies, nano-Ag have shown the ability to arrest the cell cycles during the G2/M phase in yeast, and there are many reports demonstrating G2/M phase–mediated apoptosis (Phillips et al., 2003; Kim et al., 2009). Damaged DNA needs to be restored by the cells' own repair systems, and cell division is arrested by cell-cycle checkpoint regulation to provide more time for repair before the critical phases of DNA replication (Kaufmann and Paules, 1996). To understand the effects on the cell-cycle progress of *C. albicans*, the cells were cultured in the presence or absence of nano-Ag and their DNA content was determined by staining with PI.

As shown in Figure 16.6, the yeast cells in the G2/M phase considerably increased, whereas that in the G1 phase significantly decreased in the presence of nano-Ag (Kim et al., 2009). These data suggest that nano-Ag inhibited some cellular processes that are involved in normal bud growth correlated with DNA damage. In addition to PI staining, cell-cycle arrest is also confirmed indirectly by DAPI staining (Daniel and DeCoster, 2004). Although the cell division is arrested, chromosomes remain condensed and inhibition of growth occurs. Through DAPI staining, condensation of chromosomes can be detected, indicating replication arrest, and the results strongly support that nano-Ag trigger cell-cycle arrest–mediated apoptosis in *C. albicans* cells due to the intracellular •OH accumulations.

16.8 SYNERGISTIC EFFECT OF SILVER NANOPARTICLES

To approach practical use in the medical field, the synergistic effects of silver nanoparticles were also investigated in many studies with some conventional antibiotics against

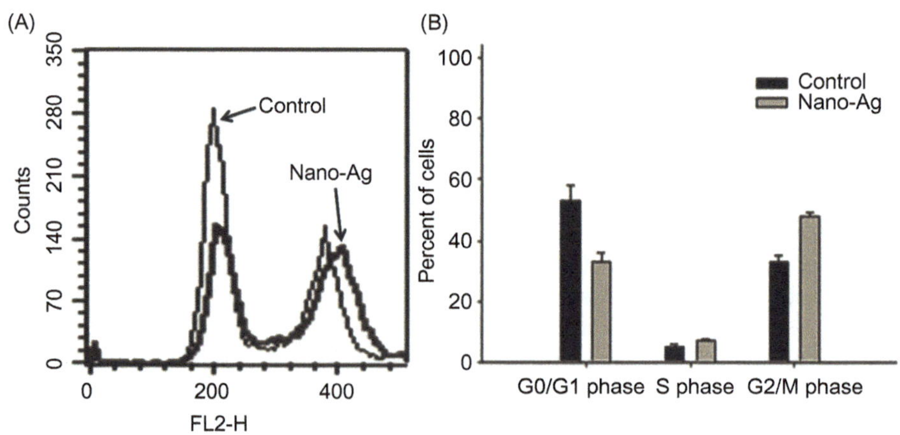

FIGURE 16.6 The effects of nano-Ag on the process of the cell cycle of *C. albicans*. (A) The FACS diagram of the cell cycle. FL2-H indicates the fluorescent intensity of PI, and the *y*-axis indicates cell number (events). (B) A histogram of the percentages of the cell cycle (Kim et al., 2009).

TABLE 16.1 Antibacterial Activity of Nano-Ag Against Human Pathogenic Bacteria (Hwang et al., 2012a)

	MIC (μg/mL)			
Species	**Nano-Ag**	**Ampicillin**	**Chloramphenicol**	**Kanamycin**
GRAM-POSITIVE				
E. faecium ATCC 19434	0.25	2	4	2
S. aureus ATCC 25923	0.5	4	2	4
S. mutans KCTC 3065	2	2	4	4
GRAM-NEGATIVE				
E. coli O-157 ATCC 43895	0.5	4	2	4
E. coli ATCC 25922	2	8	4	4
P. aeruginosa ATCC 27853	0.5	2	1	2

TABLE 16.2 Activities of Combinations of Nano-Ag and Antibiotics (Hwang et al., 2012a)

	FICI (μg/mL)		
Species	**Nano-Ag^{+} ampicillin**	**Nano-Ag^{+} chloramphenicol**	**Nano-Ag^{+} kanamycin**
GRAM-POSITIVE			
*E. faecium*ATCC 19434	0.375 (S)	0.375 (S)	0.75 (PS)
S. aureus ATCC 25923	0.5 (PS)	0.75 (PS)	0.375 (S)
S. mutans KCTC 3065	0.375 (S)	0.5 (PS)	0.375 (S)
GRAM-NEGATIVE			
E. coli O-157 ATCC 43895	0.5 (PS)	0.5 (PS)	0.5 (PS)
E. coli ATCC 25922	0.375 (S)	0.5 (PS)	0.375 (S)
P. aeruginosa ATCC 27853	0.5 (PS)	0.375 (S)	0.375 (S)

S, synergistic; PS, partially synergistic.
FICI = ($MIC_{Drug\ A}$ in combination/$MIC_{Drug\ A}$ alone) + ($MIC_{Drug\ B}$ in combination/$MIC_{Drug\ B}$ alone).

bacteria strains (Hwang et al., 2012a; Ghosh et al., 2013; Markowska et al., 2014).

Minimum inhibitory concentration (MIC) and fractional inhibitory concentration index (FICI) were used to confirm antibacterial susceptibility and synergistic effects (Table 16.1) (Hwang et al., 2012a). The results of the combination assay are presented in Table 16.2 (Hwang et al., 2012a). All combinations were shown to be effective. By reducing the viability of bacteria strains at a lower concentration, these synergistic activities of silver nanoparticles in the presence of conventional antibiotics are also possible.

Kohanski et al. (2007) suggested an intriguing mechanism of antibiotics for killing bacteria. They showed that three major bactericidal antibiotics stimulate •OH formation in bacteria, and this toxic chemical contributes to the killing efficiency of lethal antibiotics. They also suggested that bactericidal antibiotics induce bacterial cell death, exhibiting physiological and biochemical hallmarks of apoptosis in bacteria (Dwyer et al., 2012). Based on these studies, it can be concluded that generation of •OH is a common mechanism of bacterial cell death caused by antibiotics. Nano-Ag have are capable of •OH generation, and each of the two different classes of bactericidal antibiotics, including ampicillin and kanamycin, induced •OH formation in the study of synergistic effects with conventional antibiotics of nano-Ag, indicating that •OH is an important cause of synergistic effects (Figure 16.7) (Hwang et al., 2012a).

In this study, the antibacterial effects of nano-Ag and its synergistic capacity against various representative pathogenic bacteria were confirmed. Furthermore, the generation

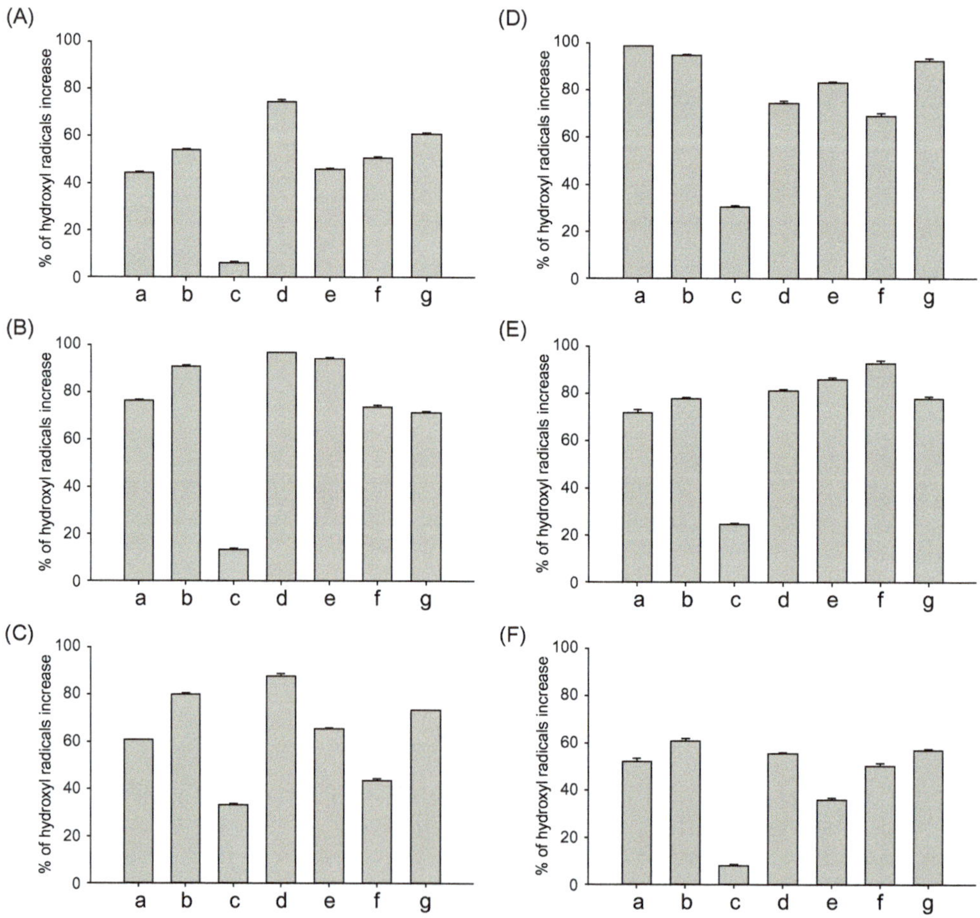

FIGURE 16.7 •OH production, determined by the HPF staining method, after addition of nano-Ag alone or in combination with other conventional antibiotics. (A) *Enterococcus faecium*, (B) *Staphylococcus aureus*, (C) *Streptococcus mutans*, (D) *Escherichia coli* O-157, (E) *E. coli* ATCC 25922, and (F) *Pseudomonas aeruginosa*. Treatments were (a) nano-Ag alone, (b) ampicillin, (c) chloramphenicol, (d) kanamycin, (e) nano-Ag^{+} ampicillin, (f) nano-Ag^{+} chloramphenicol, and (g) nano-Ag^{+} kanamycin (Hwang et al., 2012a).

of hydroxyl radicals and the malfunction of protective action are important factors in nano-Ag antibacterial effects and synergism. The findings in this study support the idea that nano-Ag have considerably more effective antimicrobial activity and deserve further investigation regarding clinical applications in pathogenic fungal cells.

16.9 CONCLUSION AND FUTURE PROSPECTS

Among the different antimicrobial agents, silver has been most extensively studied and has been used since ancient times to fight infections and prevent spoilage. The antibacterial, antifungal, and antiviral properties of silver ions, silver compounds, and silver nanoparticles have been extensively studied. Silver is also nontoxic to humans in minute concentrations. Microorganisms are unlikely to develop resistance against silver as compared with antibiotics, because silver attacks a broad range of targets in the microbes.

Recently, many studies demonstrated that nano-Ag promote apoptosis in *C. albicans* via intracellular •OH accumulation. Ultimately, nano-Ag disrupt mitochondrial membrane integrity, trigger cytochrome *c* release, activate metacaspase, and induce DNA. Although the mechanisms of nano-Ag in mitochondria-dependent apoptosis in *C. albicans* have not been fully elucidated, some reports support that nano-Ag induce programmed cell death through ROS accumulation, especially •OH generation, which deserves further study to provide elaboration of the apoptosis mechanisms of nano-Ag. Also, combination use of existing antimicrobial agents and nano-Ag can be an effective strategy for antibiotic-resistant pathogens. Many studies indicate that nano-Ag have considerable effective antimicrobial activity and are worthy of further investigation for clinical applications.

References

Barroso, G., Taylor, S., Morshedi, M., Manzur, F., Gaviño, F., Oehninger, S., 2006. Mitochondrial membrane potential integrity and plasma membrane translocation of phosphatidylserine as early apoptotic markers: a comparison of two different sperm subpopulations. Fertil. Steril. 85, 149–154.

Benaroudj, N., Lee, D.H., Goldberg, A.L., 2001. Trehalose accumulation during cellular stress protects cells and cellular proteins from damage by oxygen radicals. J. Biol. Chem. 276, 24261–24267.

Bortner, C.D., Cidlowski, J.A., 2007. Cell shrinkage and monovalent cation fluxes: role in apoptosis. Arch. Biochem. Biophys. 462, 176–188.

Carmona-Gutierrez, D., Eisenberg, T., Büttner, S., Meisinger, C., Kroemer, G., Madeo, F., 2010. Apoptosis in yeast: triggers, pathways, subroutines. Cell Death Differ. 17, 763–773.

Cerbón, J., Calderón, V., 1991. Changes of the compositional asymmetry of phospholipids associated to the increment in the membrane surface potential. Biochim. Biophys. Acta 1067, 139–144.

Cho, J., Hwang, I.S., Choi, H., Hwang, J.H., Hwang, J.S., Lee, D.G., 2012. The novel biological action of antimicrobial peptides via apoptosis induction. J. Microbiol. Biotechnol. 22, 1457–1466.

Costa, V., Moradas-Ferreira, P., 2001. Oxidative stress and signal transduction in *Saccharomyces cerevisiae*: insights in to aging, apoptosis and disease. Mol. Aspects Med. 22, 217–246.

Daniel, B., DeCoster, M.A., 2004. Quantification of sPSA-2 induced early and late apoptosis changes in neuronal cell cultures using combined TUNEL and DAPI staining. Brain Res. Protoc. 13, 144–150.

Dejean, L.M., Martinez-Caballero, S., Kinnally, K.W., 2006. Is MAC the knife that cuts cytochrome *c* from mitochondria during apoptosis?. Cell Death Differ. 13, 1387–1395.

Dwyer, D.J., Camacho, D.M., Kohanski, M.A., Callura, J.M., Collins, J.J., 2012. Antibiotic-induced bacterial cell death exhibits physiological and biochemical hallmarks of apoptosis. Mol. Cell. 46, 561–572.

Fröhlich, K.U., Madeo, F., 2000. Apoptosis in yeast—a monocellular organism exhibits altruistic behavior. FEBS Lett. 473, 6–9.

Ghosh, I.N., Patil, S.D., Sharma, T.K., Srivastava, S.K., Pathania, R., Navani, N.K., 2013. Synergistic action of cinnamaldehyde with silver nanoparticles against spore-forming bacteria: a case for judicious use of silver nanoparticles for antibacterial applications. Int. J. Nanomed. 8, 4721–4731.

Gurunathan, S., Lee, K.J., Kalimuthu, K., Sheikpranbabu, S., Vaidyanathan, R., Eom, S.H., 2009. Antiangiogenic properties of silver nanoparticles. Biomaterials 30, 6341–6350.

Haruna, S., Kuroi, R., Kajiwara, K., Hashimoto, R., Matsugo, S., Tokumaru, S., et al., 2002. Induction of apoptosis in HL-60 cells by photochemically generated hydroxyl radicals. Bioorg. Med. Chem. Lett. 12, 675–676.

Hwang, I.S., Hwang, J.H., Choi, H., Kim, K.J., Lee, D.G., 2012a. Synergistic effects between silver nanoparticles and antibiotics and the mechanisms involved. J. Med. Microbiol. 61, 1719–1726.

Hwang, I.S., Lee, J., Hwang, J.H., Kim, K.J., Lee, D.G., 2012b. Silver nanoparticles induce apoptotic cell death in *Candida albicans* through the increase of hydroxyl radicals. FEBS J. 279, 1327–1338.

Jain, P., Pradeep, T., 2005. Potential of silver nanoparticle coated polyurethane foam as an antibacterial water filter. Biotechnol. Bioeng. 90, 59–63.

Jones, S.A., Bowler, P.G., Walker, M., Parsons, D., 2004. Controlling wound bioburden with a novel silver-containing hydrofiber dressing. Wound Repair Regen. 12, 288–294.

Kapuscinski, J., 1995. DAPI: a DNA-specific fluorescent probe. Biotech. Histochem. 70, 220–233.

Kaufmann, W.K., Paules, R.S., 1996. DNA damage and cell cycle checkpoints. FASEB J. 10, 238–347.

Kim, K.J., Sung, W.S., Suh, B.K., Moon, S.K., Choi, J.S., Kim, J.G., et al., 2009. Antifungal activity and mode of action of silver nano-particles on *Candida albicans*. Biometals 22, 235–242.

Klasen, H.J., 2000. A historical review of the use of silver in the treatment of burns. II. Renewed interest for silver. Burns 26, 131–138.

Kohanski, M.A., Dwyer, D.J., Hayete, B., Lawrence, C.A., Collins, J.J., 2007. A common mechanism of cellular death induced by bactericidal antibiotics. Cell 130, 797–810.

Kumar, A., Vemula, P.K., Ajayan, M., John, G., 2008. Silver-nanoparticle-embedded antimicrobial paints based on vegetable oil. Nat. Mater. 3, 236–241.

Lee, J., Hwang, J.S., Hwang, I.S., Cho, J., Lee, E., Kim, Y., et al., 2012. Coprisin-induced antifungal effects in *Candida albicans* correlate with apoptotic mechanisms. Free Radic. Biol. Med. 52, 2302–2311.

Liu, X., Kim, C.N., Yang, J., Jemmerson, R., Wang, X., 1996. Induction of apoptotic program in cell-free extracts: requirement for dATP and cytochrome c. Cell 86, 147–157.

Liu, X., Zou, H., Widlak, P., Garrard, W., Wang, X., 1999. Activation of the apoptotic endonuclease DFF40 (caspase-activated DNase or nuclease). Oligomerization and direct interaction with histone H1. J. Biol. Chem. 14, 13836–13840.

Madeo, F., Fröhlich, E., Fröhlich, K.U., 1997. A yeast mutant showing diagnostic markers of early and late apoptosis. J. Cell Biol. 139, 729–734.

Madeo, F., Fröhlich, E., Ligr, M., Grey, M., Sigrist, S.J., Wolf, D.H., et al., 1999. Oxygen stress: a regulator of apoptosis in yeast. J. Cell Biol. 145, 757–767.

Madeo, F., Herker, E., Maldener, C., Wissing, S., Lächelt, S., Herlan, M., et al., 2002. A caspase-related protease regulates apoptosis in yeast. Mol. Cell. 9, 911–917.

Markowska, K., Grudniak, A.M., Krawczyk, K., Wróbel, I., Wolska, K.I., 2014. Modulation of antibiotic resistance and induction of a stress response in *Pseudomonas aeruginosa* by silver nanoparticles. J. Med. Microbiol. 63, 849–854.

Martin, S.J., Reutelingsperger, C.P.M., McGahon, A.J., Rader, J.A., van Schie, R.C.A.A., LaFace, D.M., et al., 1995. Early redistribution of plasma membrane phosphatidylserine is a general feature of apoptosis regardless of the initiating stimulus: inhibition by overexpression of Bcl-2 and Abl. J. Exp. Med. 182, 1545–1556.

Oberdörster, G., Oberdörster, E., Oberdörster, J., 2005. Nanotoxicology: an emerging discipline evolving from studies of ultrafine particles. Environ. Health Perspect. 113, 823–839.

Orrenius, S., Zhivotovsky, B., 2005. Cardiolipin oxidation sets cytochrome *c* free. Nat. Chem. Biol. 1, 188–189.

Parashar, U.K., Kumar, V., Bera, T., Saxena, P.S., Nath, G., Shrivastava, S.K., et al., 2011. Study of mechanism of enhanced antibacterial activity by green synthesis of silver nanoparticles. Nanotechnology 22, 415104–415117.

Pereira, C., Camougrand, N., Manon, S., Sousa, M.J., Côrte-Real, M., 2007. ADP/ATP carrier is required for mitochondrial outer membrane permeabilization and cytochrome *c* release in yeast apoptosis. Mol. Microbiol. 66, 571–582.

Perrone, G.G., Tan, S.X., Dawes, I.W., 2008. Reactive oxygen species and yeast apoptosis. Biochim. Biophys. Acta 1783, 1354–1368.

Phillips, A.J., Sudbery, I., Ramsdale, M., 2003. Apoptosis induced by environmental stresses and amphotericin B in *Candida albicans*. Proc. Natl. Acad. Sci. U.S.A. 100, 14327–14332.

Ranganathan, S., Harmison, G.G., Meyertholen, K., Pennuto, M., Burnett, B.G., Fischbeck, K.H., 2009. Mitochondrial abnormalities in spinal and bulbar muscular atrophy. Hum. Mol. Genet. 18, 27–42.

Ribeiro, G.F., Côrte-Real, M., Johansson, B., 2006. Characterization of DNA damage in yeast apoptosis induced by hydrogen peroxide, acetic acid, and hyperosmotic shock. Mol. Biol. Cell. 17, 4584–4591.

Rollet-Labelle, E., Grange, M.J., Elbim, C., Marquetty, C., Gougerot-Pocidalo, M.A., Pasquier, C., 1998. Hydroxyl radical as a potential intracellular mediator of polymorphonuclear neutrophil apoptosis. Free Radic. Biol. Med. 24 (563), 572.

Sakurada, H., Koizumi, H., Ohkawara, A., Ueda, T., Kamo, U., 1992. Use of dihydrorhodamine 123 for detecting intracellular generation of peroxides upon UV irradiation in epidermal keratinocytes. Arch. Dermatol. Res. 284, 114–116.

Sampson, S.P., Bucris, E., Horovitz-Fried, M., Parnas, A., Kahana, S., Abitbol, G., et al., 2010. Insulin increases H_2O_2-induced pancreatic beta cell death. Apoptosis 15, 1165–1176.

Setsukinai, K., Urano, Y., Kakinuma, K., Majima, H.J., Nagano, T., 2003. Development of novel fluorescence probes that can reliably detect reactive oxygen species and distinguish specific species. J. Biol. Chem. 278, 3170–3175.

Sharma, V.K., Yngard, R.A., Lin, Y., 2009. Silver nanoparticles: green synthesis and their antimicrobial activities. Adv. Colloid Interface Sci. 145, 83–96.

Smrz, D., Dráberová, L., Dráber, P., 2007. Non-apoptotic phosphatidylserine externalization induced by engagement of glycosylphosphatidylinositol-anchored proteins. J. Biol. Chem. 282, 10487–10497.

Soni, B., Visavadiya, N.P., Dalwadi, N., Madamwar, D., Winder, C., Khalil, C., 2010. Purified c-phycoerythrin: safety studies in rats and protective role against permanganate-mediated fibroblast-DNA damage. J. Appl. Toxicol. 30, 542–550.

Uren, A.G., O'Rourke, K., Aravind, L.A., Pisabarro, M.T., Seshagiri, S., Koonin, E.V., et al., 2000. Identification of paracaspases and metacaspases: two ancient families of caspase-like proteins, one of which plays a key role in MALT lymphoma. Mol. Cell. 6, 961–967.

Verhoven, B., Schlegel, R.A., Williamson, P., 1995. Mechanisms of phosphatidylserine exposure, a phagocyte recognition signal, on apoptotic T lymphocytes. J. Exp. Med. 182, 1597–1601.

Wadskog, I., Maldener, C., Proksch, A., Madeo, F., Adler, L., 2004. Yeast lacking the SRO7/SOP1-encoded tumor suppressor homologue show increased susceptibility to apoptosis-like cell death on exposure to NaCl stress. Mol. Biol. Cell. 15, 1436–1444.

Wu, X.Z., Chang, W.Q., Cheng, A.X., Sun, L.M., Lou, H.X., 2010. Plagiochin E, an antifungal active macrocyclicbis (bibenzyl), induced apoptosis in *Candida albicans* through a metacaspase-dependent apoptotic pathway. Biochim. Biophys. Acta 1800, 439–447.

Yamanaka, M., Hara, K., Kudo, J., 2005. Bactericidal actions of a silver ion solution on *Escherichia coli*, studied by energy-filtering transmission electron microscopy and proteomic analysis. Appl. Environ. Microbiol. 71, 7589–7593.

Yang, W., Shen, C., Ji, Q., An, H., Wang, J., Liu, Q., et al., 2009. Food storage material silver nanoparticles interfere with DNA replication fidelity and bind with DNA. Nanotechnology 20, 085102.

Yuan, J., Horvitz, H.R., 2004. A first insight into the molecular mechanisms of apoptosis. Cell 23, 57–59.

Zivna, L., Krocova, Z., Härtlova, A., Kubelkova, K., Zakova, J., Rudof, E., et al., 2010. Activation of B cell apoptotic pathways in the course of *Francisella tularensis* infection. Microb. Pathog. 49, 226–236.

CHAPTER

17

Silver Nanoparticles to Fight *Candida* Coinfection in the Oral Cavity

Douglas Roberto Monteiro[1], *Sónia Silva*[2], *Melyssa Negri*[3], *Luiz Fernando Gorup*[4], *Emerson Rodrigues de Camargo*[4], *Debora Barros Barbosa*[5] and *Mariana Henriques*[2]

[1]Department of Pediatric Dentistry and Public Health, Araçatuba Dental School, Univ Estadual Paulista (UNESP), São Paulo, Brazil [2]CEB—Center of Biological Engineering, LIBRO—Laboratório de Investigação em Biofilmes, Rosário Oliveira, University of Minho, Braga, Portugal [3]Faculdade INGÁ, Maringá, Paraná, Brazil [4]Department of Chemistry, Federal University of São Carlos (UFSCar), São Paulo, Brazil [5]Department of Dental Materials and Prosthodontics, Araçatuba Dental School, Univ Estadual Paulista (UNESP), São Paulo, Brazil

17.1 INTRODUCTION

The oral cavity is an environment that favors the colonization and the development of a wide variety of microorganisms, including bacteria and fungi. Among the fungi found in the oral cavity, *Candida* species are the most abundant and relevant (Rautemaa and Ramage, 2011), and they are generally identified as commensals on the tongue, palate, and buccal mucosa of healthy adults (Jewtuchowicz et al., 2007). The translation of commensal to opportunistic *Candida* may be associated with host factors and the pathogenic attributes of the microorganism itself. It is probable that no form of *Candida* infection can start in the absence of underlying pathology (Soysa et al., 2008).

Of all the *Candida* species, *C. albicans* is considered to be the most virulent and the main species associated with oral Candidosis, because of its capacity to change from blastospore to the hyphae (Samaranayake et al., 2009). Nevertheless, other non-*C. albicans Candida* (NCAC) species such as *Candida glabrata*, *Candida tropicalis*, *Candida parapsilosis*, *Candida dubliniensis*, and *Candida krusei* have also been isolated and are involved in human diseases (Weems, 1992). These *Candida* species have an attribute of virulence in common: the ability to form structured aggregates of cells called

Nanotechnology in Diagnosis, Treatment and Prophylaxis of Infectious Diseases.
DOI: http://dx.doi.org/10.1016/B978-0-12-801317-5.00017-7

biofilms (Jakubovics and Kolenbrander, 2010). Biofilms may be defined as microbial communities adhered to biotic or abiotic surfaces in an aqueous environment and surrounded by a matrix of exopolymeric material (Costerton et al., 1999). Oral pathologies such as periodontal disease, caries, endodontic infection, and oral candidosis are related to biofilm formation (Beikler and Flemmig, 2011).

In general, for *Candida* species, biofilm formation occurs as a result of a sequence of events: attachment of blastospores to the surface; cell proliferation and well-organized colony formation; matrix production with differentiation into a mature three-dimensional structure consisting of yeasts, pseudohyphae, and hyphae embedded within extracellular matrix; and, finally, detachment of biofilm cells (Chandra et al., 2001a; Seneviratne et al., 2008). Mature *Candida* biofilms exhibit a heterogeneous and complex architecture with water channels, which facilitate the transport of nutrients (Hawser and Douglas, 1994; Chandra et al., 2001a).

According to Chandra et al. (2001a), *Candida* biofilm formation was also described on the polymethylmethacrylate (PMMA) surface and occurs in three stages: (i) early (0–11 h), characterized by adhesion and development of blastospores into microcolonies; (ii) intermediate (12–30 h), represented by bilayered structure-containing yeasts and young hyphae and by the emergence of extracellular matrix production; and (iii) maturation (31–72 h), in which a dense network of yeasts, pseudohyphae, and hyphae is encased by a thick layer of extracellular matrix, forming structures with complex architectures. *Candida* biofilm formation using *in vivo* models seems to proceed similarly (Andes et al., 2004; Nett et al., 2010).

Oral candidosis related to biofilm formation may be divided into primary and secondary classes (Samaranayake and Yaacob, 1990). When the infection is confined to oral and perioral tissues, it is classified as primary; when the oral candidosis is a reflex of systemic *Candida* infections, it is categorized as secondary. Moreover, a group of diseases with multifactorial etiology called "*Candida*-associated lesions" was described and included angular cheilitis, median rhomboid glossitis, linear gingival erythema, and *Candida*-associated denture stomatitis (Samaranayake et al., 2009). Frequently, patients with these oral candidosis may have symptoms of pain, dysgeusia, oral discomfort, and repulsion to food (Samaranayake et al., 2009).

Candida-associated denture stomatitis is a well-known recurring oral disease that is characterized by biofilm development in the region between the denture base and the inflamed mucosa supporting the upper denture (Budtz-Jorgensen, 2000). This pathological condition is found in approximately 11–67% of complete denture wearers (Arendorf and Walker, 1987) and its etiology is multifactorial, including continuous use of drugs, nutritional deficiencies, precarious denture hygiene, continuous denture wearing, trauma, smoking, and oral bacteria (Budtz-Jorgensen and Bertram, 1970; Webb et al., 1998). Although *C. albicans* is the most commonly isolated yeast in denture stomatitis, *C. glabrata* has been isolated from 80% of patients with the highest degree of mucosal inflammation (Coco et al., 2008). *C. albicans* and *C. glabrata* are able to colonize acrylic surfaces and co-exist in dual species biofilms without antagonism (Silva et al., 2013). This fact may enhance the virulence of these species. Because *C. glabrata* is deprived of hyphae, it may use the hyphae of *C. albicans* as a scaffold to penetrate into host tissues (Silva et al., 2011).

In clinical terms, several antifungal drugs are available for the management of oral candidosis. These include the polyenes (nystatin [NYT] and amphotericin B), the azoles (miconazole, clotrimazole, ketoconazole, itraconazole, and fluconazole), allylaminesthiocarbamates, morpholines, DNA analog 5-fluorocytosine, and caspofungins (Soysa et al., 2008; Samaranayake et al., 2009). Chlorhexidine digluconate (CHG) (0.2 or 2%) is also used for treatment of oral *Candida* infections (Samaranayake et al., 2009). However, even with various drugs available, treatment failure and recurrence of infections are not uncommon because the effective concentration of antifungal drugs may have its availability reduced because of the diluent effect of saliva and the cleansing activity of the oral musculature (Samaranayake et al., 2009). In addition, of major clinical significance in this context is the resistance/tolerance of *Candida* biofilms to conventional therapy. Various studies have shown that *Candida* biofilms are tolerant to the commonly used antifungal drugs (Chandra et al., 2001b; Bagg et al., 2003; Hasan et al., 2009; Watamoto et al., 2009; Fonseca et al., 2014).

Many different mechanisms have been proposed to clarify *Candida* biofilm resistance, such as the robust biofilm architecture, decreased metabolic activity, altered gene expression, the extracellular matrix, the presence of "persister cells," and higher antioxidative capacities (Mah and O'Toole, 2001; Ramage et al., 2005; Seneviratne et al., 2010). Thus, this process is multifactorial and the accurate mechanism by which *Candida* species in the biofilm mode can acquire resistance remains to be understood.

17.2 SILVER NANOPARTICLES AGAINST *CANDIDA* BIOFILMS

17.2.1 Influence of Stabilizing Agent and Diameter on SN Antibiofilm Activity

The effect of silver nanoparticles (SNs) with different diameters (5, 10, and 60 nm) stabilized with two types of agents, ammonia (NH_3) and polyvinylpyrrolidone (PVP), against *C. albicans* and *C. glabrata* mature biofilms was assessed by Monteiro et al. (2012b). Although NH_3 stabilizes SNs by the formation of soluble diammine silver complexes, which trap free silver ions responsible for nanoparticle growth (Gorup et al., 2011), the polymers of the PVP group bond on the SN surfaces through the nitrogen atom in their molecule (Kvítek et al., 2008), resulting in flocculation. However, despite the flocculation, the SNs stabilized by this polymer are separated from each other through the chain of the polymer molecule, and this process allows the interaction of SNs with the cell wall of microorganisms due to their high surface energy and mobility (Kvítek et al., 2008).

It was observed that regardless of the diameter and type of stabilizing agent, SNs were very effective against *C. glabrata* biofilms, with significant total biomass reduction at SN concentrations $\geq 1.6\ \mu g/mL$ (Monteiro et al., 2012b). For this *Candida* species, all SN colloidal suspensions displayed biomass reductions of approximately 90% at SN concentration of $108\ \mu g/mL$. Contrarily, for *C. albicans* biofilms, there was some increase in biofilm biomass when compared with untreated biofilms, including different particle sizes and stabilizing agents (Monteiro et al., 2012b). It is believed that these differences in biomass reduction have occurred because of the different biofilm architectures. Regarding cell viability for both species tested, significant reductions in the number of biofilm viable cells for SN concentrations $\geq 13.5\ \mu g/mL$ were found. Moreover, using SNs (at $216\ \mu g/mL$) of 5 nm and stabilized with PVP, there were reductions of $3.36\ \log_{10}$ and $4.24\ \log_{10}$ in the number of biofilm cells for *C. albicans* and *C. glabrata*, respectively (Monteiro et al., 2012b).

However, the main finding of this study was that the nanoparticle size and the type of stabilizing agent did not interfere with SN efficacy in decreasing total biomass and cultivable cells of *Candida* biofilms. This has probably occurred because of the tendency of agglomeration of SNs when in contact with biofilms (Figure 17.1). Interactions of SNs with *Escherichia coli* biofilm promoted aggregation of particles and an increase in nanoparticle size by a factor of 40 (Choi et al., 2010). Scanning

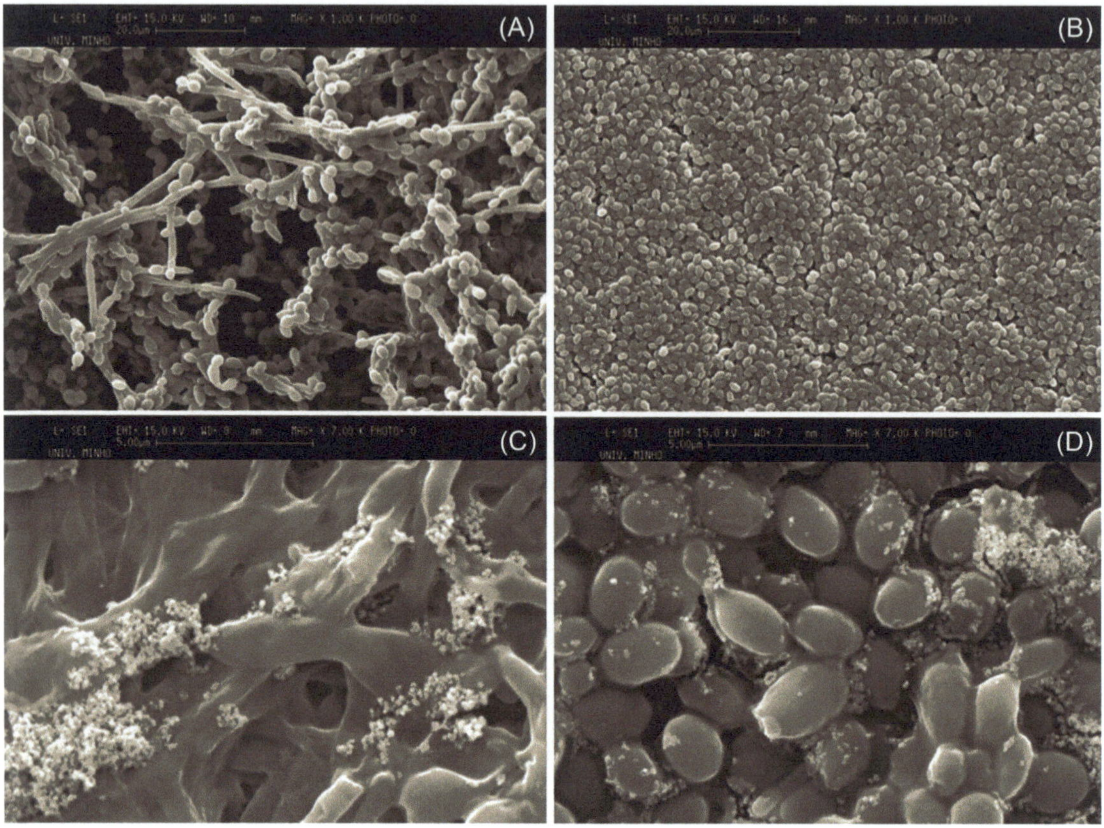

FIGURE 17.1 SEM images showing the structure of *Candida albicans* (A and C) and *C. glabrata* (B and D) mature biofilms before (A and B) and after (C and D) treatment with SN. Note clusters of SN in contact with *Candida* biofilms (C and D). Magnifications: ×1,000 (A and B); ×7,000 (C and D).

electron microscopy (SEM) observations also revealed aggregates of SN attached to the *Candida* biofilm matrices (Monteiro et al., 2013a). Biofilm traits, such as oxygen availability, substrate, pH range, and extracellular polymeric substances composition, may influence aggregation, dissolution, and diffusive transport of SNs (Stewart and Franklin, 2008). Thus, in *Candida* biofilms, SN aggregation may cause an increase in nanoparticle size, making it impossible to note an inverse relationship between SN size and antifungal activity.

Finally, these findings support the idea that the original size of the synthesized nanoparticles may be different from that observed when they are in contact with *Candida* biofilms, regardless of the type of stabilizing agent. Therefore, the size of SN and the type of stabilizing agent are not essential to their positive effect against *C. albicans* and *C. glabrata* biofilms.

17.2.2 Influence of Chemical Stability of SN on *Candida* Biofilms

The temperature and the immersion medium (ionic strength, composition, and pH) are factors that may have an effect on the size, rate of dissolution, aggregation, and, consequently, stability of SNs (Kittler et al., 2010; Gorup et al., 2011; Adegboyega et al., 2013). In view of the therapeutic potential of SN, and considering that the loss of the chemical stability might reduce their effectiveness against biofilms, a study was conducted to verify whether heating or changing the pH of a SN stock solution, as well as the treatment period, would affect the antibiofilm activity of SN against biofilms of *C. albicans* and *C. glabrata* (Monteiro et al., 2014b).

Several parameters were evaluated regarding SN (at 54 μg/mL) stability, namely temperature (50°C, 70°C, and 100°C), pH (5.0 and 9.0), and time of contact (5 and 24 h) with mature biofilms grown on acrylic resin specimens. Surprisingly, SN colloidal suspensions heated at 50°C and 70°C for 30 min did not display modifications in their absorption spectra as compared with the nonheated SN suspension. This fact pointed out that these colloidal suspensions remained stable after heating. However, the colloidal suspension heated at 100°C exhibited changes in the color and in the particle size distribution, which were evident by the shift of the absorption peak maximum toward a higher wavelength and the peak broadening of the ultraviolet/visible spectra, suggesting that this suspension was destabilized (Monteiro et al., 2014b). In addition, heating SN at 50°C, 70°C, and 100°C did not compromise the effectiveness of SN against *Candida* biofilms because, in general, there was no significant influence on total biomass or biofilms cell viability.

Regarding the experiments performed to assess the efficacy of SN colloidal suspensions with different pH, it should be mentioned that varying the pH (from 7.0 to 5.0 or 9.0) resulted in appearance of dark aggregates or flocs settled on the bottom of the vessel, evidencing instability of SN at both acidic and basic pH. Nevertheless, the findings also showed that the variations in pH did not impair or improve the effectiveness of SN in reducing the number of cultivable cells of *C. albicans* and *C. glabrata* biofilms. On the contrary, the pH modifications produced significant increases (ranging from 25% to 42.9%) in the total biofilm biomass of *C. glabrata* (Monteiro et al., 2014b). Thus, based on the results of the chemical stability through changes in temperature and pH, it seems that the use of stabilized or destabilized SNs may not be crucial to their effect against *C. albicans* and *C. glabrata* biofilms.

Finally, the comparison between the treatment periods (5 and 24 h) revealed significant differences for the biomass of *C. glabrata* biofilm and for the number of cultivable cells of *C. albicans* biofilm, with better results for the shorter treatment period for all (Monteiro et al., 2014b). A treatment period of 5 h was probably not enough to cause extensive aggregation of SN, which may have facilitated the antibiofilm action of these nanoparticles. Clinically, these findings are positive because shorter exposures to SNs may avoid problems associated with silver toxicity on human cells.

17.2.3 Effect Against Adhered Cells and Biofilms

The effect of SN against adhered cells and biofilms of *Candida* species was first reported by Monteiro et al. (2011). In this study, SNs (average diameter, 5 nm) were synthesized by silver nitrate reduction with sodium citrate and stabilized with ammonia. Then, these nanoparticles were applied (during 24 h) to adhered cells (2 h) or biofilms (48 h) of *C. albicans* and *C. glabrata* previously developed in 96-well microtiter plates.

The results for adhered cells showed a significant reduction in the total biomass and in the number of cultivable cells for SN concentrations at or higher than 3.3 μg/mL for both species. So, SNs showed the ability to inhibit biofilm formation when applied to already adhered cells. It is also important to mention that the effective SN concentration against adhered cells is lower than the toxic concentration of SNs against human cells (Carlson et al., 2008; Panácek et al., 2009).

Importantly, SN were more effective in reducing the number of cells and total biomass when applied to adhered cells than to preformed *Candida* biofilms (Monteiro et al., 2011). Because the adhesion phase contains a lower cell mass (Seneviratne et al., 2009) due to the incipient production of the extracellular matrix, the action of SNs may have been facilitated in this situation.

Regarding preformed *Candida* biofilms, the highest SN concentration tested (54 μg/mL) significantly reduced the number of cells and total biomass only for *C. glabrata* strains (Monteiro et al., 2011). In general, *C. glabrata* biofilms have reduced thickness, are less profuse, and are devoid of hyphae compared with *C. albicans* biofilms (Samaranayake et al., 2005; Seneviratne et al., 2010). Consequently, all these features can help to explain the better effect of SN against *C. glabrata* biofilms, which is still of major clinical importance because this species is known to be resistant to conventional antifungal drugs, making it difficult to eliminate.

Furthermore, the effective SN concentration against preformed *C. glabrata* biofilms (54 μg/mL) can be considered low when compared with the concentrations of conventional antifungal drugs normally tested. For instance, fluconazole at concentrations ranging from 50 to 1,250 μg/mL was ineffective against *C. glabrata* biofilms (Fonseca et al., 2014), whereas high concentrations of miconazole (2,081 μg/mL) resulted in a significant reduction in the number of cultivable cells for *C. albicans*, *C. glabrata*, *C. krusei*, *C. parapsilosis*, and *C. tropicalis* biofilms (Vandenbosch et al., 2010).

To know the antibiofilm spectrum of action of SN, the influence of the intermediate and maturation stages of biofilm development in the susceptibility of *C. albicans* and *C. glabrata* biofilms to SN was also investigated (Monteiro et al., 2014c). Here, *Candida* biofilms were developed on acrylic surfaces for 24 h (intermediate stage) and 48 h (maturation stage) and treated with SN at 54 μg/mL for 5 h. Comparisons between intermediate and mature biofilms treated with SN did not display significant differences in the reduction of total biomass and number of cultivable cells. Consequently, the stages of biofilm development analyzed (intermediate and maturation) did not interfere on the susceptibility of *Candida* biofilms to SN. This fact may guide the use of SN to prevent biofilms to reach more advanced phases of growing and to combat preformed mature biofilms, expanding the clinical applicability of SN in the control of oral *Candida* infections.

Another important finding was that the pretreatment of *Candida* biofilms with SN significantly reduced the subsequent adhesion capacity of *C. albicans* and *C. glabrata* viable cells to human epithelial cells and polystyrene surfaces (Monteiro et al., 2014a). The reductions in the capacity to adhere to HeLa cells and polystyrene ranged from 40% to 86% when both biofilms were pretreated with SN at 54 μg/mL. The adhesion of *Candida* to HeLa cells is regulated by proteins called adhesins (Miyauchi et al., 2007), whereas the adhesion to inert surfaces (polystyrene) is mainly commanded by physicochemical properties (surface hydrophobicity and electrostatic interactions) of the material and yeast cell surfaces (Rotrosen et al., 1986). In this context, it is believed that the reasons for the reductions observed are the binding of SN to adhesins present in the yeast cells and the reduction in the relative cell surface hydrophobicity promoted by the action of SN. These findings highlight the potential use of SN in controlling the dissemination of *Candida* infections because the cells released from biofilms exposed to SN may lose their ability to colonize and infect other areas.

17.2.4 Effect on Extracellular Matrix Composition and Structure of *Candida* Biofilms

One of the factors that may contribute to the resistance of *Candida* biofilms is the presence of extracellular matrix or extracellular polymeric substances (EPS) (Mah and O'Toole, 2001). The biofilm matrices are biopolymers of microbial origin that the cells are encased in; therefore, they are considered the "house of biofilm cells" (Flemming et al., 2007). The matrix is composed of several proteins, carbohydrates, phosphates (Baillie and Douglas, 2000; Blankenship and Mitchell, 2006), and extracellular DNA (Martins et al., 2010), and the proportion of each component varies with the growth medium, environmental conditions, and species (Conover et al., 2011).

Antifungal drugs may bind to constituents of the extracellular matrix of *Candida* biofilms, preventing the drug from reaching the cells in the deeper layers of the biofilm (Vediyappan et al., 2010). Therefore, strategies focusing on eradication of extracellular matrix may collaborate in controlling *Candida* infections associated with biofilms.

SNs were applied on mature *C. albicans* and *C. glabrata* biofilms for 24 h, and thereafter extracellular matrices were extracted from biofilms and analyzed chemically in terms of proteins, carbohydrates, and DNA. Moreover, the structures of *Candida* biofilms after treatment with SNs were observed by SEM and epifluorescence microscopy (Monteiro et al., 2013a).

For *C. glabrata*, it was possible to observe significant reductions in the matrix protein and DNA contents when biofilms were treated with SN at 54 μg/mL (Monteiro et al., 2013a). Because silver has affinity for proteins (Furno et al., 2004) and for phosphate groups present in the DNA molecule (Rai et al., 2009), it probably promotes cell lysis, release of intracellular constituents, and precipitation of a part of released protein and DNA. Surprisingly, the phenomena described were not noted for protein and DNA contents of *C. albicans* biofilms (Monteiro et al., 2013a).

In addition, after treatment with SN at 54 μg/mL, both *C. albicans* and *C. glabrata* displayed significant increases in total carbohydrate content compared with the untreated biofilms (Monteiro et al., 2013a). The increase observed was higher for *C. glabrata* biofilm, on which SNs are more effective in reducing total biomass and number of viable cells (Monteiro et al., 2011, 2013a). Therefore, it was hypothesized that the increases in carbohydrate content may be related to changes in the cell wall and secretion of their carbohydrates.

However, SEM observations showed that untreated *C. albicans* biofilm consisted of a mixture of yeasts and hyphae surrounded by a thick extracellular matrix. After treatment with SNs at 54 μg/mL, the resulting biofilm presented a much less compact structure and lower amounts of extracellular matrix (Monteiro et al., 2013a). For *C. glabrata*, the untreated biofilm revealed a multilayer composed entirely of yeasts, whereas the biofilm exposed to SNs displayed a more compact structure and had a thinner layer of yeasts covering the surface than the untreated biofilm (Monteiro et al., 2013a). Additionally, in both *C. albicans* and *C. glabrata* biofilms, it was possible to observe some clusters of SNs attached to the biofilm matrices and to the fungal cells. Epifluorescence microscopy images demonstrated that, in general, the amount of fluorescence and cell walls stained with Calcofluor white decreased with the increase in the SN concentration for *C. albicans* and *C. glabrata* biofilms, suggesting that SNs induced damage in the walls of biofilm cells (Monteiro et al., 2013a) (Figure 17.2). Accordingly, SN has the potential to affect the matrix composition and structure of *Candida* biofilms, and this may be associated with the mechanisms of fungicide action of SN.

Another study (unpublished data) evaluated the influence of extracellular matrix amount on

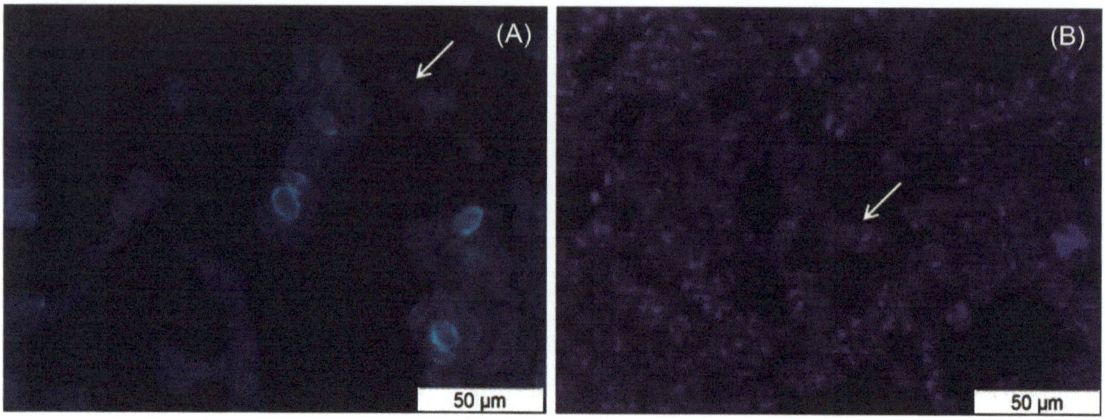

FIGURE 17.2 Epifluorescence microscopy images of *Candida albicans* (A) and *C. glabrata* (B) mature biofilms after exposure to SN. The arrows suggest that SN induced damage (pores) to cell walls of biofilm cells.

the antifungal activity of SN against single and mixed biofilms of *C. albicans* and *C. glabrata*. Single and mixed biofilms of *Candida* species were formed under agitation of 10 rpm (to form a lower amount of extracellular matrix) and 120 rpm (to form a higher amount of extracellular matrix) on the surfaces of acrylic resin specimens and then were exposed to different concentrations of SN for 24 h. The results indicated that the amount of extracellular matrix interfered in the effectiveness of SN in reducing the total biofilm biomass. However, regarding cultivable cells quantification, *Candida* biofilms formed with different incubation conditions (10 and 120 rpm) showed no differences in susceptibility to SN. Additionally, the amount of extracellular matrix may depend on the agitation speed during biofilm formation as well as the species and strains tested.

17.2.5 Combination of SN with Conventional Antifungal Drugs

Recently, an *in vitro* study (Silva et al., 2013) compared the antifungal effects of NYT and SN (both at 100 μg/mL) on preformed single-species and dual-species (*C. albicans* and *C. glabrata*) biofilms on acrylic resin surface under conditions that attempted to mimic the oral environment. The findings of this work highlighted that *C. albicans* and *C. glabrata* colonized the acrylic surfaces in the presence of artificial saliva and co-existed in dual-species biofilms without antagonism. This fact strengthens the occurrence of oral candidosis related to the presence of these two *Candida* species. Additionally, *C. albicans* displays significant differences in relation to *C. glabrata*, including cell size, morphology, and biochemistry. In general, *C. glabrata* cells are smaller (1–4 μm) than *C. albicans* cells (4–6 μm) and exhibit a narrower spectrum of carbohydrate utilization (Fidel et al., 1999). Thus, the frequent co-infection of *C. albicans* and *C. glabrata* in cases of oral candidosis may be explained in terms of these differences, seeing that they limit the competition between microorganisms, allowing both species to inhabit similar oral niches (Silva et al., 2013).

Results also revealed that NYT and SN inhibited *C. albicans* and *C. glabrata* biofilms in terms of total biomass and number of cultivable cells. Both agents promoted significant decrease in total biomass, and SN were more effective against single *C. glabrata* biofilms than NYT. Regarding the number of cultivable cells, SN were less active on single *C. albicans* biofilms and on mixed species biofilms than NYT

(Silva et al., 2013). However, because the inhibitory effects of SN were significant, its use as an alternative treatment for oral candidosis should be considered.

SNs were tested in combination with conventional antifungal drugs at low concentrations against *Candida* biofilms to increase the drug efficacy and to reduce their adverse effects (Monteiro et al., 2013b). SNs at 13.5 and 27 μg/mL were combined with NYT at 13.5 and 216 μg/mL or with CHG at 9 and 37.5 μg/mL, generating a total of eight different drug combinations. Thus, *C. albicans* and *C. glabrata* mature biofilms were grown in wells of 96-well microtiter plates in the presence of artificial saliva and after they were treated with these drug combinations.

With regard to the total biomass reduction in *C. albicans* biofilms, no synergistic effect was noted for all drug combinations, but the effects were additive and the combinations of SN at 13.5 and 27 μg/mL with CHG at 37.5 μg/mL promoted the highest reductions (~70%) in total biomass. For *C. glabrata* biofilms, all combinations of SN with NYT and the combination of SN at 13.5 μg/mL with CHG at 9 μg/mL showed synergistic activity, achieving reduction in the biofilm biomass in more than 84% (Monteiro et al., 2013b).

However, SNs at 13.5 μg/mL, when combined with NYT or CHG, proved a synergistic effect against *C. albicans* biofilm cells, and the highest decrease in the number of cells (>6 $\log_{10}$-fold) was achieved by the combination of SNs at 27 μg/mL with NYT at 216 μg/mL. Further, for *C. glabrata* biofilms, the combinations of SNs at 13.5 and 27 μg/mL with CHG at 9 μg/mL displayed synergistic activity, whereas all other drug combinations showed an additive effect with decrease in the number of biofilm cells of approximately 5 $\log_{10}$-fold (Monteiro et al., 2013b).

Interestingly, SEM observations revealed that the majority of *C. albicans* biofilm disruption was found when SNs at 13.5 μg/mL were combined with CHG at 37.5 μg/mL. This resultant biofilm showed few yeasts and deformed and elongated hyphae. With respect to the ultrastructure of *C. glabrata* biofilm, the combination of SNs at 27 μg/mL with CHG at 9 μg/mL resulted in a biofilm completely disrupted and constituted by few sparse yeasts (Monteiro et al., 2013b).

It is believed that the synergism between SNs and NYT or CHG is due to both different and similar mechanisms of drugs action. Both NYT and CHG promote disruption of the yeast cell membrane, with the consequent alteration of cell permeability and escapement of cell constituents (Kuyyakanond and Quesnel, 1992; Ellepola and Samaranayake, 2000). SNs have high surface-to-volume ratio, which improves the interaction between particles and cells (Baker et al., 2005). These nanoparticles bind to sulfur-containing proteins, resulting in defects in the cell membrane, interact with phosphorus-containing compounds like DNA (preventing cell reproduction), and attack the respiratory chain, causing the cell death (Monteiro et al., 2009; Rai et al., 2009). Therefore, the disruption of the cell membrane performed by NYT and CHG may help the entrance of SN into the cytoplasm, where these nanoparticles reach their targets.

In a clinical context, this combination of SN with NYT and CHG is very important because the concentrations of conventional antifungals tested are much lower (32–1,682 folds) than those found in commercial products. Additionally, it was not possible to observe antagonism between the combinations evaluated (Monteiro et al., 2013b).

17.2.6 Incorporation of SN into Denture Acrylic Resin

Considering that *Candida* infections involving oral mucosa represent a challenge for the dental field, the incorporation of SN into

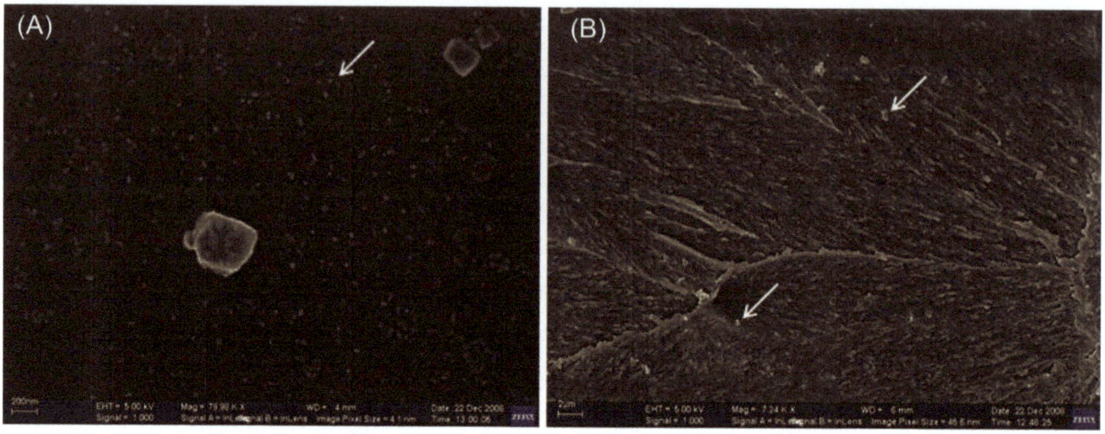

FIGURE 17.3 SEM images showing the surface (A) and the interior (B) of the PMMA/silver nanocomposite after 120 days of immersion in deionized water. The arrows indicate SN. Most of the nanoparticles are distributed in (A), and very few particles are visualized in (B). Magnifications: ×79,980 (A); ×7,240 (B).

PMMA used as denture base (Monteiro et al., 2012a) has been developed (Figure 17.3) as a method to prevent these infections.

An *in vitro* study conducted by Monteiro et al. (2012a,b) assessed a denture base acrylic resin-containing SN through morphological analysis by checking the distribution and dispersion of these particles in the polymer and by testing the silver release in deionized water at different time periods. SN were synthesized through the reduction of silver nitrate with sodium citrate and added to the monomer of the acrylic resin at 0.05%, 0.5%, and 5%, based on the polymer mass. Then, the acrylic specimens were immersed in deionized water and kept at 37°C under agitation for 7, 15, 30, 60, and 120 days. After each period of time, the amount of silver released was measured by atomic absorption spectroscopy. Moreover, SEM was used to analyze the morphology of the nanocomposites generated before and after 120 days of immersion in water.

Interestingly, the results did not reveal the presence of silver in deionized water, even after the immersion of nanocomposites for 120 days. Additional tests were performed to verify whether SN would be released in artificial saliva. Again, silver was not detected in this medium by the atomic absorption spectroscopy analysis, which is very sensitive (Monteiro et al., 2012a).

SEM images of the PMMA/SN nanocomposites showed that SN were well-adhered to the surface of the PMMA matrix, before and after storage in deionized water. It was verified that before 120 days of storage, the higher the concentration of SN incorporated into PMMA, the higher the distribution of the nanoparticles and the lower their dispersion into the polymeric mass (Monteiro et al., 2012a). However, after 120 days of storage in water, SN were found on the polymer surface, regardless of the SN concentration added to the PMMA. SN probably migrated from the interior to the surface of the polymer, where the energy is high enough to accommodate a large concentration of them.

Finally, preliminary antimicrobial and mechanical tests of the PMMA/SN nanocomposites demonstrated that they had good efficacy against *C. albicans* biofilm formation, mainly the nanocomposite containing 5% of SN. Similar values of flexural strength were also observed between the nanocomposites and the control group (PMMA without SN) (Monteiro et al., 2012a).

17.3 CONCLUSIONS AND FUTURE PERSPECTIVES

Based on available literature regarding the effect of SN against *C. albicans* and *C. glabrata*, it may be concluded that the size of SN and the type of stabilizing agent do not interfere in the antifungal activity against biofilms of these species; the temperature and pH variations of SN do not influence their effectiveness against the viable cells of mature *Candida* biofilms; SN are more effective against *Candida*-adhered cells than against their mature biofilms; SN interfere with the extracellular matrix composition of mature *Candida* biofilms in terms of protein, carbohydrate, and DNA, and with the ultrastructure of these biofilms; SN combined with either NYT or CHG show synergistic antibiofilm activity against mature *Candida* biofilms; and SN remain adhered to PMMA even after 120 days of storage in deionized water.

In view of all these attributes it is possible to consider that SN can be used as an alternative treatment for oral candidosis. However, further research is needed to elucidate the exact mechanisms by which adhesion capacity of *Candida* cells is decreased after treatment with SN, the cytotoxicity/genotoxicity of SN alone or combined with conventional antifungal drugs, the longevity of SN formulation under oral conditions, and the physical, mechanical, and microbiological properties of the nanocomposites generated by the incorporation of SN into denture acrylic resin.

References

Adegboyega, N.F., Sharma, V.K., Siskova, K., Zbořil, R., Sohn, M., Schultz, B.J., et al., 2013. Interactions of aqueous Ag + with fulvic acids: mechanisms of silver nanoparticle formation and investigation of stability. Environ. Sci. Technol. 47, 757–764.

Andes, D., Nett, J., Oschel, P., Albrecht, R., Marchillo, K., Pitula, A., 2004. Development and characterization of an *in vivo* central venous catheter *Candida albicans* biofilm model. Infect. Immun. 72, 6023–6031.

Arendorf, T.M., Walker, D.M., 1987. Denture stomatitis: a review. J. Oral. Rehabil. 14, 217–227.

Bagg, J., Sweeney, M.P., Lewis, M.A., Jackson, M.S., Coleman, D., Al, M.A., et al., 2003. High prevalence of non-*albicans* yeasts and detection of antifungal resistance in the oral flora of patients with advanced cancer. Palliat. Med. 17, 477–481.

Baillie, G.S., Douglas, L.J., 2000. Matrix polymers of *Candida* biofilms and their possible role in biofilm resistance to antifungal agents. J. Antimicrob. Chemother. 46, 397–403.

Baker, C., Pradhan, A., Pakstis, L., Pochan, D.J., Shah, S.I., 2005. Synthesis and antibacterial properties of silver nanoparticles. J. Nanosci. Nanotechnol. 5, 244–249.

Beikler, T., Flemmig, T.F., 2011. Oral biofilm-associated diseases: trends and implications for quality of life, systemic health and expenditures. Periodontology 2000, 87–103.

Blankenship, J.R., Mitchell, A.P., 2006. How to build a biofilm: a fungal perspective. Curr. Opin. Microbiol. 9, 588–594.

Budtz-Jorgensen, E., 2000. Ecology of *Candida*-associated denture stomatitis. Microb. Ecol. Health Dis. 12, 170–185.

Budtz-Jorgensen, E., Bertram, U., 1970. Denture stomatitis I. The etiology in relation to trauma and infection. Acta Odontol. Scand. 28, 71–92.

Carlson, C., Hussain, S.M., Schrand, A.M., Braydich-Stolle, L.K., Hess, K.L., Jones, R.L., et al., 2008. Unique cellular interaction of silver nanoparticles: size-dependent generation of reactive oxygen species. J. Phys. Chem. B. 112, 13608–13619.

Chandra, J., Kuhn, D.M., Mukherjee, P.K., Hoyer, L.L., McCormick, T., Ghannoum, M.A., 2001a. Biofilm formation by the fungal pathogen *Candida albicans*: development, architecture, and drug resistance. J. Bacteriol. 183, 5385–5394.

Chandra, J., Mukherjee, P.K., Leidich, S.D., Faddoul, F.F., Hoyer, L.L., Douglas, L.J., et al., 2001b. Antifungal resistance of candidal biofilms formed on denture acrylic *in vitro*. J. Dent. Res. 80, 903–908.

Choi, O., Yu, C.P., Esteban Fernández, G., Hu, Z., 2010. Interactions of nanosilver with *Escherichia coli* cells in planktonic and biofilm cultures. Water Res. 44, 6095–6103.

Coco, B.J., Bagg, J., Cross, L.J., Jose, A., Cross, J., Ramage, G., 2008. Mixed *Candida albicans* and *Candida glabrata* populations associated with the pathogenesis of denture stomatitis. Oral Microbiol. Immunol. 23, 377–383.

Conover, M.S., Mishra, M., Deora, R., 2011. Extracellular DNA is essential for maintaining *Bordetella* biofilm integrity on abiotic surfaces and in the upper respiratory tract of mice. PLoS One 6, e16861.

Costerton, J.W., Stewart, P.S., Greenberg, E.P., 1999. Bacterial biofilms: a common cause of persistent infections. Science 284, 1318–1322.

Ellepola, A.N., Samaranayake, L.P., 2000. Oral candidal infections and antimycotics. Crit. Rev. Oral Biol. Med. 11, 172–198.

Fidel, P.L., Vazquez, J.A., Sobel, J.D., 1999. *Candida glabrata*: review of epidemiology, pathogenesis, and clinical disease with comparison to *C. albicans*. Clin. Microbiol. Rev. 12, 80–96.

Flemming, H.C., Neu, T.R., Wozniak, D.J., 2007. The EPS matrix: the "house of biofilm cells". J. Bacteriol. 189, 7945–7947.

Fonseca, E., Silva, S., Rodrigues, C.F., Alves, C.T., Azeredo, J., Henriques, M., 2014. Effects of fluconazole on *Candida glabrata* biofilms and its relationship with ABC transporter gene expression. Biofouling 30, 447–457.

Furno, F., Morley, K.S., Wong, B., Sharp, B.L., Arnold, P.L., Howdle, S.M., et al., 2004. Silver nanoparticles and polymeric medical devices: a new approach to prevention of infection? J. Antimicrob. Chemother. 54, 1019–1024.

Gorup, L.F., Longo, E., Leite, E.R., Camargo, E.R., 2011. Moderating effect of ammonia on particle growth and stability of quasi-monodisperse silver nanoparticles synthesized by the Turkevich method. J. Colloid Interface Sci. 360, 355–358.

Hasan, F., Xess, I., Wang, X., Jain, N., Fries, B.C., 2009. Biofilm formation in clinical *Candida* isolates and its association with virulence. Microbes Infect. 11, 753–761.

Hawser, S.P., Douglas, L.J., 1994. Biofilm formation by *Candida* species on the surface of catheter materials *in vitro*. Infect. Immun. 62, 915–921.

Jakubovics, N.S., Kolenbrander, P.E., 2010. The road to ruin: the formation of disease-associated oral biofilms. Oral Dis. 16, 729–739.

Jewtuchowicz, V.M., Brusca, M.I., Mujica, M.T., Gliosca, L. A., Finquelievich, J.L., Lovannitti, C.A., et al., 2007. Subgingival distribution of yeast and their antifungal susceptibility in immunocompetent subjects with and without dental devices. Acta. Odontol. Latinoam. 20, 17–22.

Kittler, S., Greulich, C., Diendorf, J., Koller, M., Epple, M., 2010. Toxicity of silver nanoparticles increases during storage because of slow dissolution under release of silver ions. Chem. Mater. 22, 4548–4554.

Kuyyakanond, T., Quesnel, L.B., 1992. The mechanism of action of chlorhexidine. FEMS Microbiol. Lett. 79, 211–215.

Kvítek, L., Panácek, A., Soukupová, J., Kolár, M., Vecerová, R., Prucek, R., et al., 2008. Effect of surfactants and polymers on stability and antibacterial activity of silver nanoparticles (NPs). J. Phys. Chem. C. 112, 5825–5834.

Mah, T.F., O'Toole, G.A., 2001. Mechanisms of biofilm resistance to antimicrobial agents. Trends Microbiol. 9, 34–39.

Martins, M., Uppuluri, P., Thomas, D.P., Cleary, I.A., Henriques, M., Lopez-Ribot, J.L., et al., 2010. Presence of extracellular DNA in the *Candida albicans* biofilm matrix and its contribution to biofilms. Mycopathologia 169, 323–331.

Miyauchi, M., Giummelly, P., Yazawa, S., Okawa, Y., 2007. Adhesion of *Candida albicans* to HeLa cells: studies using polystyrene beads. Biol. Pharm. Bull. 30, 588–590.

Monteiro, D.R., Gorup, L.F., Takamiya, A.S., Ruvollo-Filho, A.C., de Camargo, E.R., Barbosa, D.B., 2009. The growing importance of materials that prevent microbial adhesion: antimicrobial effect of medical devices containing silver. Int. J. Antimicrob. Agents 34, 103–110.

Monteiro, D.R., Gorup, L.F., Silva, S., Negri, M., de Camargo, E.R., Oliveira, R., et al., 2011. Silver colloidal nanoparticles: antifungal effect against *Candida albicans* and *Candida glabrata* adhered cells and biofilms. Biofouling 27, 711–719.

Monteiro, D.R., Gorup, L.F., Takamiya, A.S., de Camargo, E.R., Filho, A.C., Barbosa, D.B., 2012a. Silver distribution and release from an antimicrobial denture base resin containing silver colloidal nanoparticles. J. Prosthodont. 21, 7–15.

Monteiro, D.R., Silva, S., Negri, M., Gorup, L.F., de Camargo, E.R., Oliveira, R., et al., 2012b. Silver nanoparticles: influence of stabilizing agent and diameter on antifungal activity against *Candida albicans* and *Candida glabrata* biofilms. Lett. Appl. Microbiol. 54, 383–391.

Monteiro, D.R., Silva, S., Negri, M., Gorup, L.F., de Camargo, E.R., Oliveira, R., et al., 2013a. Silver colloidal nanoparticles: effect on matrix composition and structure of *Candida albicans* and *Candida glabrata* biofilms. J. Appl. Microbiol. 114, 1175–1183.

Monteiro, D.R., Silva, S., Negri, M., Gorup, L.F., de Camargo, E.R., Oliveira, R., et al., 2013b. Antifungal activity of silver nanoparticles in combination with nystatin and chlorhexidine digluconate against *Candida albicans* and *Candida glabrata* biofilms. Mycoses 56, 672–680.

Monteiro, D.R., Negri, M., Silva, S., Gorup, L.F., de Camargo, E.R., Oliveira, R., et al., 2014a. Adhesion of *Candida* biofilm cells to human epithelial cells and polystyrene after treatment with silver nanoparticles. Colloids Surf. B Biointerfaces 114, 410–412.

Monteiro, D.R., Takamiya, A.S., Feresin, L.P., Gorup, L.F., de Camargo, E.R., Delbem, A.C.B., et al., 2014b. Silver colloidal nanoparticle stability: influence on *Candida* biofilms formed on denture acrylic. Med. Mycol. 52, 627–635.

Monteiro, D.R., Takamiya, A.S., Feresin, L.P., Gorup, L.F., de Camargo, E.R., Delbem, A.C.B., et al., 2014c. Susceptibility of *Candida albicans* and *Candida glabrata* biofilms to silver nanoparticles in intermediate and mature development phases. J. Prosthodont. Res. Available from: <http://dx.doi.org/10.1016/j.jpor.2014.07.004>.

Nett, J.E., Marchillo, K., Spiegel, C.A., Andes, D.R., 2010. Development and validation of an *in vivo Candida albicans* biofilm denture model. Infect. Immun. 78, 3650–3659.

Panácek, A., Kolár, M., Vecerová, R., Prucek, R., Soukupová, J., Krystof, V., et al., 2009. Antifungal activity of silver nanoparticles against *Candida* spp. Biomaterials 30, 6333–6340.

Rai, M., Yadav, A., Gade, A., 2009. Silver nanoparticles as a new generation of antimicrobials. Biotechnol. Adv. 27, 76–83.

Ramage, G., Saville, S.P., Thomas, D.P., López-Ribot, J.L., 2005. *Candida* biofilms: an update. Eukaryot. Cell 4, 633–638.

Rautemaa, R., Ramage, G., 2011. Oral candidosis—clinical challenges of a biofilm disease. Crit. Rev. Microbiol. 37, 328–336.

Rotrosen, D., Calderone, R.A., Edwards Jr., J.E., 1986. Adherence of *Candida* species to host tissues and plastic surfaces. Rev. Infect. Dis. 8, 73–85.

Samaranayake, L.P., Yaacob, H.B., 1990. Classification of oral candidosis. In: Samaranayake, L.P., MacFarlane, T.W. (Eds.), Oral Candidosis. Wright-Butterworth, London, pp. 124–132.

Samaranayake, L.P., Keung Leung, W., Jin, L., 2009. Oral mucosal fungal infections. Periodontol. 2000. 49, 39–59.

Samaranayake, Y.H., Ye, J., Yau, J.Y., Cheung, B.P., Samaranayake, L.P., 2005. *In vitro* method to study antifungal perfusion in *Candida* biofilms. J. Clin. Microbiol. 43, 818–825.

Seneviratne, C.J., Jin, L., Samaranayake, L.P., 2008. Biofilm lifestyle of *Candida*: a mini review. Oral Dis. 14, 582–590.

Seneviratne, C.J., Silva, W.J., Jin, L.J., Samaranayake, Y.H., Samaranayake, L.P., 2009. Architectural analysis, viability assessment and growth kinetics of *Candida albicans* and *Candida glabrata* biofilms. Arch. Oral Biol. 54, 1052–1060.

Seneviratne, C.J., Wang, Y., Jin, L., Abiko, Y., Samaranayake, L.P., 2010. Proteomics of drug resistance in *Candida glabrata* biofilms. Proteomics 10, 1444–1454.

Silva, S., Henriques, M., Hayes, A., Oliveira, R., Azeredo, J., Williams, D.W., 2011. *Candida glabrata* and *Candida albicans* co-infection of an *in vitro* oral epithelium. J. Oral. Pathol. Med. 40, 421–427.

Silva, S., Pires, P., Monteiro, D.R., Negri, M., Gorup, L.F., Camargo, E.R., et al., 2013. The effect of silver nanoparticles and nystatin on mixed biofilms of *Candida glabrata* and *Candida albicans* on acrylic. Med. Mycol. 51, 178–184.

Soysa, N.S., Samaranayake, L.P., Ellepola, A.N., 2008. Antimicrobials as a contributory factor in oral candidosis—a brief overview. Oral Dis. 14, 138–143.

Stewart, P.S., Franklin, M.J., 2008. Physiological heterogeneity in biofilms. Nat. Rev. Microbiol. 6, 199–210.

Vandenbosch, D., Braeckmans, K., Nelis, H.J., Coenye, T., 2010. Fungicidal activity of miconazole against *Candida* spp. biofilms. J. Antimicrob. Chemother. 65, 694–700.

Vediyappan, G., Rossignol, T., d'Enfert, C., 2010. Interaction of *Candida albicans* biofilms with antifungals: transcriptional response and binding of antifungals to beta-glucans. Antimicrob. Agents Chemother. 54, 2096–2111.

Watamoto, T., Samaranayake, L.P., Jayatilake, J.A., Egusa, H., Yatani, H., Seneviratne, C.J., 2009. Effect of filamentation and mode of growth on antifungal susceptibility of *Candida albicans*. Int. J. Antimicrob. Agents 34, 333–339.

Webb, B.C., Thomas, C.J., Willcox, M.D., Harty, D.W., Knox, K.W., 1998. *Candida*-associated denture stomatitis. Aetiology and management: a review. Part 1. Factors influencing distribution of *Candida* species in the oral cavity. Aust. Dent. J. 43, 45–50.

Weems Jr., J.J., 1992. *Candida parapsilosis*: epidemiology, pathogenicity, clinical manifestations, and antimicrobial susceptibility. Clin. Infect. Dis. 14, 756–766.

CHAPTER

18

Nanomedical Therapeutic and Prophylaxis Strategies Against Intracellular Protozoa in the Americas

Maria Jose Morilla and Eder Lilia Romero

Programa de Nanomedicinas, Departamento de Ciencia y Tecnología, Universidad Nacional de Quilmes, Buenos Aires, Argentina

18.1 INTRODUCTION

Therapeutics and prophylaxis of tumor and infections using three-dimensional nanoobjects (metal and oxides nanoparticles [NPs], polymeric and lipid NPs, nanocapsules, dendrimers, micelles, liposomes, and other vesicles up to 200–300 nm size, referred to here as NPs) are the most important applications of nanomedicine. Bare NPs or NP-loaded active principles (apis) (also known as nonbiological complex drugs [NBCDs]) constitute the so-called nanomedicines. Pharmacokinetics and biodistribution of api in NBCDs do not depend on the chemical structure, but rather on the structural features of the NPs. NPs of adequate size, shape, and surface can surpass anatomical and phenomenological barriers, protecting and enabling the passage of apis across gastrointestinal mucus, skin, and blood–brain barriers. Because eukaryotic cells can recognize and take-up NPs by means of endocytic mechanisms, nanomedicines are used for targeted delivery of foreign, highly charged, and/or high-molecular-weight apis such as proteins and nucleic acids to cells, modifying their subsequent intracellular processing pathway. The targeted delivery of proteins to antigen-presenting cells (APC), however is the core of NP-mediated prophylactic strategies. Overall, NPs enable cell targeting of poorly bioavailable, labile in circulation, poorly cell-penetrating, and/or nonselectively biodistributed apis. More specifically, nanomedicines provide an advantage for standard low-molecular-weight (LMW) apis by reducing their renal excretion and/or hepatic degradation, leading to prolonged circulation times. Nanomedicines also reduce the volume of distribution of apis, leading to less accumulation in healthy nontarget tissues ("site-avoidance drug delivery"), improvement in the ability of apis to accumulate at pathological sites

Nanotechnology in Diagnosis, Treatment and Prophylaxis of Infectious Diseases.
DOI: http://dx.doi.org/10.1016/B978-0-12-801317-5.00018-9

("site-specific drug delivery") and improvement in the therapeutic index of apis by increasing their accumulation at the target site and/or reducing their localization in potentially endangered healthy organs. In addition, NBCDs assist LMW (chemo-)therapeutic agents in overcoming several additional barriers such as vascular endothelium, perivascular space, cellular membrane, nuclear membrane, and blood–brain barrier, which impair the access to pathological sites. In sum, nanomedicines help to overcome the so-called clinical barriers such as low efficacy, high toxicity, the need for hospitalization, and/or frequent administration, as well as low cost-effectiveness (Rizzo et al., 2013). Excluding the targeting to intravascular sites, the selective distribution of intravenously administered nanomedicines necessitates a suitable anatomopathological environment with characteristics such as increased vascular permeability (accompanied or not by impaired lymphatic drainage). The presence of the target site having determined tissue/cell activity or ligand expression does not help improve selectivity, but the cell uptake of nanomedicines can be improved. Infections are usually accompanied by inflammation that, as in solid tumors, generally present leaky blood vessels. This fact, as opposed to the vasculature in the majority of healthy tissues, allows for the extravasation of NPs with sizes of up to 400 nm. Nanomedicines are generally able to accumulate both effectively and selectively at such pathological sites by means of a mechanism known as the enhanced permeability and retention (EPR) effect. Besides the enhanced leakiness of tumorous and inflamed blood vessels, the EPR effect is also based on the fact that solid tumors tend to lack functional lymphatics, thereby limiting the removal of extravasated NPs from the target site. Because EPR-mediated drug targeting exclusively relies on the pathological properties of the target tissue, that is, enhanced leakiness and poor lymphatic drainage, it is generally referred to as passive drug targeting. Active drug targeting, however, relies on the use of antibodies, peptides, or sugar moieties, which are physically or chemically incorporated into nanomedicine formulations to facilitate localization to and/or uptake by target cells.

Various vascular mediators commonly found in inflammation and cancer contribute to the EPR effect. These mediators include vascular endothelial growth factor (VEGF), which enhances the vascular permeability of normal blood vessels as well as that of tumor vessels, with this process partly involving endothelial nitric oxide synthase (eNOS) and thus NO; activated cyclooxygenase-1, interferon-γ (IFN-γ), inducible NOS (iNOS), and other cytokines such as tumor necrosis factor-α (TNF-α), transforming growth factor-β (TGF-β), interleukin-2 (IL-2), and IL-8 also affect normal blood vessels.

Among the various requirements for and factors influencing the EPR effect, the most important is having a molecular size larger than 40 kDa (larger than the renal clearance threshold). However, this requirement is only partly valid, because size alone is not sufficient for the EPR effect to occur. Certain heavy macromolecular complexes have a short half-life in circulation and do not passively accumulate despite the EPR context. Hence, the circulating nanomedicines must avoid the interaction with blood components (causing, for instance, cell lysis) or blood vessels and minimize antigenicity to reduce clearance by the reticulo-endothelial system. The surface charge of nanomedicines has to be weakly negative to near neutral. The time required to achieve accumulation has to be longer than several hours in systemic circulation in mice, with distinct accumulation seen at 30 min; this leads to a retention time of days to weeks. This is in great contrast to passive targeting of LMW api, which are rapidly cleared and enter the systemic circulation in a few minutes. This is the case of LMW contrast agents used in angiography, which are taken-up in the tumor tissue by passive targeting but not retained.

Similarly, arterial injection of LMW anticancer agents, although they hit the tumor by the first path effect, is not retained in tumor tissue; therefore, there is not much clinical benefit. Unfortunately, a major difference between tumors and inflammation related to infections is the clearance rate of extravasated macromolecules such as plasma proteins, lipid particles, or nanomedicine, resulting in prolonged retention time in tumor tissue compared with that in inflamed normal tissue (Maeda et al., 2013).

Currently, nearly 120 therapeutic and prophylactic nanomedicines are in clinical trials; there are already 70 products on the market for cancer, cardiovascular, neurodegenerative, musculoskeletal, and inflammatory diseases. In this chapter, the question posed by the treatment and prophylaxis of the main parasitosis of the Americas—leishmaniasis and Chagas disease—using nanomedicines are addressed.

18.2 LEISHMANIASIS

The different clinical manifestations of leishmaniasis are cutaneous leishmaniasis (CL; with ulcerative skin lesions), mucocutaneous leishmaniasis (MCL; destructive mucosal inflammation), and visceral leishmaniasis (VL), and they are dependent on the infecting leishmania genus (Table 18.1).

Once injected in the skin by the bite of a vector sand fly of *Lutzomia* or *Phlebotomus* genus, leishmania promastigotes invade local phagocytic host cells. Inside the cells, the promastigotes transform into amastigotes, which survive in the harsh environment of phagolysosomes. After a brief period during which amastigotes multiply, promastigotes are released in a cell burst and dissemination to local or distant phagocytes occurs. Liver, spleen, and bone marrow macrophages are colonized in

TABLE 18.1 Characteristics of Leishmaniasis and Chagas Disease

Disease	Causative agent	Numbers	Drugs used and adverse reactions
Leishmaniasis	*Leishmania* spp.	1.5 million and 500,000 new cases of CL and VL occur each year, respectively.	CL heals by reepithelization with scarring, but it is treated to avoid disfiguring lesions; MCL can produce potentially life-threatening inflammatory disease and must be treated; VL is mortal if untreated.
	L. donovani and *L. infantum* (old world); *L. chagasi* (new world) cause VL.	Endemic in 98 countries in Asia, Africa, South and Central America and southern Europe.	Pentavalent antimonials (Sb^V) (meglumine antimoniate and sodium stibogluconate) first line for some forms. Disadvantages: resistance, administered parenterally daily for 3 weeks.
	L. major, *L. tropica*, and *L. aethiopica* (old world); *L. mexicana*, *L. amazonensis*, *L. braziliensis*, *L. panamensis*, and *L. guyanensis* (new world) cause CL	Opportunistic infection in HIV patients, potentially fatal. HIV infection increase 10–100 folds the risk of development of VL	Miltefosine, paromomycin, pentamidine and AmB (Fungizone, micellar formulation of AmB and deoxycholate) used as second-line drugs, also are highly toxic and need parenteral administration. Lipid-based AmB formulations have reduced nephrotoxicity of AmB.

(Continued)

TABLE 18.1 (Continued)

Disease	Causative agent	Numbers	Drugs used and adverse reactions
Chagas disease	*T. cruzi*	10−12 million people infected in endemic areas of Latin America. 15,000 deaths each day.	BNZ and NFX active in acute (up to 80% of cures) and early chronic phases (up to 60% cures), are contraindicated in pregnancy.
		Infection becoming increasingly prevalent in Europe and United States.	Two randomized trials are in the process to provide evidence regarding the evolution of mild or advanced heart disease in chronic patients treated with BNZ.
			BNZ adverse effects: dermatitis with cutaneous eruptions, myalgias, arthralgias and lymphadenopathy; polyneuropathy, paresthesias and polyneuritis; bone marrow disorders, such as thrombopenic purpura, and agranulocytosis; and genotoxicity.

VL, and skin macrophages and dendritic cells, including Langerhans cells, are colonized in CL, together with lymph nodes and mucosas cells in MCL. The intracellular location of amastigotes within phagolysosomes is the main structural and phenomenological barrier to overcome by leishmanicidal drugs.

Therapy for leishmaniasis depends on its clinical manifestation (Table 18.1). Today, the intravenous infusion of liposomal amphotericin (AmB; AmBisome) has become the standard treatment for the lethal VL and the basis for new short course treatments.

18.2.1 Nanomedical Therapeutic Strategies

18.2.1.1 Visceral Leishmaniasis

18.2.1.1.1 NANOMEDICINES BASED ON AmB

18.2.1.1.1.1 PARENTERAL ROUTE The high activity in animal models of VL of intravenous liposomal formulations of Sb^V and AmB was known since the mid-1970s because of their direct uptake by liver and spleen macrophages. AmBisome—a brand name for liposomal AmB—was reported to eliminate liver parasites to the same extent, but faster than free Sb^V. AmBisome is effective in cases of Sb^V unresponsiveness in *Leishmania infantum* and *Leishmania donovani* foci and is recommended for the treatment of VL in immunosuppressed patients (Abeer et al., 2012). Although presenting significant regional variation, AmBisome is more efficacious and less toxic than other AmB lipid-based formulations (Table 18.2) (Bern et al., 2006). The main limitation for the use of AmBisome, however, is its high cost (Olliaro et al., 2009); even a single dose (5 mg/kg 97.5% cure rate in VL patients in India) is still considered expensive. AmBisome requires a cold chain and is not stable at temperatures above 25°C (Croft and Olliaro, 2011). Even minor alterations in the molar ratios of the drug to phospholipids or the process of manufacture may affect the performance of liposomal AmB formulations. Cheaper, nonbioequivalent generic versions of AmBisome have been

TABLE 18.2 Lipidic AmB Formulations Used in the Clinics

Name	Composition	Size	LD50 mice (mg/kg)	Manufacture
AmBisome	Hydrogenated soy phosphatidylcholine (HSPC): cholesterol (chol): distearoylphos-phatidylglycerol (DSPG): AmB (2:1:0.8:0.4 molar ratio) (unilamellar liposomes)	80 nm	>175	Gilead Sciences, USA
Abelcet	Dimyristoylphosphatidylcholine (DMPC): Dimyristoylphosphatidylglycerol (DMPG): AmB (7:3:10 molar ratio) (ribbon-like complex)	1.6–11 μm, with 90% < 6 μm	>40	Enzon Pharmaceutical, USA
Amphocil amphotec	Cholesteryl sulfate: AmB (1 :1 molar ratio) (lipid complex, colloidal dispersion)	100 ± 22 nm	36	Three Rivers Pharmaceuticals, USA
Fungisome	Phosphatidylcholine: chol (7:3 molar ratio) AmB/lipid 2.2% (multilamellar liposome)	Convert to small unilamellar vesicles through 45 min sonication 0.0884–3 μm	17.6	Lifecare Innovations, India
Amphomul	Similar to Amphocil			Bharat Serums and Vaccines Limited, India
Phosome	No data			CIPLA, India
Anfogen	Similar to AmBisome, but manufactured differently	111 nm	10	Genpharma, Argentina

produced. For instance, Anfogen has a chemical composition similar to AmBisome, but its different manufacture produces liposomes with significantly different efficacy and toxicity (Olson et al., 2008).

Aiming to increase its activity and reduce its toxicity, the performance of AmB within different NPs has been recently screened in preclinical models (Table 18.3).

AmB loaded in poly(lactide-co-glycolide-polyethilenglycol [PLGA-PEG]) (Kumar et al., 2014) and in phosphoserine-coated gelatin NPs for specific targeting to macrophages (Khatik et al., 2014), showed higher activity than free AmB. In both cases the inhibition of parasite burden was not higher than 93% and the activities were not compared with AmBisome.

Increased phagocytosis, production of proinflammatory cytokines (IL-6, IL-1b, and TNF-α), and oxidate species (reactive oxygen species [ROS] and nitric oxide species) in *Leishmania*-infected phagocytes accompany parasites elimination. AmB-loaded chitosan nanocapsules showed higher *in vivo* activity than AmBisome and Fungizone because of the upregulation of TNF-α, IL-12, and iNOS, along with chitosan-mediated downregulation of TGF-β, IL-10, and IL-4 (Asthana et al., 2013). In the same sense, lipo-polymerosomes of glycol chitosan–stearic acid copolymer-containing AmB showed higher

TABLE 18.3 Nanomedicines Based on AmB

Formulation	Composition	Efficacy and toxicity	Reference
Lipo-polymerosomes	Self-assembly of synthesized glycol chitosan–stearic acid copolymer and cholesterol. 240 nm, +27 mV Z potential	*L. donovani*-infected hamsters	Gupta et al. (2014a)
		Inhibited 78.66% splenic parasite burden. Fungizone and AmBisome caused 56.54% and 66.46% inhibition, respectively, in *L. donovani*-infected hamsters. Less toxicity *in vitro* than Fungizone and AmBisome	Gupta et al. (2014b)
Gelatine NP	1,2-Diacyl-sn-glycero-3-phospho-L-serine (PS)-coated gelatin NP	Inhibited 85.3% splenic parasitic burden in *L. donovani*-infected hamsters. Noncoated NPs and AmB caused 71.0% and 50.5% inhibition, respectively.	Khatik et al. (2014)
PLGA-PEG NP	30 nm	Inhibited 93% splenic parasite burden, AmB caused 74.42% inhibition	Kumar et al. (2014)
Chitosan nanocapsule	AmB-containing Tween 80 and soya lecithin as surfactants and with soya bean oil as core covered by chitosan. 146 nm, +29 mV Z potential	Inhibited 86% splenic parasite burden in *L. donovani*-infected hamsters (1 mg/kg for 5 consecutive days i.p.). AmBisome and Fungizone caused 69.8% and 55.5% inhibition, respectively. Nanocapsules were less toxic than Fungizone and AmBisome on J774A macrophages and erythrocytes.	Asthana et al. (2013)
AmB deoxycholate NP	AmB NP. 10–20 nm	Inhibited 92.18% splenic parasite burden in *L. donovani*-infected hamster (5 mg/kg for 5 consecutive days i.p.). Fungizone caused 74.57% inhibition.	Manandhar et al. (2008)
		ED50 for intracellular amastigotes in peritoneal macrophage and J774 cells was significantly less than the required dose of Fungizone (0.0027 vs. 0.0426 μg/mL and 0.0038 vs. 0.0196 μg/mL).	Manandhar et al. (2014)
Carbon nanotubes	AmB attached to amino-terminated carbon nanotubes diameter ~40–70 nm, length ~2–8 μm	Inhibited 89.85% splenic parasite burden in *L. donovani*-infected hamsters (5 mg/kg for 5 consecutive days i.p.). AmB caused 68.97% inhibition. No hepatic and renal toxicity was observed up to 20 mg/kg	Prajapati et al. (2011)
		Inhibited 98.2% splenic parasite burden in *L. donovani*-infected hamster (15 mg/kg for 5 consecutive days orally). Oral miltefosine and AmBisome (5 mg/kg i.p.) caused 80.6% and 97.6% inhibition, respectively.	Prajapati et al. (2012)

(Continued)

TABLE 18.3 (Continued)

Formulation	Composition	Efficacy and toxicity	Reference
Poly(HPMA)–polymer conjugates	HPMA-AmB	Inhibited 93.8% and 99.6% hepatic parasite burden at 1 and 3.0 mg/kg i.v. administered respectively, in *L. donovani*-infected Balb/c mice. AmBisome inhibited 99.9% at both doses. Favorable cytotoxicity profile compared with AmB and Fungizone, and comparable with AmBisome.	Nicoletti et al. (2009)
	HPMA-AmB-alendronic acid	Conjugates of AmB (1.0 mg/kg) containing 0% and 1.8% of alendronic acid displayed 77% and 67.9% hepatic parasite burden inhibition, respectively	Nicoletti et al. (2010)

in vivo activity than AmBisome, upregulation of TNF-α, IL-12, IFN-γ, and iNOS, and downregulation of TGF-β, IL-10, and IL-4 (Gupta et al., 2014b). However, lectin-functionalized lipo-polymerosomes showed higher accumulation in macrophage-rich organs (liver, spleen, lung) and higher activity as compared with nonfunctionalized lipo-polymerosomes, AmBisome, and Fungizone (Gupta et al., 2014a).

Converting AmB deoxycholate (1–2 μm) into NPs by applying high-pressure milling homogenization showed greater efficacy than conventional AmB (Manandhar et al., 2008). However, only *in vitro* leishmanicidal activity has been reported (Manandhar et al., 2014).

Intraperitoneal and oral administration of AmB covalently bound to functionalized carbon nanotubes showed leishmanicidal activity *in vivo*. However, toxicity, biopersistence, and biodegradability carbon nanotubes remain a matter of major concern.

In a different approach, AmB attached to poly (*N*-(2-hydroxypropyl)methacrylamide, HPMA)-polymer through a degradable GlyPheLeuGly linker (cleaved by cathepsin B in phagolysosomes) showed potent leishmanicidal activity *in vivo*, although it was less active than AmBisome (Nicoletti et al., 2009). Conjugation of poly(HPMA)-AmB with the bisphosphonate alendronic acid did not increase the leishmanicidal activity (Nicoletti et al., 2010).

18.2.1.1.1.2 ORAL ROUTE Except miltefosine, all current leishmanicidal drugs are administered by parenteral routes because of their poor aqueous solubility and bioavailability. New approaches aimed to change the administration route have recently been described. For instance, to enhance its oral absorption and minimize its side effects, AmB was formulated in self-emulsions loaded in PLGA or polymer–lipid NPs.

Self-emulsifying AmB formulations (triglyceride [Peceol]), surfactant (Gelucire 44/14) and co-surfactant (vitamin E-polyethylene glycol succinate [vitamin E-TPGS]), improved the oral absorption (Wasan et al., 2009a). The AmB self-emulsion at 10 mg/kg twice daily for 5 days inhibited 99.5% of parasitemia (Wasan et al., 2009b), whereas five daily doses of 3 mg/kg of oral miltefosine inhibited 47.5% of liver parasites, and a single intravenous dose of AmBisome of 2 mg/kg completely eradicated liver parasites (Wasan et al., 2010). Moreover, the cytotoxicity against renal cells of the AmB self-emulsion was significantly lower than those of AmBisome and Fungizone (Leon et al., 2011), whereas *in vivo* the renal toxicity was avoided (Ibrahim et al., 2012). The formulations showed good stability at 30°C and 43°C.

AmB-PLGA NPs stabilized with vitamin E-TGPS reduced AmB nephrotoxicty by oral and intravenous routes as compared with Fungizone. AmB-NPs exhibited an eight-fold increase in oral bioavailability compared with Fungizone and reduced liver parasite burdens (Italia et al., 2012).

AmB loaded into polymer–lipid hybrid NPs (lecithin anionic core coated with cationic type A gelatin to impart stability in the gastrointestinal tract [AmB-LAGNPs]) exhibited a 4.69-fold increased oral bioavailability as compared with free AmB (Jain et al., 2012). AmB-LAGNPs showed significantly lesser blood urea nitrogen and plasma creatinine levels as compared with free AmB and Fungizone, but comparable with that of Fungisome, after intravenous administration. However, AmB-LAGNPs has not been tested in the VL model.

Finally, a cochelate complex of AmB (precipitation of AmB with phosphatidylserine and Ca^{+2}, bioral amphotericin B) highly effective in treating murine candidiasis and aspergillosis by oral route has completed phase I trials (Delmas et al., 2002). Moreover, the *in vitro* activity of AmB cochleate against *Leishmania chagasi* was found to be similar to that of Fungizone (Sesana et al., 2011), but it has not been tested in the VL model.

18.2.1.1.2 NANOMEDICINES BASED ON Sb^V

Dogs are the main urban reservoirs of *L. infantum* and represent the major source of contagion for the vectors by virtue of the high prevalence of infection and intense cutaneous parasitism. Achieving the complete cure of dogs with VL, or at least the blockade of infectivity to the sand flies vector, is a great challenge. Treatment of dogs naturally infected by *L. infantum* with four or six intravenous doses of liposomal Sb^V (distearoylphosphatidylcholine [DSPC]/chol/dicetylphosphate [DCP] 5:4:1 molar ratio, 400–1,200 nm; 6.5 mg Sb^V/kg at 4-day intervals) promoted both long-term parasite suppression and reduction of infectivity to sand flies (Da Silva et al., 2012). However, parasites remained in the spleen, bone marrow, and skin of treated dogs. Mixed formulations of Sb^V-containing plain (200 nm) and PEGylated liposomes (DSPC/chol/DCP/distearoylphosphatidyl-ethanolamine-polyethylene glycol 2000 [DSPE-PEG] 4.53:4:1:0.47 molar ratio; 180 nm) showed improved targeting of Sb^V to the bone marrow of dogs after intravenous administration. The mixed formulation promoted parasite suppression to a higher extent in both spleen and bone marrow of *L. infantum*–infected Balb/C mice compared with PEGylated or conventional liposomes alone (Azevedo et al., 2014). However, the activity of these formulations in infected dogs remains to be tested.

A different approach showed that only one intravenous administration of cationic liposomes (egg phosphatidylcholine: stearylamine [SA]) containing Sb^V suppressed liver (93–97%), spleen (96–98%), and bone marrow (84–86%) parasitic load in Balb/C mice infected with Sb^V-resistant *L. donovani* strains, which were comparable with Fungizone (2 mg/kg) (Roychoudhury et al., 2011). Interestingly, in comparison with AmB, mice treated with cationic liposomes were more resistant to reinfection and treated mice showed enhanced T-cell proliferation, persistent IgG1 levels, increased IgG2a, and upregulated IL-12, IFN-γ, and TNF-α production in leishmania-pulsed splenocytes.

18.2.1.1.3 NANOMEDICINES BASED ON DRUGS OTHER THAN AmB AND Sb^V

Nonapproved leishmanicidal drugs such as furazolidone, buparvaquone, and bisnaphthalimidopropyldiaaminooctane improved their performance compared with free drugs when loaded in liposomes and PLGA NPs, respectively (Tempone et al., 2010, Reimão et al., 2012; Costa Lima et al., 2012). The antitumoral doxorubicin loaded into nanocapsules (oil core of soyabean oil, Span 80/Tween 80, covered by protamine sulfate and sodium alginate and grafted with phosphatidylserine) accumulated in macrophage-rich organs (liver and spleen) in Wistar rats and inhibited splenic parasitic burden by 85.23% in *L. donovani*–infected hamsters after intraperitoneal administration (250 mg/kg/day, 4 consecutive days). Noncoated nanocapsules and free doxorubicin caused only 72.88% and 42.85% parasite inhibition, respectively (Kansal et al., 2012). Further, doxorubicin loaded into alginate-coated nanocapsules despite upregulated T-cell responses, NO production, and enhanced levels of iNOS, TNF-α, IFN-γ, and IL-12 only caused 72.88% inhibition of splenic parasitic burden in infected hamsters (Kansal et al., 2013).

18.2.1.2 *Cutaneous Leishmaniasis*

18.2.1.2.1 NANOMEDICINES BASED ON AmB BY PARENTERAL ROUTE

Parenteral administration of Sb^{V} to treat severe CL caused by Leishmania species with the potential to disseminate and of MCL CL are recommended to expedite healing, reduce the risk of scarring, prevent parasite dissemination, and reduce chance of relapse (Sundar and Chakravarty, 2013).

Although short-term treatments with AmBisome (3 mg/kg for 5 consecutive days and at day 10) were shown to be effective and better tolerated than Sb^{V} against CL caused by *Leishmania braziliensis* (Solomon et al., 2013), there is no available data regarding optimum dosage regimen and extension of treatment.

Only one work has showed that AmB encapsulated in PLGA NPs reduced the number of parasites on the paw of *L. amazonensis*–infected C57BL/6 mice after four intraperitoneal administrations (6 mg/kg on days 1, 4, and 7, and 2 mg/kg on day 10) to the same extent as free AmB administered for 10 days (2 mg/kg/day) (De Carvalho et al., 2013). Mixed PLGA NP with magnetic NP (maghemite) to control the release of AmB by magnetohyperthermia did not improve the results.

18.2.1.2.2 NANOMEDICINES BASED ON PAROMOMYCIN BY TOPICAL ROUTE

For uncomplicated "Old World" CL, the WHO recommends local chemotherapy (e.g., intralesional Sb^{V}, topical paromomycin, cryotherapy, or thermotherapy). For "New World" CL, systemic treatment is preferred because self-healing is rare and the evolution of the disease is potentially severe. Local therapy should not only be able to eliminate the parasite and prevent the risk of dissemination but also be able to reduce the scar formation and disfigurement.

An overview of 14 randomized controlled trials concluded that the activity of 15% paromomycin ointment associated with the permeation enhancer methylbenzethonium chloride (12%) was similar to that of intralesional ointment Sb^{V} in *Leishmania major* infections (Kim et al., 2009). However, local side effects are frequently observed because of the permeation enhancer. A cream containing paromomycin and gentamicin (WR279,396) showed high cure rates (94% versus 71% placebo) in patients with *L. major* with only mild local irritation (Ben Salah et al., 2009). However, efficacy of both treatments was inferior in treating New World CL. Topically applied dispersion of Amphocil in 5% ethanol has been successful in treatment of *L. major*–infected patients in Israel (Zvulunov et al., 2003). However, clinical trials addressed with topical AmB did not provide sufficient evidence for their use in CL.

The use of NPs for the topical treatment of CL has been recently reviewed by Moreno et al. (2014). NPs can increase the amount of the loaded drug arriving at the dermis and modulate the permeation rate, can target drugs near the parasite, and can have immunomodulatory properties or wound-healing capabilities.

Only flexible liposomes have shown significant reduction of lesion size in CL models. For instance, topical application of liposomes containing paromomycin (SoyPc: chol: propilenglycol: vitamin E, 15:27:0.3 w/w ~500 nm) twice per day for 4 weeks showed significant reduction of lesion size in *L. major*–infected mice that were completely cured 8 weeks later (Jaafari et al., 2009). Currently, this formulation is in clinical trial to test its efficacy in combination with systemic Sb^{V} for the treatment of CL caused by *Leishmania tropica*. Recently, topical application of transfersomes containing paromomycin (SoyPc: sodium cholate: ethanol 20:2:5 w/w, 200 nm) twice per day for 4 weeks caused significant reductions in the lesion sizes and lower parasite burden in spleen in *L. major*–infected mice compared with paromomycin cream (Bavarsad et al., 2012). Using a similar approach (Carneiro et al., 2010) showed that 300–500 nm fluid liposomes containing paromomycin enhanced *in vitro* drug permeation across stripped skin and improved the *in vivo* antileishmanial activity in *L. major*–infected mice.

18.2.1.2.3 NANOMEDICAL PHOTODYNAMIC THERAPY

Photodynamic therapy (PDT) is based on the use of photosensitizers that are excited by light to produce ROS in the presence of oxygen. Because of the accessibility of skin to irradiation from laser or incoherent light sources, PDT has been used with variable outcomes in experimental and clinical settings against CL (Van Der Snoek et al., 2008). Analysis of 49 trials confirmed the efficacy of a weekly session of PDT with 5-aminolevulinic acid (5-ALA) or its derivative methyl-ALA (MAL) for 4 weeks against *L. major* infections. A comparison of PDT with topical treatment of paromomycin daily for 28 days in Iran showed higher cure rates in the PDT group and the prevention of scarring was similar between them, with absence of recurrences and good cosmetic results. A recent study has corroborated the efficacy of PDT based on methylene blue to treat CL caused by *Leishmania amazonensis* (Song et al., 2011).

Regarding the use of NPs to increase the delivery of photosensitizer to infected macrofages, *in vivo* studies are still lacking. Our research group has shown that anti-amastigote activity of hydrophobic Zn phthalocyanine against intracellular *L. braziliensis* was significantly increased when loaded in ultradeformable liposomes (SoyPc: sodium cholate, 6:1 w/w) because of the increased skin penetration and uptake by infected cells of these liposomes (Montanari et al., 2010). An improved formulation containing total polar lipids (TPL) extracted from the hyperhalophile archaea *Halorubrum tebenquichense* that were more extensively taken-up by macrophages eliminated intracellular *L. braziliensis* amastigotes, without reducing the viability of host cells, keratinocytes, and bone marrow–derived dendritic cells on visible light illumination (0.2 J/cm^2) (Perez et al., 2014). *In vivo* topical application of these liposomes for 5 days on *L. braziliensis*–infected Balb/c mice followed by 15 min of sunlight irradiation produced a significant reduction of lesion size (unpublished results).

Incorporation of another photosensitizer, chloroaluminum phthalocyanine, in ultradeformable liposomes also increased the anti-*Leishmania panamensis* amastigote activity on illumination (17 J/cm^2) (Hernández et al., 2012). However, the treatment was equally toxic for macrophages and amastigotes. Other liposomal formulation (DMPC: chol: DSPG 2:1:0.8 molar ratio) also increased the effectiveness of carbaporphyrin dimethyl ketal on intracellular amastigotes of *L. amazonensis* on 2 h of visible light illumination (Taylor et al., 2011).

18.2.1.2.4 METALLIC AND METAL OXIDE NPs

Macrophages produce a high level of ROS to kill microbial agents. However, ROS production is inhibited in infected macrophages by *Leishmania* parasites, which leads to its survival inside macrophages. Metallic NPs are able to produce ROS; hence, metallic NPs were proposed to overcome the inhibition of ROS production by *Leishmania* in macrophages.

Only two works tested *in vivo* activity of metallic NPs. The first work showed that topically applied silver NPs were ineffective in reducing mean size of lesions of *L. major*–infected Balb/c mice (Nilforoushzadeh et al., 2012). Their low penetration across the skin could explain their inefficacy against *Leishmania in vivo.* However, intraperitoneally administered *Bacillus* sp. biosynthesized Se NPs at nontoxic doses (5 or 10 mg kg/day) for 14 days before infection and delayed the development of lesions in *L. major*–infected Balb/c mice (Beheshti et al., 2013). Daily intraperitoneal administration of Se NPs for 14 days to infected animals not only significantly decreased the lesion sizes compared with the control group but also completely abolished these lesions.

In vitro, AgNP inhibited the survival of *L. tropica* amastigotes, an effect that was more significant in the presence of UV light and in the absence of macrophages cytotoxicity (Allahverdiyev et al., 2011). TiO_2 NP doped with Ag (to pull light absorption into visible light) inhibited survival of *L. tropica* and *L. infantum* amastigotes 2-fold and 2.5-fold in the dark, whereas visible light-exposed TiAg-NPs inhibited them 4-fold and 4.5-fold, respectively (Allahverdiyev et al., 2013). The authors propose that TiAg-NP and visible light can be further used for treatment of CL, whereas TiAg-NPs alone can be used for VL treatment. Finally, Jebali and Kazemi (2013) showed that different NPs (Ag, Au, TiO_2, ZnO, and MgO) presented leishmanicidal activity against intracellular amastigotes of *L. major* in the dark that was increased under UV illumination. However, all NPs showed toxicity on host macrophages.

18.2.2 Nanomedical Prophylactic Strategies

A mass vaccination of the population of an endemic area would be the most cost-effective tool to diminish *Leishmania* burden. The feasibility of developing a vaccine against CL has been sustained by the fact that long-lasting protection was observed on recovery from CL or by leishmanization (LZ; inducing the infection by injecting live virulent parasites).

The protection is chiefly mediated through upregulation of protective cytokines (IFN-γ, IL-12, IL-2, and TNF-α), with concomitant down-regulation of disease-promoting cytokines such as TGF-β, IL-10, and IL-4. The first generation of leishmania vaccines consisting of whole-cell autoclave-killed *L. major* mixed with BCG have reached phase 3 trials but failed to show enough efficacy (54%) (Mutiso et al., 2010). However, recombinant second-generation vaccines and third-generation DNA vaccines achieved mean parasite load reductions of 68% and 59%, respectively, in animal models, but their success in field trials has not been reported yet (Palatnik-De-Sousa, 2008).

One limitation of recombinant antigens (Ag) is their poor immunogenicity that requires adjuvants to enhance the immune responses. Adjuvants approved for human use (alum, monophosphoryl lipid A [MPL], AS04, and MF59) enhance antibody production but are limited in their ability to induce cellular immunity. In that sense, NPs can enhance and/or facilitate the uptake of Ag by APC (dendritic cells or macrophages), may serve as a depot for controlled release of Ag, can protect the integrity of Ag against degradation, and can potentially cross-present Ag to generate cytotoxic T lymphocytes against intracellular pathogens (Oyewumi et al., 2010).

The performance of different NPs was thoroughly described recently (Doroud and Rafati, 2012). Here, we describe novel approaches to improve the protective immune response by loading Ag within NPs to further develop a Th1 response (Table 18.4). For instance, Bhowmick et al. (2010) showed that cationic SA containing MLV loaded with *L. donovani* promastigote membrane Ag induced almost complete protection, as well as significantly high delayed-type hypersensitivity (DTH) (index of cell-mediated immunity), IgG2a antibodies, and IFN-γ. Liposomal Ag demonstrated durable cell-mediated immunity, and mice challenged 10 weeks after vaccination could resist infection. Ravindran et al. (2010) also showed that SA liposomes loaded with soluble leishmania Ag (SLA) provided significant protection against murine VL compared with BCG and PPL plus trehalose dicorynomycolate (MPL-TDM toll-like receptor agonists) as adjuvants. Liposomes reduced parasite loads by 93% and 98% in liver and spleen, respectively, and produced the highest levels of IgG2a, IFN-γ, and DTH responses, together with lowest IL-4.

SA liposomes loaded with SLA plus MPL-TDM subcutaneously or intraperitoneally administered induced high levels of short-term (89%) and long-term (86%) protection in the liver and spleen (87% and 83%, respectively) (Ravindran et al., 2012). SLA entrapped in liposomes alone or SLA mixed MPL-TDM elicited partial protection. Highest levels of DTH were exhibited by mice immunized with liposomal Ag injected subcutaneously in combination with MPL-TDM. Protection was sustained up to 12 weeks, and infection was controlled for at least 4 months of the challenge. Immunization increased IFN-γ and IgG2a production even 4 months after the challenge infection and decreased IL-4 production. Taken together, results suggest these represent a good vaccine formulation for the induction of durable protection against *L. donovani*, but if administered to humans through a more friendly route.

However, cationic SLN (containing the synthetic quaternary ammonium DOTAP—an active stimulator of dendritic cells resulting in extracellular signal-regulated kinase activation and β-chemokine induction) loaded with pDNA encoding CpG motifs (able to trigger plasmacytoid dendritic cells) and *L. major* cysteine proteinases showed high protection levels with specific Th1 immune response (Doroud et al., 2011a). Mice vaccinated with cationic SLN-containing cysteine proteinases without the C-terminal extension increased the specific Th1 immune response and caused parasite inhibition (Doroud et al., 2011b). Interestingly, PLGA nanospheres loaded with autoclaved *L. major* induced strong protection against a challenge with small increase in footpad thickness. Surprisingly, a reverse effect on protective immune response was seen in the presence of Quillaja saponins as inmunomodulator (Tafaghodi et al., 2010). Alginate microspheres loaded with CpG-ODN and autoclaved *L. major* induced the lowest lesion development with significantly high IgG2a/IgG1 ratio and IFN-γ (Tafaghodi et al., 2011).

DOTAP containing liposomes loaded with nuclease-sensitive phosphodiester CpG-ODN and SLA caused smaller footpad swelling and parasite loads, higher IFN-γ, together with an increase of two-fold in the IgG2a/IgG1 ratio as compared with controls (Shargh et al., 2012). After immunization, an early IL-4 response in the spleen cells was increased, but it decreased significantly 9 weeks after infection. Authors claimed that the early IL-4 production did not hinder the Th1 response, which is crucial for the priming of long-term $CD8^{+}$ T-cell memory responses. The same cationic liposomes but complexed with cationic polymer (protamine) condensing CpG-ODN mixed with leishmania parasites induced a milder leishmania lesion and a minimum number of *L. major* in the spleen and lymph nodes, accompanied by a Th1 type of immune response with a preponderance of IgG2a isotype (Alavizadeh et al., 2012).

TABLE 18.4 Nanomedical Prophylaxis Strategies Against Leishmaniasis

Carrier type, composition, size and Z potential	Antigen	Immunization protocol /animal model/dose	Route	Challenge	Reference
Liposomes eggPC:chol: SA 7:2:2 molar ratio, 306 nm	SLA *L. donovani*	Three doses at 2-week intervals, Balb/c mice 20 μg SLA	i.p.	i.v. 2×10^7 promastigotes, 10 d after last booster	Ravindran et al. (2010)
Liposomes eggPC:chol: SA 7:2:2 molar ratio mixed with MPL-TDM, 330 nm ~40 mV	SLA *L. donovani*	Days 0 and 22, 15 μg SLA	s.c. i.p.	i.v. 2×10^7 promastigotes, 10 days or 12 weeks after last booster	Ravindran et al. (2012)
Liposomes DSPC:chol:SA, 7:2:2 molar ratio, MLV, 40–50 mV	SLA *L. donovani*	3 doses at 2 week intervals, Balb/c mice at 20 μg	i.p.	i.v. 2×10^7 promastigotes, 10 days after last booster	Bhowmick et al. (2010)
	Type I, II, and III cysteine proteinase	2 dose at 2-week interval, hamster at 2.5 μg of each protein	s.c.	Intracardially 2.5×10^7 promastigotes 10 d after last booster	Das and Ali (2014)
Liposomes, DOTAP:chol 1:1 molar ratio with CpG-ODN, 200–500 nm ~60 mV	SLA *L. major*	3 times at 2-week intervals, Balb/c mice, 25 μg SLA-10 μg CpG-ODN	s.c.	SC, in the left footpad, 1×10^6 promastigotes, 2 weeks after the last booster	Shargh et al. (2012)
Liposomes, DOTAP:chol with protamine CpG, 58 mV	live *L. major* parasites	1×10^6 parasites, Balb/c mice	s.c.	None	Alavizadeh et al. (2012)
Liposomes, DOTAP:chol 1:1 molar ratio, 100 nm, 32 mV	SLA	3 doses at 3-week intervals Balb/c mice, 50 μg	s.c.	SC in the right footpad with 1×10^6 *L. major* promastigotes	Firouzmand et al. (2013)
Chitosan NPs 250–300 nm	SODB1	3 doses at 3-week intervals Balb/c mice	s.c.		Danesh-Bahreini et al. (2011)
PLGA NPs with DOTAP 300–450 nm, 20–30 mV	KMP-11 and pDNA encoding KMP-11	Days 0, 14, and 28, Balb/c mice 100 μg pDNA or prime with 30 μg pDNA NPs and d 21 10 μg KMP-11 NPs with 25 μg CPG-ODN	i.m. intra der mal	Intradermal *L. braziliensis* with sand fly saliva	Santos et al. (2012)
Alginate microspheres and CPG-ODN, 1.8 μm	Autoclaved *L. major*	3 times at 3-week intervals, Balb/c mice at 180 μg parasites	s.c.	SC into the footpad, promastigotes 3 weeks after last booster	Tafaghodi et al. (2011)
Cationic SLN DOTAP 0.4%w/v, cetyl palmitate, chol, Tween 80, 3.2:1 molar ratio, 240 nm, ~25 mV	pDNA encoding cysteine proteinase	Days 0 and 21, 50 μg each pDNA	Right-hind foot pad	3 weeks after booster	Doroud et al. (2011a,b)

DPPC, dipalmitoylphosphatidylcholine; chol, cholesterol; eggPC, eggphosphatidylcholine; DOTAP, di-octadecenoyl-trimethylammonium-propane; SA, stearylamine; DSPC, distearoylphosphatidylcholine; DDAB, dimethyldioctadecylammonium (bromide salt); DCP, dicetylphosphate.

These results suggested that immune modulation using NPs might be a practical approach to improve the safety of LZ.

Interestingly, mice immunized with DOTAP liposomes without any additive and containing SLA showed significantly smaller footpad swelling and the lowest spleen and footpad parasite burden after the challenge, accompanied by the high IFN-γ, lower IL-4 level, and higher IgG2a antibody titer than that of control groups (Firouzmand et al., 2013).

DOTAP containing PLGA NP loaded with pDNA encoding for 11 kDa kinetoplastid membrane protein (KMP-11, a strong inducer of IFN-γ production by cells from cured patients; pDNA/DOTAP/PLGA) were used to immunize mice by heterologous prime-boost regimen (pDNA/DOTAP/PLGA NPs followed by a boost with PLGA NPs-loaded KMP-11 protein plus CpG-ODN) (Santos et al., 2012). An increased level of IL-2 and IFN-γ ascribed to DOTAP and TNF-α production due to CpG motifs were found. Mice challenged with *L. braziliensis* in the presence of sand fly saliva, mimicking the context of natural infection, showed a significant reduction in parasite load as well as an increased IFN-γ expression; however, immunization did not prevent lesion development.

However, chitosan NP-containing recombinant *Leishmania* superoxide dismutase B1 (SODB1) was used to obtain sustained release of the Ag, aiming to develop a single-dose vaccine (Danesh-Bahreini et al., 2011). It was found that both single-dose and triple-dose vaccinations were equally effective for inducing cell-mediated immunity compared with the control group (soluble SODB1).

Finally, SA liposomes containing a cocktail of recombinant type I, type II, and type III cysteine proteinase mixed with MPL-TDM reduced the organ parasite burden by 10^{13}-fold to 10^{16}-fold and increased the disease-free survival to 80% for hamsters for at least up to 6 months after *L. donovani* infection (Das and Ali, 2014). Robust secretion of IFN-γ and IL-12, along with concomitant downregulation of Th2 cytokines, was observed in cocktail vaccinates, even 3 months after infection. Although the mouse serves as a good animal model for dissecting the protective immune responses with the available immunological reagents, murine infection is usually self-curing and differs from human VL. In contrast, hamsters closely mimic clinical symptoms of human VL, characterized by severe immunosuppression and development of progressive fatal infection when challenged with *L. donovani*.

18.3 CHAGAS DISEASE

The flagellate protozoan *Trypanosoma cruzi* is transmitted to humans by the feces of hematophagous reduviidae bugs vectors. After entering the wounded skin or mucosal membrane, the circulating extracellular forms of *T. cruzi*, trypomastigotes, may invade different cell types. Inside the cell cytoplasm, the trypomastigotes transform into *amastigotes* that multiply by binary fission producing cell lysis, releasing new *trypomastigotes* into the blood stream that can invade any nucleated cell. After a generally asymptomatic acute phase (characterized by high parasitemia, with the presence of *trypomastigotes* in blood) lasting from a few weeks to several months, the infection is well-controlled by the host immune system. In the absence of specific treatment, the infection may remain asymptomatic. However, nearly one-third of the infected persons may develop clinical manifestations of different degrees of severity, which include cardiomyopathy, fatal heart failure, and digestive tract abnormalies such megacolon and megaesophagus. The irreversible structural damage to the heart, the esophagus, and the colon and severe disorders of nerve conduction in these organs, is caused by the intracellular amastigotes. Similar to leishmania, these cytoplasmatic amastigotes represent the major structural and phenomenological barriers that antichagasic drugs have to overcome.

Benznidazole (BNZ) and nifurtimox (NFZ) are the only drugs used to treat the acute phase of Chagas (Table 18.1), with the BNZ being less toxic and also capable of reducing the severity of the associated inflammatory processes of patients with chronic disease. However, the toxicity of BNZ arising from its metabolization and the incapacity to completely eliminate the intracellular parasites—failure to cure the chronic phase—remain unsolved challenges of treatment for adults. Despite its extensive oral bioavailability (>90%), BNZ exerts relatively poor *in vivo* antichagasic activity. Effectively, if well more than 40% of the drug is bound to plasma proteins exhibiting a half-life of 12–15 h, during which it acts against the trypomastigotes, its apparent volume of distribution (Vap) is 0.56 L/kg. Such low Vap value indicates a poor tendency to penetrate the body tissues and to remain in the vascular compartment. Moreover, because BNZ is a class III (high solubility and low permeability) drug, its low permeability would explain its low antiparasitic activity in the chronic phase, which is the most prevalent presentation. To overcome its poor permeation across the plasma membrane, a huge concentration gradient is required for BNZ to gain access to the cytoplasmic amastigotes. Both low Vap and permeability, together with extensive intracellular metabolization resulting in toxic metabolites for the host, are major drawbacks of BNZ.

18.3.1 Nanomedical Therapeutic Strategies

18.3.1.1 Nanomedicines Based on BNZ

The first attempts to increase the efficacy of BNZ by loading it into liposomes failed. For instance, BNZ loaded in liposomes (HSPC: chol: DPPG, 2:2:1 molar ratio) by intravenous administration (0.4 mg/kg, twice per week for 3 weeks) did not decrease parasitemia of mice infected with *T. cruzi* RA strain (Morilla et al., 2004). The hydrophobic BNZ in the bilayer of the uptaken liposomes was trapped within the endo-lysosomal pathway, which impaired its access to the cell cytoplasm. To escape from the endo-lysosomes, the hydrophilic 2-nitroimidazole etanidazole loaded into the inner space of pH-sensitive liposomes (dioleoylphosphatidylethanolamine [DOPE]: cholesteryl hemisuccinate [CHEMS], 6:4 mol:mol, ~400 nm) was used. At a pH less than 5 or 6, a phase transition from bilayer to inverted hexagonal phase II is triggered, which is responsible for the fusion of liposomes with the endo-lysosomal bilayers. As a result, the etanidazole is released to the cytoplasm. Intravenous administration of pH-sensitive liposomes containing etanidazole (3 days per week for 3 weeks) significantly decreased the parasitemia of *T. cruzi* RA strain–infected Balb/c mice, whereas 180-fold higher dose of free etanidazole failed to reduce the parasitemia (Morilla et al., 2005).

18.3.1.2 Nanomedicines Based on AmB

Early studies by Yardley and Croft (1999) showed that AmBisome at a single dose of 25 mg/kg suppressed the acute infection of mice infected with *T. cruzi* Tulahuen strain. Twelve years later, the efficacy of AmBisome was tested again in a different experimental setting in which immunosuppression with cyclophosphamide was induced to assess the cure rate and the efficacy in chronically infected animals.

In one of the first studies, Balb/CJ mice infected with 1,000 trypomastigotes of the *Tulahuen* strain received six intraperitoneal injections of AmBisome at 25 mg/kg administered on alternate days starting on the first day after infection (dpi 1) either during the acute phase (dpi 10) or during the chronic phase (dpi 45), or during both phases (dpi 10 and 45) (Cencig et al., 2011). AmBisome prevented death in the acute phase and drastically reduced parasite loads in heart, liver, spleen, skeletal muscle, and adipose tissues in acute and chronic infection, but it failed to completely cure animals. Importantly, the treatment given during the chronic phase only significantly reduced parasitic loads in cardiac tissue. However, treatment during the acute and chronic phases did not present significant advantages over the chronic treatment.

The lack of complete cure was ascribed to the preferential tropism of the *Tulahuen* strain for muscle tissues. The authors predict a beneficial effect of AmBisome treatment in *T. cruzi* congenital infection, in which parasites are preferentially targeted to the liver by the fetal circulation, because the early treatment was able to drastically reduce parasite loads in liver and spleen and allowed survival of all infected animals. However, AmBisome is preferentially taken-up by liver, spleen, and lungs as long as it is intravenously administered, but not if it is intraperitoneally injected.

More recently, Balb/CJ mice acutely or chronically infected (1,000 trypomastigotes) of either a BNZ-susceptible (Tulahuen) or a partially resistant BNZ (Y) strain of *T. cruzi* were treated with suboptimal doses of Ambisome (five intraperitoneal administrations 25 mg/kg every day for the chronic phase and on alternate days for the acute phase) in combination with BNZ (100 mg/kg for 20 days in acute phase and 10 days in chronic phase; suboptimal received half the doses) (Cencig et al., 2012). It was found that the combination did not cure acutely or chronically infected mice. Despite these discouraging results, the therapeutic failure of the two approaches using Ambisome was probably attributable to the nonoptimal route of administration because the intraperitoneal instead of the intravenous route was chosen.

18.3.1.3 Nanomedicines Based on Sesquiterpene Lychnopholide

Lychnopholide (LYC) is a chemically unstable natural product with low aqueous solubility, high lipophilicity, and *in vitro* anti-*T. cruzi* activity. LYC loaded in poly(lactic acid)-co-polyethylene glycol (PLA-PEG) nanocapsules (105 nm) were tested on an acute model of murine infection. Swiss mice infected with the Y strain and treated with 20 consecutive intravenous doses starting at 7 dpi (pre-patent period) at 2 mg/kg/day achieved 100% of cure, whereas those treated with BNZ (50 mg/kg/day) achieved 75% of cure (Branquinho et al., 2014). Free LYC reduced the parasitemia and improved mice survival, but no mice were cured. Authors suggested that LYC-PLA-PEG nanocapsules would exert at the same time for sustained (drug-associated plasma half-life is prolonged) and passive targeted delivery (nanocapsules extravasate to infected tissues and are taken up by infected macrophages at the inflammatory sites). However, the sustained delivery could also be achieved by a patch or subcutaneous depot, avoiding the problem of the PEGylated formulation that could cause complement activation-related pseudo allergy effect (CARPA effect) once intravenous injected. However, treatment is more effective when the highest number of blood trypomastigotes is exposed to drug and immediately after the rupture of highest number of pseudocists (in the pre-patent period) at the longest period of treatment (20 days). Hence, the role of a passive targeted delivery was likely negligible.

18.3.2 Nanomedical Prophylactic Strategies

A preventive Chagas vaccine would be cost-effective even when transmission to humans and prevalence of infection are low, and even for a vaccine of moderate protective efficacy (Lee et al., 2010). A therapeutic vaccine administered alone or in combination with BNZ may greatly improve the prognosis for Chagasic patients by increasing treatment efficacy, reducing its duration and cost, or at least delaying disease progression to advanced stages and heart failure, and could potentially prevent congenital Chagas disease if used during pregnancy (Dumonteil et al., 2012).

A vaccine against *T. cruzi* requires the activation of a Th1 immune profile with the stimulation of $CD8^+$ T cells. In addition, a very stable formulation is required for a vaccine to be efficiently distributed in remote rural areas.

A wide range of prophylactic and therapeutic vaccines have been evaluated in mice, from the use of whole parasites, to purified or recombinant proteins, to viral vectors and DNA vaccines; however, few of these candidates have been recently evaluated in dogs (animal model of chronic disease). With respect of the use of NPs, only proteoliposomes and archeosomes have been tested in acute animal models.

PS proteoliposomes containing multiple parasite proteins solubilized in SDS (dipalmitoylphosphatidylcholine: dipalmitoyl-PS: chol 5:2:15 w:w, 200 nm) were once intraperitoneally administered to Balb/c mice, which for 4 weeks were challenged with 300 trypomastigotes of Y strain (Migliaccio et al., 2008). Results show that PS proteoliposomes do not induce complete protection from death in mice, but they are able to significantly delay their death.

Archeosomes are vesicles enclosed by one or more bilayers prepared with TPL extracted from microorganisms that belong to the Archaea domain. These lipids are structurally very different from lipids of organisms from the Eukarya and bacteria domain: the glycerol backbone is ether-linked to saturated isoprenoid chains, mainly phytanyls and diphytanils, in an *sn*-2,3 enantiomeric configuration. Archaeosomes are more avidly internalized by macrophages and APC than liposomes (Krishnan et al., 2001; Perez et al., 2014). After subcutaneous administration in mice, archeosomes are potent adjuvants for the induction of Th1, Th2, and $CD8^+$ T-cell responses to the entrapped soluble Ag (Krishnan and Sprott, 2008). We showed that three administrations of archeosomes containing 12.5 μg of soluble *T. cruzi* proteins (days 0, 14, 21) in C3H/HeN mice generated higher levels of circulating antibodies than those measured in the sera from animals receiving the Ag alone, with a dominant IgG2a isotype associated with Th1-type immunity (Higa et al., 2013). Immunized mice displayed reduced parasitemia during infection and were protected against the lethal challenge (intraperitoneal administration of 150 trypomastigotes of Tulahuen strain; 100% survival of animals immunized with archeosomes at 30 dpi versus 100% mortality of control groups at 25 dpi). From the point of view of possible large-scale production, archeosomes can be produced by scalable techniques from sustainable sources and are highly stable (the isoprenoid chains are resistant to peroxidation, the ether linkages are resistant to hydrolysis, and the *sn*-2,3 stereoisomers are resistant to hydrolysis by stereospecific phospholipases).

18.4 CONCLUSIONS

The evolution of therapeutic and prophylactic nanomedical strategies against neglected diseases such as leishmaniasis and Chagas disease is undoubtedly slow. Despite its efficacy against VL, AmBisome is unaffordable for underdeveloped countries. Extensive efforts to replace AmBisome with new AmB formulations have not accessed the clinic yet. Nanonization of Fungizone could be the more rapid strategy to reach patients, assuming the NPs are highly cost-effective and structurally stable. However, the activities of NPs used in most of the preclinical studies were not compared with AmBisome, and animal models and doses lacked standardization. Some of them were administered intraperitoneally, which is an unsuitable route of administration for humans. Other less invasive formulations, such as oral self-emulsifying AmB, seem to be promising because of their efficacy, cost-effectiveness, and stability. A potential issue could be the toxic and immunological responses to NPs, which have not been addressed yet. The development of delivery strategies based on drugs different than AmB will take more time, because their use as leishmanicidals is not approved and because the drugs could be highly toxic, such as the cardiotoxic doxorubicin (maximal recommended accumulated dose 450 mg/m^2).

Overall, no preclinical approaches using NPs—except the clinical outcomes of topical liposomal paromomicyn on CL that remain to be published—have entered clinical trials yet. However, safety and penetration into the blood stream of metallic NPs and effectivity of PDT with NPs in animal models remain to be shown. Preclinical vaccination strategies against leishmaniasis show encouraging results, such as subcutaneous application of SA liposomes against VL and the use of pDNA within SLN against CL.

Finally, Chagas disease remains the most neglected of the parasitosis, with scarce preclinical approaches testing the efficacy of nanomedicines against acute phase murine models. Except our own research, a single approach using nanoparticulate material as an adjuvant that provided modest results has been tested so far. In this scenario, the poor interest in using nanomedicines against neglected diseases is attributable to affordability issues, lack of technical knowledge, and the absence of strategic health plans by the governments of the endemic countries.

References

Abeer, H.A., Ahmed, M., Brocchini, S., Croft, S.L., 2012. Recent advances in development of amphotericin B formulations for the treatment of visceral Leishmaniasis. Curr. Opin. Infect. Dis. 25, 695–702.

Alavizadeh, H., Badiee, A., Khamesipour, A., Jalali, S.A., Firouzmand, H., Abbasi, A., et al., 2012. The role of liposome–protamine–DNA nanoparticles containing CpG oligodeoxynucleotides in the course of infection induced by *Leishmania major* in BALB/c mice. Exp. Parasitol. 132, 313–319.

Allahverdiyev, A.M., Abamor, E.S., Bagirova, M., Ustundag, C.B., Kaya, C., Kaya, F., et al., 2011. Antileishmanial effect of silver nanoparticles and their enhanced antiparasitic activity under ultraviolet light. Int. J. Nanomed. 6, 2705–2714.

Allahverdiyev, A.M., Abamor, E.S., Bagirova, M., Baydar, S.Y., Ates, S.C., Kaya, F., et al., 2013. Investigation of antileishmanial activities of TiO_2@Ag nanoparticles on biological properties of *L. tropica* and *L. infantum* parasites, *in vitro*. Exp. Parasitol. 55–63.

Asthana, S., Jaiswal, A.K., Gupta, P.K., Pawar, V.K., Dube, A., Chourasia, M.K., 2013. Immunoadjuvant chemotherapy of visceral leishmaniasis in hamsters using amphotericin B-encapsulated nanoemulsion template-based chitosan nanocapsules. Antimicrob. Agents Chemother. 57, 1714–1722.

Azevedo, E.G., Ribeiro, R.R., Da Silva, S.M., Ferreira, C.S., De Souza, L.E., Ferreira, A.A., et al., 2014. Mixed formulation of conventional and PEGylated liposomes as a novel drug delivery strategy for improved treatment of visceral leishmaniasis. Expert Opin. Drug Deliv. 11, 1551–1560.

Bavarsad, N., Bazzaz, B.S.F., Khamesipour, A., Jaafari, M. R., 2012. Colloidal, *in vitro* and *in vivo* anti-leishmanial properties of transfersomes containing paromomycin sulfate in susceptible BALB/c mice. Acta Trop. 124, 33–41.

Beheshti, N., Soflaei, S., Shakibaie, M., Yazdi, M.H., Ghaffarifar, F., Dalimi, A., et al., 2013. Efficacy of biogenic selenium nanoparticles against Leishmania major: *in vitro* and *in vivo* studies. J. Trace. Elem. Med. Biol. 27, 203–207.

Ben Salah, A., Buffet, P.A., Morizot, G., Ben Massoud, N., Zâatour, A., Ben Alaya, N., et al., 2009. WR279 a third generation aminoglycoside ointment for the treatment of *Leishmania major* cutaneous leishmaniasis: a phase 2, randomized, double blind, placebo controlled study. PLoS Negl. Trop. Dis. 3, e432.

Bern, C., Adler-Moore, J., Berenguer, J., Boelaert, M., Den Boer, M., Davidson, R.N., et al., 2006. Liposomal amphotericin B for the treatment of visceral leishmaniasis. Clin. Infect. Dis. 43, 917–924.

Bhowmick, S., Mazumdar, T., Sinhá, R., Ali, N., 2010. Comparison of liposome based antigen delivery systems for protection against *Leishmania donovani*. J. Control Release 141, 199–207.

Branquinho, R.T., Mosqueira, V.C., Oliveira-Silva, J.C., Simoes-Silva, M.R., Saude-Guimaraes, D.A., et al., 2014. Sesquiterpene lactone in nanostructured parenteral dosage form is efficacious in experimental Chagas disease. Antimicrob. Agents Chemother. 58, 2067–2075.

Carneiro, G., Santos, D.C., Oliveira, M.C., Fernandes, A.P., Ferreira, L.S., Ramaldes, G.A., et al., 2010. Topical delivery and *in vivo* antileishmanial activity of paromomycin loaded liposomes for treatment of cutaneous leishmaniasis. J. Liposome Res. 20, 16–23.

Cencig, S., Coltel, N., Truyens, C., Carlier, Y., 2011. Parasitic loads in tissues of mice infected with *Trypanosoma cruzi* and treated with AmBisome. PLoS Negl. Trop. Dis. 5, e1216.

Cencig, S., Coltel, N., Truyens, C., Carlier, Y., 2012. Evaluation of benznidazole treatment combined with nifurtimox, posaconazole or AmBisome(R) in mice infected with *Trypanosoma cruzi* strains. Int. J. Antimicrob. Agents 40, 527–532.

Costa Lima, S.A., Resende, M., Silvestre, R., Tavares, J., Ouaissi, A., et al., 2012. Characterization and evaluation of BNIPDaoct-loaded PLGA nanoparticles for visceral leishmaniasis: *in vitro* and *in vivo* studies. Nanomedicine (Lond). 7, 1839–1849.

Croft, S.L., Olliaro, P., 2011. Leishmaniasis chemotherapy—challenges and opportunities. Clin. Microbiol. Infect. 17, 1478–1483.

Da Silva, S.M., Amorim, I.F., Ribeiro, R.R., Azevedo, E.G., Demicheli, C., Melo, M.N., et al., 2012. Efficacy of combined therapy with liposome-encapsulated meglumine antimoniate and allopurinol in treatment of canine visceral leishmaniasis. Antimicrob. Agents Chemother. 56, 2858–2867.

Danesh-Bahreini, M.A., Shokri, J., Samiei, A., Kamali-Sarvestani, E., Barzegar-Jalali, M., Mohammadi-Samani, S., 2011. Nanovaccine for leishmaniasis: preparation of chitosan nanoparticles containing Leishmania superoxide dismutase and evaluation of its immunogenicity in BALB/c mice. Int. J. Nanomed. 6, 835–842.

Das, A., Ali, N., 2014. Combining cationic liposomal delivery with MPL-TDM for cysteine protease cocktail vaccination against *Leishmania donovani* protection. PLoS Negl. Trop. Dis. 8 (8), e3091.

De Carvalho, R.F., Ferreira Ribeiro, I., Miranda-Vilela, A.L., De Souza Filho, J., Martins, O.P., Cintra e Silva Dde, O., et al., 2013. Leishmanicidal activity of amphotericin B encapsulated in PLGA–DMSA nanoparticles to treat cutaneous leishmaniasis in C57BL/6 mice. Exp. Parasitol. 135, 217–222.

Delmas, G., Park, S., Chen, Z., Tan, F., Kashiwazaki, R., Zarif, L., et al., 2002. Efficacy of orally delivered cochleates containing amphotericin B in a murine model of aspergillosis. Antimicrob. Agents Chemother. 46, 2704–2707.

Doroud, D., Rafati, S., 2012. Leishmaniasis: focus on the design of nanoparticulate vaccine delivery systems. Expert Rev. Vaccines 11, 69–86.

Doroud, D., Zahedifard, F., Vatanara, A., Najafabadi, A.R., Taslimi, Y., Vahabpour, R., et al., 2011a. Delivery of a cocktail DNA vaccine encoding cysteine proteinases type I, II and III with solid lipid nanoparticles potentiate protective immunity against *Leishmania major* infection. J. Control Release 153, 154–162.

Doroud, D., Zahedifard, F., Vatanara, A., Taslimi, Y., Vahabpour, R., Torkashvand, F., et al., 2011b. C-terminal domain deletion enhances the protective activity of cpa/cpb loaded solid lipid nanoparticles against *Leishmania major* in BALB/c mice. PLoS Negl. Trop. Dis. 5, e1236.

Dumonteil, E., Bottazzi, M.E., Zhan, B., Heffernan, M.J., Jones, K., Valenzuela, J.G., et al., 2012. Accelerating the development of a therapeutic vaccine for human Chagas disease: rationale and prospects. Expert Rev. Vaccines 11, 1043–1055.

Firouzmand, H., Badiee, A., Khamesipour, A., Heravi Shargh, V., Alavizadeh, S.H., Abbasi, A., et al., 2013. Induction of protection against leishmaniasis in susceptible BALB/cmice using simple DOTAP cationic nanoliposomes containing soluble Leishmania antigen (SLA). Acta Trop. 128, 528–535.

Gupta, P.K., Asthana, S., Jaiswal, A.K., Kumar, V., Verma, A.K., Shukla, P., et al., 2014a. Exploitation of lectinized lipo-polymerosome encapsulated amphotericin B to target macrophages for effective chemotherapy of visceral leishmaniasis. Bioconjug. Chem. 18, 1091–1102.

Gupta, P.K., Jaiswal, A.K., Kumar, V., Verma, A., Dwivedi, P., Dube, A., et al., 2014b. Covalent functionalized self-assembled lipo-polymerosome bearing amphotericin B for better management of Leishmaniasis and its toxicity evaluation. Mol. Pharm. 11 (3), 951–963.

Hernández, I.P., Montanari, J., Valdivieso, W., Morilla, M. J., Romero, E.L., Escobar, P., 2012. *In vitro* phototoxicity of ultradeformable liposomes containing chloroaluminum phthalocyanine against New World Leishmania species. J. Photochem. Photobiol. B. 117, 157–163.

Higa, L.H., Corral, R.S., Morilla, M.J., Romero, E.L., Petray, P.B., 2013. Archaeosomes display immunoadjuvant potential for a vaccine against Chagas disease. Hum. Vaccin. Immunother. 9.

Ibrahim, F., Gershkovich, P., Sivak, O., Wasan, E.K., Bartlett, K., Wasan, K.M., 2012. Efficacy and toxicity of a tropically stable lipid-based formulation of amphotericin B (iCo-010) in a rat model of invasive candidiasis. Int. J. Pharm. 436, 318–323.

Italia, J.L., Kumar, M.N., Carter, K.C., 2012. Evaluating the potential of polyester nanoparticles for per oral delivery of amphotericin B in treating visceral leishmaniasis. J. Biomed. Nanotechnol. 8, 695–702.

Jaafari, M.R., Bavarsad, N., Bazzaz, B.S.F., Samiei, A., Soroush, D., Ghorbani, S., et al., 2009. Effect of topical liposomes containing paromomycin sulfate in the course of Leishmania major infection in susceptible BALB/c mice. Antimicrob. Agents Chemother. 53, 2259–2265.

Jain, S., Valvi, P.U., Swarnakar, N.K., Thanki, K., 2012. Gelatin coated hybrid lipid nanoparticles for oral delivery of amphotericin B. Mol. Pharm. 9, 2542–2553.

Jebali, A., Kazemi, B., 2013. Nano-based antileishmanial agents: a toxicological study on nanoparticles for future treatment of cutaneous leishmaniasis. Toxicol. In Vitro 27, 1896–1904.

Kansal, S., Tandon, R., Dwivedi, P., Misra, P., Verma, P.R., Dube, A., et al., 2012. Development of nanocapsules bearing doxorubicin for macrophage targeting through the phosphatidylserine ligand: a system for intervention in visceral leishmaniasis. J. Antimicrob. Chemother. 67, 2650–2660.

Kansal, S., Tandon, R., Verma, A., Dube, A., Mishra, P.R., 2013. Development of doxorubicin loaded novel core shell structured nanocapsules for the intervention of visceral leishmaniasis. J. Microencapsul. 30, 441–450.

Khatik, R., Dwivedi, P., Khare, P., Kansal, S., Dube, A., Mishra, P.R., et al., 2014. Development of targeted 1,2-diacyl-sn-glycero-3-phospho-l-serine-coated gelatin nanoparticles loaded with amphotericin B for improved *in vitro* and *in vivo* effect in leishmaniasis. Expert Opin. Drug Deliv. 11, 633–646.

Kim, D.H., Chung, H.J., Bleys, J., Ghohestani, R.F., 2009. Is paromomycin an effective and safe treatment against cutaneous leishmaniasis? A meta-analysis of 14 randomized controlled trials. PLoS Negl. Trop. Dis. 3, e381.

Krishnan, L., Sad, S., Patel, G.B., Sprott, G.D., 2001. The potent adjuvant activity of archaeosomes correlates to the recruitment and activation of macrophages and dendritic cells *in vivo*. J. Immunol. 166 (3), 1885–1893.

Krishnan, L., Sprott, G.D., 2008. Archaeosome adjuvants: immunological capabilities and mechanism(s) of action. Vaccine 26, 2043–2055.

Kumar, R., Sahoo, G.C., Pandey, K., Das, V., Das, P., 2014. Study the effects of PLGA-PEG encapsulated amphotericin B nanoparticle drug delivery system against *Leishmania donovani*. Drug Deliv. (Epub ahead of print).

Lee, B.Y., Bacon, K.M., Connor, D.L., Willig, A.M., Bailey, R.R., 2010. The potential economic value of a *Trypanosoma cruzi* (Chagas disease) vaccine in Latin America. PLoS Negl. Trop. Dis. 4, e916.

Leon, C.G., Lee, J., Bartlett, K., Gershkovich, P., Wasan, E. K., Zhao, J., et al., 2011. *In vitro* cytotoxicity of two novel oral formulations of amphotericin B (iCo-009 and iCo-010) against *Candida albicans*, human monocytic and kidney cell lines. Lipids Health Dis. 10, 144.

Maeda, H., Nakamura, H., Fang, J., 2013. The EPR effect for macromolecular drug delivery to solid tumors: improvement of tumor uptake, lowering of systemic toxicity, and distinct tumor imaging *in vivo*. Adv. Drug Del. Rev. 65, 71–79.

Manandhar, K., Yadav, T., Prajapati, V., Kumar, S., Rai, M., Dube, A., et al., 2008. Antileishmanial activity of nano-amphotericin B deoxycholate. J. Antimicrob. Chemother. 62, 376–380.

Manandhar, K., Yadav, T., Prajapati, V., Basukala, O., Aganja, R.P., Dude, A., et al., 2014. Nanonization increases the antileishmanial efficacy of amphotericin B: an *ex vivo* approach. Adv. Exp. Med. Biol. 808, 77–91.

Migliaccio, V., Santos, F.R., Ciancaglini, P., Ramalho-Pinto, F.J., 2008. Use of proteoliposome as a vaccine against *Trypanosoma cruzi* in mice. Chem. Phys. Lipids 152, 86–94.

Montanari, J., Maidana, C., Esteva, M.I., Salomon, C., Morilla, M.J., Romero, E.L., 2010. Sunlight triggered photodynamic ultradeformable liposomes against *Leishmania braziliensis* are also leishmanicidal in the dark. J. Control Release 147, 368–376.

Moreno, E., Schwartz, J., Fernandez, C., Sanmartın, C., Nguewa, P., Irache, J.M., et al., 2014. Nanoparticles as multifunctional devices for the topical treatment of cutaneous leishmaniasis. Exp. Opin. Drug Deliv. 11, 579–597.

Morilla, M.J., Montanari, J.A., Prieto, M.J., Lopez, M.O., Petray, P.B., Romero, E.L., 2004. Intravenous liposomal benznidazole as trypanocidal agent: increasing drug delivery to liver is not enough. Int. J. Pharm. 278, 311–318.

Morilla, M.J., Montanari, J., Frank, F., Malchiodi, E., Corral, R., Petray, P., et al., 2005. Etanidazole in pH-sensitive liposomes: design, characterization and *in vitro/in vivo* anti-*Trypanosoma cruzi* activity. J. Control Release 103, 599–607.

Mutiso, J.M., Macharia, J.C., Gicheru, M.M., 2010. A review of adjuvants for vaccine candidates. J. Biomed. Res. 24, 16–25.

Nicoletti, S., Seifert, K., Gilbert, I.H., 2009. *N*-(2-hydroxypropyl)methacrylamide-amphotericin B (HPMA-AmB) copolymer conjugates as antileishmanial agents. Int. J. Antimicrob. Agents 33, 441–448.

Nicoletti, S., Seifert, K., Gilbert, I.H., 2010. Water-soluble polymer–drug conjugates for combination chemotherapy against visceral leishmaniasis. Bioorg. Med. Chem. 18, 2559–2565.

Nilforoushzadeh, M.A., Shirani-Bidabadi, L.A., Zolfaghari-Baghbaderani, A., Jafari, R., Heidari-Beni, M., Siadat, A. H., et al., 2012. Topical effectiveness of different concentrations of nanosilver solution on *Leishmania major* lesions in Balb/c mice. J. Vector Borne Dis. 49, 249–253.

Olliaro, P., Darley, S., Laxminarayan, R., Sundar, S., 2009. Cost-effectiveness projections of single and combination therapies for visceral leishmaniasis in Bihar, India. Trop. Med. Int. Health 14, 918–925.

Olson, J.A., Adler-Moore, J.P., Jensen, G.M., Schwartz, J., Dignani, M.C., Proffitt, R.T., 2008. Comparison of the physicochemical, antifungal, and toxic properties of two liposomal amphotericin B products. Antimicrob. Agents Chemother. 52, 259–268.

Oyewumi, M.O., Kumar, A., Cui, Z., 2010. Nano-microparticles as immune adjuvants: correlating particle sizes and the resultant immune responses. Expert Rev. Vaccines 9, 1095–1107.

Palatnik-De-Sousa, C.B., 2008. Vaccines for leishmaniasis in the fore coming 25 years. Vaccine. 26, 1709–1724.

Perez, A.P., Casasco, A., Defain Tesoriero, M.V., Pappalardo, J.S., Altube, M.J., Duempelmann, L., et al., 2014. Enhanced photodynamic leishmanicidal activity of hydrophobic zinc phthalocyanine within archaeolipids containing liposomes. Int. J. Nanomedicine. 9, 3335–3345.

Prajapati, V., Awasthi, K., Gautam, S., et al., 2011. Targeted killing of Leishmania donovani in vivo and in vitro with amphotericin B attached to functionalized carbon nanotubes. J. Antimicrob. Chemother. 66, 874–879.

Prajapat, V., Awasthi, K., Yadav, T., et al., 2012. An oral formulation of amphotericin B attached to functionalized carbon nanotubes is an effective treatment for experimental visceral leishmaniasis. J. Infect. Dis. 205, 333–336.

Ravindran, R., Bhowmick, S., Das, A., Ali, N., 2010. Comparison of BCG, MPL and cationic liposome adjuvant systems in leishmanial antigen vaccine formulations against murine visceral leishmaniasis. BMC Microbiol. 10, 181.

Ravindran, R., Maji, M., Ali, N., 2012. Vaccination with liposomal leishmanial antigens adjuvanted with monophosphoryl lipid – trehalose dicorynomycolate (MPL-TDM) confers long-term protection against visceral leishmaniasis through a human administrable route. Mol. Pharm. 9, 59–70.

Reimão, J.Q., Colombo, F.A., Pereira-Chioccola, V.L., Tempone, A.G., 2012. Effectiveness of liposomal buparvaquone in an experimental hamster model of *Leishmania (L.) infantum chagasi*. Exp. Parasitol. 130, 195–199.

Rizzo, L.Y., Theek, B., Storm, G., Kiessling, F., Lammers, T., 2013. Recent progress in nanomedicine: therapeutic, diagnostic and theranostic applications. Curr. Opin. Biotechnol. 24, 1159–1166.

Roychoudhury, J., Sinhá, R., Ali, N., 2011. Therapy with sodium stibogluconate in stearylamine-bearing liposomes confers cure against SSG-resistant *Leishmania donovani* in BALB/c mice. PLoS One. 6, e17376.

Santos, D.M., Carneiro, M.W., De Moura, T.R., Fukutani, K., Clarencio, J., Soto, M., et al., 2012. Towards development of novel immunization strategies against leishmaniasis using PLGA nanoparticles loaded with kinetoplastid membrane protein-11. Int. J. Nanomed. 7, 2115–2127.

Sesana, A.M., Monti-Rocha, R., Vinhas, S.A., Morais, C.G., Dietze, R., Lemos, E.M., 2011. *In vitro* activity of amphotericin B cochleates against *Leishmania chagasi*. Mem. Inst. Oswaldo Cruz. 106, 251–253.

Shargh, V.H., Jaafari, M.R., Khamesipour, A., Jaafari, I., Jalali, S.A., Abbasi, A., et al., 2012. Liposomal SLA co-incorporated with PO CpG ODNs or PS CpG ODNs induce the same protection against the murine model of leishmaniasis. Vaccine. 30, 3957–3964.

Solomon, M., Pavlotzky, F., Barzilai, A., Schwartz, E., 2013. Liposomal amphotericin B in comparison to sodium stibogluconate for *Leishmania braziliensis* cutaneous leishmaniasis in travelers. J. Am. Acad. Dematol. 68, 284–289.

Song, D., Lindoso, J.A., Oyafuso, L.K., Kanashiro, E.H., Cardoso, J.L., Uchoa, A.F., et al., 2011. Photodynamic therapy using methylene blue to treat cutaneous leishmaniasis. Photomed. Laser Surg. 29, 711–715.

Sundar, S., Chakravarty, J., 2013. Leishmaniasis: an update of current pharmacotherapy. Expert Opin. Pharmacother. 14, 53–63.

Tafaghodi, M., Eskandari, M., Kharazizadeh, M., Khamesipour, A., Jaafari, M.R., 2010. Immunization against leishmaniasis by PLGA nanospheres loaded with an experimental autoclaved *Leishmania major* (ALM) and Quillaja saponins. Trop. Biomed. 27, 639–650.

Tafaghodi, M., Eskandari, M., Khamesipour, A., Jaafari, M. R., 2011. Alginate microspheres encapsulated with autoclaved *Leishmania major* (ALM) and CpG-ODN induced partial protection and enhanced immune response against murine model of leishmaniasis. Exp. Parasitol. 129, 107–114.

Taylor, V.M., Cedeño, D.L., Muñoz, D.L., Jones, M.A., Lash, T.D., Young, A.M., et al., 2011. *In vitro* and *in vivo* studies of the utility of dimethyl and diethyl carbaporphyrin ketals in treatment of cutaneous leishmaniasis. Antimicrob. Agents Chemother. 55, 4755–4764.

Tempone, A.G., Mortara, R.A., De Andrade Jr., H.F., Reimão, J.Q., 2010. Therapeutic evaluation of free and liposome-loaded furazolidone in experimental visceral leishmaniasis. Int. J. Antimicrob. Agents 36, 159–163.

Van Der Snoek, E.M., Robinson, D.J., Van Hellemond, J.J., Neumann, H.A., 2008. A review of photodynamic therapy in cutaneous leishmaniasis. J. Eur. Acad. Dermatol. Venereol. 22, 918–922.

Wasan, E.K., Bartlett, K., Gershkovich, P., Sivak, O., Banno, B., Wong, Z., et al., 2009a. Development and characterization of oral lipid-based amphotericin B formulations with enhanced drug solubility, stability and antifungal activity in rats infected with *Aspergillus fumigatus* or *Candida albicans*. Int. J. Pharm. 372, 76–84.

Wasan, K.M., Wasan, E.K., Gershkovich, P., Zhu, X., Tidwell, R.R., Werbovetz, K.A., et al., 2009b. Highly effective oral amphotericin B formulation against murine visceral leishmaniasis. J. Infect. Dis. 200, 357–360.

Wasan, E.K., Gershkovich, P., Zhao, J., Zhu, X., Werbovetz, K., Tidwell, R.R., et al., 2010. A novel tropically stable oral amphotericin B formulation (iCo-010) exhibits efficacy against visceral leishmaniasis in a murine model. PLoS Negl. Trop. Dis. 4, e913.

Yardley, V., Croft, S.L., 1999. *In vitro* and *in vivo* activity of amphotericin B-lipid formulations against experimental *Trypanosoma cruzi* infections. Am. J. Trop. Med. Hyg. 61, 193–197.

Zvulunov, A., Cagnano, E., Frankenburg, S., Barenholz, Y., Vardy, D., 2003. Topical treatment of persistent cutaneous leishmaniasis with ethanolic lipid amphotericin B. Pediatr. Infect. Dis. J. 22, 567–569.

Index

Note: Page numbers followed by "*f*" and "*t*" refer to figures and tables, respectively.

FSC
www.fsc.org
MIX
Papier | Fördert
gute Waldnutzung
FSC® C083411

Kolibri 360 GmbH
Ferdinand-Jühlke-Straße 7
99095 Erfurt, Deutschland
produktsicherheit@kolibri360.de